Complexity and Synergetics

Stefan C. Müller · Peter J. Plath
Günter Radons · Armin Fuchs
Editors

Complexity and Synergetics

Springer

Editors
Stefan C. Müller
Institute of Experimental Physics
Otto-von-Guericke University Magdeburg
Magdeburg
Germany

Peter J. Plath
Department of Physical Chemistry
Fritz-Haber-Institute of the MPG
Berlin
Germany

Günter Radons
Institute of Physics, Complex Systems
Technical University Chemnitz
Chemnitz
Germany

Armin Fuchs
Center for Complex Systems and Brain Sciences
Florida Atlantic University
Boca Raton, FL
USA

ISBN 978-3-319-64333-5 ISBN 978-3-319-64334-2 (eBook)
https://doi.org/10.1007/978-3-319-64334-2

Library of Congress Control Number: 2017947483

Printed on acid-free paper

This Springer imprint is published by Springer Nature
The registered company is Springer International Publishing AG
The registered company address is: Gewerbestrasse 11, 6330 Cham, Switzerland

Preface

The two notions of Complexity and Synergetics appear equally weighted in the title of this book. The word "complexity" has been used in a quite general way for a long time before it appeared more and more often in a scientific context. The initiator of deciphering the hieroglyphes, Jean-Francois Champillon, wrote already in the year 1824: *The hieroglyphic scripture is a complex system, it is pictorial, symbolic and phonetic at the same time*. In biology, the evolution of insect communities can be described by relatively simple mathematical equations, about which the biologist Robert May stated in 1976 that we are dealing here with *simple mathematical models with very complicated dynamics*. Two seemingly far separated disciplines come to a similar statement about complex behavior, which is, in fact, determining the temporal and spatial evolution of many systems in a very large number of dynamical situations.

An explanation of such system behavior calls for intensive experimentation as well as intelligent modelling, based on physical principles and closely supported by mathematical methods, to solve the nonlinear equations that govern complexity. Thus, the field of nonlinear dynamics comes into play, which comprises a number of mechanistic approaches, among which the concept of "synergetics" has played a major role since its introduction by Hermann Haken in the 1970s. (A description of the history is given in the book "Beiträge zur Geschichte der Synergetik" by H. Haken, P.J. Plath, W. Ebeling, and Y. Romanovsky, Springer Spektrum, Wiesbaden, 2016.)

Going further, we notice that complexity and synergetics are intimately connected with the unfolding of ordered and irregular structures and their temporal evolution, which is an omnipresent phenomenon in our natural world. One may think of the shape of galaxies, various cloud formations, vortices that form in mountain creeks, or the wonderful patterns on snail shells and tropical fish. In addition to spatio-temporal self-organization acting here on the basis of system immanent mechanisms, there are other aspects to be considered: For instance, the far reaching coordination of a multitude of small individuals (subsystems) within a much larger entity, typically found in the surprisingly coherent motion of large swarms of birds or fish. Equally impressive is the interaction among individuals that form large systems and can accomplish global synchronization, as manifested in the periodic blinking of many hundreds of fireflies. If we consider the most complex biological system, the human brain, we can state that from measurements of electroencephalograms (EEG) quite different dynamical states can be deduced. One of the most pronounced transitions between such states is observed when an epileptic seizure occurs: distinct periodicities with large amplitude develop in the normally quite irregular EEG, an indication for a large scale coordinated activity of many thousands of neurons. Moreover, another "dynamical disease", migraine, is frequently connected with the appearance of a visual aura, which moves in a scintillating and patterned way through the visual field and hints at a strong reduction of excitability of the affected brain tissue.

In view of such a diversity of natural phenomena, for which an "invisible" ordering hand must be involved, the scientist will ask the question, whether a basic principle stands behind all these patterning phenomena to form a common basis for this diversity. Among other approaches, the introduction of the slaving principle by Hermann Haken, originating from laser physics, has achieved this goal to a large extent. For a laser to function, an appropriate optical material has to be "pumped" with light. Then, above a certain threshold of light intensity, i.e., in a state far from thermal equilibrium, a sharp change (a non-equilibrium phase transition) occurs, beyond which intensive and phase coherent light is emitted from the optically active material. A bifurcation has taken place from irregular fluctuations with incoherent phase relations to a highly organized state characterized by coherent light radiating with a well-defined phase. The numerous fluctuating modes of light emission have disappeared (the suppressed slaves) and a master dominates the event, namely a light beam generated by stimulated emission. If the initial pump intensity is restored, the laser light disappears and the system returns to an equilibrium state with no phase coherence.

This picture of enslaving of modes by a dominant master mode (the coherently emitted laser light) has shaped the idea of synergetics since its "inauguration" by Hermann Haken.

There are many other well-known examples of spontaneous structure formation that can be explained by this approach, e.g., the pioneering experiments performed by Henri Bérnard around 1900, shortly later analyzed theoretically by Lord Rayleigh (1916). A liquid layer is heated from below. While for a small temperature difference between the layer's bottom and surface heat conduction determines the upward energy flux, there is a sharp threshold (a bifurcation) at a critical value of the temperature gradient. Beyond this point the heat transfer in the liquid is mostly facilitated by convection and characteristic convective rolls or hexagonal cells are formed—in a spontaneous process the formation of patterns with a well-defined spatial wavelength has emerged. As in the laser, a dominant spatial mode gains the upper hand and enslaves all the others.

A great variety of theoretical treatments based on the synergistic concept and other approaches of complex systems theory can be found in the literature and will not be discussed here. Generally, the book focusses on the complex behavior of macroscopic and mesoscopic systems, composed of numerous interacting subsystems existing under non-equilibrium conditions. One goal is to find low-dimensional systems of differential equation with appropriate initial and boundary conditions to describe the dynamical properties satisfactorily. This is a challenging and difficult endeavor of many research groups all over the world. By virtue of the interdisciplinary character of their common goals, fruitful contacts between various study groups have been established and are constantly growing.

To strengthen international and interdisciplinary connections, a Symposium with the title of this book was held in Summer 2015 in Hannover-Herrenhausen under the roof of the Volkswagen Stiftung, organized by S.C. Müller, P.J. Plath and G. Radons. The scientific program of the Symposium forms the basis of this volume, and a few more contributions have been added.

Symposium “Complexity and Synergetics” Schloß Herrenhausen, Hannover, July 2015

Totally 30 articles have been assembled in eight chapters representing modern research in various fields around the keywords “complexity” and “synergetics”. Contributions by international experts are complemented by articles from young investigators who appreciated the opportunity to contribute to a book of this kind. The authors made efforts to write a generally comprehensible introduction and to present their actual findings in an understandable way. Therefore, we believe that this volume offers stimulating and appropriate readings for a rather general scientific audience, including students and the interested community.

This book starts with an introductory article by Werner Ebeling and Rainer Feistel on information, the evolution of life and its relation to synergetics. There follows a wonderful overview written by Hermann Haken showing how the synergetic approach can be successfully applied to a multitude of phenomena, covering systems "From Laser Light to Cognition".

Many of the following chapters can be assigned to specific disciplines, where for each case an effort is made to build interdisciplinary bridges. There are chapters on physics, chemistry, biology, economy, brain science, coordination dynamics, and more. This enumeration underlines the wide range of disciplines dealing with topics that are related to the theme of this book. As an example we take a closer look at the field of chemistry.

Chemistry and Physical Chemistry have their own prototypes of self-organization. The famous "Liesegang rings", already discovered in 1896, with patterned precipitation of insoluble salts, have been investigated until today, especially with a view to various comparable processes in geology. Another pattern forming prototype in chemistry is the generation of rotating spirals. At the basis of these phenomena are the purely temporal dynamics of some chemical reactions which, due to the interplay between activation (often by autocatalysis) and inhibition, exhibit sustained oscillations, as long as they remain in a state far from thermodynamic equilibrium. Such pattern formation does not only occur in "liquid" chemistry, but also in electrochemical systems. Whereas in liquid systems the coupling of reaction and diffusion (a local activation mechanism) plays a major role to generate propagating chemical waves, in heterogeneous systems, as in catalysis or in electrochemistry, a global coupling mechanism acting through the gas phase or through electrical effects has to be taken into account which leads to a significant increase in the variability of the observed patterns. Especially renowned in this field is the "Belousov-Zhabotinsky reaction" (known since the 1950s), because it is excellently suited for experiments with a large significance beyond pure chemistry, for instance to biology.

On the heart muscle rotating activity may lead to life-threatening fibrillations; by controlling spiral dynamics using external forcing (electric fields or light), important knowledge for cardiology can be obtained. Revealing the excitability concept is of similar importance for a treatment of dynamical diseases such as migraine, epilepsy and an understanding of cerebral cortical dynamics.

Complex system theories and synergetics provide comprehensive approaches for many other domains of science, some of which have found their place in chapters of this volume. An important application lies in various studies about the finance sector. The article on "Financial Market Models" is an attractive introduction to this field, explaining perhaps, why it is so difficult to interpret stock prices. An appropriate counterpart is found in the article on structural changes in dynamic econometric models.

Finally, we venture a big jump to brain research. Chapter "Brain and Coordination Dynamics" shows in several articles how important the concepts discussed above are for the most complex organ known to man. We see the description of specifics about brain diseases and collect the experience of researchers on coordination dynamics (with the example of a ballet dancer). We learn how movement coordination can be used to probe the brain of individual subjects and how coordination modes are reflected in brain signals. Furthermore, how humans with their brains and bodies probe for neural, behavioral and social scales, is investigated using human-human and human-machine interactions.

A short chapter is added to present additional topics. Here we find thoughts on the interplay between art and science; the usefulness of ultra high-speed cameras for observations of fast phenomena (e.g., crack dynamics, microbubbles) is demonstrated.

The last chapter is devoted to a special contribution by Otto Rössler. We read here about a most interesting approach between deterministic thermodynamics and cryodynamics, which is promoted as new fundamental discipline based on long-range attraction. We think that this article will lead to many exciting discussions.

In summary, this book aims to reflect some of the progress made in today's research in the fields of complexity and synergetics, as well as to initiate discussions on the current state and future directions of this research domain.

Acknowledgements

The editors thank all authors for their constructive efforts in preparing the manuscripts and contributing to the rich variety of topics included in this volume. Special thanks are due to Thomas Ditzinger and Holger Schaepe from Springer-Verlag, Heidelberg, for their continuous commitment, efficient support, and skillful technical assistance. We also thank Ulrike Bischler and Margot Jädick-Jäckel from the VW Foundation for their support in organizing the Symposium in Hannover-Herrenhausen in summer 2015 that laid the basis for the layout and chapters of this book. Furthermore, the assistance of Kinko Tsuji and Patricia Dähmlow in preparing this volume, as well as the efficient secretarial and technical support by Katja Guttmann and Dorothea Erndt are gratefully acknowledged.

Magdeburg, Germany — Stefan C. Müller
Berlin, Germany — Peter J. Plath
Chemnitz, Germany — Günter Radons
Boca Raton, USA — Armin Fuchs
May 2017

Contents

Contributors

Takashi Amemiya Graduate School of Environment and Information Sciences, Yokohama National University, Yokohama, Japan

Mehrnaz Anvari ForWind, University of Oldenburg, Oldenburg, Germany

Gerold Baier Cell and Developmental Biology, University College London, London, UK

Carsten Ballandis Universität Bremen, Physikalische Chemie, Bremen, Germany

Thimo Böhl Division of Theoretical Chemistry, University of Tübingen, Tübingen, Germany

Lisa Borland Cerebellum Capital, San Francisco, CA, USA

Demetris T. Christopoulos Department of Economics, National and Kapodistrian University of Athens, Athens, Greece

Patricia Dähmlow Institute of Experimental Physics, Otto-von-Guericke University Magdeburg, Magdeburg, Germany

Nirmali Prabha Das Indian Institute of Technology Guwahati, Guwahati, India

Hans H. Diebner Faculty of Medicine Carl Gustav Carus, Institute for Medical Informatics and Biometry, Department of Medical Information, Technical University Dresden, Dresden, Germany

Sabine Dietrich Department of Earth Sciences, Technical University, Berlin, Germany

Guillaume Dumas Human Genetics and Cognitive Functions Unit, Institut Pasteur, Paris, France; Center for Complex Systems and Brain Sciences, Florida Atlantic University, Boca Raton, FL, USA

Sumana Dutta Indian Institute of Technology Guwahati, Guwahati, India

Werner Ebeling Institute of Physics, Humboldt University, Berlin, Germany

Rainer Feistel Leibniz Institute for Baltic Sea Research, Warnemünde, Germany

Ulrike Feudel Institute of Chemistry and Biology of the Marine Environment, Carl-von-Ossietzky University Oldenburg, Oldenburg, Germany

Jan Friedrich Institut für Theoretische Physik I, Ruhr-Universität Bochum, Bochum, Germany

Dieter Fröhlich Division of Theoretical Chemistry, University of Tübingen, Tübingen, Germany

Armin Fuchs Center for Complex Systems & Brain Sciences and Department of Physics, Florida Atlantic University, Boca Raton, USA

Rainer Grauer Institut für Theoretische Physik I, Ruhr-Universität Bochum, Bochum, Germany

Hermann Haken Center of Synergetics, Institut für Theoretische Physik 1, Universität Stuttgart, Stuttgart, Germany

Ernst-Christoph Haß Institut für Angewandte und Physikalische Chemie, Arbeitsgruppe Synergetik, Universität Bremen, Bremen, Germany

Marcus J. B. Hauser Biophysics Group, Institute of Experimental Physics, Department of Biometry and Medical Informatics, Otto-von-Guericke University Magdeburg, Magdeburg, Germany

Karl-Heinz Jacob Department of Earth Sciences, Technical University, Berlin, Germany

Oliver Kamps Center for Nonlinear Science, University of Münster, Münster, Germany

J. A. Scott Kelso Center for Complex Systems & Brain Sciences and Department of Physics, Florida Atlantic University, Boca Raton, FL, USA; Intelligent Systems Research Centre, Ulster University, Derry Londonderry, Northern Ireland

Christian Kleiber Faculty of Business and Economics, University Basel, Basel, Switzerland

Katharina Knicker Institut für Angewandte und Physikalische Chemie, Arbeitsgruppe Synergetik, Universität Bremen, Bremen, Germany

Lisa Kolb Carl Zeiss Microscopy GmbH, Munich, Germany

Susanne Krömker Interdisziplinäres Zentrum für Wissenschaftliches Rechnen – IWR, Universität Heidelberg, Heidelberg, Germany

Rei Kurita Department of Physics, Tokyo Metropolitan University, Tokyo, Japan

Frank Kuske Division of Theoretical Chemistry, University of Tübingen, Tübingen, Germany

Anastasia I. Lavrova Medical Faculty, Saint-Petersburg State University, Saint-Petersburg, Russia; Saint-Petersburg State Research Institute of Phthisiopulmonology, Saint-Petersburg, Russia

Aline Lefebvre Human Genetics and Cognitive Functions Unit, Institut Pasteur, Paris, France; Department of Child and Adolescent Psychiatry, Assistance Publique-Hôpitaux de Paris, Robert Debré Hospital, Paris, France

Christophe Letellier Physics Department, University of Rouen CORIA, Saint-Étienne du Rouvray, France

Chaiya Luengviriya Department of Physics, Kasetsart University, Jatujak, Bangkok, Thailand

Jiraporn Luengviriya Department of Industrial Physics and Medical Instrumentation, Lasers and Optics Research Center, King Mongkut's University of Technology North Bangkok, Bangkok, Thailand

Dhriti Mahanta Indian Institute of Technology Guwahati, Guwahati, India

Alexei G. Makeev Faculty of Computational Mathematics and Cybernetics, Lomonosov Moscow State University, Moscow, Russia

Yoshihito Mori Graduate School of Humanities and Sciences, Ochanomizu University, Tokyo, Japan

David Müller Institute of Physics, Chemnitz University of Technology, Chemnitz, Germany

Stefan C. Müller Institute of Experimental Physics, Otto-von-Guericke University Magdeburg, Magdeburg, Germany

Satoshi Nakata Graduate School of Science, Hiroshima University, Higashi-Hiroshima, Japan

Frank W. Ohl Leibniz Institute for Neurobiology, Magdeburg, and Institute for Biology, Otto-von-Guericke University, Magdeburg, Germany

Noriko Oikawa Department of Physics, Tokyo Metropolitan University, Tokyo, Japan

Andreas Otto Institute of Physics, Chemnitz University of Technology, Chemnitz, Germany

Joachim Peinke ForWind, University of Oldenburg, Oldenburg, Germany

Nikolai V. Peskov Faculty of Computational Mathematics and Cybernetics, Lomonosov Moscow State University, Moscow, Russia

Metinee Phantu Department of Physics, Kasetsart University, Jatujak, Bangkok, Thailand

Peter J. Plath Universität Bremen, Physikalische Chemie, Bremen, Germany; Fritz-Haber-Institut der Max-Planck-Gesellschaft, Berlin-Dahlem, Germany

Porramain Porjai Department of Physics, Kasetsart University, Jatujak, Bangkok, Thailand

Eugene B. Postnikov Kursk State University, Kursk, Russia

Yury Prokazov Special Laboratory for Electron and Laserscanning Microscopy, Leibniz Institute for Neurobiology, Magdeburg, Germany

Günter Radons Institute of Physics, Chemnitz University of Technology, Chemnitz, Germany

M. Reza Rahimi Tabar Department of Physics, Sharif University of Technology, Tehran, Iran

Frank Rietz Institute of Experimental Physics, Otto-von-Guericke University, Magdeburg, Germany

Richard Rosch Wellcome Trust Centre for Neuroimaging, University College London, London, UK

Otto E. Rossler Division of Theoretical Chemistry, University of Tübingen, Tübingen, Germany

Tatsunari Sakurai Department of Creative Arts & Media, Yamaguchi College of Arts, Yamaguchi, Japan

Katrin Schmietendorf Institute for Physics, Carl-von-Ossietzky University, Oldenburg, Germany

Matthias Schulz Fraunhofer-Institut für Keramische Technologien und Systeme IKTS, Stationäre Energiespeicher, Hermsdorf, Germany

Marina M. Slinko Institute of Chemical Physics, Russian Academy of Science, Moscow, Russia

Ralf Stannarius Institute of Experimental Physics, Otto-von-Guericke University, Magdeburg, Germany

Nobuhiko J. Suematsu Graduate School of Advanced Mathematical Sciences, Meiji Institute of Advanced Study of Mathematical Sciences, Meiji University, Tokyo, Japan

Malee Sutthiopad Department of Physics, Kasetsart University, Jatujak, Bangkok, Thailand

Uwe Sydow Fraunhofer-Institut für Keramische Technologien und Systeme IKTS, Stationäre Energiespeicher, Hermsdorf, Germany

Kentaroh Takagaki Leibniz Institute for Neurobiology, Magdeburg, and Institute for Biology, Otto-von-Guericke University, Magdeburg, Germany

Peter Neal Taylor Institute of Neuroscience, Newcastle University, Newcastle, UK

Emmanuelle Tognoli Center for Complex Systems and Brain Sciences, Florida Atlantic University, Boca Raton, FL, USA

Kinko Tsuji Shimadzu Europa GmbH, Duisburg, Germany

Tohru Tsujikawa Faculty of Engineering, University of Miyazaki, Miyazaki, Japan

Daisuke Umeno Department of Applied Chemistry and Biotechnology, Chiba University, Inage-ku, Chiba, Japan

Yujiang Wang ICOS, School of Computing Science, Newcastle University, Newcastle, UK

Matthias Wächter ForWind, University of Oldenburg, Oldenburg, Germany

André Weber Biophysics Group, Institute of Experimental Physics, Otto-von-Guericke University Magdeburg, Magdeburg, Germany; Special Laboratory for Electron and Laserscanning Microscopy, Leibniz Institute for Neurobiology, Magdeburg, Germany

Mengsen Zhang Center for Complex Systems and Brain Sciences, Florida Atlantic University, Boca Raton, FL, USA

Werner Zuschratter Special Laboratory for Electron and Laserscanning Microscopy, Leibniz Institute for Neurobiology, Magdeburg, Germany

Part I
Introduction

About Self-organization of Information and Synergetics

Werner Ebeling and Rainer Feistel

Abstract The self-organization of information belongs to the basic topics of Haken's synergetics. The basic statement of this paper is following Eigen and Haken-Krell that information is a product of evolution and: there is no information processing without life, and there is no life without information processing. The origin of information processing is based on the self-organization of life. In our view, information is not a physical quantity but its transfer is always related to transfer of a universal physical quantity, the entropy. Information cannot be reduced to physics alone, it is basically a non-physical term and bound to the evolution of life. Information-processing systems exist only in the context of life and its descendants: animal behaviour, human sociology, science, technology etc.

1 Introduction: The Role of Information in Synergetics

The science of synergetics was founded by Hermann Haken and the fundaments were given in the book "Synergetics—An Introduction". We refer here to the second enlarged edition [1]. This book, which has just appeared in a new revised Russian translation, starts with the three chapters: Aims, Probability, Information. This shows us already how basic the role is that information plays in synergetics. All the basic ideas about the role of information we find already here; in later books, these ideas were worked out and refined in several respects [2], in particular regarding the relation between information and life [3]. We all know that the term information is basic for our everyday life and that it is used in articles and books published in so different disciplines such as physics, biology, computer technology or philosophy. On the other hand, a generally accepted and universally valid

W. Ebeling (✉)
Institute of Physics, Humboldt University, 12489 Berlin, Germany
e-mail: ebeling@physik.hu-berlin.de

R. Feistel
Leibniz Institute for Baltic Sea Research, 18119 Warnemünde, Germany
e-mail: rainer.feistel@io-warnemuende.de

S.C. Müller et al. (eds.), *Complexity and Synergetics*,
https://doi.org/10.1007/978-3-319-64334-2_1

definition of information is still subject to ongoing fundamental discussion [4, 5]. Following the views of Haken and Krell [6], we do not associate information processing with simple physical systems such as a gas or a pendulum, but consider complex systems like humans, animals and computers being able to store, exchange and accumulate information. What we have to discuss in this context is where the separation line lies between those two classes of systems, and what the physical nature of the transition between them is.

Our plain and basic statement is: there is no information processing without life, and there is no life without information processing. In this context, we understand technology as a part of the human culture that belongs to the realm of life. We share the view of Haken and Haken-Krell: "Die Frage nach der Entstehung der biologischen Information ist mit dem Konzept der Evolution untrennbar verbunden" [6]. Similarly, Manfred Eigen [7] formulated that "a living entity can be described as a complex adaptive system which differs from any, however complex, chemical structure by its capability of functional self-organization based on processing of information. If one asks where does this information come from and what is its primary semantics, the answer is: information generates itself in feedback loops via replication and selection, the objective being 'to be or not to be'." Information processing, we may conclude, is a key process in the struggle for the existence of living beings. The search for the origin and the physics of information takes us to the self-organization and evolution of life.

We emphasize that there are still many unresolved questions. A particularly difficult problem is the definition of the value of information [8, 9, 2, 10, 11]. However, values are a different concept which is to be considered separately and, although closely related, is not our topic here (see Ebeling et al. [12, 13].

2 Relation of Information to Physics and Entropy

The question to be discussed in this section is: Can information be reduced to physical terms like entropy? Our answer is a clear NO [13]. Information is not a physical term, this concept does not belong to the axiomatics of physics, but on the other hand the statement is: The transfer of information is necessarily accompanied by a transfer of energy and entropy.

Entropy is an emergent physical quantity with respect to the microscopic dynamics. Thermodynamic energy can be expressed as a function of the coordinates and momenta of the particles involved, while entropy cannot [14]. The irreversible growth of entropy represents a different symmetry, namely that of a semigroup, compared to that of the reversible microscopic motion, mathematically described by a group. Entropy has become a key concept to so many disciplines of science [15] that it may be worth to quote here the original arguments of Clausius [16] for his invention of the word: "If one looks for a term to denote S, one could, similar to what is said about the quantity U that it is the heat and work content of the body, say about the quantity S that it is the transformation content of the body.

Since I prefer to borrow the name for quantities so important for science from the old languages to permit their unaltered use in all languages, I suggest to name the quantity S the entropy of the body, after the Greek word η $\tau\rho o\pi\acute{\eta}$, the transformation. Intentionally I have formed the word entropy as similar as possible to the word energy because the two quantities denoted by those words are so closely related with respect to their physical meaning that a certain similarity in their nomenclature appears appropriate to me" (English translation from [17]). In the sense of this reasoning, entropy has been regarded by Clausius as the value of energy [18].

The close relation between thermodynamic entropy and information was first revealed in the investigations of Maxwell [19], Szilard [20] and Brillouin [21]. Only recently, Eigen [22] returned to the question of whether an equation "information = entropy" may be justified, and came to an affirmative conclusion, at least as far as the amount of information is concerned. The consequential question for the *information value* is subject to ongoing scientific discussions [11, 10, 18] and aims at a suitable quantitative measure of the impact of information on dynamical processes. Already Brillouin [22] had stated that (quotation from Eigen [22]): "The present theory of information completely ignores the value of the information handled, transmitted or processed… At any rate, one point is immediately obvious: any criterion for value will result in an evaluation of the information received. This is equivalent to selecting the information according to a certain figure of merit". Here, mainly the evolutionary aspects of the problem are considered, where the "merit" consists in winning the struggle for survival. Thus, from our perspective, the origin of information processes is intimately connected with the evolution of life [18, 23, 24, 25, 26, 27, 28].

We have to notice at this place that there are alternative views of information. Mathematicians speak about a mathematical Universe, imbedding the physical world, which is just a realization of mathematical patterns. Konrad Zuse developed the concept "Rechnender Raum". Stephen Wolfram and Nobel Laureat Gerard t' Hooft consider the real Universe as a digital computer. Similar views were developed by Bekenstein [29] and in books by Stonier [30], as well as in the Internet forum FIS [31]. Our view is different and more related to the pivotal approaches of scientific pioneers such as Chernavsky, Eigen, Haken, Klix, Nicolis, Stratonovich, Tembrock, or Volkenstein.

3 The Origin of Information

We know that the existence of all living beings is intimately connected with information processing and valuation. These concepts are considered here as the central aspect of life. On the other hand, considering Shannon's work, the concept of information is a product of technical sciences and was devised as a technical "signal theory" [32] or a "communication theory" [33]. However, this does not mean that information is a merely technical concept. Information appeared on our planet first in the context of life at a time when no technicians were available yet who may have designed it. We define *a living system as a natural, ordered and information-processing macroscopic system with an evolutionary history.* In our view, this criterion may even be used as a basis for decisions. Imagine the staff on a space ship far from our home planet which observes unknown objects moving in space, sending signals and doing maneuvers; should the crew meet them with the respect for living objects? We consider information processing as a special high form of self-organization. Symbolic information is an emergent property, but we see several open problems here. How did information emerge by self-organization? Genuine information is symbolic information, needing a source that creates signals or symbols, a carrier to store or transport it, and finally a receiver that knows about the meaning of the message and transforms it into the structure or function the text is a blueprint for. This way symbolic information is always related to an ultimate purpose connected with valuation [17].

Information-processing systems exist only in the context of life and its descendants, such as animal behavior, human sociology, science, technology. Historically, the very first such system was the genetic expression machinery of early life [7, 11, 25, 34]. However, the details how life appeared, which way symbolic information developed out of non-symbolic, native one, are hidden behind the veils of an ancient history. Other, later examples for the self-organization of information are much easier to study, and this was done first by Julian Huxley in the beginning of the last century in behavioral biology. The evolutionary process of the transition from use activities of animals to signal activities he called "ritualization". In our approach to the problem, the qualitative transition associated with "ritualization" or "symbolization" is a central point. A more detailed view onto this transition process reveals rather general features which we consider as a universal way to information processing in the course of evolution [17, 35]. When a process or a structure becomes symbolized, its original full form of appearance is successively reduced to a representation by symbols, together with a building-up of its processing machinery, which is able to react on the symbol as if its complete original was still there. At the end of the transition, the physical properties of the symbolic representation are no longer dependent on the physical properties of its origin, and this new symmetry (termed "coding invariance" or "code symmetry" by [13, 35] makes drift and diversification of symbols possible because of their neutral stability. Ritualization is the transition by which emergent symbolic information gradually separates from the original structural information in the course of evolution. In the form of their structural information, the symbols preserve traces of their evolution history.

In all processes of transferring and processing information, necessarily some amount of entropy is flowing. While the quantity of the related entropy flux may be extremely small compared to the thermal background entropy [36], it is far from being irrelevant. As an example in a similar context, Max Planck [14] had emphasized that relative to the vast amount of Einstein's relativistic rest energy of a macroscopic body, merely the tiny changes related to heat or work are of any practical importance in thermodynamics. Who likes to claim that the waves on an ocean or the surface structures of the terrestrial crust are "without relevance" because of their smallness compared to the planet's dimension, may endeavour to cross the sea or a mountain range without modern technology. Not quantity matters here, but quality! Also, while the entropy related to the ink distribution of a printed letter may be small compared to the thermal entropy of the paper sheet, the change of Shannon's information entropy is a relevant quantity.

There must have been a point in history at which pre informational processes and structures smoothly transformed into rudimentary genuine information processing for the very first time. That transition had a number of properties in common with later, similar events in the course of biological, social and technical evolution. The historically very first such system was the genetic expression machinery of early life. The process how life appeared on Earth, which way symbolic information developed out of non-symbolic, native one, was extremely complicated and is connected with ritualization. Note that behavior biologists and ethologists were apparently the first who observed and analyzed the general character of that phenomenon. Sir Julian Huxley [37] described it as "the gradual change of a useful action into a symbol and then into a ritual; or in other words, the change by which the same act which first subserved a definite purpose directly comes later to subserve it only indirectly (symbolically) and then not at all." Later, the transition process was termed *ritualization* [38, 39, 40]. In the more general approach taken here, the origin of life is understood as the first ritualization transition that converted an abiotic complex physico-chemical structure into a primitive living organism. From then on, that kind of transition process repeated over and over again, at very different levels of organisation, under very different circumstances [3, 17, 26, 40], recently culminating in the digitalization of the human society on a global scale [41].

References

1. H. Haken, *Synergetics—An Introduction*, 2nd enlarged edn. (Springer, Berlin, Heidelberg, New York, 1978) [Russian Edition, URSS Moskva 2014]
2. H. Haken, *Information and Selforganization* (Springer, Berlin, 1988/2000/2006)
3. I. Jahn, A. Wessel (eds.), *For a Philosophy of Biology* (Kleine, München, 2010)
4. M. Burgin, *Theory of Information: Fundamentality, Diversity and Unification* (World Scientific, New York, NY, USA; London, UK; Singapore, 2010)
5. M. Burgin, R. Feistel, Structures and Structural Information. Proceedings 1, 217 (2017). https://doi.org/10.3390/IS4SI-2017-04036
6. H. Haken, M. Haken-Krell, *Entstehung biologischer Information und Ordnung* (Wiss. Buchgesellschaft, Darmstadt, 1989/1995)
7. M. Eigen, The origin of genetic information. Origins. Life. Evol. B. 24, 241–262 (1994). https://doi.org/10.1007/BF02627944
8. R.L. Stratonovich, On information value (in Russian). Izv AN SSSR, Ser Tekhn Kibern 5, 3–12 (1965)
9. R.L. Stratonovich, *Teoriya Informatsii (Information Theory)* (Sovietskoye Radio, Moskva, 1975)
10. R.L. Stratonovich, On the problem of the valuability of information, in *Thermodynamics and Regulation of Biological Processes*, ed. by L. Lamprecht, A.I. Zotin (De Gruyter, Berlin, New York, 1984)

11. M.V. Volkenstein, *Entropy and Information* (Birkhäuser, Basel, 2009)
12. W. Ebeling, J. Freund, F. Schweitzer, *Komplexe Strukturen: Entropie und Information* (B.G, Teubner, Berlin, Stuttgart, Leipzig, 1998)
13. R. Feistel, W. Ebeling, Entropy and the selforganization of information and value. Entropy 18 (2016). https://doi.org/10.3390/e170x000x
14. M. Planck, Über Neuere Thermodynamische Theorien. (Nernstsches Wärmetheorem und Quanten-Hypothese). Berichte der Deutschen Chemischen Gesellschaft 45, 5–23 (1912)
15. E. Schrödinger, *What is Life?* (Cambridge University Press, Cambridge, 1944)
16. R. Clausius, Ueber verschiedene für die Anwendung bequeme Formen der Hauptgleichungen der mechanischen Wärmetheorie. Annalen. der. Physik. 201, 353–400 (1865)
17. R. Feistel, W. Ebeling, *Physics of Self-Organization and Evolution*, 1st edn. (Wiley, Weinheim, Germany, 2011)
18. W. Ebeling, R. Feistel, Theory of selforganization: The role of entropy, information and value. J. Non-Equil. Thermodyn. 17, 303–332 (1992)
19. J.C. Maxwell, *Theory of Heat* (Longmans, Green and Co., London, UK; New York, NY, USA, 1888)
20. L. Szilard, Über die Entropievermehrung in einem thermodynamischen System bei Eingriffen intelligenter Wesen. Z. Physik 53, 840–856 (1929)
21. L. Brillouin, Negentropy principle of information. J. Appl. Phys. 24, 1152–1163 (1953)
22. M. Eigen, *From Strange Simplicity to Complex Familiarity* (Oxford University Press, Oxford, UK, 2013)
23. J. Avery, *Information Theory and Evolution* (World Scientific, Singapore, 2003)
24. R.U. Ayres, *Information, Entropy, and Progress—A New Evolutionary Paradigm* (AIP Press, Woodbury, 1994)
25. W. Ebeling, R. Feistel, *Physik der Selbstorganisation und Evolution* (Akademie-Verlag, Berlin, 1982)
26. W. Ebeling, R. Feistel, *Chaos und Kosmos: Prinzipien der Evolution* (Spektrum-Verlag, Heidelberg, 1994)
27. R. Feistel, Ritualisation und die Selbstorganisation der Information, in Jahrbuch für Komplexität Selbstorganisation und Determination, ed. by U. Niedersen, L. Pohlmann (Duncker & Humboldt, Berlin, 1990), pp. 83–98
28. H.P. Yockey, *Information Theory, Evolution and the Origin of Life* (Cambridge University Press, Cambridge, 2005)
29. J.D. Bekenstein, Black holes and information theory. Contemp. Phys. 45, 31-43 (2003)
30. T. Stonier, *Information und die innere Struktur des Universums* (Springer, Berlin, 1991)
31. P.C. Marijuan et al., *Foundations of Information Science* (FIS) (2016), fis@listas.unizar.es
32. H. Haken, J. Portugali, *Information Adaptation: The Interplay Between Shannon Information and Semantic Information in Cognition* (Springer, Heidelberg, 2015)
33. C.E. Shannon, A mathematical theory of communication. Bell Syst. Tech. J. 27(379–423), 623–656 (1948)
34. M. Eigen, The selforganization of matter and the evolution of biological macromolecules. Naturwissenschaften 58, 465 (1971)
35. R. Feistel, Self-organisation of symbolic information. Eur. Phys. J. Special. Topics. 226, 207–228 (2017). https://doi.org/10.1140/epjst/e2016-60170-9
36. W. Ebeling, On the relation between various entropy concepts and the valoric interpretation. Physica. A. 182, 108–120 (1992)
37. J. Huxley, The courtship-habits of the great crested grebe (Podiceps cristatus); with an addition to the theory of sexual selection. J. Zool. 84, 491–562 (1914)
38. K. Lorenz, *Das sogenannte Böse* (Borotha-Schoeler, Wien, 1963)
39. G. Tembrock, *Grundlagen des Tierverhaltens* (Akademie-Verlag, Berlin, 1977)
40. F. Klix, *Erwachendes Denken* (Deutscher Verlag der Wissenschaften, Berlin, 1989)
41. J. Rifkin, *The Zero Marginal Cost Society: The Internet of Things, the Collaborative Commons, and the Eclipse of Capitalism* (Palgrave Macmillan, New York, 2014)

Some Aspects of Synergetics. From Laser Light to Cognition

Hermann Haken

Abstract While the study of the selforganized formation of spatial, temporal or even functional structures has a long history in *individual* scientific disciplines, ranging from physics over medicine till economy, the goal of synergetics is to unearth the underlying principles of selforganization. The explicit example of a physical device, the laser, allows the development and illustration of general concepts such as order parameters, the slaving principle and circular causality. I show how these concepts and the related mathematical approach can be applied to brain processes connected with visual perception. At the level of order parameters, hysteresis in perception and ambivalent figures are dealt with. A bridge between this phenomenological level and that of real neurons is provided by the Synergetic Computer based on the analogy between pattern formation and pattern recognition. A more recently established link with information theory is outlined,—including a brief discussion of Shannon-, pragmatic and semantic information.

1 Introduction. The Aim of Synergetics

In my contribution, I want to elucidate the scope of Synergetics, an interdisciplinary endeavor that I initiated in 1969 (for some historical aspects (Kröger [1] and Haken [2])). Synergetics is a search for unifying principles for the selforganized formation of spatial, temporal or functional structures in complex systems. The exploration of such structures has a long history. A few hints may suffice:

Chemistry (cf. the nice contribution by P. Plath in this issue)
Fluid dynamics (Bénard [3])
Psychology/Gestalt theory (Köhler [4])
Medicine, Synergy of muscles (Sherrington [5])
Economy (Nash [6]).

H. Haken (✉)
Center of Synergetics, Institut für Theoretische Physik 1, Universität Stuttgart,
Pfaffenwaldring 57, 70550 Stuttgart, Germany
e-mail: cos@itp1.uni-stuttgart.de

S.C. Müller et al. (eds.), *Complexity and Synergetics*,
https://doi.org/10.1007/978-3-319-64334-2_2

So why Synergetics? My late Russian friend Yuri Klimontovich (who was also the thesis advisor of Werner Ebeling, a pioneer of complexity research) brought it to the point: He said, present days'science resembles a situation where miners working in *different* mines are digging deeper and deeper, while they are not aware of what is found in other mines. So Klimontovich considered Synergetics as an enterprise to foster interdisciplinary cooperation. I stress right at the beginning that Synergetics is a *system* science that not only refers to the "hard sciences" such as physics and chemistry, but also to the "soft sciences" such as sociology or even epistemology. Synergetics focusses its attention on a widespread phenomenon that has marvelled scientists since the antique: the formation of structures by *selforganization,* i.e. without interference by means of any kind of "sculptor" [7]. By fortunate circumstances I came across such a phenomenon not in biology, where selforganization happens all the time, but in physics, i.e. the inanimated world.

Because a detailed study of the concrete physical system, the then novel light source has revealed so many basic insights into the "mechanism" of selforganization, you surely will forgive me, when I briefly recapitulate basic aspects.

2 The Laser Paradigm

The ruby laser is a nice example of a laser device (Maiman [8]). Ions embedded in a crystal of the shape of a rod may emit light. At the end faces of the rod mirrors are mounted, one of them semitransparent. Thus light emitted in axial direction is reflected several times and can interact with the ions more strongly than light emitted in other directions (Fig. 1).

The electrons of the ions are excited from the outside by *pumplight* so strongly that more electrons are in the upper, excited state than in the lower state (ground state). Denoting the occupation numbers by N_2, N_1, respectively, a positive *inversion* $D = N_2 - N_1$ results. The predominant reason for the construction of a laser was to produce a light source with high spectral purity (Prochorov [9], Schawlow and Townes [10]). They based their suggestion on the process of *stimulated emission* that had been invoked by Einstein 1915 to derive Planck's formula for black body radiation. This process had been used in masers ("*M*icrowave *A*mplification by *S*timulated *E*mission of *R*adiation"). There were, however, two essential

Fig. 1 Scheme of a laser

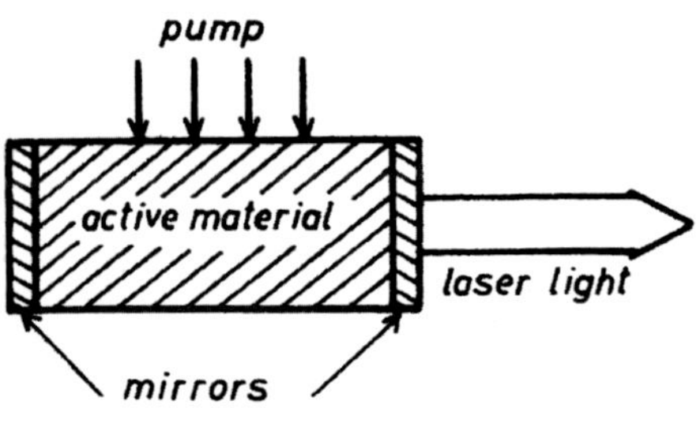

obstacles for the extension into the *optical* region with its much shorter wavelengths: In the laser rod, many axial waves (called "modes") can coexist, and the emission line is very broad so that many modes can be excited. We (Haken and Sauermann [11], as well as Tang et al. [12]) could show that in a ring laser a competition between modes leads to the amplification of only a *single mode* according to the principle: "winner takes all" or, using an expression of the evolution theory of biological species "survival of the fittest". Actually, later we found a close analogy between our equations and those of Eigen [13]. So my story could have ended here, weren't there a real suprise. As I could show in 1964 [14], laser light has a further fundamental property besides spectral purity (which had been considered by the experts as the only typical feature of laser light). To explain this and to show why all this has to do with selforgnization, I have to consider the laser process in more detail, but invoking a simple model for visualization (Fig. 2).

In an atom (or impurity ion in a crystal), a negatively charged electron is bound to a positively charged nucleus so that at atom j a dipole moment p_j results. When an atom is excited so that $d_j > 0$, p_j begins to oscillate spontaneously and emits a light wave (for a survey, cf. Table 1 and following text)

1. When the p_js oscillate independently, the fields E_j are uncorrelated. This happens in all conventional lamps. But in the laser, something new happens:

$$\text{stimulated emission.} \tag{1}$$

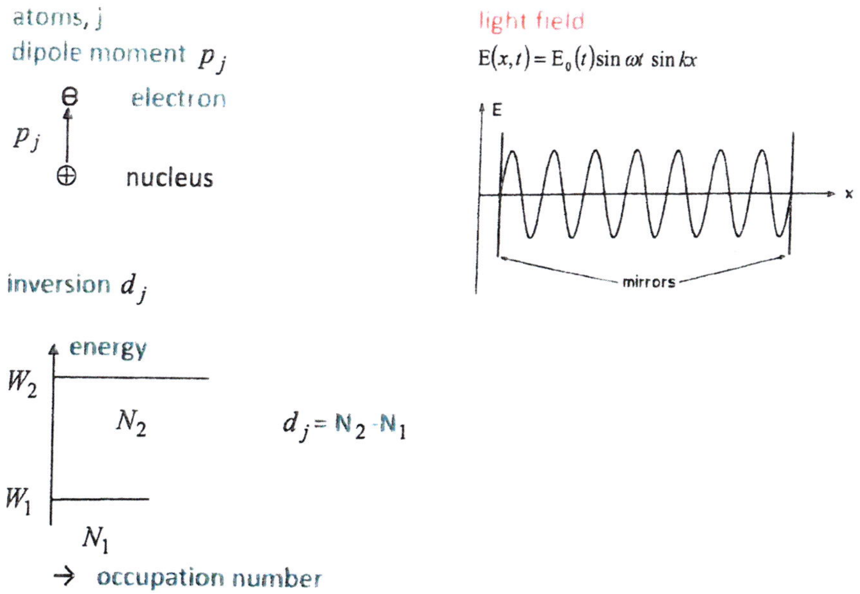

Fig. 2 Upper part, l.h.s.: dipole moment of atom *j*, *upper part*, r.h.s.: standing wave between two mirrors. ω: frequency, k: wave number. $E_0(t)$ envelope, *lower part* energy levels and inversion

Table 1 Basic elementary processes

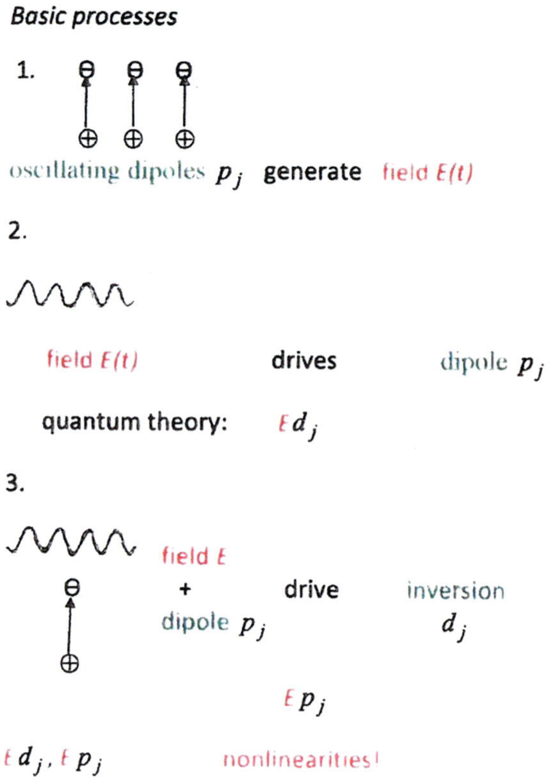

2. The resulting field $E = \sum E_j$ acts on the individual dipoles by a driving force Ed_j.

$$E \to p_j \tag{2}$$

3. E jointly with p_j causes also a change of the inversion d_j of each atom.

$$E \to d_j \tag{3}$$

Putting (1), (2), and (3) together, we obtain a loop

$$\begin{array}{c} \swarrow \quad \mathrm{E} \quad \nwarrow \\ (p_j, d_j) \end{array} \tag{4}$$

that I have called circular causality. At this stage of our approach (4) results in a complete mess concerning the temporal behavior of E. However, our laser device is coupled to its surround: the atoms (ions) are embedded in a matrix that perturbs the dipole oscillations leading to their damping and fluctuations. In addition, the pump light tries to restore the inversion within a typical relaxation time, or, equivalently, with a specific damping. In analogy to the notation of spin resonance two types of damping constants come into play, inversion: γ_1, dipole: γ_2. Furthermore, the lightfield is damped also, because it can escape through the semitransparent mirror. I denote the corresponding damping constant by κ. In typical lasers we have a time scale separation.

$$\kappa \ll \gamma_1 < \gamma_2 \tag{5}$$

This means that the atoms react quickly to the rather slowly decaying field E_0 after splitting off the rapidly oscillating factor exp($i\omega t$), ω light frequency.

$$E = E_0 \sin \omega t \sin kx$$

This allowed me to express the dipole moments p_j and in the quantum mechanical approach the inversion d_j by the amplitude of the field strength E_0. Thus, I found a simple equation for E_0

$$\frac{dE_0(t)}{dt} = (g - \kappa)E_0(t) - sE_0(t)^3 + F(t) \tag{6}$$

Actually, for my presentation, (6) is somewhat simplified but captures the most essential features of laser light. In (6) g is the "gain" that results from the continuous excitation of the atoms by means of pumplight. K is the just introduced damping constant. The next nonlinear term sE_0^3 is most important and serves for the stabilization of the laser light amplitude. The last term $F(t)$ describes the impact of fluctuations. This effect becomes very small if $g - \kappa > 0$. The meaning of (6) can be easily explained when we realize that it can be expressed by means of a potential $V(E_0)$:

$$\frac{dE_0}{dt} = -\frac{\partial V}{\partial E_0} + F(t) \tag{7}$$

$V(E_0)$ is plotted in Fig. 3. For $g < \kappa$, i.e. weak excitations, the parabola with only one minimum of V applies: The field amplitude is mainly zero and fluctuates due to the random "kicks" of $F(t)$.

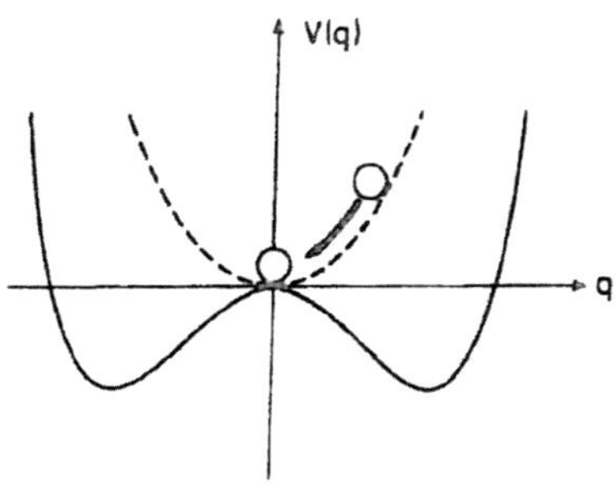

Fig. 3 cf. text, $q \equiv E_0$

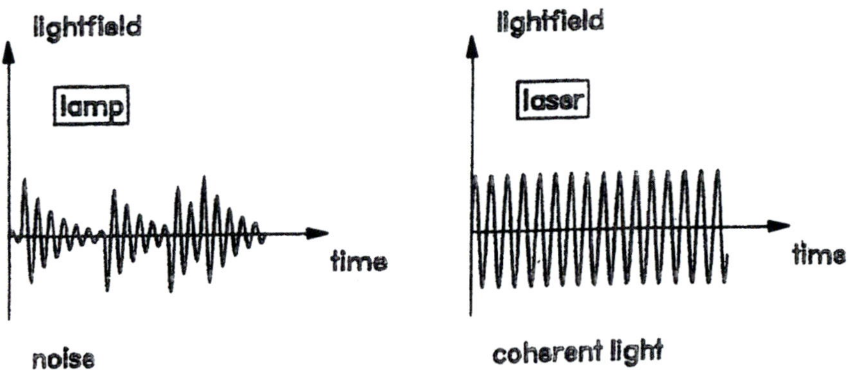

Fig. 4 L.h.s.: light of a lamp; r.h.s.: laser light

For g > κ evidently stable minima result indicating a stable field amplitude. Figure 3 is, of course, familiar to all those scientists who have been concerned with phase transitions of systems in thermal equilibrium such as freezing of water, ferromagnetism, superconducting a.s.o. Thus, the highly ordered laser light emerges from the (chaotic) light of conventional lamps by a phase transition! But here we are dealing with a driven system far from thermal equilibrium. Furthermore, we are dealing with a macroscopic quantum system subject to quantum noise $F(t)$ (actually in my original paper (6) is an *operator* equation). Laser light is brought about by the *selforganization* of the dipoles that must oscillate in phase. The many individual wave tracks of the light of lamps are replaced by a single, practically infinitely long laser wave (Fig. 4).

3 The Road to Selforganization

My laser example allows us to gain insight into basic features of selforganization that can be generalized to far more complicated systems. First of all, we deal with a system composed of many elements that interact with each other (the laser atoms interacting via the light field) and the system is coupled to its surround (Fig. 5).

The system receives inputs of energy, matter and/or information (e.g., pump light) and may dispose of "waste", e.g., heat (entropy). The coupling(s) of the system is (are) described by control parameter(s), e.g., pump strength.

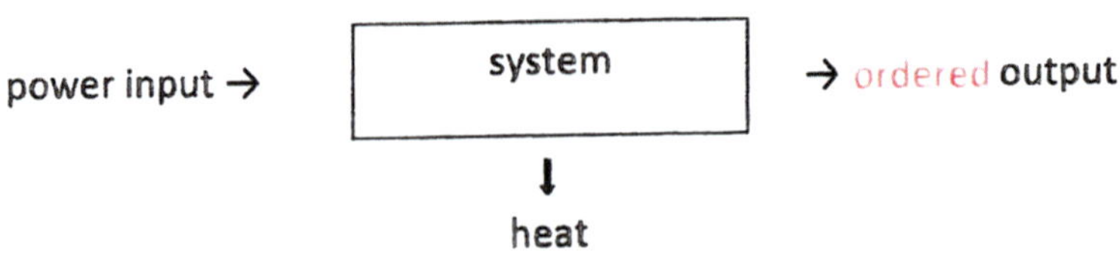

Fig. 5 Scheme of an open system

At a critical value of a control parameter, the system changes its state qualitatively. Its state becomes unstable.

In such a situation, one (or several) new collective variables appear, e.g. in the laser: the coherent field E. Generalizing Landau's notion we call these variables *order parameters*. These order parameters *enslave* the individual parts (the field E fixes the behavior of p_j, d_j). (The slaving principle is more general than the center manifold or inertial manifold theorem by the inclusion of fluctuations). Order parameters and enslaved parts condition each other by

"circular causality" (cf.4)).

The concept of order parameters and the slaving principle lead to a remarkable

complexity reduction.

Instead of dealing with the numerous variables of the parts (e.g. p_j, d_j), their behavior is determined and described by only few order parameters (in the single mode laser described above by *only one*). By use of dynamic systems theory, a number of order parameter equations can be classified leading to fixed points, limit cycles, tori and deterministic chaos.

How far can we go when we try to apply these concepts (and the underlying mathematical tools that I haven't presented here) to really complex systems? Probably the most complex system we know is the human brain. So let's try!

4 The Brain as a Synergetic System

Figure 6 shows the human brain as seen from above. Its white-gray matter of about 1.5 kg has the shape of a walnut. When studying it under a microscope, we observe small, complicated structures, the neurons of which Fig. 7 shows two examples. There are about 10^{11} neurons in the brain, and a neuron can receive inputs from up to 10^4 other neurons. Figure 8 shows a typical structure of a neuron. Its center is the cell body. Synapses receive (usually) voltage "spikes" from other neurons, convert them into electric currents in dendrites. The cell body sums them up and "fires" if the sum is beyond a threshold (This model picture forms the basis of many neural network approaches since McCulloch and Pitts [15]). The cell body then sends out a sequence of short pulses using pulse code modulation. The great challenge to neuroscience is the question: Who or what steers the neurons so that the brain may recognize objects, steer our movements, and may produce so many of its other complex activities. The answer I have proposed in 1983 (Haken [16]) is: The brain is a synergetic system, i.e. in particular, that it is selforganizing. To be concrete, I suggest an analogy:

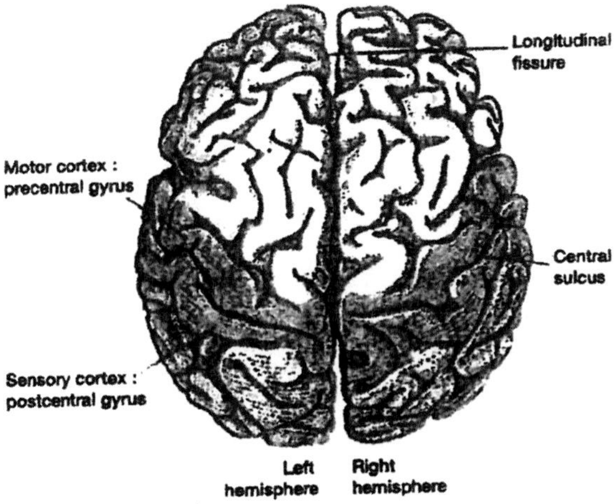

Fig. 6 The brain seen from above

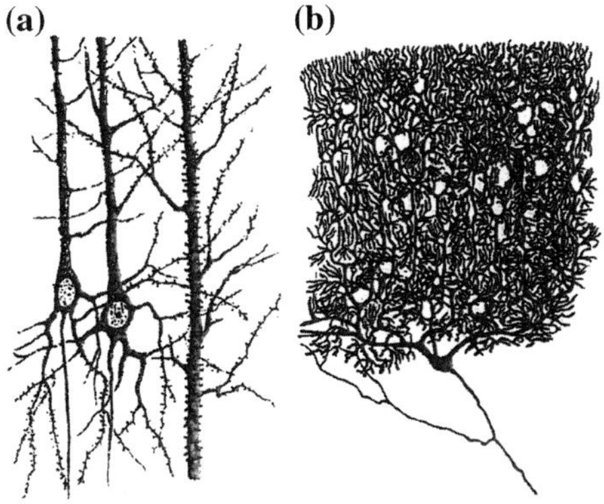

Fig. 7 **a** Pyramidal cells, **b** Purkinje cells

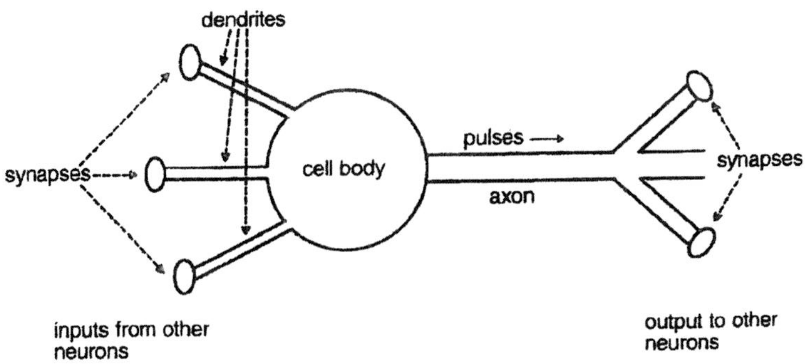

Fig. 8 Scheme of a neuron

	laser		brain
macro:	laser light	order parameters	percepts
	↕	↕	↕
micro:	atoms	parts	neurons

Over the past decades, I have devoted one line of my activities to the relation between neurons and percepts, or in other words, to pattern recognition. To this end, I proceeded in three steps which I can discuss here only briefly.

(4.1) phenomenological approach based on order parameters
(4.2) mesocopic approach based on the synergetic computer in which model neurons produce order parameters
(4.3) microscopic approach that treats the dynamics of neurons

4.1 Phenomenological Approach

If there is only 1 order parameter, ξ, the dynamics is prescribed by a potential $V(\xi)$, where V depends on a control parameter that causes deformations of V such as in Fig. 9. We ascribe to the minima of V specific percepts. We start with the upper row from the left. When the control parameter is increased (middle part), the system is still in its original state. But finally, a new minimum will be occupied. Now follow the lower part in the reverse direction. In the middle part the system is still kept in the right minimum. Though the control parameter is the same as before in the middle part of Fig. 9, the states occupying the minima are different. Clearly, here I have visualized the hysteresis effect. But the reader may verify this effect in visual perception, when in Fig. 10 s/he first looks at the upper row from the left to the right and then at the lower row from the right to the left. Did you notice that your perception switches at different positions depending on what you have seen before?

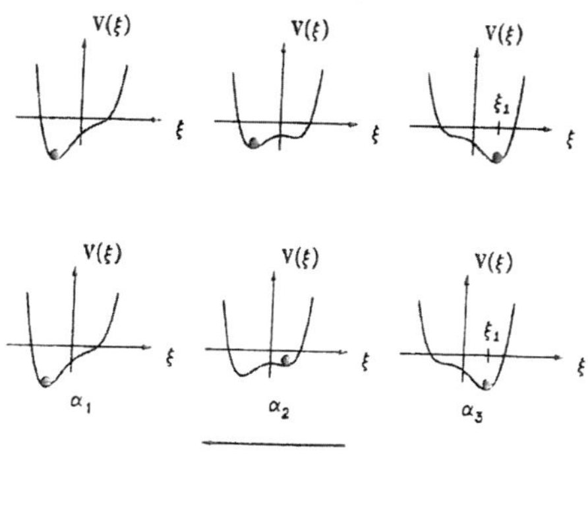

Fig. 9 cf. text

I have "invented" a further example shown in Fig. 11. First look at the upper line, hiding the other lines. Then proceed step by step to the lower lines. Then do the same in the opposite direction. Could you notice the hysteresis?

The effect of ambivalent figures (in German: Kippfiguren) on perception has been modelled by means of the oscillatory dynamics of two order parameters (cf. Figures 12 and 13) (Ditzinger and Haken [17]).

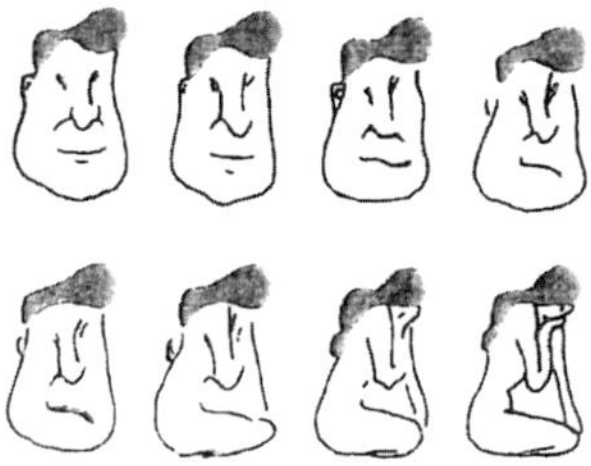

Fig. 10 Hysteresis in visual perception

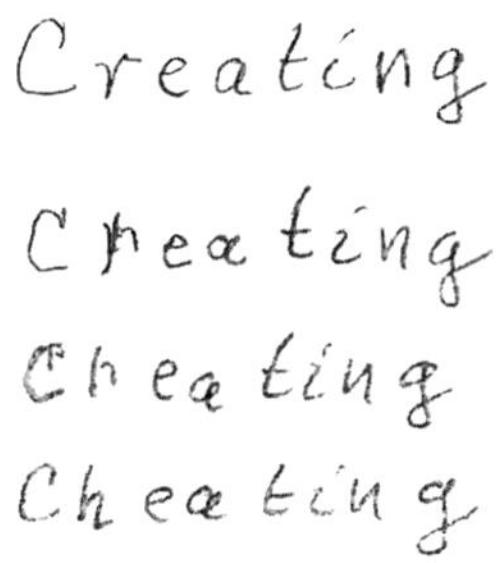

Fig. 11 Hysteresis in understanding (cf. text)

Fig. 12 Vase or faces?

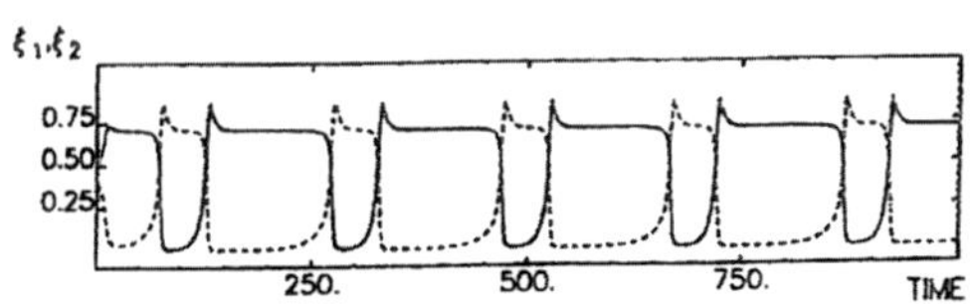

Fig. 13 Plot of the order parameters ξ_1, ξ_2 corresponding to two interpretations, for instance vase/faces, versus time [17]

4.2 Mesoscopic Approach

Here I modelled perception by means of the Synergetic Computer. (My algorithm was implemented on a computer by my Ph.D. student Armin Fuchs). The basic concept is based on the analogy between pattern formation (e.g. in fluids) and pattern recognition by humans or computers. A main ingredient of pattern is associative memory. A telephone book is an example: a person's name is supplemented by his/her telephone number. Thus, based on memory, incomplete data are complemented. This is achieved in the Synergetic Computer by a specific dynamics using order parameters (Fig. 14).

In pattern formation an initially only partly ordered pattern calls up several order parameters which compete with each other. Eventually one order parameter wins and enslaves the total system so that it is now in its fully ordered state. A similar process happens in pattern recognition. An only partly offered picture is calling up several order parameters that compete with each other. Again, the winning order parameter restores the complete picture Fig. 15 (face recognition).

More recently, we (Haken and Portugali, to be published) found a nice new application of the Synergetic Computer. Some time ago, so-called hybrid images were "constructed" by Oliva and Schyns [18]. A typical example is shown in Fig. 16.

When you look at it from a short distance, you will recognize Einstein, but when looking from a larger distance, or with half-closed eyes, you will recognize Marílyn Monroe. The "secret" of the construction principle is as follows: the total picture consists of a superposition of two pictures, Einstein and Monroe. While Einstein's face is composed of fine lines (i.e. high band pass filtered) that of Monroe is based on smooth shades. To see, how the Synergetic Computer can deal with hybrid images, we first stored the corresponding *unfiltered* images in the computer. Then

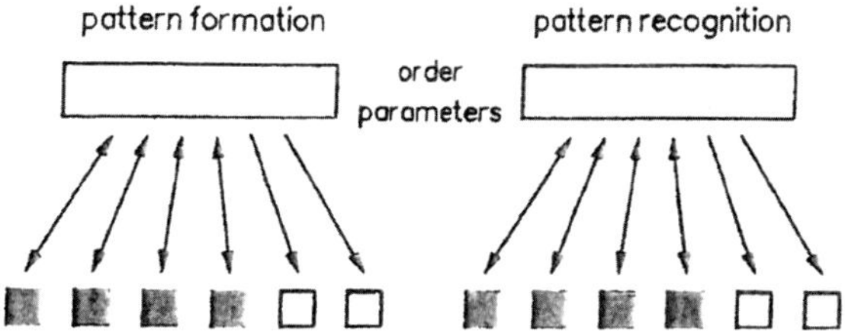

Fig. 14 Analogy between pattern formation and pattern recognition

Fig. 15 Example of a recognition process where part of a face is prescribed as initial state

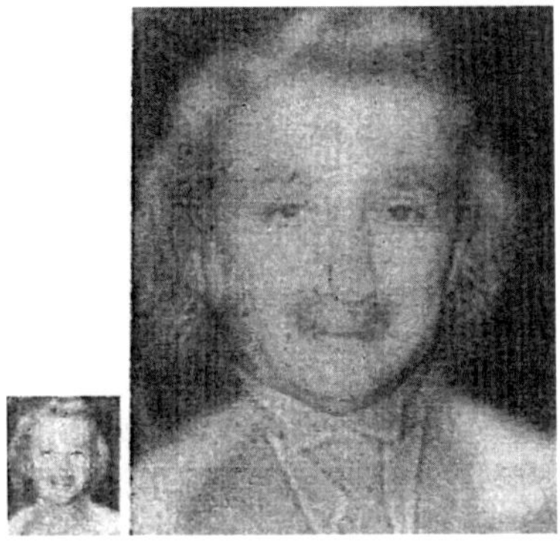

Fig. 16 The same picture is interpreted in two different ways

the hybrid image was offered to the computer which "recognized" Einstein. Then we studied the effect of *blurring* the hybrid image by forming its convolution with a Gaussian. We found that the fine lines were more affected ("smeared out") than the soft shades. So when the blurred hybrid image is offered to the Synergetic Computer, the "Monroe" interpretation wins. We expect that for a well-tuned blurring, a hybrid image may cause oscillations in perception as in the case of the "traditional" ambivalent figures.

4.3 Microscopic Approach—Spiking Neurons. Light House Model [19]

A survey is provided by Table 2 and an explanation on the next page.

As we know, the cell body of a neuron emits voltage spike trains along its axon. To model this effect, we consider a light house that emits a rotating light beam (upper part of Table 2, l.h.s.). Seen from above, (middle part): the light beams form an angle ϕ with respect to a fixed axis. Whenever $\phi = 0, 2\pi, l2\pi, l$: integer, an observer sees a flash. To apply this model to the action of a real neuron, we recall (cf. Fig. 8) that a neuron with label n receives its input from another neuron k in the form of pulses, where ϕ_k corresponds to ϕ in the upper middle part of Table 2, δ is Dirac's δ-function. These pulses p_k generate an electric current j_m in a dendrite m of neuron n. Note that according to measurements, the exponent α is a fraction in between 1 and 2. γ is a damping constant, and a a conversion factor. Finally, the cell body sums the incoming currents in a nonlinear fashion up (cf. F). The resulting F together with an incoming external signal s_n act as "driving force" for the overdamped rotation of the angle $\phi(t)$. κ is a damping constant, and the operation "mod 2π" serves the purpose that ϕ_n mod 2π remains restricted to $0 \leq {}_n < 2\pi$. It can be shown that ϕ_n mod 2π mod is just the action potential in suitable units. In my book [19], I have treated a network of such "light house" neurons and shown that it covers interesting special cases. On the one hand, it allows the

Table 2 Scheme of the light house model. For details cf. text

Spiking neurons: light house model

neuron n receives input from neuron k

$$P_k = c\sum_m \delta(\phi_k - 2\pi m) \quad , \quad m = 1, 2, \ldots$$

⇓

generates electric current j_m in dendrite m

$$\left(\frac{d}{dt} + \gamma\right)^{\alpha} j_m = a P_k$$

⇓

j_m causes „rotation" of phase angle ϕ_n in neuron n

$$\frac{d\phi_n}{dt} + \kappa\phi_n \bmod 2\pi = F(j_m) + s_j$$

external signal

$\phi_n \bmod 2\pi \propto$ action potential, $F = \sum_m j_m + \sum_{ml} \cdot\cdot j_m j_l$

↙ ↘

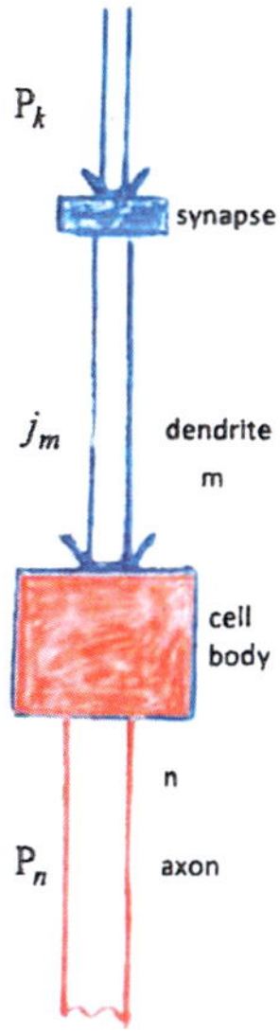

synergetic computer
pattern recognition

synchronization
phase-locking

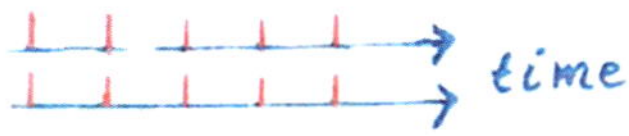

Gray, Singer: feature binding
Parkinson: Peter Tass
coordinated reset

derivation of the equations of the synergetic computer for pattern recognition; on the other hand it captures synchronization between neurons, in particular phase locking between groups of neurons as observed by Gray and Singer [20], who invoke this effect to explain feature binding. Other synchronization effects play a fundamental role in Parkinson tremor, where Tass [21] has developed efficient methods of treatment by "coordinated reset".

5 Concluding Remarks

I am very glad to see that the "Synergetics Endeavor" is still flourishing as is witnessed by this meeting at Herrenhausen Castle, by the proceedings of a meeting at Delmenhorst [22] (both meetings sponsered by the Volkswagen Foundation) as well as by numerous other publications. In a way, Synergetics rests on three columns:

dynamic systems theory dealing with deterministic processes

the theory of stochastic processes dealing with chance events

phase transition concepts of physics.

Needless to say, that Synergetics strongly profits from the Synergy between these columns as well as from the traditionally tight ties between experiment and theory. There is, however, a fourth colomn, that relates the topic *selforganization* with *information* (Haken [23]). More recently, jointly with Juval Portugali, I elaborated this relationship more closely [24]. Our starting point is Shannon's definition of information, $S = -\sum_j p_j log_2 p_j$, where p_j is the probability (or relative frequency in a different interpretation) of an event labeled by j. A well known "drawback" of this definition is that "meaning is exorcized". So there have been attempts to introduce concepts on "meaningful information". In my book [23] I defined it as such signals that have an effect on their receiver. For a historical account of the corresponding concepts of "pragmatic" and "semantic information" cf. [25]. Portugali and I have been intrigued by the question whether there is any relation between Shannonian information (SHI) on the one hand and pragmatic/semantic information (PI/SI) on the other. We found by the concrete example of visual perception based on the model of the Synergetic Computer, jointly with Jaynes' maximum (information) entropy principle, that SHI and PI/SI condition each other. Furthermore, we found that in cognition the human brain "deflates" or "inflates" SHI depending on the situation. While usually in pattern recognition the brain must select one specific percept out of many other interpretations ("deflation"), in other cases our brain "produces" several interpretations. An example is provided by Fig. 17 (from [24]). Its middle part shows a drawing by Picasso.

Lack of space does not allow me to present our results in more detail, but I can refer readers who are interested in this rather entertaining story to our book as well as to a forthcoming issue on Information and Selforganization in the journal "Entropy".

Fig. 17 Is the drawing at the center a bull? a mountain goat? a gnou? a buffalo?

References

1. B. Kröger, H. Haken, *From the Laser Light to Synergetics* (Springer International Publishing, Switzerland, 2015)
2. H. Haken, Entwicklungslinien der Synergetik, in *Beiträge zur Geschichte der Synergetik*, ed. by H. Haken, P.J. Plath, W. Ebeling, Y.M. Romanovsky (Springer Spektrum, Heidelberg, 2016)
3. H. Bénard, Les tourbillons cellulaires dans une nappe liquid. Rev. Gén. Sci. Pures Appl. **11** (1261–1271), 1309–1328 (1900)
4. W. Köhler, *Die Physischen Gestalten in Ruhe und im Stationären Zustand* (Vieweg, Braunschweig, 1920)
5. C.S. Sherrington, *The Integrative Action of the Nervous System (Electronic Resource)* (Yale University Press, New Haven, 1911)
6. J. Nash, Equilibrium points in n-person games. Proc. Natl. Acad. Sci. U.S.A. **36**, 48–49 (1950)
7. R. Paslack, *Urgeschichte der Selbstorganisation: zur Archäologie eines wissenschaftlichen Paradigmas* (Friedrich Vieweg & Sohn Verlagsgesellschaft mbH, Braunschweig, Wiesbaden, 1991)
8. T.H. Maiman, Stimulated optical radiation in ruby. Nature **147**, 493–494 (1960)
9. A.M. Prokhorov, J. Exptl. Theor. Phys. USSR **34**, 1658 (1958)
10. A.L. Schawlow, C. Townes, Infrared and optical masers. Phys. Rev. **112**, 1940–1949 (1958)
11. H. Haken, H. Sauermann, Nonlinear interaction of laser modes. Z. Physik **173**, 261–275 (1963)
12. C.L. Tang, H. Statz, G.A. de Mars, Spectral output and spiking behavior of solid-state lasers. J. Appl. Phys. **34**, 2289 (1963)

13. M. Eigen, Selforganization of matter and the evolution of biological macromolecules. Naturwiss **58**, 465–523 (1971)
14. H. Haken, A nonlinear theory of laser noise and coherence I. Z. Physik **181**, 96–124 (1964)
15. W.S. McCulloch, W. Pitts, A logical calculus of the ideas immanent in nervous activity. Bull. Math. Biophys. **5**, 115–133 (1943)
16. H. Haken, Synopsis and Introduction, in *Synergetics of the Brain,* ed. by E. Bazar, H. Flohr, H. Haken, A.J. Mandell (Springer Verlag, Berlin, 1983)
17. T. Ditzinger, H. Haken, Oscillations in the perception of ambiguous patterns. Biol. Cybern. **61**, 279–287 (1989)
18. A. Oliva, P.F. Schyns, Coarse blobs or fine edges? Evidence that information diagnosticity changes the perception of complex visual stimuli. Cognit. Psychol. **14**, 72 (1997)
19. H. Haken, *Brain Dynamics. An Introduction to Models and Simulation* (Springer, Berlin, 2002/2007)
20. C.M. Gray, W. Singer, Stimulus-dependent neuronal oscillations in the cat visual cortex area 17. IBRO Abstr. Neurosci. Lett. Suppl. **22**, 1301 (1987)
21. P. Tass, *Phase Resetting in Medicine and Biology-stochastic Modelling and Data Analysis* (Springer, Berlin, 1999)
22. A. Pelster, G. Wunner (eds.), *Selforganization in Complex Systems: The Past, Present, and the Future of Synergetics* (Springer International Publishing, Switzerland, 2016)
23. H. Haken, *Information and Selforganization: A Macroscopic Approach to Complex Systems* (Springer, Heidelberg, 1988/2000/2006)
24. H. Haken, J. Portugali, *Information Adaptation: The Interplay Between Shannon Information and Semantic Information in Cognition, Springer Briefs in Complexity* (Springer, Heidelberg, 2015)
25. P. beim Graben, H. Atmanspacher, Special issue on pragmatic information. Mind Matter **4**, 131–140 (2006)

Part II
Physics

Dynamical Systems with Time-Varying Delay: Dissipative and More Dissipative Systems

David Müller, Andreas Otto and Günter Radons

Abstract We investigate the dynamical properties of time-delay systems with time-varying delay, where we focus on the influence of the functional structure of the delay. Two universality classes of systems with time-varying delays are presented which lead to fundamental differences in the dynamics of the related systems, as for example the scaling behavior of the Lyapunov spectrum. The classification is connected to the well-known existence or non-existence of topological conjugacies of circle maps to constant rotations. It is independent of the specific delay system.

1 Introduction

Dynamical systems with time-delay arise in various fields [1] such as control theory [2, 3], climate dynamics [4, 5], life sciences [6–9], synchronization of networks [10, 11] and engineering [12, 13]. Time-varying delays are introduced for example for better controlling systems in time-delayed feedback-control [14, 15] and for taking into account environmental fluctuations or the age structure in population models [6–8, 16]. They lead to interesting phenomena in synchronization [17, 18] and are induced into engineering processes like turning and milling [19, 20] for stabilization purposes.

For our present work we consider general systems of delay differential equations (DDE) with one time-varying delay $\tau(t)$, which can be defined by

D. Müller · A. Otto · G. Radons (✉)
Institute of Physics, Chemnitz University of Technology, 09107 Chemnitz, Germany
e-mail: guenter.radons@physik.tu-chemnitz.de

D. Müller
e-mail: david.mueller@physik.tu-chemnitz.de

A. Otto
e-mail: andreas.otto@physik.tu-chemnitz.de

S.C. Müller et al. (eds.), *Complexity and Synergetics*,
https://doi.org/10.1007/978-3-319-64334-2_3

$$\dot{\mathbf{z}}(t) = \mathbf{f}(\mathbf{z}(t), \mathbf{z}(R(t)), t), \quad \text{where} \quad R(t) := t - \tau(t) \tag{1}$$

is called *retarded argument*. Especially, we are interested in the influence of the functional structure of the delay on the dynamics of the delay system. We identify two universality classes of time-varying delays, which we call *conservative* and *dissipative* delays. Some systems implicitly require the delay to be conservative. Examples are *threshold delays* in biological models [7, 8] and *variable transport delays* [21], where the time-varying delay is implicitly defined by the time to reach adulthood by a time-varying speed of development or the time needed to cover a fixed distance by a time-varying velocity, respectively. As we will see in the following, one should be aware of the fact that the modeling of such a system with a general time-varying delay may cause unphysical situations.

2 Time-Varying Delays in Applications

At first, we present typical applications of DDEs with time-varying delay, where the delay has to be conservative by definition. We start with an abstract model for a process inducing a time-varying delay which can be found in more complex and varied forms in biological models, especially population models, and in systems involving the transport over a finite distance. The model is schematically illustrated by a conveyor belt in Fig. 1a, which describes the transport of the output of a dynamical system over the finite distance c by the time-varying velocity $v(t)$. After the length c of the transport line is covered, the output of the dynamical system is fed back into the dynamical system. Hence, if the process is modeled by a differential equation and $\tau(t)$ is defined to be the time to cover the distance c, the derivative of the current state $\dot{\mathbf{z}}(t)$ depends not only on the current state $\mathbf{z}(t)$ as one expects for a model given by an ordinary differential equation, but also on the state in the past $\mathbf{z}(t - \tau(t))$ where $\tau(t)$ is time-varying due to the time-varying velocity $v(t)$. So the system can be described by a DDE with time-varying delay $\tau(t)$ as in Eq. (1). Systems with such a type of transport can be found for example in automotive engineering applications, especially in the control of the fuel-air ratio in combustion engines [22, 23], where the delay is caused by the transport of the exhaust gas from the combustion chamber to the FAR-sensor and the velocity $v(t)$ is time-varying due to the variation of the exhaust pressure.

A further example is given by FIFO buffers in electrical engineering [14], where FIFO stands for "first in first out". The delay $\tau(t)$ is determined by the time needed for the transport of the saved information from the first to the last memory area, the "distance" is given by the total amount of memory and the time-variation of the transport velocity can be induced by the modulation of the clock frequency. The mentioned transport mechanism is also applicable to biological systems in a slightly modified way. For these systems the transport line represents the grade of maturity and the delay is given by the time needed to reach a threshold of maturity which must be exceeded for reaching the adults population, i.e. the point of maturity at which a

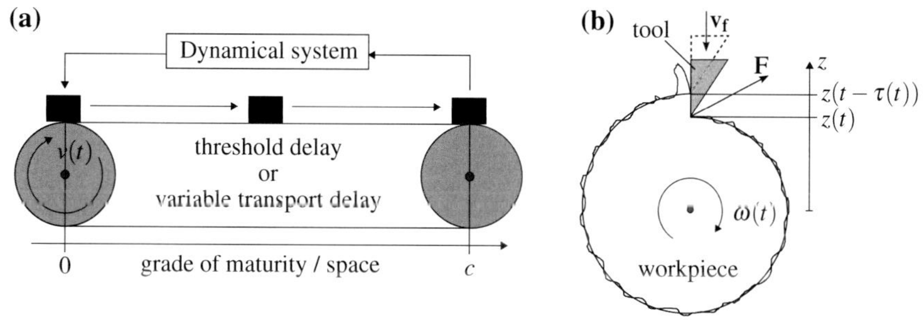

Fig. 1 **a** Threshold delay in biological models and variable transport delay. **b** Simple model for regenerative chattering in turning processes

cell is able to proliferate etc. Such type of systems can be also modeled by DDEs as Eq. (1), whereby the delay is time-varying since the rate of evolution is in general time-dependent. Environmental fluctuations or the age-structure of the population [7, 8, 16] cause the rate of evolution, represented by the transport velocity $v(t)$, to be time-varying.

Now let us consider a simple model from the field of engineering, in detail a model for regenerative chatter in turning. Stability in turning and milling processes is a problem of great interest since suppressing chatter enables the improvement of the efficiency of industrial processes due to lower degradation of the involved cutting tools and lower strain of the machines. A reduced noise exposure is also a positive side-effect. The process is schematically illustrated in Fig. 1b. The workpiece rotates with a time-varying angular velocity $\omega(t)$, whereby it is penetrated by the tool moving with feed velocity $\mathbf{v}_f$. The force acting on the tool depends on the thickness $h(t)$ of the uncut chip which is determined by the difference of the current position of the tool $z(t)$ and the position $z(t-\tau(t))$, where $\tau(t)$ is the time needed to perform one rotation. Typically, the force is defined as

$$F(t) = K\,b\,h(t) = K\,b\,(z(t-\tau(t)) - z(t)), \tag{2}$$

where K denotes the cutting force coefficient and b denotes the chip width. A simple model for regenerative chatter, which is analyzed for example in [19, 24], is obtained if one assumes that the tool has only one vibrational mode and is given by a harmonic oscillator driven by the force $F(t)$. By introducing dimensionless parameters and defining $\mathbf{z}(t) = (z(t), \dot{z}(t))^{\mathrm{T}}$ we obtain the first order DDE

$$\dot{\mathbf{z}}(t) = \mathbf{A}\,\mathbf{z}(t) + \mathbf{B}\,\mathbf{z}(t-\tau(t)). \tag{3}$$

The matrices $\mathbf{A}$ and $\mathbf{B}$ are given by

$$\mathbf{A} = \begin{pmatrix} 0 & 1 \\ -(1+w) & -2\xi \end{pmatrix}, \quad \mathbf{B} = \begin{pmatrix} 0 & 0 \\ w & 0 \end{pmatrix}, \tag{4}$$

where the damping of the tool is represented by ξ and the cutting force coefficient K, the chip width b as well as the frequency of the dominant mode of the system are taken into account by w [24]. All in all, the cause of the time-varying delay $\tau(t)$ of this system is analogous to the examples above and is given by the modulation of the angular velocity $\omega(t)$ of the workpiece.

Now let $\boldsymbol{\Phi}(t)$ be the integral of the velocity $v(t)$ for the transport models or $\omega(t)$ for the turning model, in other words let $\boldsymbol{\Phi}(t)$ be the covered distance or angle, respectively. The delay $\tau(t)$ is implicitly defined by the connection between the length of the transport line c and $\boldsymbol{\Phi}(t)$

$$c = \int_{t-\tau(t)}^{t} dt'\, v(t') = \boldsymbol{\Phi}(t) - \boldsymbol{\Phi}(t - \tau(t)), \quad \text{where} \quad \dot{\boldsymbol{\Phi}}(t) = v(t), \tag{5}$$

where for the turning model the distance c represents the angle between one turn of the workpiece, i.e. $c = 2\pi$, and the velocity $v(t)$ is replaced by the angular velocity $\omega(t)$. The specific structure of the physical process contains an inherent constant delay, which is related to an intrinsic "timescale" depending on the specific problem and equals the length of the transport line c. So even if the evolution of a dynamical system in the physical time is a more natural representation, it has the disadvantage of dealing with a time-varying delay. Since systems with constant delay are extensively studied in the literature it is advantageous to find a representation where the delay becomes constant. The existence of such a representation was already recognized and applied [25–28] and it can be easily derived from Eq. (5). If we assume the velocity $v(t)$ to be strictly positive for all times, the covered distance $\boldsymbol{\Phi}(t)$ is invertible and defines a bijective nonlinear transformation between the physical timescale and an "internal" timescale which represents for the specific problem the covered distance, the grade of maturity or the covered angle, respectively, at time t. Let φ be the new timescale, which is connected to the physical time t by $\varphi = \boldsymbol{\Phi}(t)$. Using the definition of the retarded argument with constant delay in the new timescale

$$R_c = \varphi - c \tag{6}$$

and by inserting $t = \boldsymbol{\Phi}^{-1}(\varphi)$ into Eq. (5), one obtains

$$R_c = \boldsymbol{\Phi} \circ R \circ \boldsymbol{\Phi}^{-1}, \tag{7}$$

where $\circ$ denotes function composition. This shows that the definition of a transport delay, as in Eq. (5), is nothing but the time domain formulation of a topological conjugacy, defined by Eq. (7). To transform the representation related to the physical time, given by Eq. (1), of the system to the representation related to the timescale φ we introduce the state $\mathbf{y}(\varphi)$, which is connected to the state in the physical time $\mathbf{z}(t)$ by

$$\mathbf{y} = \mathbf{z} \circ \boldsymbol{\Phi}^{-1}. \tag{8}$$

The substitution of Eq. (8) into Eq. (1) leads, after some rearrangements and the utilization of Eq. (7), to the following system with constant delay

$$\mathbf{y}'(\varphi) = \frac{1}{\dot{\Phi}(\Phi^{-1}(\varphi))} \cdot \mathbf{f}(\mathbf{y}(\varphi), \mathbf{y}(R_c(\varphi)), \Phi^{-1}(\varphi)). \tag{9}$$

As an example, the simple model of regenerative chatter in turning in angular domain, where the constant delay is given by the angle 2π and the modulation of the angular velocity is taken into account by a time-varying prefactor [27, 29], reads

$$\mathbf{y}'(\varphi) = \frac{1}{\dot{\Phi}(\Phi^{-1}(\varphi))}(\mathbf{A}\,\mathbf{y}(\varphi) + \mathbf{B}\,\mathbf{y}(\varphi - 2\pi)). \tag{10}$$

So we have seen that systems with time-varying delay, where the delay is caused by the transport over a finite and constant distance, are characterized by an inherent constant delay. In other words these systems are equivalent to systems with constant delay in the sense that they show the same dynamics, since they are connected by a well-behaved nonlinear timescale transformation which maps the trajectories of the system with time-varying delay one-to-one to the trajectories of a system with constant delay. Immediately the question raises, whether all systems with time-varying delay can be transformed to systems with constant delay, or not. In general this is not true and based on this criterion, DDEs with time-varying delay can be separated into two universality classes. The systems considered in this section are examples for the class of systems which we call systems with conservative delay. The reason is explained in the next section.

3 Conservative and Dissipative Delays

At this point we derive the two classes of systems with time-varying delay from the equivalence or non-equivalence to systems with constant delay as done in [30, 31]. We consider only periodic delays which define an invertible retarded argument and without any loss of generality we set the delay period to one. The question, whether a system is equivalent to a system with constant delay can be reduced to the question, whether there exists a somehow well-behaved transformation function $\Phi(t)$ which solves Eq. (7) for a given retarded argument $R(t)$. Here "well-behaved" means that the timescale transformation maps all solutions of the original DDE one-to-one to solutions of the resulting DDE, such that dynamical quantities like the Lyapunov exponents are well defined and are preserved under the transformation. Equation (7) takes the structure of a conjugacy equation, which is known from the theory of iterated maps, especially circle maps [32]. In our case it defines the well-understood topological conjugacy between a general circle map $R(t)$ and the pure, constant rotation $R_c(\varphi)$. The maps are defined by the retarded arguments and are given by

$$t_k = R(t_{k-1}) = R^k(t_0) \quad \text{and} \quad \varphi_k = R_c(\varphi_{k-1}) = R_c^k(\varphi_0) = \varphi_0 - k\,c, \tag{11}$$

where $R^k(t)$ denotes the composition of $R(t)$ with itself k-times, where $R^0(t) = t$ and $R_c^0(\varphi) = \varphi$, and the orbits of the maps are connected by $\varphi_k = \Phi(t_k)$. Topological conjugate maps exhibit the same dynamics and dynamical quantities are invariant under the conjugacy. One invariant quantity is given by the rotation number [32]

$$\rho = \lim_{k\to\infty} \frac{R^k(t) - R^0(t)}{k} \stackrel{!}{=} -c, \tag{12}$$

which can be understood as the mean drift per iteration of the map $R(t)$. Trivially, the rotation number of the pure rotation R_c is $\rho = -c < 0$ and consequently, the rotation number of a topological conjugate map $R(t)$ is also equal to $-c$.

Another invariant quantity is given by the Lyapunov exponent μ of the map $R(t)$, which describes the rate of the evolution of small perturbations at a reference orbit of the map [33]. Since the pure rotation R_c preserves initial perturbations, the Lyapunov exponent equals zero. Hence the Lyapunov exponent of a topological conjugate map $R(t)$, where the first derivatives of $\Phi(t)$ and $\Phi^{-1}(\varphi)$ exist and are bounded, is also equal to zero. These properties of a map which is topological conjugate to the pure rotation are automatically fulfilled if the criterion

$$\lim_{n\to\infty} R^{q_n}(t) + p_n = t \tag{13}$$

holds, where $\frac{p_n}{q_n}$ are the convergents of the continued fraction expansion of the constant delay c [34]. It can be shown that a transformation function $\Phi(t)$ exists and can be computed by

$$\Phi(t) = \lim_{n\to\infty} \frac{1}{n} \sum_{k=0}^{n-1} R^k(t) =: \langle R^k(t)\rangle, \tag{14}$$

if the map defined by the retarded argument $R(t)$ fulfills Eq. (13) [35]. Another quantity which is invariant under this conjugacy is the mean distance between two orbits t_k and $\tilde{t}_k$ or φ_k and $\tilde{\varphi}_k$, respectively. With Eq. (14) and the connection between the orbits of $R_c(\varphi)$ and $R(t)$, it is easy to derive that the mean distance between orbits of the pure rotation $\langle \Delta\varphi_k\rangle = \langle \tilde{\varphi}_k - \varphi_k\rangle$ is equal to the mean distance $\langle \Delta t_k\rangle = \langle \tilde{t}_k - t_k\rangle$ between two orbits of $R(t)$, i.e.

$$\langle \Delta\varphi_k\rangle = \Delta\varphi_0 = \Phi(\tilde{t}_0) - \Phi(t_0) = \langle R^k(t_0 + \Delta t_0) - R^k(t_0)\rangle = \langle \Delta t_k\rangle. \tag{15}$$

Hence, the maps defined by this type of delays are conservative systems and we call this class of delays *conservative delays*. The reduced version of maps, obtained by taking Eq. (11) modulo one, which are not topological to a pure rotation, possess attractive fixpoints or periodic orbits due to Poincaré's classification [32]. Consequently, the mean distance between close orbits vanishes and except for some pathological cases the related Lyapunov exponent is negative, $\mu < 0$. DDEs which are

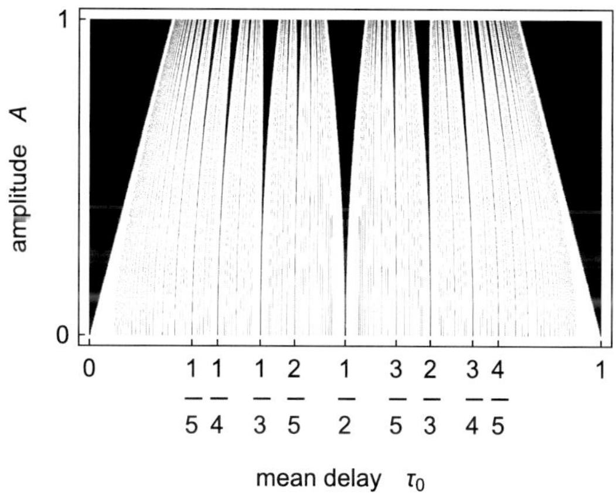

Fig. 2 Parameter regions of the delay classes for a sine-shaped delay with amplitude A and mean delay τ_0. The set of parameters corresponding to dissipative delays (*black*) take the structure of Arnold tongues and the parameter set corresponding to conservative delays (*white*) is a Cantor set. The Arnold tongues related to circle maps possessing periodic orbits with a period of 1–20 are shown

characterized by a delay defining a map of this type are not equivalent to a system with constant delay. Since the corresponding circle maps are dissipative dynamical systems, we call the related delays *dissipative delays*.

As mentioned in the last section, systems with variable transport delays or threshold delays are equivalent to systems with constant delay by definition. With our preceding derivation of the delay classification it is clear that this is the case, because the implicit definition of the time-varying delay by Eq. (5), which is equivalent to Eq. (7), implies a topological conjugacy between a pure rotation defined by the intrinsic constant delay and the circle map defined by the time-varying delay. Hence variable transport delays and threshold delays are conservative delays by definition. So let us consider a DDE equipped with a common parameter family of time-varying delays

$$\tau(t) = \tau_0 + \frac{A}{2\pi}\sin(2\pi t), \tag{16}$$

with the mean delay τ_0 and the amplitude A, and let us check which set of parameters defines which type of delay. In Fig. 2 the sets corresponding to conservative and dissipative delays are represented by the white regions and the black regions, respectively. The parameters related to conservative delays form a Cantor set and the related circle maps are characterized by quasiperiodic and marginally stable dynamics, $\mu = 0$. The parameters related to dissipative delays are characterized by Arnold tongues and the related circle maps are characterized by mode-locking with $\mu < 0$ [33, 36]. An interesting fact is that circle maps which are topological conjugate to the pure rotation are structurally unstable, whereas circle maps which are characterized by mode-locking are structurally stable [32]. This means that a small parameter change or a small perturbation of a conservative delay likely leads to a dissipative delay. Moreover, if a system is modeled by a DDE with time-varying delay, where the delay is given by a parameter family as defined above and one parameter is continuously varied starting with a conservative delay, the system crosses infinitely many regions of dissipative and conservative delays due to the fractal structure of the parameter space. Thus, modeling a system, which is characterized by a threshold or variable transport delay, with a DDE with a general delay parameter family may leads to unphysical situations even for small parameter changes, since having a conservative delay is an implicit requirement for these systems.

The influence of the delay class on the dynamics of time-delay systems can be exposed by the analysis of the Lyapunov spectrum. Hence, we consider the evolution of small perturbations of a reference trajectory. For a given solution $\tilde{\mathbf{z}}(t)$ of Eq. (1), the linearized system describing the dynamics of small perturbations $\mathbf{x}(t) = \mathbf{z}(t) - \tilde{\mathbf{z}}(t)$ can be given by

$$\dot{\mathbf{x}}(t) = \mathbf{A}(t)\,\mathbf{x}(t) + \mathbf{B}(t)\,\mathbf{x}(R(t)), \tag{17}$$

where $\mathbf{A}(t) = \partial_{\mathbf{z}(t)}\mathbf{f}$ and $\mathbf{B}(t) = \partial_{\mathbf{z}(R(t))}\mathbf{f}$ denote the Jacobians of $\mathbf{f}$ with respect to the state $\mathbf{z}(t)$ and the delayed state $\mathbf{z}(R(t))$ taken at the unperturbed solution $\tilde{\mathbf{z}}(t)$. The Lyapunov spectrum $\{\lambda_n : n \in \mathbb{N}\}$ is given by the relaxation rates, called Lyapunov exponents, of perturbations related to a specific direction of the initial perturbation. Since DDEs are infinite dimensional systems, the Lyapunov spectrum consists of infinitely many Lyapunov exponents, which are typically ordered from the largest to the smallest [37]. Delay systems are strongly dissipative since for large index the Lyapunov exponents tend to $-\infty$. In Fig. 3 the Lyapunov spectra of the three systems

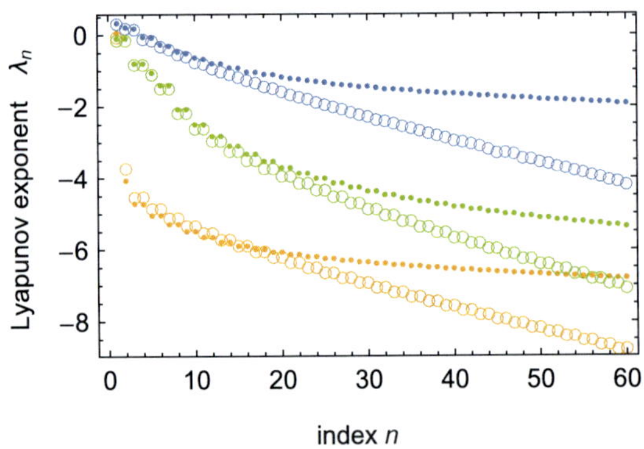

Fig. 3 Comparison of the Lyapunov spectra between systems with conservative delay (dots) and dissipative delay (circles) for the systems, from top to bottom, Eqs. (18a), (18b) and (18c). We have chosen the delay $\tau(t) = \tau_0 + (0.9/2/\pi)\sin(2\pi\, t)$ with the mean delay $\tau_0 = 1.51$ and $\tau_0 = 1.54$ for dissipative and conservative delay, respectively. The spectra where computed by the *semi-discretization* method [12, 38]

$$\dot{z}(t) = 2z(t)\,(1 - z(R(t)))\,, \tag{18a}$$

$$\dot{z}(t) = \frac{10z(R(t))}{1 + z(R(t))^{10}} - 5z(t), \tag{18b}$$

$$\ddot{z}(t) + \dot{z}(t) + 4\pi^2 z(t) = 8z(R(t)), \tag{18c}$$

related to an exemplary dissipative and an exemplary conservative delay are shown, where Eq. (18a) is the Hutchinson equation, modeling population dynamics [6], Eq. (18b) is the Mackey-Glass equation, a model for blood production [39] and Eq. (18c) is the simple turning model from the last section. DDEs with conservative delays are equivalent to systems with constant delays. Hence characteristic properties of the Lyapunov spectrum for DDEs with constant delay are characteristic for DDEs with conservative delay as well. The asymptotic scaling for exponents with large index is logarithmic for systems with constant delay [37].

In Fig. 3 the spectra for systems with conservative and dissipative delay are represented by dots and empty circles, respectively. We notice that the Lyapunov exponents related to dissipative delay decay much faster with the index than the exponents of systems with conservative delay. Moreover, it can be shown that the asymptotic scaling for systems with dissipative delay is linear [30, 31]. Hence the dissipative dynamics of the circle map related to dissipative delay leads to an *additional* dissipation in the dynamics of the delay system in contrast to conservative delays, where the related circle map is a conservative system and preserves the logarithmic scaling behavior known from systems with constant delay. All in all, we have shown that the dynamical properties of the access to the past by the delay itself influences the dynamics of the delay system, whereby systems with dissipative delay are more dissipative than systems with conservative delay which are dissipative as well.

4 Conclusion

We have presented two universality classes of delay systems with time-varying delay. The classification is independent of the specific form of the delay system and depends only on the functional structure of the delay. Systems with conservative delay are equivalent to systems with constant delay and are characterized by a logarithmic scaling behavior of the Lyapunov spectrum. In contrast, systems with dissipative delay are not equivalent to systems with constant delay and the dissipative dynamics related to the delay access induces an additional dissipation into the dynamics of the delay system, which leads to a linear scaling behavior of the Lyapunov spectrum. As an exemplary application we have shown that the delay type becomes relevant in the case of systems involving transport, where a conservative delay is an implicit requirement of the physical system, due to the implicit definition of the time-varying delay.

References

1. W. Just, A. Pelster, M. Schanz, E. Schöll, Delayed complex systems: an overview. Phil. Trans. R. Soc. A **368**, 303–304 (2010)
2. J.P. Richard, Time-delay systems: an overview of some recent advances and open problems. Automatica **39**, 1667–1694 (2003)
3. E. Schöll, H.G. Schuster, *Handbook of Chaos Control* (Wiley-VCH, Weinheim, 2008)
4. M. Ghil, I. Zaliapin, S. Thompson, A delay differential model of ENSO variability: parametric instability and the distribution of extremes. Nonlinear Processes Geophys. **15**, 417–433 (2008)
5. E. Tziperman, L. Stone, M.A. Cane, H. Jarosh, El Niño chaos: overlapping of resonances between the seasonal cycle and the Pacific ocean-atmosphere oscillator. Science **264**, 72–74 (1994)
6. K. Gopalsamy, *Stability and Oscillations in Delay Differential Equations of Population Dynamics* (Kluwer Academic, Dordrecht, 1992)
7. Y. Kuang, *Delay Differential Equations: With Applications in Population Dynamics* (Academic Press, San Diego CA, 1993)
8. H.L. Smith, Reduction of structured population models to threshold-type delay equations and functional differential equations: a case study. Math. Biosci. **113**, 1–23 (1993)
9. H.L. Smith, *An Introduction to Delay Differential Equations with Applications to the Life Sciences*, vol. 57 (Springer, New York, 2010)
10. F.M. Atay, J. Jost, A. Wende, Delays, connection topology, and synchronization of coupled chaotic maps. Phys. Rev. Lett. **92**, 144101 (2004)
11. M. Lakshmanan, D.V. Senthilkumar, *Dynamics of Nonlinear Time-Delay Systems* (Springer, Berlin, 2011)
12. T. Insperger, G. Stépán, *Semi-Discretization for Time-Delay Systems: stability and Engineering Applications* (Springer, New York, 2011)
13. Y.N. Kyrychko, S.J. Hogan, On the use of delay equations in engineering applications. J. Vibr. Control **16**, 943–960 (2010)
14. T. Jüngling, A. Gjurchinovski, V. Urumov, Experimental time-delayed feedback control with variable and distributed delays. Phys. Rev. E **86**, 046213 (2012)
15. Y. Sugitani, K. Konishi, N. Hara, Experimental verification of amplitude death induced by a periodic time-varying delay-connection. Nonlinear Dyn. **70**, 2227–2235 (2012)
16. M. Kloosterman, S. Campbell, F. Poulin, An NPZ model with state-dependent delay due to size-structure in juvenile zooplankton. SIAM J. Appl. Math. **76**, 551–577 (2016)
17. G. Ambika, R.E. Amritkar, Anticipatory synchronization with variable time delay and reset. Phys. Rev. E **79**, 056206 (2009)
18. W.H. Kye, M. Choi, M.S. Kurdoglyan, C.M. Kim, Y.J. Park, Synchronization of chaotic oscillators due to common delay time modulation. Phys. Rev. E **70**, 046211 (2004)
19. A. Otto, G. Radons, Application of spindle speed variation for chatter suppression in turning. CIRP J. Manuf. Sci. Technol. **6**, 102–109 (2013)
20. M. Zatarain, I. Bediaga, J. Muñoa, R. Lizarralde, Stability of milling processes with continuous spindle speed variation: analysis in the frequency and time domains, and experimental correlation. CIRP Ann. Manuf. Technol. **57**, 379–384 (2008)
21. F. Zhang, M. Yeddanapudi, Modeling and simulation of time-varying delays, in *Proceedings of TMS/DEVS*, San Diego CA, (2012) pp 34:1–34:8
22. D. Bresch-Pietri, N. Petit, Implicit integral equations for modeling systems with a transport delay. In: (Eds.), E. Witrant, E. Fridman, O. Sename, L. Dugard, Recent results on time-delay systems, no. 5 in Advances in delays and dynamics, (Springer International Publishing, Cham, 2016), pp 3–21
23. D. Bresch-Pietri, J. Chauvin, N. Petit, Prediction-based stabilization of linear systems subject to input-dependent input delay of integral-type. IEEE Trans. Autom. Control **59**, 2385–2399 (2014)
24. T. Insperger, G. Stépán, Stability analysis of turning with periodic spindle speed modulation via semidiscretization. J. Vib. Control **10**, 1835–1855 (2004)

25. R. Bellman, K.L. Cooke, On the computational solution of a class of functional differential equations. J. Math. Anal. Appl. **12**, 495–500 (1965)
26. J.P. Seddon, R.A. Johnson, The simulation of variable delay. IEEE Trans. Comput. **1**, 89–94 (1968)
27. T.C. Tsao, M.W. McCarthy, S.G. Kapoor, A new approach to stability analysis of variable speed machining systems. Int. J. Mach. Tools Manuf. **33**, 791–808 (1993)
28. E.I. Verriest, Inconsistencies in systems with time-varying delays and their resolution. IMA J. Math. Control Inf. **28**, 147–162 (2011)
29. A. Otto, G. Radons, The influence of tangential and torsional vibrations on the stability lobes in metal cutting. Nonlinear Dyn. **82**, 1989–2000 (2015)
30. D. Müller, A. Otto, G. Radons, From dynamical systems with time-varying delay to circle maps and Koopman operators. Phys. Rev. E **95**, 062214 (2017)
31. A. Otto, D. Müller, G. Radons, Universal dichotomy for dynamical systems with variable delay. Phys. Rev. Lett. **118**, 044104 (2017)
32. A. Katok, B. Hasselblatt, *Introduction to the Modern Theory of Dynamical Systems*, vol. 54 (Cambridge University Press, Cambridge, 1997)
33. E. Ott, *Chaos in Dynamical Systems* (Cambridge University Press, Cambridge, 2002)
34. A.Y. Khinchin, *Continued Fractions* (Dover Publications, Mineola, 1997)
35. M.R. Herman, Sur la conjugaison différentiable des difféomorphismes du cercle a des rotations. Inst. Hautes Études Sci. Publ. Math. **49**, 5–233 (1979)
36. V.I. Arnold, *Collected Works* (Springer, Berlin, 2009)
37. J. Farmer, Chaotic attractors of an infinite-dimensional dynamical system. Physica D **4**, 366–393 (1982)
38. T. Insperger, G. Stépán, Updated semi-discretization method for periodic delay-differential equations with discrete delay. Int. J. Numer. Method Eng. **61**, 117–141 (2004)
39. M.C. Mackey, L. Glass, Oscillation and chaos in physiological control systems. Science **197**, 287–289 (1977)

Markov Property of Velocity Increments in Burgers Turbulence

Jan Friedrich and Rainer Grauer

Abstract We investigate the intermittency properties of a turbulent flow without pressure described by the Burgers equation. To this end, we make use of a phenomenogical description devised by R. Friedrich and J. Peinke [Phys. Rev. Lett. 78, 863 (1997)] that interprets the turbulent energy cascade as a Markov process in scale. The impact of Burgers-shocks on the Markov property of the velocity increments is discussed and compared to numerical simulations. Furthermore, we give a brief outlook on the use of the Markov property as a possible closure of a hierarchy of multi-increment probability density functions derived directly from the Burgers equation.

1 Introduction

Despite tremendous efforts to establish a statistical description of turbulence over the past century, the problem is still far from being solved and we yet have to identify the basic physical mechanisms underlying the phenomenon [1]. The presence of a variety of interlacing vortical structures of different sizes and the complex nonlinear transfer of energy through the corresponding scales forms a system, which is far from equilibrium and therefore not amenable to any of our current perturbative methods [2]. Especially the phenomenon of intermittency manifesting itself by the occurrence of very large velocity fluctuations at smaller scales and the resulting anomalous statistics is ambitious. Accordingly, descriptions of intermittency effects are of rather phenomenological nature [3, 4].

J. Friedrich (✉) · R. Grauer
Institut für Theoretische Physik I, Ruhr-Universität Bochum,
Universitätsstraße 150, 44801 Bochum, Germany
e-mail: jaf@tp1.rub.de

R. Grauer
e-mail: grauer@tp1.rub.de

S.C. Müller et al. (eds.), *Complexity and Synergetics*,
https://doi.org/10.1007/978-3-319-64334-2_4

Nevertheless, the last decades of turbulence research have seen several promising methods that were mainly borrowed from field theory such as the so-called operator product expansion [5, 6] or the instanton method [7, 8]. Another interesting approach via the theory of stochastic processes was provided by Friedrich and Peinke [9]. Here, the concept of the energy cascade, i.e., the transfer of energy from large to small scales boils down to a Markov process of the velocity increments in scale. It could be shown experimentally that the Markov property is fulfilled over a broad range of scales and is broken only at smaller scale separations [10]. This signifies a considerable reduction of the complexity of the problem, since the n-increment probability density function (PDF) can be factorized into a product containing only transition probabilities. Furthermore, it follows that the evolution in scale of the one-increment PDF as well as the transition PDF are described by the same so-called Kramers-Moyal expansion, which allows for a quite intuitive description of intermittency effects.

However, since this is a purely phenomenological theory, no contact to the basic fluid dynamical equation, the Navier-Stokes equation, has been made up to this point. In this context, a quite promising approach seems to be the Lundgren-Monin-Novikov (LMN) hierarchy that derives PDF equations directly from the basic fluid dynamical equations [11–13]. Its inherent difficulty lies in the occurrence of unclosed terms that couple to statistical quantities of higher order [14]. The n-increment PDF equation, for instance, involves terms containing the $n + 1$-increment PDF and we are left with an infinite hierarchy of PDF equations that is formally similar to the BBGKY-hierarchy from statistical physics. The aim of this contribution is to give a brief outlook on a possible reconciliation of the Markov phenomenology and the LMN hierarchy at the example of the Burgers equation, a simplified version of the Navier-Stokes equation. Here, intermittency effects are caused by shocks that eventually form due to the nonlinear and compressible character of the Burgers equation [15]. In Sect. 2, we describe those effects via the Markov phenomenology and investigate its validation. Section 3 mentions work in progress that addresses a possible closure of the Burgers-LMN hierarchy via the Markov property of the increment statistics.

2 Description of the Intermittency of Burgers Velocity Increments by a Markov Process in Scale

The Burgers equation applies for a turbulent flow with neglected pressure and reads

$$\frac{\partial}{\partial t}u(x,t) + u(x,t)\frac{\partial}{\partial x}u(x,t) = \nu\frac{\partial^2}{\partial x^2}u(x,t) + F(x,t), \tag{1}$$

where the forcing is assumed to be white noise in time $\langle F(x,t)F(x',t)\rangle = \chi(x - x')\delta(t - t')$. A typical realization of the velocity field obtained from numerical simulations is depicted in Fig. 1a: Smooth ramps of positive velocity increments

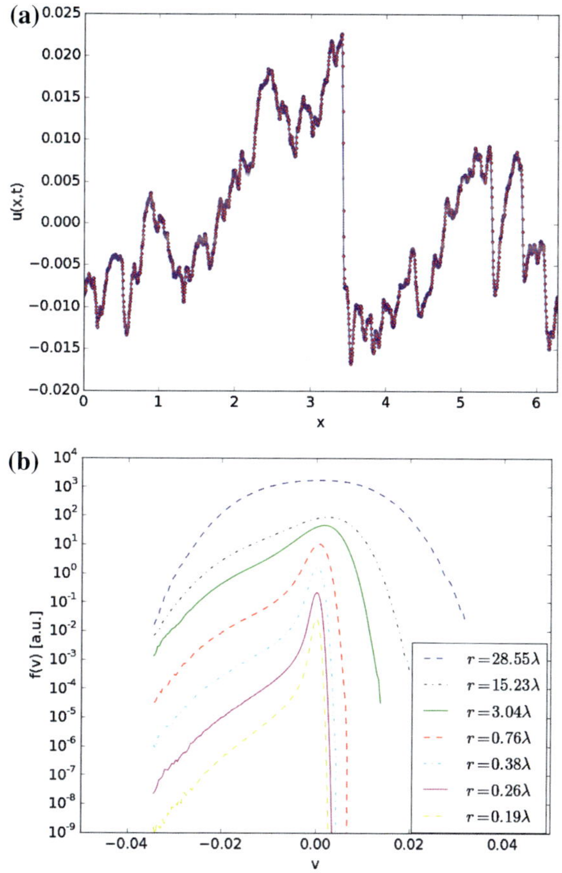

Fig. 1 **a** Typical realization of the velocity field obtained from numerical simulations of forced Burgers turbulence. Large negative velocity field gradients such as the dominant shock in the center are build up by the nonlinearity in Eq. (1) and are then subject to an increased rate of energy dissipation. The characteristic parameters of the simulation are assembled in Table 1. The spatial correlation of the forcing is $\sim k^{-1}$ in Fourier space [16]. **b** Evolution of the one-increment PDF $f_1(v, r)$ in scale obtained from the same numerical simulations. The PDFs are shifted vertically for a better view. The scale is given in units of the Taylor length λ (see Table 1 for definitions). On large scales, the PDF is close to Gaussian whereas it develops a pronounced asymmetry as it tends to smaller scales due to the presence of shocks in the left tail. Finally, it approaches the gradient PDF

are interrupted by violent shocks of negative velocity increments that are eventually smeared out by viscosity. The latter manifest themselves by the extreme scaling of the velocity structure functions [15]:

$$\langle v(R,r)^n \rangle = \langle (u(R+r) - u(R))^n \rangle \sim r^{\zeta_n} \qquad \text{with} \quad \zeta_n = \begin{cases} n & \text{for } n < 1, \\ 1 & \text{for } n \geq 1. \end{cases} \tag{2}$$

Here, the first scaling exponent is due to regions where the velocity field is smooth whereas the latter exponent corresponds to shock events similar to Fig. 1a. A full statistical description of the fluctuating velocity field in Eq. (1) is tantamount to the knowledge of the n-increment PDF

$$f_n(v_n, r_n; v_{n-1}, r_{n-1}; ...; v_1, r_1) = \prod_{i=1}^{n} \langle \delta(v_i - v(R, r_i)) \rangle. \tag{3}$$

where the increments are taken at the same reference point R whose dependence along with t has been dropped under the assumption of homogeneity and stationarity.

The n-increment PDF can be simplified significantly under the assumption that the increments possess a Markov property in scale for the ordering $r_n \ll r_{n-1} \ll r_1$

$$p(v_n, r_n | v_{n-1}, r_{n-1}; ...; v_1, r_1) = \frac{f_n(v_n, r_n; ...; v_1, r_1)}{f_{n-1}(v_{n-1}, r_{n-1}; ...; v_1, r_1)} \overset{Markov}{=} p(v_n, r_n | v_{n-1}, r_{n-1}) \tag{4}$$

which means that the n-increment PDF can be factorized according to

$$f_n(v_n, r_n; v_{n-1}, r_{n-1}; ...; v_1, r_1) = p(v_n, r_n | v_{n-1}, r_{n-1}) ... p(v_2, r_2 | v_1, r_1) f_1(v_1, r_1). \tag{5}$$

For a given distribution $f_1(v_1, r_1)$ on large scales, typically a Gaussian, the knowledge of the transition PDF $p(v_n, r_n | v_{n-1}, r_{n-1})$ is thus sufficient for an entire statistical description. Furthermore, a central notion of a Markov process is that the one-increment PDF as well as the transition PDF follow the same Kramers-Moyal expansion

$$-\frac{\partial}{\partial r} f_1(v, r) = \sum_{n=1}^{\infty} (-1)^n \frac{\partial^n}{\partial v^n} D^{(n)}(v, r) f_1(v, r) = \hat{L}_{KM} f(v, r) \tag{6}$$

$$-\frac{\partial}{\partial r} p(v, r | v', r') = \sum_{n=1}^{\infty} (-1)^n \frac{\partial^n}{\partial v^n} D^{(n)}(v, r) p(v, r | v', r') = \hat{L}_{KM} p(v, r | v', r') \tag{7}$$

where the minus sign indicates that the process occurs from large to small scales and the Kramers-Moyal coefficients are defined as

Table 1 Characteristic parameters of the numerical simulations: root mean square velocity $u_{rms} = \sqrt{\langle u^2 \rangle}$, viscosity ν, averaged rate of local energy dissipation $\langle \varepsilon \rangle = \nu \left\langle \left(\frac{\partial u}{\partial x} \right)^2 \right\rangle$, grid spacing dx, dissipation length $\eta = \left(\frac{\nu^3}{\langle \varepsilon \rangle} \right)^{1/4}$, Taylor length $\lambda = u_{rms} \sqrt{\frac{\nu}{\langle \varepsilon \rangle}}$, Taylor-Reynolds number $\mathrm{Re}_\lambda = \frac{u_{rms} \lambda}{\nu}$, integral length scale $L = \frac{u_{rms}^3}{\langle \varepsilon \rangle}$, large-eddy turn-over time $T_L = \frac{L}{u_{rms}}$ and number of grid points N

u_{rms}	ν	$\langle \varepsilon \rangle$	dx	η	λ	Re_λ	L	T in T_L	N
0.01	0.0001	1.02×10^{-6}	0.003	0.0315	0.099	9.96	0.989	25,815	2048

Table 2 Phenomenological models of turbulence (we refer the reader to [17] for further discussions) with scaling exponents ζ_n and the corresponding reduced Kramers-Moyal coefficients from Eq. (8): Kolmogorov's mean field theory from 1941 (K41), Kolmogorov-Oboukhov theory from 1962 (K62), Burgers phenomenology, She-Leveque phenomenology, Yakhot's model and anti-de Sitter/conformal field theory random geometry model [18]. Note that the K41 as well as the Burgers-ramps do not exhibit intermittency corrections due to the absence of multiplicative noise represented by $D^{(2)}(v, r)$. The K62 phenomenology is the only intermittency model that can be reproduced with just two Kramers-Moyal coefficients

Model	ζ_n	Reduced Kramers-Moyal coefficients
K41	$n/3$	$C_1 = 1/3$, no higher orders
K62[a]	$n/3 - \mu n(n-3)/18$	$C_1 = (3+\mu)/9$, $C_2 = \mu/9$, no higher orders
Burgers-ramps	n	$C_1 = 1$, no higher orders
Burgers-shocks	1	$C_n = 1$
She-Leveque[d]	$\frac{n}{9} + 2\left(1 - \left(\frac{2}{3}\right)^{n/3}\right)$	$C_n = \frac{1}{9}\left(n\,{}_1F_0(1-n;;1) + 18\left(1 - \sqrt[3]{\frac{2}{3}}\right)^n\right)$
Yakhot[b]	$\zeta_{2n} = \frac{2(1+3\beta)n}{3(1+2\beta n)}$	$C_n = \frac{\Gamma[n+1]}{\Gamma[n+1+\frac{1}{\beta}]}\left(\Gamma[1+\frac{1}{\beta}] + \frac{1}{3\beta^2}\Gamma[\frac{1}{\beta}]\right)$
ADS/CFT[c]	$\frac{\left((1+\gamma^2)^2+4\gamma^2(\frac{n}{3}-1)\right)^{\frac{1}{2}}+\gamma^2-1}{2\gamma^2}$	no analytic expression

Experimental values for intermittency parameters: [a]$\mu \approx 0.227$, [b]$\beta = 0.05$ and [c]$\gamma^2 = 0.161$. [d]Here, ${}_vF_q(a; b; z)$ is the generalized hypergeometric function

$$D^{(n)}(v', r') = \frac{1}{n!} \lim_{r \to r'} \frac{1}{r - r'} \int dv (v - v')^n p(v, r|v', r'). \tag{8}$$

In comparison, Fig. 1b shows a typical evolution in scale of the one-increment PDF obtained directly from the numerical simulation. The right part corresponds to smooth ramps, whereas the shocks enter in the left revealing a pronounced left tail of the PDF at small scales. The correct choice of the Kramers-Moyal coefficients in Eq. (6) should thus enable us to reproduce this behaviour. To this end, we take the moments $\langle v^n \rangle$ from Eq. (6). The scaling of the structure functions (2) can now be guaranteed by the choice

$$D^{(1)}(v, r) = -\frac{v}{r}\ , \quad D^{(n)}(v, r) = 0 \ \text{ for } \ n > 1\ , \qquad \text{for positive increments,} \tag{9}$$

$$D^{(n)}(v, r) = \frac{(-1)^n}{n!} \frac{v^n}{r}\ , \qquad \text{for negative increments.} \tag{10}$$

In order to compare this result with other existing phenomenological models (K41, K62, She-Leveque etc.), we have assembled the corresponding Kramers-Moyal coefficients in Table 2.

Remarkably, any of our current phenomenological models can be reproduced by the Kramers-Moyal expansion (6) and the proper choice of the Kramers-Moyal coefficients

$$D^{(n)}(v,r) = C_n \frac{(-1)^n}{n!} \frac{v^n}{r}, \tag{11}$$

which corresponds to scaling solutions for the moments $\langle v^n \rangle$. As it can be seen from Table 2 as well as from the plots of the reduced Kramers-Moyal coefficients C_n in Fig. 2, the Burgers phenomenology exhibits the most extreme intermittency effects of all the considered models. This sheds also some new light on the use of the so-called Pawula theorem [19] for the Kramers-Moyal expansion (6). The theorem states that if $D^{(4)}(v,r) = 0$, then all higher order coefficients are zero and the expansion can be terminated at $n = 2$, leaving us with an ordinary Fokker-Planck equation. In [9], the use of Pawula's theorem was justified from the free jet experiment, since $D^{(4)}(v,r) \approx 0$ in this case. We have plotted the reduced Kramers-Moyal coefficients in Fig. 2. From the ordinary plot in Fig. 2a, it seems that $C_4 \approx 0$ for all models except for the Burgers model. However, the same logarithmic plo reveals the correct asymptotics of the reduced coefficients.

We are now in the position to calculate the transition PDFs for positive and negative increments from the Kramers-Moyal expansion (7) with the corresponding Kramers-Moyal coefficients from Eqs. (9) or (10). For the ramps, we only have to deal with a drift coefficient and the solution can be determined by the method of characteristics [20] according to

$$p(v,r|v',r') = \delta\left(v - \frac{r}{r'}v'\right). \tag{12}$$

Here, we used the initial condition $p(v,r'|v',r') = \delta(v-v')$ as a consequence of the coincidence property of the n-increment PDF (3). Note that although this initial condition seems to be the proper choice with regard to the coincidence property, it is also known [10] that the Markov property (4) breaks down at the so-called Markov-Einstein coherence length, i.e., if $|r - r'| \ll \lambda_{ME}$. Therefore, one might also seek for other initial conditions that explicitly take into account finite λ_{ME}. Turning to the Kramers-Moyal expansion of the shocks (10), a solution can be obtained from its Dyson series representation [19] according to

$$\begin{aligned}
p(v,r|v',r') &= \delta(v-v') + \int_{r'}^{r} \mathrm{d}r_1 \hat{L}_{KM}(v,r_1)\delta(v-v') \\
&\quad + \int_{r'}^{r} \mathrm{d}r_1 \int_{r'}^{r_1} \mathrm{d}r_2 \hat{L}_{KM}(v,r_1)\hat{L}_{KM}(v,r_2)\delta(v-v') + \ldots \\
&= \delta(v-v') + \int_{r'}^{r} \mathrm{d}r_1 \frac{\hat{L}}{r_1}\delta(v-v') + \int_{r'}^{r} \mathrm{d}r_1 \int_{r'}^{r_1} \mathrm{d}r_2 \frac{\hat{L}^2}{r_1 r_2}\delta(v-v') + \ldots \\
&= \delta(v-v') + \ln\frac{r}{r'}\hat{L}\delta(v-v') + \frac{1}{2!}\left(\ln\frac{r}{r'}\right)^2 \hat{L}^2\delta(v-v') + \ldots \\
&= \exp\left[\ln\frac{r}{r'}\hat{L}\right]\delta(v-v')
\end{aligned} \tag{13}$$

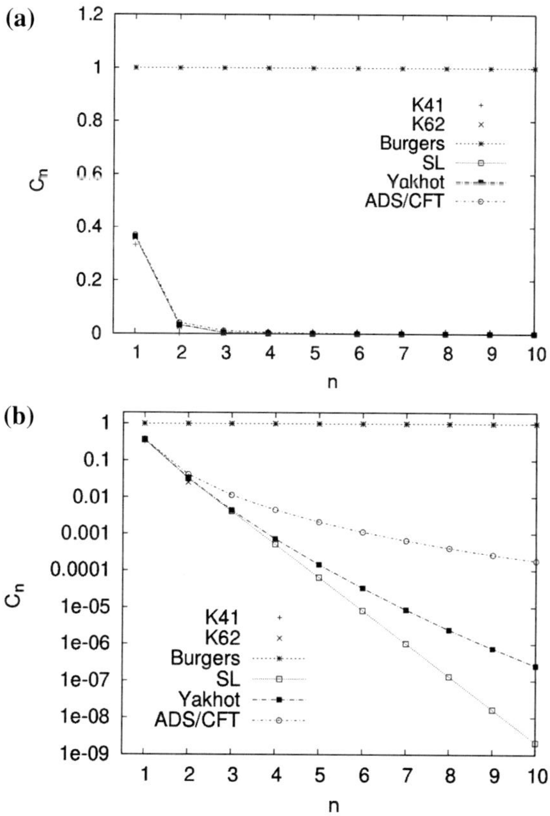

Fig. 2 **a** Reduced Kramers-Moyal coefficients C_n for different phenomenological models according to Table 2. Note that the K41 and the K62 coefficients end after $n = 1$ and $n = 2$, respectively. Only the Burgers phenomenology seems to depend on an infinite number of coefficients whereas the use of Pawula's theorem seems to be adequate for all other models, since the reduced Kramers-Moyal coefficient of order four appears to vanish, $C_4 \approx 0$. However, this can be misleading as it is shown in the logarithmic plot **b** that reveals the correct asymptotics. Clearly, coefficients of higher order are needed to reproduce the correct intermittency effects. This also guarantees that the scaling exponents ζ_n are monotonically increasing functions of n, which is not the case for intermittency of the K62 type

where we introduced the operator $\hat{L} = -\sum_{n=1}^{\infty} \frac{1}{n!}\frac{\partial^n}{\partial v^n} v^n$. Letting this operator act on the delta function yields

$$\hat{L}\delta(v - v') = -\sum_{n=1}^{\infty} \frac{1}{n!}\frac{\partial^n}{\partial v^n} v^n \delta(v - v') = -\sum_{n=1}^{\infty} \frac{1}{n!}\frac{\partial^n}{\partial v^n} (v')^n \delta(v - v') \tag{14}$$

where we made use of the sifting property of the delta function. We can now write the delta function in its Fourier representation and obtain

$$\hat{L}\delta(v-v') = -\sum_{n=1}^{\infty}\frac{1}{n!}\frac{\partial^n}{\partial v^n}v'^n\int\frac{\mathrm{d}u}{2\pi}e^{iu(v-v')} = -\sum_{n=1}^{\infty}\int\frac{\mathrm{d}u}{2\pi}\frac{(iuv')^n}{n!}e^{iu(v-v')}$$
$$= \int\frac{\mathrm{d}u}{2\pi}e^{iu(v-v')} - \int\frac{\mathrm{d}u}{2\pi}e^{iuv'}e^{iu(v-v')} = \delta(v-v') - \delta(v). \quad (15)$$

Another application of this operator yields

$$\hat{L}^2\delta(v-v') = \hat{L}\delta(v-v') - \hat{L}\delta(v) = \delta(v-v') - \delta(v). \quad (16)$$

By inserting these relations into Eq. (13), we obtain

$$p(v,r|v',r') = \delta(v-v') + \left[\ln\frac{r}{r'} + \frac{1}{2!}\left(\ln\frac{r}{r'}\right)^2 + \ldots\right](\delta(v-v') - \delta(v)) \quad (17)$$
$$= e^{\ln\frac{r}{r'}}\delta(v-v') + (1-e^{\ln\frac{r}{r'}})\delta(v) = \frac{r}{r'}\delta(v-v') + \left(1-\frac{r}{r'}\right)\delta(v).$$

The transition probability for positive (12) and negative increments (18) thus reads

$$p(v,r|v',r') = \begin{cases} \delta\left(v - \frac{r}{r'}v'\right) & \text{for } v' \geq 0, \\ \frac{r}{r'}\delta(v-v') + \left(1-\frac{r}{r'}\right)\delta(v) & \text{for } v' \leq 0. \end{cases} \quad (18)$$

By inserting Eq. (18) into the definition of the Kramers-Moyal coefficients (8) it can be verified that the corresponding coefficients in Eqs. (9) and (10) are recovered. Figure 3 shows the validity of the Markov property for two different scale separations.

3 Multi-increment PDF Hierarchy and a Possible Closure Scenario

The aim of this section is to derive a of multi-increment PDFs directly from the Burgers equation (1) and to outline a possible closure on the basis of the Markov property (3) insinuated in [21]. To this end, we derive the one-increment PDF with respect to time t according to

$$\frac{\partial}{\partial t}f(v_1,r_1,R_1,t) = \frac{\partial}{\partial t}\langle\delta(v_1 - v(R_1,r_1,t))\rangle = -\frac{\partial}{\partial v_1}\langle\delta(v_1 - v(R_1,r_1,t))\,\mathring{v}\,(R_1,r_1,t)\rangle \quad (19)$$

By the use of the chain rule in the last step, we are able to make contact to the temporal derivative of the velocity increment indicated by the dot. The latter, however, is determined by Eq. (1) and we are able to derive an evolution equation 'from first principles' [21], namely

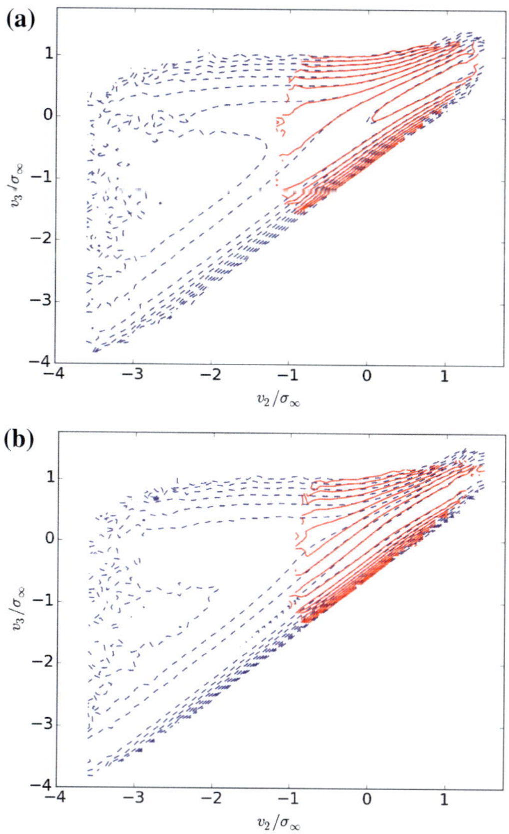

Fig. 3 **a** Contour plot of the Markov property $p(v_3, L/2 - r|v_2, L/2; v_1, L/2 + r) = p(v_3, L/2 - r|v_2, L/2)$ for $r = 1.48\lambda$, where $v_1 = -\sigma_\infty$ and $\sigma_\infty = \sqrt{2}u_{rms}$ has been introduced in order to obtain a uniform increment-representation. The Markov property is found to be valid. Only the boundaries of the two-times conditional PDF (straight lines) deviate from the transition PDF. Furthermore, it can be seen that the analytic result (18) captures the main features of the transition PDF (dashed lines). More precisely the (broadened) part due to $\delta(v_3)$ as well as the transition at $v_2 = 0$ can be observed in the contour plot. **b** Validation of the Markov property for $r = 0.88\lambda$. The Markov property is starting to get violated in the upper part. This tendency becomes even more pronounced for smaller r. At this point it is not clear whether the Taylor length λ is the right quantity for the determination of the Markov-Einstein length λ_{ME}, where the Markov property is starting to deteriorate significantly

$$\frac{\partial}{\partial t} f_1(v_1, r_1, t) + v_1 \frac{\partial}{\partial r_1} f_1(v_1, r_1, t) = -2 \int_{-\infty}^{v_1} \mathrm{d}v \frac{\partial}{\partial r_1} f_1(v, r_1, t)$$
$$- \nu \frac{\partial}{\partial v_1} \int_0^\infty \mathrm{d}r_2 \left[\delta(r_2 - r_1) - 2\delta(r_2)\right] \frac{\partial^2}{\partial r_2^2} \int \mathrm{d}v_2 v_2 f_2(v_2, r_2; v_1, r_1, t)$$
$$+ [\chi(0) - \chi(r_1)] \frac{\partial^2}{\partial v_1^2} f_1(v_1, r_1, t). \tag{20}$$

Interestingly, the terms that stem from the nonlinearity in Eq. (1) are closed and only the viscous term involves the two-increment PDF confronting us with an infinite chain of evolution equations [14]. Note that the cumulative one-increment PDF enters in Eq. (20) as a consequence of the compressibility of the velocity field. For the case of the incompressible Navier-Stokes equation, this term is replaced by pressure contributions involving the two-increment PDF [22]. Furthermore, letting $\nu \to 0$ in Eq. (20) will also not resolve the closure problem, since the rate of energy dissipation $\langle \varepsilon \rangle$ remains finite, a phenomenon termed as dissipation anomaly [5]. So far possible closures of the hierarchy (20) have been restricted to Gaussian approximations [23] of the unclosed terms or to methods borrowed from field theory [5, 6]. Their inherent difficulty lies in the fact that the nonlinear contributions in Eq. (1) are too strong to be accurately described by close-to-equilibrium distributions. Furthermore, they are based on parameterized families of PDFs, i.e., certain quantities have to be specified in advance (e.g., the structure function of second order for a Gaussian approximation). It is therefore of great importance to seek for ways to truncate the hierarchy that preserve the intermittent character of the velocity field. Our approach consists in truncating the hierarchy at the level of the two-increment PDF via the Markov property, which corresponds to a three-point closure. We approximate the unclosed terms in the evolution equation of the two-increment PDF in scale according to

$$f_3(v_3, r_3; v_2, r_2; v_1, r_1) = p(v_3, r_3 | v_2, r_2) f_2(v_2, r_2; v_1, r_1) \qquad \text{for} \quad r_3 \leq r_2 \leq r_1 \tag{21}$$

which leads to a a quite complex but closed nonlinear partial integro-differential equation for f_2 whose properties are examined carefully at present. Obviously, this can only be a first approximation, since in the limit procedures in (20), we inevitably approach the Markov-Einstein length λ_{ME}, where Eq. (4) is violated. However, since the closure allows for a description of intermittency effects and contains no additional parametric function, we are quite hopeful that a deeper insight in the intermittency properties of turbulent flows can be revealed.

4 Conclusion

We have presented a phenomenological description of Burgers turbulence by a Markov process and its corresponding Kramers-Moyal expansion in scale. The transition probabilities have been explicitly calculated and compared to numerical simulations. The influence of the shock structures on the transition PDFs and the breakdown of the Markov property has been analyzed. Furthermore, we briefly mentioned a possible closure scenario of the LMN hierarchy of velocity increments in Burgers turbulence. It remains a task for the near future to further quantify the proposed closure and to examine whether it is an appropriate approximation of the hierarchy. If the latter is true this also opens up a possible way to a pseudo-perturbative treatment of the Navier-Stokes equation beyond the Markov-Einstein length in successively taking into account a larger 'scale history' of the system.

References

1. M. Nelkin, In what sense is turbulence an unsolved problem? Science **255**, 566 (1992)
2. A.A. Migdal, *Turbulence as Statistics of Vortex Cells*, hep-th arXiv:9306152 (1993)
3. A.S. Monin, A.M. Yaglom, *Statistical Fluid Mechanics* (MIT Press, Cambridge, 1975)
4. U. Frisch, *Turbulence* (Cambridge University, Cambridge, 1995)
5. A.M. Polyakov, Turbulence without pressure. Phys. Rev. E **52**, 6183 (1995)
6. M. Lässig, Dynamical anomalies and intermittency in Burgers turbulence. Phys. Rev. Lett. **84**, 2618–2621 (2000)
7. E. Balkovsky, G. Falkovich, I. Kolokolov, V. Lebedev, Intermittency of Burgers' turbulence. Phys. Rev. Lett. **78**, 1452–1455 (1997)
8. T. Grafke, R. Grauer, T. Schäfer, The instanton method and its numerical implementation in fluid mechanics. J. Phys. A: Math. Theor. **48**, 333001 (2015)
9. R. Friedrich, J. Peinke, Description of a turbulent cascade by a Fokker-Planck equation. Phys. Rev. Lett. **78**, 863–866 (1997)
10. S. Lück, C. Renner, J. Peinke, R. Friedrich, The Markov-Einstein coherence length—a new meaning for the Taylor length in turbulence. Phys. Lett. A **359**, 335–338 (2006)
11. T.S. Lundgren, Distribution functions in the statistical theory of turbulence. Phys. Fluids **10**, 969–975 (1967)
12. A.S. Monin, Equations of turbulent motion. Prikl. Mat. Mekh. **31**, 1057 (1967)
13. E.A. Novikov, Kinetic equations for a vortex field. Sov. Phys. Dokl. **12**, 1006–1008 (1968)
14. R. Friedrich, A. Daitche, O. Kamps, J. Lülff, M. Voßkuhle, M. Wilczek, The Lundgren-Monin-Novikov hierarchy: kinetic equations for turbulence. C. R. Phys. **13**, 929–953 (2012)
15. J. Bec, K. Khanin, Burgers turbulence. Phys. Rep. **447**, 1 (2007)
16. A. Chekhlov, V. Yakhot, Kolmogorov turbulence in a random-force-driven Burgers equation. Phys. Rev. E **51**, R2739–R2742 (1995)
17. D. Nickelsen, Markov Processes linking Thermodynamics and Turbulence, arXiv:1510.06281, (2015)
18. C. Eling, Y. Oz, The anomalous scaling exponents of turbulence in general dimension from random geometry, JHEP 09 (2015)
19. H. Risken, *The Fokker-Planck Equation: Methods of Solutions and Applications*, 2nd edn. (Springer, Heidelberg, 1996)
20. R. Courant, D. Hilbert, *Methods of Mathematical Physics II* (Wiley, New Jersey, 1962)
21. O. Kamps, M. Wilczek, Statistical description of turbulent flows—a short review in memory of Rudolf Friedrich, in *Progress in Turbulence V, Springer Proceedings in Physics* vol. 149, p. 1 (2014)
22. F.R. Ulinich, B.Y. Lyubimov, The statistical theory of turbulence of an incompressible fluid at large reynolds numbers. JETP **28**, 494 (1969)
23. M. Wilczek, A. Daitche, R. Friedrich, On the velocity distribution in homogeneous isotropic turbulence: correlations and deviations from Gaussianity. J. Fluid Mech. **676**, 191–217 (2011)

Beads in a Rotating Box

Frank Rietz and Ralf Stannarius

Abstract Granular materials contain, even though they have been an integral part of human technology since ancient times, numerous mysteries even today. Dynamic phenomena observed in such materials depend on a multitude of parameters, and a general description is still missing. Here, a simple experiment is presented that extends the diversity of these pattern forming phenomena. A flat container (Hele-Shaw cell) is filled with a mixture of granular material and slowly horizontally rotated. Depending on the height of the filling the beads either stripe-wise demix or circulate in convection rolls that are known from hydrodynamic instabilities. Irrespective of some superficial similarity to convection in other granular systems, the driving mechanism is completely different, and only partially understood. Particularly interesting are traveling wave structures where demixed stripes move periodically along the axial direction.

1 Introduction

Rotating containers of different geometries filled with grain mixtures are among the standard systems in the investigation of granular matter physics. Since the first description of pattern formation in the horizontally rotating cylindrical mixer [1] the experiment has kept its attractiveness for researchers. A list of references can be found in a resource paper by Kakalios [2], in reviews and books (e.g. [3, 4]). Different scenarios of spontaneous pattern formation have been detected in thin rec-

This text has been published earlier as "Pattern formation in a flat rotating box" by Frank Rietz and Ralf Stannarius in AIP Conf. Proc. 1542, 763 (2013). It is reproduced by permission of AIP, with minor updates.

F. Rietz (✉) · R. Stannarius
Institute of Experimental Physics, Otto-von-Guericke University,
39106 Magdeburg, Germany
e-mail: frank.rietz@gmx.net

R. Stannarius
e-mail: ralf.stannarius@physik.uni-magdeburg.de

S.C. Müller et al. (eds.), *Complexity and Synergetics*,
https://doi.org/10.1007/978-3-319-64334-2_5

tangular cells that rotate about their horizontal axis [5–8]. Only on first glance, the reduction of the system to a quasi-2D geometry seems to simplify the dynamics and greatly reduce the variety of observed patterns. It turns out that this cell still produces several types of spontaneous pattern. The elucidation of the pattern forming mechanisms remains to be a challenge for theoreticians. In this paper, we restrict to experiments with bidisperse mixtures of spherical grains.

Two regimes have to be treated separately in dependence of the fill level C (volume of grains divided by whole cell volume). The first regime is found when the cell is filled with a subcritical amount of granulate, the critical value is $C_{\mathrm{crit}} \approx 0.6$. Then, chute flow is dominant. One observes segregation stripes analogous to the well-known bands in cylindrical containers. These patterns can coarsen when neighboring stripes of the same grain species merge. Peculiarities of the flat cell geometry are discussed below. In particular, we focus on quasi-periodic dynamic solutions.

Even more interesting are the dynamics at fill levels above C_{crit}. The segregation stripe patterns are completely absent there, the limited free volume above the grains largely inhibits individual grain dynamics. The granulate starts to move collectively in slow convection rolls. Segregation and transport patterns are coupled in these structures [5, 6], which will be described in the second part of the Results section. Convection in cells with aspect ratios close to one exhibits some unique features that will be discussed separately.

2 Experimental

The basic geometry is sketched in Fig. 1. If not indicated otherwise, the flat cell has dimensions of 500 mm in width and 5 mm in thickness, the height can be varied, it is usually between 50 and 110 mm. The cell is rotated about a horizontal axis with an angular velocity of the order of $2\,\mathrm{s}^{-1}$. In the standard experiment, the cell rotates with the upper edge towards the camera. The side of the cell seen by the camera will be called front view in the following. Of course, both sides of the cell are experimentally equivalent after 180° revolution. If not specified otherwise, we used a mixture of (0.3 ± 0.05) mm and (0.9 ± 0.1) mm diameter soda-lime glass beads of density 2.49 g/cm^3 in 1:1 volume ratio.

3 Results

3.1 Chute Flow Regime

In the subcritical filling regime, at $C < C_{\mathrm{crit}}$, the initial instability is usually towards vertical segregation stripes, in which the grains follow a slow circulation in the same sense as the container rotates (Fig. 2, right), i.e. when the material slides down the tilted cell it is slightly faster on the top side. The segregation stripes can be subject

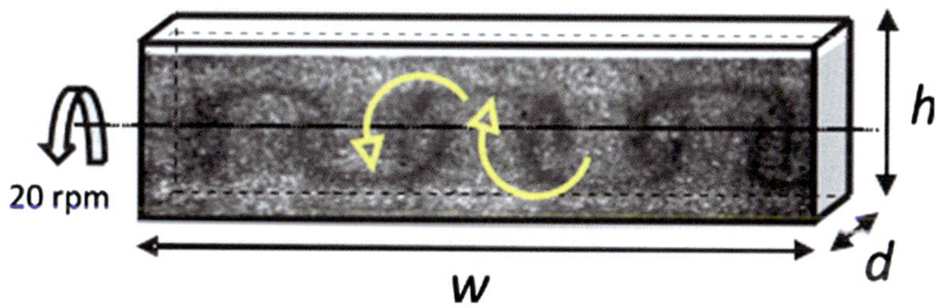

Fig. 1 Experimental geometry. The sketch has been decorated with a convection roll pattern typical for high filling fractions. Images of one side are taken in regular intervals by a CCD camera. *Arrows* indicate the sense of convective flow. The cell is illuminated uniformly from the back. Reproduced from AIP Conf. Proc. 1542, 763 (2013), with the permission of AIP Publishing

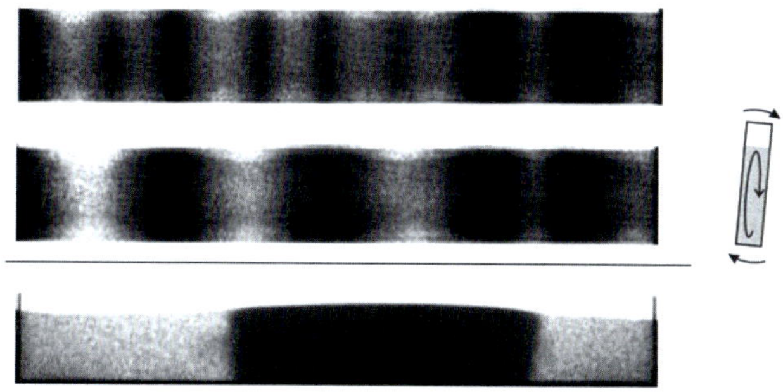

Fig. 2 Rotating cell filled with $C = 0.581 < C_{\mathrm{crit}}$, after 1.5 K rotations (*top*) and 15 K rotations (*middle*) at 20 rpm. The *bottom* image shows a state reached at $C = 0.4$ after 50 K rotations. Darker bands are composed mainly of small beads, brighter bands contain larger beads. The drawing at the *right* is the *side view* of the rotating cell and a sketch of the effective granulate circulation. Reproduced from AIP Conf. Proc. 1542, 763 (2013), with the permission of AIP Publishing

to coarsening, so that finally a single region with small beads is enclosed by two regions of segregated large beads.

More interesting is the situation shown in Fig. 3, with quasi-periodic structures. In the bottom-left inset of the image, snapshots of the cell show that segregation stripes travel to both sides of the cell, but after approximately 2200 rotations, the initial state is almost re-established. The profiles of the cell view along the central axis were stacked to a space-time plot of the complete experiment, shown at the top of Fig. 3. Dark areas indicate an excess of the smaller spheres, brighter areas contain more large spheres.

Interestingly, the quasi-periodic patterns survive for a certain number of periods, but finally the cell locks into another quasi-periodicity with different temporal and spatial pattern. Both the number of stripes, their width and the traveling velocities can vary non-systematically between the quasi-periodic phases. We exclude that the transitions are related to some potential ageing effects of the material, since some of the observed traveling wave patters were found to reappear after several ten thousand

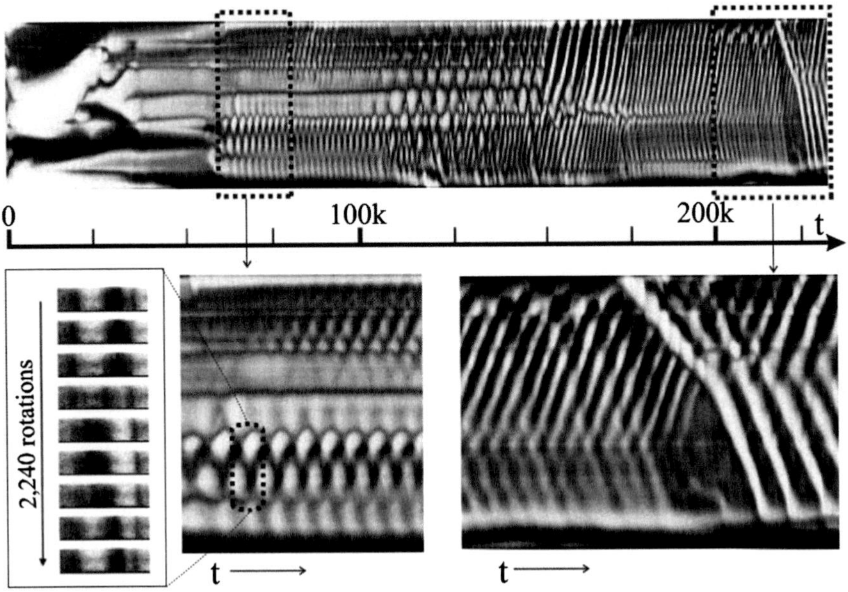

Fig. 3 Inset *bottom left* snapshots of a quasi-periodic structure in the cell. The *top* is a space-time plot of the evolution during about 230 K cell rotations, two sections are magnified in the *bottom* images. Each *vertical line* is a complete cross-section of the cell's *middle* axis. Reproduced from AIP Conf. Proc. 1542, 763 (2013), with the permission of AIP Publishing

revolutions of the container [7]. Obviously this system has several attractors between which the solutions are chosen. This could be a hint that in such experiments, one has to be very careful to derive general statements from only few experiments or limited observation periods. A periodic pattern observed for short time periods (several thousand rotations) does not necessarily prove that a final asymptotic cycle is reached.

3.2 Convection Roll Arrays

When the cell is filled above a critical level, it develops a qualitatively different type of patterns, viz convection rolls. The flow field is quasi-2D, i.e., identical on front and back sides of the cell. These structures were found for all tested combinations of granulates, even in one-component samples with relatively narrow size distributions (in real monodisperse granulate crystallization suppresses such structures). Figure 4 shows examples of the convection and segregation patterns in different types of granular mixtures. A peculiarity of the system (c), which is a mixture of materials of a large density mismatch, is the appearance of oblique stripes. Segregation occurs not primarily within individual rolls, but between neighboring rolls.

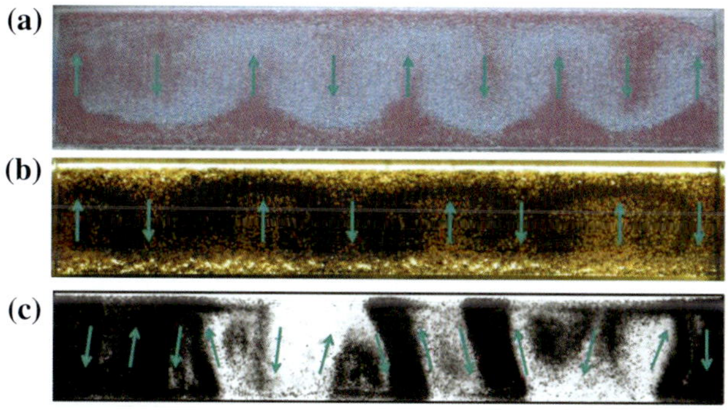

Fig. 4 Convection in **a** decoration sand, *dark red*, $\varnothing \approx 50 \ldots 250\,\mu$m and *bright blue*, $1.0 \ldots 1.4$ mm, **b** mustard seeds (*yellow*, $\rho \approx 0.75\,\text{g/cm}^3$, $\varnothing \approx 2 \ldots 2.5$ mm) and poppy seeds (*dark*, $\rho \approx 0.6\,\text{g/cm}^3$, $\approx 1.0 \times 0.7 \times 0.5$ mm, **c** glass spheres (bright, $\rho = 2.50\,\text{g/cm}^3$, $(650 \pm 100)\,\mu$m) and poppy seeds. Reproduced from AIP Conf. Proc. 1542, 763 (2013), with the permission of AIP Publishing

Fig. 5 Influence of the rotation rate: at 20 rpm (*top*, Froude number $Fr \approx 0.01$) the convection pattern is formed, whereas after continuation at 250 rpm (*bottom*, $Fr \approx 1.75$), banding in radial direction sets in. Cell height 50 mm. Reproduced from AIP Conf. Proc. 1542, 763 (2013), with the permission of AIP Publishing

The onset of convection is relatively insensitive to the rotation rate, as long as the Froude number remains small, $Fr \ll 1$. Figure 5 shows that at larger Froude numbers, the pattern topology changes. Small beads are partially sieved into two mixed bands at the outer edges, convection ceases.

The influence of the interstitial fluid was tested in experiments in vacuum and in water. The reduction of ambient air pressure down to a few Pascal did not change the convection scenario noticeably. Similarly, the preparation of the experiment with water as embedding fluid did not reveal qualitative influences on convection [6].

A detailed analysis of the granulate structure reveals the existence of three zones. In the central part of the cells (cf. Fig. 4 and top of Fig. 5), a zone is formed where particles move collectively in slowly convecting structures, while on top and bottom two layers of rapidly moving grains are located. Their width scales with the cell depth. In these two regions, it is difficult to track the individual particles and to determine velocities.

In Fig. 6a, b the composition and the assumed profile of the granular top surface after two subsequent half rotations is sketched. There is a clear correlation with the roll positions. The edge zones are enriched with small beads in sections where flow

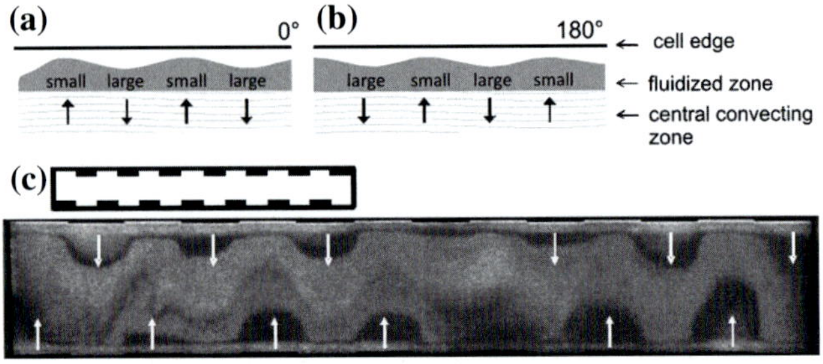

Fig. 6 Profile of the *upper* granular surface and *vertical* flow component **a** and the same after a 180° rotation, **b** the undulations are considerably exaggerated. Artificial modulation of the cell profile **c** and imprinted convection roll array. Reproduced from AIP Conf. Proc. 1542, 763 (2013), with the permission of AIP Publishing

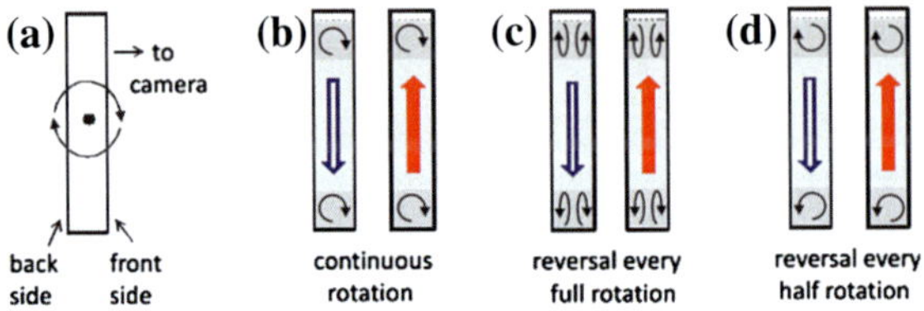

Fig. 7 Circulation of grains in the fluidized edge zones, *side view*. **a** Definition of cell rotation sense, **b–d** different excitation schemes (see text). Reproduced from AIP Conf. Proc. 1542, 763 (2013), with the permission of AIP Publishing

enters the zones and with large beads where the flow leaves them. The upper edge of the granular layer is not uniformly flat, it also correlates with the convection structure. One can make use of this relation by modification of the cell edges. Figure 6c shows that the pattern wavelength can be controlled with a step-modulated edge [6]. Notches correlate with flow from the edge towards the cell axis, similar to Fig. 6a, b, where the depressions correlate with inward flow.

A further detail is the dynamics within the fluidized edge zones. Primarily, there is some axial transport with the convection rolls. But in addition, the grains perform cyclic motions, as sketched in Fig. 7. When different rotation schemes of the cell are applied, this has almost no influence on the convection patterns but it changes the dynamics in the fluidized zones. The figure shows three situations, (b) is the standard experiment, (c) is a cradle-like ±360° swinging and (d) a ±180° swinging [6]. In the latter, the symmetry of front and back sides is broken. This is seen in the circulation pattern in the fluidized edges. In each subfigure, the sketches correspond to regions with upward and downward convection, respectively.

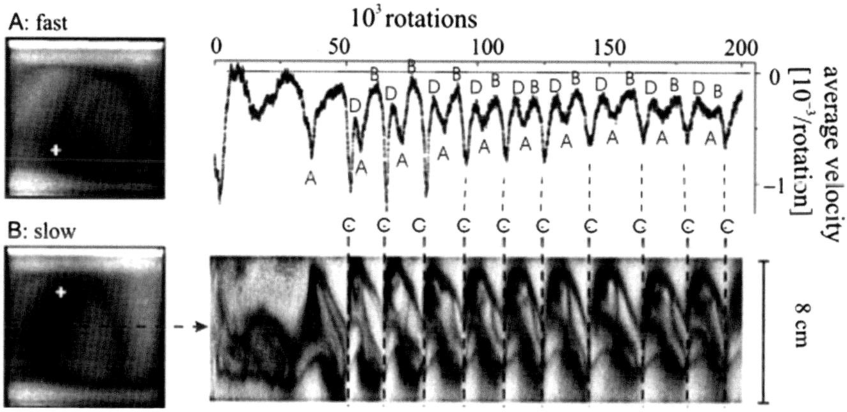

Fig. 8 Single roll in a cell with aspect ratio one ($w = d = 8$ cm, $C = 0.650$). The convection velocity is not constant as in multi-roll patterns. The flow oscillates with a period of half an orbit of the roll and a superimposed four-fold periodicity (peaks A to D). The *top* graph shows the time dependent flow amplitude, and the space-time plot below, obtained from the transmission profiles along the central cell axis (*dashed line*) evidences the correlation between segregation pattern and orbit velocity. Snapshots A and B of the cell were taken in the corresponding extrema of the velocity curve. Reproduced from AIP Conf. Proc. 1542, 763 (2013), with the permission of AIP Publishing

3.3 Single Convection Rolls

A special case are cells with aspect ratios close to unity. The square cell develops a single roll as seen in Fig. 8, left. In contrast to the multi-rolls which are characterized by a relatively uniform rotation velocity when the pattern is fully developed, the single-roll pattern can oscillate, stop, and even reverse the rotation direction. Figure 8 shows one of the scenarios observed in the single roll cell. The periodicity of the dynamics arises from a coupling of the segregation pattern and the revolution velocity [8], as seen in the velocity characteristics and the space-time plot in Fig. 8. The snapshots show typical distributions of small beads along the orbit in the phases A and B of velocity extrema. On longer observation periods, the behavior can be much more complex. The orbits can cease, the revolution direction can change, and an irregular dynamics can be observed, as seen in Fig. 9.

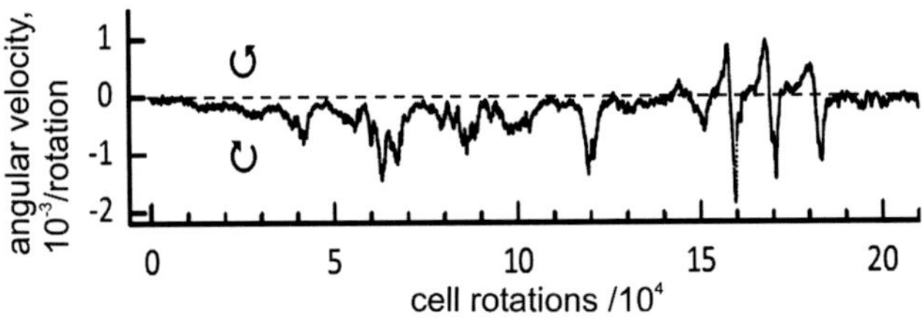

Fig. 9 Angular velocity of the single roll pattern in a cell of aspect ratio ≈1, observed over ≈210 K rotations. *Arrows* indicate the sense of rotation for positive and negative angular velocities, resp. Reproduced from AIP Conf. Proc. 1542, 763 (2013), with the permission of AIP Publishing

4 Discussion and Summary

4.1 Traveling Stripe Patterns

The traveling wave patterns observed in cells filled with a low, subcritical amount of granulate are not yet understood. There is no clear dispersion relation between the velocities and wavelengths of the patterns nor any other evident relation between the dynamics of the stripes and their geometry. Stripes can move in both directions in the cell, they can pass each other. They can be reflected at the container ends, but not necessarily. Often, they are absorbed at the walls. Generators can appear somewhere in the cell. In most cases, the traveling wave patterns obey a quasi-periodicity in time, up to several dozen periods. The characteristics of the traveling wave patterns is that they are not long-term stable but the cell undergoes transitions between different oscillating patterns. The reason for this behavior is not clear. We took particular care to keep experimental conditions constant over the full lifetime of each experiment. Ageing of the glass beads can be excluded as source of the effect.

Traveling wave patterns in cylindrical mixers have been observed earlier [9, 10], but in these cases, the dynamic patterns appeared only as transients or were recorded for time periods much shorter than in our experiments, with exception of traveling wave patterns in a nearly completely filled mixer [11]. The latter develops traveling stripes that can show a clear periodicity.

4.2 Multi-roll Convection Patterns

Even though some basic understanding of the prerequisites for convection in the almost filled flat cells has been reached, the mechanism is far from being explained satisfactorily. The effect is robust, it is found in various material combinations, in water, atmospheric air and vacuum. The pattern can be controlled to a certain extend by modifications of the container edges. Multi-roll patterns reach a stationary convection state, in which they are fundamentally different from the single rolls described below.

4.3 Oscillations, Cessations and Flow Reversal in Single-Roll Convection

The single roll dynamics in cells of aspect ratio close to unity is more complex that of the multi-roll patterns. Even cessations and the reversal of flow are observed. Obviously, the synchronization of multiple rolls tends to suppress these temporal variations of the convection efficiency. Interestingly, the behavior is qualitatively quite similar to Rayleigh-Bénard convection rolls at high Rayleigh numbers [12], where similar phenomena have been reported. In particular, the flat cuboid cell geometry [13, 14] exhibits very similar time series of the convection roll dynamics. Irrespective of the completely different mechanisms for both types of instabilities and the different characteristics of the materials, this may indicate a closer relation between both systems. The prominent feature of the granulate is the grainy microscopic structure. In the Rayleigh-Bénard experiment, a continuous fluid has been used, but it is not unrealistic to assume that the highly turbulent nature of flow at high Rayleigh numbers mimics a granular structure.

References

1. Y. Oyama, Sci. Pap. Inst. Phys. Chem. Res. (Jpn.) **6**, 600–639 (1939). Mixing of binary system of two sizes by ball mill motion (transl.) **37**, 17–29 (1940)
2. J. Kakalios, Granular physics or nonlinear dynamics in a sandbox. Am. J. Phys. **73**, 8–22 (2005)
3. J.M. Ottino, Granular matter as a window into collective systems far from equilibrium, complexity, and scientific prematurity. Chem. Eng. Sci. **61**, 4165–4171 (2006)
4. I. Aranson, L. Tsimring, *Granular Patterns* (Oxford University Press, Oxford, 2009)
5. F. Rietz, R. Stannarius, On the brink of jamming: granular convection in densely filled containers. Phys. Rev. Lett. **108**, 228–302 (2008)
6. F. Rietz, R. Stannarius, Convection and segregation in a flat rotating sandbox. New J. Phys. **14**, 015001 (2012)
7. F. Rietz, R. Stannarius, Transitions between multiple attractors in a granular experiment. Phys. Rev. E. **85**, 040302(R) (2012)
8. F. Rietz, R. Stannarius, Oscillations, cessations and circulation reversals of granular convection in a densely filled rotating container. Phys. Rev. Lett. **108**, 118001 (2012)
9. K. Choo, T.C.A. Molteno, S. Morris, Traveling granular segregation patterns in a long drum mixer. Phys. Rev. Lett. **79**, 2975–2978 (1997). K. Choo, M. W. Baker, T. C. A. Molteno, S. W. Morris, Dynamics of granular segregation patterns in a long drum mixer. Phys. Rev. E **58**, 6115–6123 (1998)
10. S.J. Fiedor, J.M. Ottino, Dynamics of axial segregation and coarsening of dry granular materials and slurries in circular and square tubes. Phys. Rev. Lett. **91**, 244–301 (2003)
11. S. Inagaki, K. Yoshikawa, Traveling wave of segregation in a highly filled rotating drum. Phys. Rev. Lett. **105**, 118001 (2010). S. Inagaki, H. Ebata, K. Yoshikawa, Steadily oscillating axial bands of binary granules in a nearly filled coaxial cylinder. Phys. Rev. E **91**, 010201(R) (2015)

12. E. Brown, G. Ahlers, Rotations and cessations of the large-scale circulation in turbulent Rayleigh-Bénard convection. J. Fluid Mech. **568**, 351–386 (2006). H.-D. Xi, S.-Q. Zhou, K.-Q. Xia, Azimuthal motion of the mean wind in turbulent thermal convection. Phys. Rev. E. **73**, 056312 (2006)
13. K. Sugiyama, R. Ni, R.J.A.M. Stevens, T.S. Chan, S.-Q. Zhou, H.-D. Xi, C. Sun, S. Grossmann, K.-Q. Xia, D. Lohse, Flow reversals in thermally driven turbulence. Phys. Rev. Lett. **105**, 034503 (2010)
14. A.Y. Vasil'ev, P.G. Frick, Reversals of large-scale circulation in turbulent convection in rectangular cavities. JETP Lett. **93**, 330–334 (2011)

Power Grids as Synergetic Systems

Oliver Kamps and Katrin Schmietendorf

Abstract In this article we study power grids from the viewpoint of Synergetics. We show that the typical behavior of self-organizing systems like phase transitions and critical fluctuations can be observed in models for the dynamics of power grids. Therefore we numerically investigate a model, where the phase and voltage dynamics are represented by Kuramoto-like equations. For the topology of the grid we use real world data from the northern Europe high voltage transmission grid.

1 Introduction

Self-organization of complex systems composed of many interacting parts is an ubiquitous phenomenon in nature and can be understood within the framework of Synergetics [1]. The self-organized formation of spatial, temporal or spatiotemporal patterns is often the base for the functioning of a system. An example are the spatial patterns of chemical concentrations needed in the process of morphogenesis [1]. Also in technical systems, self-organization can play a key role for the resilient operation of the system. The power grid composed of producing and consuming units connected by a network of power lines is a complex system. On time scales up to several seconds the frequency stability, which is the prerequisite for the utility of the system, is guaranteed by self-organized synchronization. Regarding the consumers (motors) and producers (generators) as synchronous machines the dynamics of the power grid can be described by a model that is close to Kuramoto's model for synchronization. This is the point where Synergetics meets an important application

O. Kamps (✉)
Center for Nonlinear Science, University of Münster, Münster, Germany
e-mail: okamp@uni-muenster.de

K. Schmietendorf
Institute for Physics, Carl-von-Ossietzky University, Oldenburg, Germany
e-mail: katrin.schmietendorf@uni-oldenburg.de

S.C. Müller et al. (eds.), *Complexity and Synergetics*,
https://doi.org/10.1007/978-3-319-64334-2_6

from engineering. In this article we use a model that describes the coupled dynamics of the phase and the voltage of synchronous machines in a realistic way together with a real world grid topology to show that in such models the typical signatures of self-organization can be found.

2 Kuramoto-Like Power Grid Modeling

Kuramoto-like power grid modeling was initiated in 2008, when Filatrella et al. uncovered the relationship between Kuramoto's well-known model for coupled oscillators and a simple power grid consisting of synchronous machines [2]. The original Kuramoto model (KM) [3] describes the phase dynamics of a population of N coupled oscillators with phases $\theta_i(t)$, natural frequencies ω_i (usually drawn from a unimodal distribution) and coupling matrix $\mathrm{K} = \{K_{ij}\}$:

$$\dot{\theta}_i = \omega_i + \sum_{j=1}^{N} K_{ij} \sin(\theta_j - \theta_i). \tag{1}$$

One commonly introduces the complex order parameter

$$r e^{i\Phi} = \frac{1}{N} \sum_{j=1}^{N} e^{i\theta_j}, \tag{2}$$

whose absolute $r \in [0, 1]$ can be interpreted as a coherence measure. Complete incoherence is indicated by $r = 0$, whereas $r = 1$ corresponds to a fully synchronized state with zero phase shifts. In between there are partially synchronized states [4]. Kuramoto's pioneering work refers to an infinite population with global all-to-all coupling. Equation (1) can, of course, also be applied to finite populations, for various types of topologies and distributions of natural frequencies. The order parameter maintains its function as a measure for the degree of synchronization. Systems with finite population size feature finite size fluctuations [5].

The Kuramoto model has become a canonical model for the investigation of synchronization phenomena in a variety of contexts. Synchronization also plays a crucial role in power system dynamics. Here, a synchronized state with constant system frequency, voltages and steady power transfer is the desired mode of operation. Within the KM-like approach, reduced power grid models are considered. They consist of synchronous generators as prototypal power injecting units, and synchronous motors as power consumers respectively, which together with the set of transmission lines form a network. The system dynamics depends on the interplay between the node dynamics and the underlying network structure. Most research on KM-like power grids focusses on the interplay of stability and topology [6–9]. Furthermore, stability assessment [10] and extension and diversification of the modeling framework [11] were studied.

2.1 Network Topology

Each transmission line of a power network is weighted by a complex admittance **Y**. Its topological information is provided by the nodal admittance matrix **Y** corresponding to the network's Laplacian matrix. **Y** connects the vectors of nodal currents and voltages in the *nodal network equations* [12]:

$$\underline{\mathbf{I}} = \mathbf{Y}\underline{\mathbf{I}}. \tag{3}$$

Equation (3) is a compact formulation of Kirchhoff's law. By means of *Kron reduction* only the N active nodes, i.e. the power-injecting or consuming nodes (here: generators and motors), remain [13]. This yields a $N \times N$ matrix $\mathbf{Y} = \{Y_{ij}\}$ with $Y_{ij} = G_{ij} + \mathrm{i}B_{ij}$ (G_{ij} being the conductance, which accounts for transfer losses, and B_{ij} the susceptance).

The KM-like approach is intended for large networks with size $N \sim 100$. On the one hand, standard reference topologies can be used like the Northern grid [7], the British power grid [6] or an IEEE topology. On the other hand, investigations often aim at relating specific dynamical features or stability properties to the characteristics of the underlying network structure. Therefore, algorithms for synthetic networks are needed in order to provide sets of network realizations with specific features. The standard network algorithms for random, smart-world and scale-free networks have proven to be insufficient, since they do not display the characteristics of real networks adequately [14] and in particular neglect spatial embedding. On this account, the *random growth model* was proposed in [15]. Designed to meet the needs of power grid modeling, it features real transmission grid properties taking into account cost and redundancy.

2.2 The Dynamical Models

The dynamical equations presented in the following all base upon established models from engineering literature. Assume a network of N synchronous machines. Each machine $i \in \{1, \ldots, N\}$ is associated with a complex nodal voltage $\boldsymbol{E}_i = E_i \mathrm{e}^{\mathrm{i}\delta_i}$ with magnitude E_i and phase angle δ_i. The system is standardly transferred into a reference frame rotating with nominal frequency (corresponding to 50 Hz or 60 Hz). The most reduced model ansatz is the *classical model*, which primarily takes into account the mechanical dynamics of the machine (given by Newton's law for rotating masses) plus real power transmission among the nodes [12, 16]:

$$m_i \ddot{\delta}_i = -\gamma_i \dot{\delta}_i + P_{\mathrm{m},i} + \sum_{j=1}^{N} B_{ij} E_i E_j \sin(\delta_j - \delta_i). \tag{4}$$

Here γ_i comprises all sources of mechanical damping and $P_{\mathrm{m},i}$ is the mechanical input ($P_{\mathrm{m},i} > 0$ for generators) or output power ($P_{\mathrm{m},i} < 0$ for motors). Under the assumption of a loss less grid ($G_{ij} = 0 \,\forall\, i,j$), the electrical power transferred between two nodes i and j depends on the phase angle difference: $P_{ij} = B_{ij}E_iE_j \sin(\delta_j - \delta_i)$. The nodal voltages E_i and E_j are assumed to be constant parameters in the classical model, so one can replace $B_{ij}E_iE_j$ by the constant K_{ij}. Equations (4) correspond to the KM modified by additional inertia terms. The classical model's degree of reduction, especially the assumption of constant voltages, limits its range of unconfined validity to time scales of $\sim$1 s.

The *extended model* [11, 16] complements the classical model with dynamical equations for the nodal voltages:

$$\dot{\delta}_i = -\gamma_i\dot{\delta}_i + P_{m,i} + \sum_{j=1}^{N} B_{ij}E_iE_j \sin(\delta_j - \delta_i) \tag{5}$$

$$\alpha_i\dot{E}_i \quad = E_{\mathrm{f},i} - E_i + X_i \sum_{j=1}^{N} B_{ij}E_j \cos(\delta_j - \delta_i) \tag{6}$$

$E_{\mathrm{f},i}$ is proportional to the synchronous machine's rotor field voltage, X_i and the time constant α_i inclose specific synchronous machine parameters. The interaction term in Eq. (6) refers to reactive power. $B_{ij}E_i(t)E_j(t)$ can be interpreted as a time-dependent coupling matrix $K_{ij}(t)$. According to Eqs. (5) and (6), the voltage and angle dynamics are coupled—a known feature of real power grids! The *extended model* is a non-trivial extension of Kuramoto's model causing a richer dynamical behavior. Its range of validity is in the order of 10 s. In [11] it was shown, that the extended model can lead to dynamical behavior and hence stability predictions different from the classical model.

2.3 Current Agenda

Up to now, the dynamical models and the common network topologies as described above only apply for high-voltage transmission networks. The stability investigations so far involved disturbances like transmission line failures or local load changes. Recent attempts will also take into account the distribution grid and implement fluctuating feed-in units mimicking renewable power plants. Therefore, the dynamic models have to be supplemented by transfer losses and inverter-like units. Furthermore, algorithms for the low and medium voltage levels are being prepared. KM-like power grid modeling may contribute to meet the challenges of the transition from fossil-fuel to renewable supply.

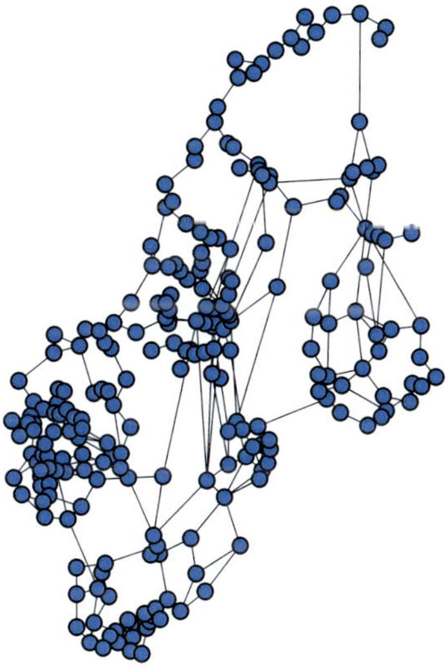

Fig. 1 Visualization of the topology of the northern European high voltage transmission grid

3 Numerical Simulations

To investigate the model we numerical integrate Eqs. (5) and (6) using the topology of the northern grid [7] (see Fig. 1). The numerical integration is done with a Runge-Kutta method of 4th order with a time step $dt = 0.01$. The number of time steps for the different runs varies between 25,000 and 60,000. The B_{ij} are given by $B_{ij} = Ba_{ij}$ where the a_{ij} are the entries of the adjacency matrix ($a_{ij} = 1$ if connected, $a_{ij} = 0$ otherwise) of the northern European high voltage transmission grid with 236 nodes and 320 links. B plays the role of the control parameter. The other parameters are chosen as $\alpha = 2.0$, $\gamma = 0.2$, $C = 0.993$ and $\chi = 0.1$. The $P_{m,i}$ except for $P_{m,N}$ are drawn from a Gaussian distribution with $\sigma = 0.5$. The term $P_{m,N}$ balances the power consumption and production so that $\sum_i^N P_{m,i} = 0$.

3.1 Phase Transition and Hysteresis

To investigate the phase transition of the system we performed simulations with different B for the case of increasing and decreasing coupling values. Since the order parameter displays finite size fluctuations, we use the time average $\langle r \rangle$ for fixed B. In Fig. 2 one can see the typical pattern of a transition from an incoherent state with small $\langle r \rangle$ to an synchronized state. Also the hysteretic behavior, which is found in Kuramoto models with inertia, can be seen.

For increasing coupling strength the system follows a *weakly synchronized branch*, for decreasing coupling a *strongly synchronized branch* [17]. It is apparent, that the phase transition is not as pronounced as in larger systems, and the $r(B)$ behavior dif-

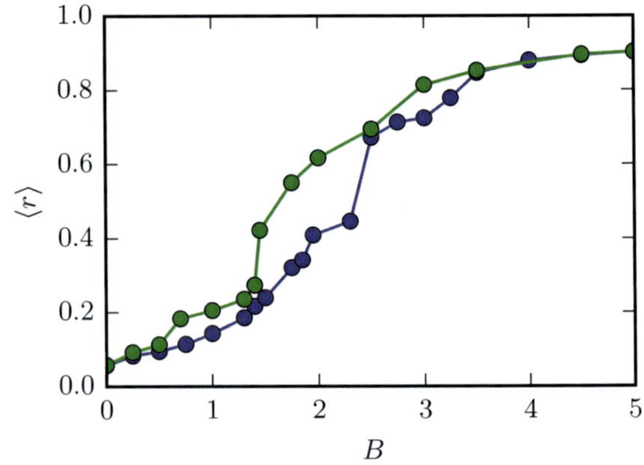

Fig. 2 *Blue dots* denote the branch for increasing B while *green dots* denote the branch with decreasing control parameter

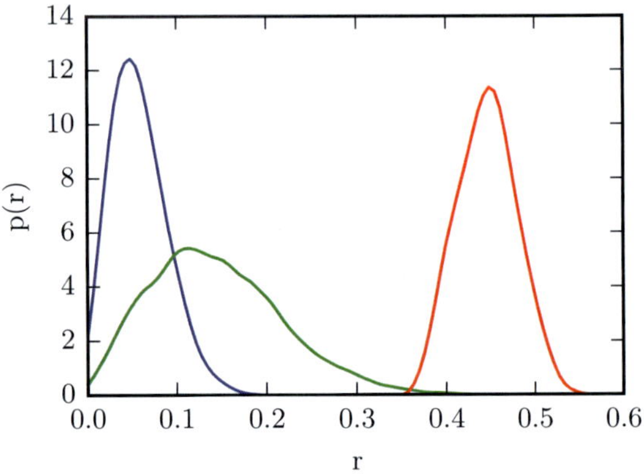

Fig. 3 PDF of coherence r for $B = 0.0$ (*left* maximum), $B = 1.0$ and $B = 2.3$ (*right* maximum). The PDFs are estimated via kernel density estimation with an Epanechnikov kernel and a bandwidth of 0.025

fers from [17]. The reason is that the size of $N = 236$ is rather small for Kuramoto oscillator populations. The system displays quite large finite N fluctuations, which makes the phase transition become blurred. Furthermore, we simulate one specific system, with a specific topology and distribution of $P_{m,i}$. If we modify one of this features, $r(B)$ will also change. In addition, taking the voltage dynamics Eq. (6) into account, makes the system dynamics more complex. We observe multi-stability with several coexisting fixed points and limit cycles. We ensured that the system did not jump onto one of these limit cycles while varying B. This explains the irregular step size in B.

3.2 Critical Fluctuations

One very important signature of self-organization is the occurrence of critical fluctuations close to the phase transition in the system. A very prominent example where

Fig. 4 Plot of $\sigma_r(B)$

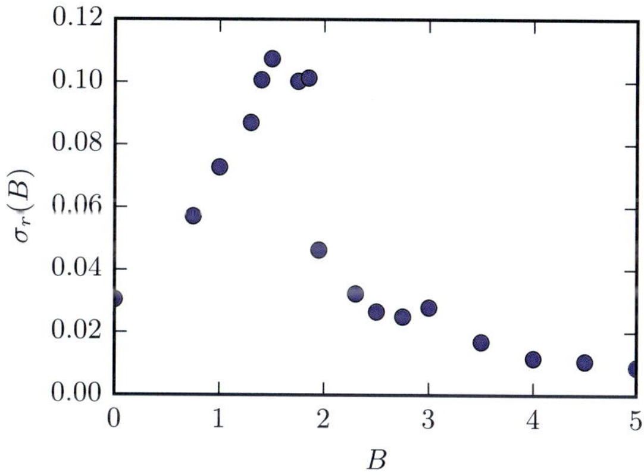

the observation of critical fluctuations led to the conclusion that the process is driven by self-organization rather than by a kind of program is the Haken-Kelso-Bunz model [18, 19]. Figure 3 shows the probability density function (PDF) of the order parameter r for three fixed B values: below, around and above the phase transition. In each case the PDF has a finite width but close to the phase transition the PDF broadens significantly. To quantify this effect we also plotted the standard deviation $\sigma_r(B)$ (see Fig. 4). This plot shows a clear maximum around at the critical coupling value $B \approx 1.5$.

4 Conclusion and Outlook

In this article we have shown that reduced (KM-like) power grid models show typical features of self-organizing systems. It is important to emphasize that self-organization is in this case not only a nice artifact but the central mechanism that keeps the system in a stable operational mode within a time scale of several seconds. After this time automatic control mechanisms and human control stabilize the system e.g. by matching demand and production. Further decentralization and fluctuating feed-in from wind and solar power production pose a big challenge for power grid stability. A future question is whether ideas from Synergetics can help redesign the power grid in a way that also under this influences the synchronized state is stable. A nice example where Synergetics gave the answer to stabilize or to return to the desired state is the destruction of, in this case unwanted, synchronization of neurons causing Parkinson's disease or essential tremor [20].

References

1. H. Haken, *Synergetics—An Introduction* (Springer, Heidelberg, 1983)
2. G. Filatrella, A.H. Nielsen, N.F. Pedersen, Analysis of a power grid using a Kuramoto-like model. Eur. Phys. J. B **61**, 485–491 (2008)
3. Y. Kuramoto, Self-entrainment of a population of coupled non-linear oscillator, in *International Symposium on Mathematical Problems in Theoretical Physics Lecture Notes*, pp. 420–422, 1975
4. S.H. Strogatz, From Kuramoto to Crawford: exploring the onset of synchronization in populations of coupled oscillators. Physica D **143**, 1–202 (2000)
5. J.A. Acebrón, L.L. Bonilla, C.J. Pérez Vicente, F. Ritort, R. Spigler, The Kuramoto model: a simple paradigm for synchronization phenomena. Rev. Mod. Phys. **77**, 137–185 (2005)
6. M. Rohden, A. Sorge, M. Timme, D. Witthaut, Self-organized synchronization in decentralized power grids. Phys. Rev. Lett. **109**, 064101 (2012)
7. P.J. Menck, J. Heitzig, J. Kurths, H.J. Schellnhuber, How dead ends undermine power grid stability. Nat. Commun. **5**, 3969 (2014)
8. D. Witthaut, M. Timme, Braess's paradox in oscillator networks, desynchronization and power outage. New J. Phys. **14**, 083036 (2012)
9. Y. Susuki, I. Mezic, T. Hikihara, Global swing instability of multimachine power systems, in *47th IEEE Conference on Decision and Control*, pp. 2487–2492, 2008
10. P. Menck, J. Heitzig, N. Marwan, J. Kurths, How basin stability complements the linear-stability paradigm. Nat. Phys. **9**, 89–92 (2013)
11. K. Schmietendorf, J. Peinke, O. Kamps, R. Friedrich, Self-organized synchronization and voltage stability an networks of synchronous machines. Eur. Phys. J. **223**, 2577–2592 (2014)
12. P. Kundur, N.J. Balu, M.G. Lauby, *Power System Stability and Control* (McGraw-Hill, New York, 1994)
13. F. Dörfler, F. Bullo, Synchronization and transient stability in power networks and non-uniform Kuramoto oscillators, in *American Control Conference*, pp. 930–937, 2010
14. P. Hines, S. Blumsack, E. Cotilla Sanchez, C. Barrows, The topological and electrical structure of power grids, in *43rd Hawaii International Conference on System Sciences*, pp. 1–10, 2010
15. P. Schultz, J. Heitzig, J. Kurths, A random growth model for power grids and other spatially embedded infrastructure networks. Eur. Phys. J. **223**, 12 (2014)
16. J. Machowski, J. Bialek, D.J. Bumby, *Power System Dynamics: Stability and Control* (Wiley, New York, 2008)
17. H.-A. Tanaka, A.J. Lichtenberg, S. Oishi, First order phase transition resulting from finite inertia in coupled oscillator systems. Phys. Rev. Lett. **78**, 2104–2107 (1997)
18. H. Haken, J.A.S. Kelso, H. Bunz, A theoretical model of phase transitions in human hand movements. Biol. Cybern. **51**, 347–356 (1985)
19. G. Schöner, H. Haken, J.A.S. Kelso, A stochastic theory of phase transitions in human hand movement. Biol. Cybern. **53**, 247–257 (1986)
20. A.P. Tass, Desynchronization of brain rhythms with soft phase-resetting techniques. Biol. Cybern. **87**, 102–115 (2002)

Part III
Chemistry

Nonlinear Phenomena in Heterogeneous Catalysis

Marina M. Slinko, Alexei G. Makeev and Nikolai V. Peskov

Abstract The manuscript demonstrates with a few selected examples how the nonlinear phenomena can change the way in which heterogeneous reactions proceed. Firstly it is shown that the kinetic oscillations and synergetic effect can originate due to the diffusion coupling of two different catalysts. Secondly it is demonstrated that due to the nonlinearity of the system and the ability to form spatial structures two catalysts can produce the same catalytic activity as one catalyst.

1 Introduction

Heterogeneous catalytic reactions are highly nonlinear systems, which can operate under conditions far from thermodynamic equilibrium, where temporal and spatiotemporal organization becomes possible. The difficulties in the study of the nonlinear phenomena in heterogeneous catalytic systems are connected with their multilevel structure shown in Fig. 1. The first level would be an element on a single crystal surface which can be described by a microscopic kinetic model considering the interaction of individual atoms and molecules. The second level is represented by an element of a polycrystalline metallic surface or a single metal cluster bounded by differently indexed planes. The next level for supported catalysts may be considered as a pellet. The fourth level for massive catalysts is represented by an extended polycrystalline surface and for supported catalysts by a catalyst bed. Finally the last level is a reactor, where the global reaction rate can be estimated.

M.M. Slinko (✉)
Institute of Chemical Physics, Russian Academy of Science,
Kosygina Str. 4, Moscow 117334, Russia
e-mail: marinaslinko@gmail.com

A.G. Makeev · N.V. Peskov
Faculty of Computational Mathematics and Cybernetics,
Lomonosov Moscow State University, Moscow 119991, Russia

S.C. Müller et al. (eds.), *Complexity and Synergetics*,
https://doi.org/10.1007/978-3-319-64334-2_7

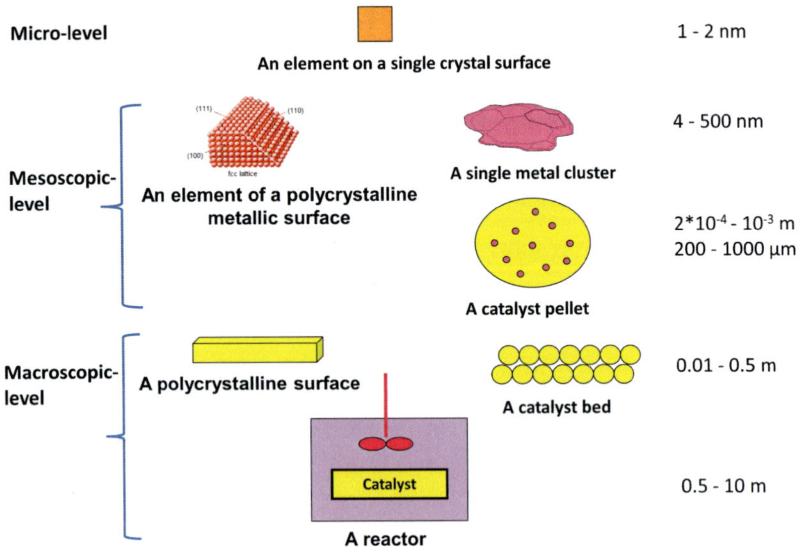

Fig. 1 The multilevel structure of a heterogeneous catalytic system

The nonlinearity of the system originates at the first microscopic level, where the processes of adsorption, surface diffusion, chemical transformation of adsorbed species and desorption proceed on the element of a single crystal surface. The causes of the nonlinearity at the microscopic level are the lateral interactions between the adsorbed particles, adsorbate-induced surface restructuring, faceting of the metallic surface and phase transitions in the subsurface layer [1, 2]. Knowledge of the kinetics of the elementary processes of surface reactions on the atomic scale has to be the basis for modeling of the macroscopic kinetics. However understanding of the role of microscopic nonlinearity in the macroscopic dynamics and the nature of the transition from the micro to the macro level remain basically unsolved problems.

From the second level up, each level can be represented by an array of local oscillators. The nature of the local oscillator depends on the level of the system. It may be different single crystal planes on the polycrystalline surface, various particles in a pellet or different pellets in a catalyst bed. Coupling between the local oscillators may occur through surface diffusion, by heat transfer and/or through the gas phase diffusion. The mode of coupling depends upon the level under study and experimental conditions. In UHV studies using single crystal planes, thermal coupling is not efficient while surface diffusion operates over relatively short distances. At atmospheric pressure in the case of supported catalysts and highly exothermic reactions, heat transfer may be important for the coupling of various particles in the pellet and pellets in a catalyst bed. However at all levels, in both high and low pressure experiments the local oscillators are coupled via gas-phase diffusion, where local reaction rate variations are communicated via variations in the partial pressure of reactants in the gas-phase above the surface.

Different types of the global nonlinear behaviour may occur, including periodic, aperiodic and chaotic oscillations depending upon the nature and the strength of coupling between local oscillators. The purpose of this article is not to give an overview of various types of nonlinear phenomena of the global reaction rate in heterogeneous catalytic systems—this topic is well reflected in many reviews [1–4]—but to demonstrate with a few selected examples how the nonlinear phenomena can change the way in which heterogeneous reactions proceed.

2 The Coupling of Two Polycrystalline Surfaces via Surface Diffusion. Can Two Steady State Surfaces Produce Reaction Rate Oscillations Due to the Coupling?

At each level of a heterogeneous catalytic system an array of local elements may behave differently from a single separated unit due to the coupling. Zhdanov and Kasemo [5] showed theoretically that the dynamic behaviour of catalytic reactions occurring on nm-sized metal particles, exposing different crystalline facets, can be dramatically different from the superposition of the dynamics of the individual facets. These differences can be connected with the coupling effects between different orientations present on the metallic particle. The application of Field Ion Microscopy demonstrated that dynamic coupling effects between the individual nm-sized planes may lead to different dynamic behaviour than that observed in single crystal experiments. Thus it was found out that the oscillatory behaviour during H_2 oxidation could be observed only on a monocrystalline Pt field emitter tip with many different planes on its surface, while no oscillations have been found on separated Pt single crystal planes [6]. Below it will be demonstrated how the coupling of two non-oscillating surfaces of Pt and Rh may lead to the appearance of kinetic oscillations during CO oxidation.

Carbon monoxide oxidation on noble metal catalysts is a seemingly simple reaction, but it can have strongly non-linear kinetics leading to the intriguingly rich dynamic behavior observed during the last 25 years [1, 7, 8]. The nonlinearity of CO oxidation causes the appearance of self-sustained oscillations of the reaction rate, spatial structures and chaotic regimes (chemical turbulence) at various scale levels of the catalytic system [2]. An important result of the nonlinearity is bistable kinetics of CO oxidation occurring over Pt and Rh metallic surfaces. The well-known Langmuir-Hinshelwood mechanism of CO oxidation on Pt and Rh surfaces contains the following stages:

$$\begin{aligned}
&1.\ \ CO + [M] \leftrightarrow [M-CO] \\
&2.\ \ O_2 + [M] \rightarrow 2[M-O] \qquad (1)\\
&3.\ \ [M-CO] + [M-O] \rightarrow CO_2 + 2[M]
\end{aligned}$$

where [M] is an active site of a Pt and Rh surface.

The bistability of the reaction rate originates essentially from the inequivalence of the adsorption of carbon monoxide and oxygen. Oxygen adsorbs dissociatively (requiring 2 free sites) and CO molecularly (requiring only one active site), with CO having a strongly inhibiting effect on oxygen adsorption [9].

Bimetallic Pt/Rh catalysts have been studied experimentally in details by Kalinkin et al. [10]. They demonstrated that a synergistic effect between Pt and Rh in the CO_2 production reaction was observed only for the Pt/Rh composite catalyst, while the activity of the alloy was intermediate between the activities of the individual metals. The bimetallic composite surface represented the array of platinum particles of size 10–20 nm, which were supported onto the flat surface of polycrystalline Rh. The surface fraction occupied by Pt patches was about 50%.

The dynamic behavior of surface coverages over Pt (i = 1) and Rh (i = 2) surfaces can be described by the following equations:

$$\begin{aligned}
\frac{d\theta_{CO,i}}{dt} &= P_{CO}S_{CO,i}k^{ads}_{CO,i}\theta_{*,i} - k^{des}_{CO,i}\theta_{CO,i} - k^r_i\theta_{CO,i}\theta_{O,i} = f_{CO,i};\\
\frac{d\theta_{O,i}}{dt} &= 2P_{O_2}S_{O_2,i}k^{ads}_{O_2,i}(\theta_{*,i})^2 - 2k^{des}_{O2,i}(\theta_{O,i})^2 - k^r_i\theta_{CO,i}\theta_{O,i} = f_{O,i};
\end{aligned} \tag{2}$$

where P_{CO} and P_{O_2} are the partial pressures of CO and O_2; $\theta_{CO,1}, \theta_{O,1}, \theta_{CO,2}, \theta_{O,2}$ are CO and O coverages over Rh and Pt respectively; $S_{CO,i}$ and $S_{O_2,i}$—the initial sticking coefficients of CO and O_2; $k^{ads}_{CO,i}$ and $k^{ads}_{O2,i}$—the adsorption rate constants of CO and O_2; $k^{des}_{CO,i}$ and $k^{des}_{O2,i}$—the desorption rate constants of CO and O_2: $k^{des}_{CO,i} = \tilde{k}^{des}_{CO,i}\exp(-E^{des}_{CO,i}/(RT))$, $k^{des}_{O_2,i} = \tilde{k}^{des}_{O_2,i}\exp(-E^{des}_{O,i}/(RT))$; k^r_i—the reaction rate constant, $k^r_i = \tilde{k}^r_i\exp(-E^r_i/(RT))$. The coverages of free active sites for Rh and Pt are defined as: $\theta_{*,1} = 1 - \theta_{CO,1} - \theta_{O,1}$, $\theta_{*,2} = 1 - \theta_{CO,2} - \theta_{O,2}$. The mathematical model, describing the dynamic behavior over the Pt-Rh composite surface is represented by the following system of differential-algebraic equations:

$$\begin{aligned}
&\alpha\frac{d\theta_{CO,1}}{dt} + (1-\alpha)\frac{d\theta_{CO,2}}{dt} = \alpha f_{CO,1} + (1-\alpha)f_{CO,2};\\
&\alpha\frac{d\theta_{O,1}}{dt} + (1-\alpha)\frac{d\theta_{O,2}}{dt} = \alpha f_{O,1} + (1-\alpha)f_{O,2};\\
&0 = \exp(\frac{\Delta E_{CO}}{RT})\theta_{CO,1}\theta_{*,2} - \theta_{CO,2}\theta_{*,1}\\
&0 = \exp(\frac{\Delta E_{O}}{RT})\theta_{O,1}\theta_{*,2} - \theta_{O,2}\theta_{*,1}
\end{aligned} \tag{3}$$

where α is the surface area fraction of Rh and $(1-\alpha)$ is the part of Pt.

The model (3) supposes the quasi-equilibrium conditions due to the equality of chemical potentials of adsorbed CO and O atoms on two parts of the composite catalyst. Fast equilibration of the CO and O coverages is assumed due to rapid surface diffusion of both species. The algebraic equations contain the important parameters ΔE_{CO} and ΔE_O, denoting the difference in the binding energies of CO_{ads}

Table 1 The kinetic parameters of the models (2) and (3)

Definition	Parameter	$i=1$ (Rh)	$i=2$ (Pt)
CO adsorption rate constant	$k^{ads}_{CO,i}$ [$s^{-1}mbar^{-1}$]	2×10^5	2×10^5
CO sticking coefficient	$S_{CO,i}$ [–]	0.5	1
O_2 adsorption rate constant	$k^{ads}_{O_2,i}$ [$s^{-1}mbar^{-1}$]	2×10^5	2×10^5
O_2 sticking coefficient	$S_{O_2,i}$ [–]	0.25	0.2
CO desorption rate constant	$\tilde{k}^{des}_{CO,i}$ [s^{-1}]	10^{15}	2×10^{15}
CO desorption activation energy	$E^{des}_{CO,i}$ [kJ/mol]	140	140
O_2 desorption rate constant	$\tilde{k}^{des}_{O_2,i}$ [s^{-1}]	0	0
Reaction rate constant	$\tilde{k}^{r}_{i}$ [s^{-1}]	10^{13}	10^{12}
Reaction activation energy	E^{r}_{i} [kJ/mol]	100	100

and O_{ads}, respectively, on two parts of the composite catalyst. They can be defined as: $\Delta E_{CO}=E^{des}_{CO,2}-E^{des}_{CO,1}$, $\Delta E_O=(E^{des}_{O,2}-E^{des}_{O,1})/2$.

Simulations were carried out using the mathematical models (2) and (3) with the values of the parameters for Rh and Pt, taken from the literature and shown in Table 1. Oxygen desorption from both metals is neglected, since experiments have shown that this process does not occur at temperatures lower than 600 K. The kinetic parameters for desorption of CO were chosen to be equal for Rh and Pt. The equal activation energies for desorption of CO from Rh and Pt indicate the equal binding energy of CO on both metals and $\Delta E_{CO}=0$. On the contrary the large difference in the binding energy of atomic oxygen on Rh and Pt equal to $\Delta E_O=-35$ kJ/mol can be estimated by combining the initial heat of adsorption of O_2 on Pt(111), 300 kJ/mole [11], with the activation energy of O_2 desorption from Rh(111) in the limit of zero coverage, 370 kJ/mole [12]. The external parameters, including CO and O_2 partial pressures correspond to the experimental data [10]. The details of the choice of parameters can be found in Ref. [13].

Figure 2 demonstrates bifurcation diagrams and the corresponding dependence of the reaction rate upon P_{CO} on Rh ($\alpha=1$) and Pt ($\alpha=0$) catalysts. It can be seen that at low P_{CO} the surface is predominantly oxygen-covered with the CO_2 production rate increasing with the increase of P_{CO}. At a certain CO partial pressure CO adsorption becomes higher than oxygen adsorption, the surface becomes covered by CO molecules leading to the low activity steady state. As shown in bifurcation diagrams there is a region of bistability between two saddle-node bifurcations. This region is larger for the Rh catalyst. Here during the variation of P_{CO} in the "up" and "down" directions at a fixed oxygen pressure and temperature the CO_2 production rate follows the hysteresis loop with an upper branch of high reactivity and a lower branch of low reactivity (Fig. 2b). The oscillatory region is not detected for both catalysts.

Figure 3a, b demonstrate the analogous calculations for the composite Rh/Pt catalyst. The bifurcation diagram and the dependence of the reaction rate upon CO partial pressure reveal more complicated character. There are two points of Andronov-Hopf bifurcation, h_1 and h_2, and also four points of a saddle-node

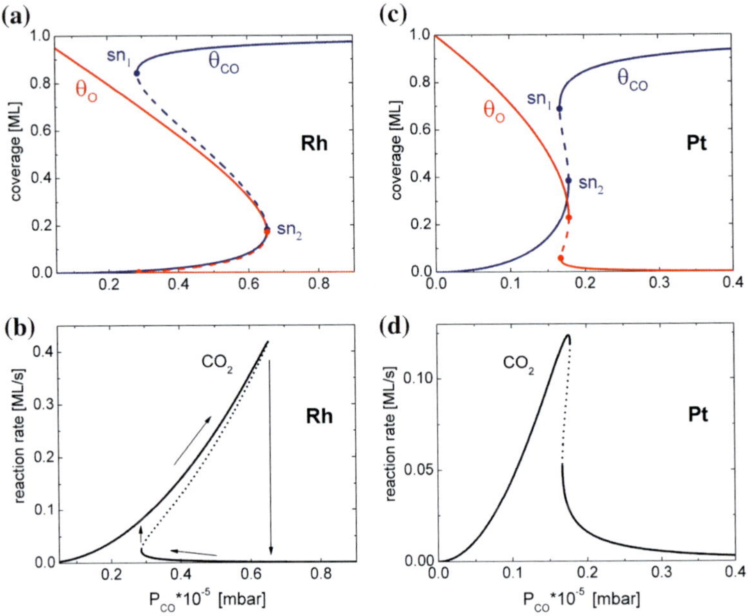

Fig. 2 Bifurcation diagrams and corresponding dependences of the reaction rate of CO oxidation upon CO partial pressure for Rh (**a**, **b**) and Pt (**c**, **d**) at $P_{O_2} = 10^{-5}$ mbar and T = 440 K. *Solid* (*dashed*) *lines* denote stable (unstable) branches. θ_{CO} is shown in *blue* and θ_O is shown in *red*; sn_1 and sn_2 denote saddle-node bifurcations

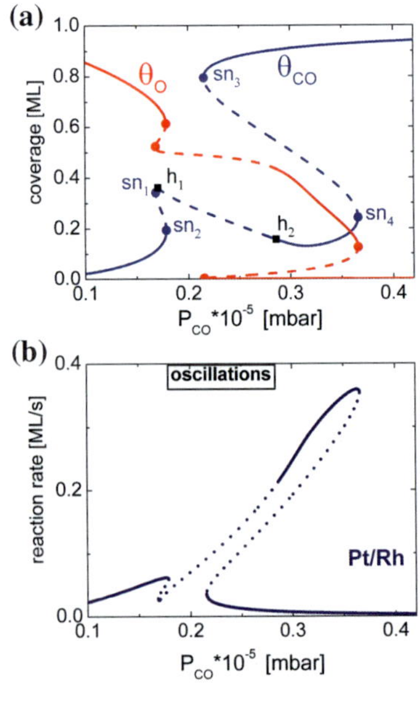

Fig. 3 **a** Bifurcation diagram and **b** corresponding dependence of the reaction rate upon CO partial pressure for Pt-Rh catalyst at $P_{O_2} = 10^{-5}$ mbar and T = 440 K. *Solid* (*dashed*) *lines* denote stable (unstable) branches. Two points of Andronov-Hopf bifurcation, h_1 and h_2 are marked by *filled squares* and four points of saddle-node bifurcation, sn_1, sn_2, sn_3 and sn_4 are marked by *filled circles*

bifurcation, sn_1, sn_2, sn_3 and sn_4. For P_{CO} values from 1.7861×10^{-6} (point sn_2) to 2.8635×10^{-6} mbar, the system exhibits oscillations in the reaction rate. Near the right boundary of the oscillatory region, a hard loss of stability occurs as P_{CO} is decreased, and a limit cycle with a large amplitude emerges from the subcritical Andronov-Hopf bifurcation h_2. At the left boundary, a limit cycle collides with a saddle-node point sn_2 leading to an oscillation with the "infinite" period via a "saddle-node infinite period" or a "SNIPER" bifurcation. The point h_1 represents a supercritical Andronov-Hopf bifurcation, where oscillations are born with a small amplitude. However, near the point h_1 the stable limit cycle exists only in a very narrow range of P_{CO} values (the range width is $\approx 2 \times 10^{-10}$ mbar).

In order to understand the mechanism of oscillations one period has been divided in 3 stages and variation of surface coverages, different reaction rates and diffusion fluxes were shown in Fig. 4. The diffusion fluxes were calculated as:

$$
\begin{aligned}
W_{CO}^{Pt \to Rh} &= \alpha \frac{d\theta_{CO,1}}{dt} - \alpha f_{CO,1} \\
W_{O}^{Pt \to Rh} &= \alpha \frac{d\theta_{O,1}}{dt} - \alpha f_{O,1}
\end{aligned}
\tag{4}
$$

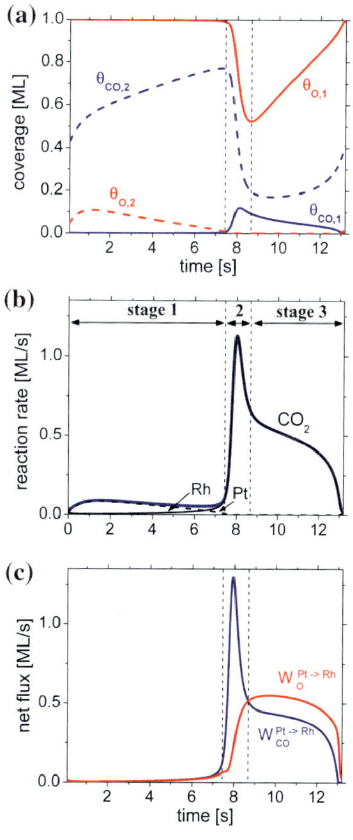

Fig. 4 One cycle of kinetic oscillations on the composite Rh/Pt catalyst at fixed $P_{O_2} = 10^{-5}$ mbar, T = 440 K. A cycle is subdivided into three distinct stages. Shown are the time dependencies of: **a** adsorbate coverages on Rh (*solid lines*) and Pt (*dashed lines*); **b** the total reaction rate (*thick blue line*), and the partial reaction rates on Rh (*thin line*) and on Pt (*dashed line*); **c** diffusion fluxes of CO (*blue*) and O (*red*) from Pt to Rh. (Reprinted from Ref. [13])

The oscillations in the model result from the interplay between the reaction rates over both metals and the diffusion fluxes. Due to the larger binding energy of atomic oxygen on Rh practically all oxygen atoms, adsorbed on the Pt surface quickly spillover onto the Rh surface if vacant sites are available. Therefore $W_O^{Pt \to Rh}$ is always positive despite the small concentration of oxygen on Pt as compared to Rh. In spite of $\Delta E_{CO} = 0$, the flux $W_{CO}^{Pt \to Rh}$ is also always positive during oscillations, because the CO coverage is constantly larger on Pt than that on Rh.

Figure 4 shows that at stage 1 the Rh part of the composite Pt-Rh catalyst is covered by oxygen ($\theta_{O,1} > 0.99$ ML) although for the isolated Rh surface the oxygen coverage would be $\theta_O \approx 0.7$ ML.

The total reaction rate is low and it coincides mainly with the reaction rate over the Pt part. The high oxygen coverage on Rh is fueled by the diffusion of oxygen from Pt although the values of fluxes are low due to the small amount of free sites over Rh. On Pt the CO coverage gradually increases (under these conditions isolated Pt would be covered by CO), while O coverage gradually decreases. The first stage ends, when the oxygen coverage over Pt reaches its critical minimal value and the flux of CO from Pt to Rh starts to exceed the flux of O from Pt to Rh. At stage 2 the oxygen concentration on Rh cannot be sustained at the high value and it begins to decrease. The free sites appearing on Rh due to the CO_2 formation reaction are quickly occupied by adsorbed CO species diffusing from Pt. The supply of adsorbed CO from Pt results in a sharp autocatalytic increase of the reaction rate on Rh. The maximum of the global reaction rate coincides with the maximum of CO coverage on Rh and the maximum value of $W_{CO}^{Pt \to Rh}$. During stage 2 the CO coverage on Pt and oxygen coverage on Rh sharply decreases. This stage ends when both fluxes become practically equal and the decrease of oxygen coverage on Rh stops. At the third stage the flux of oxygen atoms from Rh exceeds the flux of adsorbed CO and the surface coverage of oxygen on Rh gradually increases. Finally the whole surface of Rh becomes covered by oxygen and the system returns to the beginning of the new cycle.

It has been well-established that kinetic oscillations during CO oxidation over Pt and Rh originate as a result of the coupling of adsorption–desorption processes with the reaction rates [2]. The oscillations on Pt-Rh catalyst occur under conditions at which no kinetic oscillations exist over isolated Pt and Rh catalyst. They originate due to the coupling of surface diffusion between two catalysts and the reaction rates. The spillover of adsorbed CO species from Pt to Rh (due to a concentration gradient) and the spillover of adsorbed oxygen atoms from Pt to Rh (due to the large difference in binding energy of oxygen over two metals) are the driving forces for these "synergetic" oscillations. The results of simulations demonstrate that the model (3) generates oscillations in a very large range of the parameters values. In particular the model exhibits oscillations even for $k_2^r = 0$.

In summary the results of mathematical modelling revealed the origin of the synergetic effect during CO oxidation over Pt-Rh composite catalyst, which was observed experimentally by Kalinkin et al. [10]. At elevated temperature O atoms and CO molecules diffused to Rh immediately after adsorption on Pt surface leaving it adsorbate free. The CO_2 formation occurred only over the Rh part, while the Pt part of the composite catalyst served only as a source of adsorbed CO and O species. The larger CO sticking coefficients on Pt provided a positive synergistic effect due to the larger supply of CO to the active Rh centers in Pt-Rh composite catalyst in comparison with the isolated Rh surface.

3 Coupling of 2 Nonlinear Oscillators at a Reactor Level. Can One Catalyst Produce the Same Activity as Two Catalysts?

The gas-phase coupling, where local reaction rate variations are communicated via variations in the partial pressure of reactants in the gas phase above the surface is the very important mechanism of communication of kinetic oscillations at all levels of a heterogeneous catalytic system. For kinetic oscillations in CO oxidation under normal pressure conditions it was shown that this is a dominant mechanism operated during the coupling of individual grains on polycrystalline catalysts [14], during the coupling of various parts of the Pt wire [15] and during the coupling of various parts of the catalyst bed [16]. In all these experiments "in-phase" synchronization of local oscillations has been detected under the proper conditions.

The "in-phase" synchronization for equal oscillators in a CSTR has been detected experimentally and theoretically also for methane oxidation over two Ni foils [17].

The temperatures of similar Ni foils oscillated always in-phase even for the largest distance between two foils. Therefore, it was surprising to find out that depending upon the experimental conditions two equal Pd foils could generate both 'in-phase' and 'anti-phase' oscillations shown in Fig. 5 [18].

In order to understand the origin of "antiphase" oscillations of two coupled catalytic oscillators mathematical modelling has been done. Each oscillator was simulated with the simplest mathematical model, describing oscillations during methane oxidation over Pd catalysts [19]. The reaction steps and parameters of the mathematical model are shown in Table 2. Here Pd_s and Pd_v denote free sites on the catalyst surface and in the subsurface layer, and Pd_v-O and Pd_v-C indicate the subsurface O and C atoms, P_{CH_4} and P_{O_2} are the partial pressures of CH_4 and O_2; $\theta_{CH_3}, \theta_{CH_2}, \theta_{CH}, \theta_O, \theta_C, \theta_H$ are the surface coverages of adsorbed species, θ_{O_ν} is the concentration of the dissolved oxygen in the subsurface layer, θ_{C_ν} is the concentration of the dissolved carbon in the subsurface layer, $\theta_* = 1 - (\theta_{CH_3} + \theta_{CH_2} + \theta_{CH} + \theta_O + \theta_C + \theta_H)$ is the concentration of free adsorption sites on the catalyst surface; $\theta_r = 1 - (\theta_{O_\nu} + \theta_{C_\nu})$ is the concentration of free sites in the first subsurface layer. The reaction mechanism includes stages shown in Table 2. Agreeing to the experimental data the carbon atoms can penetrate into Pd bulk to a depth of up to hundred subsurface atomic layers [20]. Taking into account this fact we assume that the dissolved carbon atoms C_ν are uniformly distributed over n_z(= 100) atomic layers. However, only the carbon atoms from the first subsurface layer participate in stage 7. According to the proposed reaction mechanism, the dynamic behaviour of the system can be simulated by the following equations:

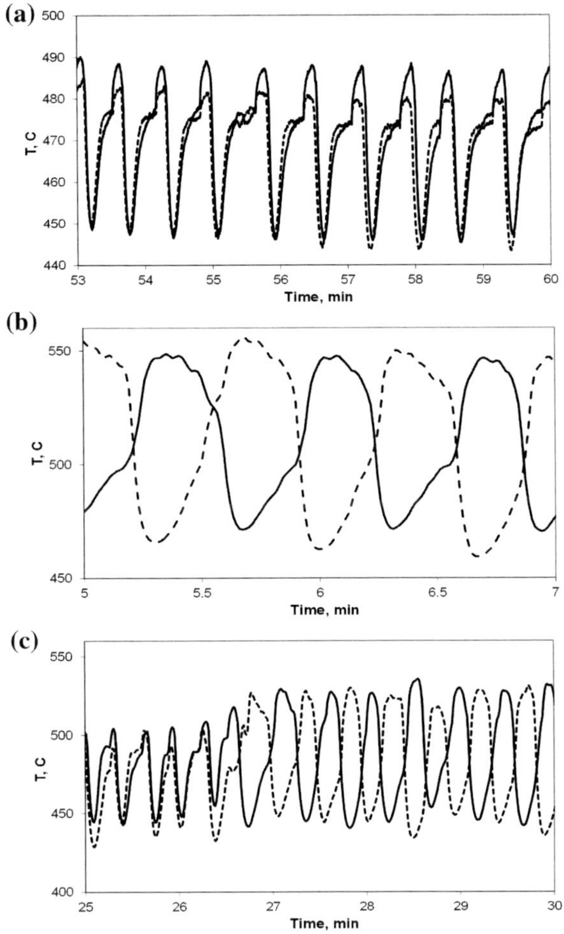

Fig. 5 Different dynamic behavior, which was observed over two Pd foils inserted in one reactor: **a** "in-phase" synchronization, **b** "anti-phase" synchronization, **c** the transition from "in-phase" oscillations to "anti-phase" oscillations during the increase of the temperature. Reaction mixture: 15% $O_2 + CH_4$; feed flow rate, 50 mL/min. The *thick solid* (*dashed*) *line* refers to the first (second) oscillator. (Reprinted from Ref. [19])

$$
\begin{aligned}
&\theta'_{CH3} = r_1 - r_{1.2},\\
&\theta'_{CH2} = r_{1.2} - r_{1.3},\\
&\theta'_{CH} = r_{1.3} - r_{1.4},\\
&\theta'_{H} = r_1 + r_{1.2} + r_{1.3} + r_{1.4} - 2r_{-1} - 2r_4 - 2r_8,\\
&\theta'_{C} = r_{1.4} - r_3 - r_6,\\
&\theta'_{O} = 2r_2 - 2r_{-2} - 2r_3 - r_4 - 2r_7 - r_5,\\
&\theta'_{Cv} = (r_6 - r_7)/n_z,\\
&\theta'_{Ov} = r_5 - r_8,\\
&T' = \alpha(T^0 - T) + \beta \sum_{i=1}^{12} r_i \Delta H_i,\\
&P'_{CH4} = (P^0_{CH4} - P_{CH4})/\tau - \gamma r_1,\\
&P'_{o2} = (P^0_{O2} - P_{O2})/\tau - \gamma(r_2 - r_{-2}),
\end{aligned}
\tag{5}
$$

Table 2 Temperature dependent reaction parameters and heat effects of various stages of methane oxidation over Pd

i	Reaction step	The rates of r_i	Prefactor k_{oi} (s^{-1})	Activation energy (kJ/mol)	Heats of reaction steps (kJ/mol)
1.	$CH_4 + 2Pd_s \rightarrow Pd_s\text{-}CH_3 + Pd_s\text{-}H$	$r_1 = k_1 P_{CH4}\theta_*^2$	$s = 2 \times 10^{-3}$	20	−103
1.2	$Pd_s\text{-}CH_3 + Pd_s \rightarrow Pd_s\text{-}CH_2 + Pd_s\text{-}H$	$r_{1.2} = k_{1.2}\theta_{CH3}\theta_*$	10^{13}	20	−103
1.3	$Pd_s\text{-}CH_2 + Pd_s \rightarrow Pd_s\text{-}CH + Pd_s\text{-}H$	$r_{1.3} = k_{1.3}\theta_{CH2}\theta_*$	10^{13}	20	−103
1.4	$Pd_s\text{-}CH + Pd_s \rightarrow Pd_s\text{-}C + Pd_s\text{-}H$	$r_{1.4} = k_{1.4}\theta_{CH}\theta_*$	10^{13}	20	−103
−1	$Pd_s\text{-}H + Pd_s\text{-}H \rightarrow H_2 + 2Pd_s$	$r_{-1} = k_{-1}\theta_H^2$	3.8×10^7	80	−102.5
2	$O_2 + 2Pd_s \rightarrow 2Pd_s\text{-}O$	$r_2 = k_2 P_{O2}\theta_*^2 e^{-\delta O_v}$	$s = 1$	0	221
−2	$Pd_s\text{-}O + Pd_s\text{-}O \rightarrow O_2 + 2Pd_s$	$r_{-2} = k_{-2}\theta_O^2$	10^{13}	170	−221
3.	$Pd_s\text{-}C + 2Pd_s\text{-}O \rightarrow CO_2 + 3Pd_s$	$r_3 = k_3\theta_C\theta_O^2$	0.9×10^8	60	722
4.	$2Pd_s\text{-}H + Pd_s\text{-}O \rightarrow H_2O + 3Pd_s$	$r_4 = k_4\theta_O\theta_H^2$	5.1×10^6	29	30.2
5.	$Pd_s\text{-}O + Pd_v \rightarrow Pd_v\text{-}O + Pd_s$	$r_5 = k_5\theta_O\theta_r$	1.0×10^5	90	0
6.	$Pd_s\text{-}C + Pd_v \rightarrow Pd_v\text{-}C + Pd_s$	$r_6 = k_6\theta_C\theta_r$	4.0×10^{10}	107	0
7.	$Pd_v\text{-}C + 2Pd_s\text{-}O \rightarrow CO_2 + 2Pd_s + Pd_v$	$r_7 = k_7\theta_{Ov}\theta_O^2$	1.2×10^6	50	722
8.	$Pd_v\text{-}O + 2Pd_s\text{-}H \rightarrow H_2O + 2Pd_s + Pd_v$	$r_8 = k_8\theta_{Ov}\theta_H^2$	0.6×10^6	80	30.2

Here $\alpha = (hS_O)/(WC_p)$; $\beta = (S_O N)/(WC_p)$; $\gamma = (SNRT_0)/V$; $\tau = V/F$, where: h is the heat transfer coefficient, S_O—the catalyst surface area, W—the catalyst weight, C_p—the catalyst heat capacity; N—the adsorption capacity of Pd; R—the universal gas constant, V—the reactor volume; F—the flow rate of the reactant mixture, ΔH_i—the heat effects of the corresponding studies of the reaction mechanism. The numerical values of these parameters correspond to the conditions of the experimental study of the oscillatory behaviour during methane oxidation in a CSTR over the Pd foil presented in Ref. [20]: $h = 0.003$ cal/(s × K), $S_O = 0.27$ cm^2, $W = 0.065$ g, $C_p = 0.057$ cal/(g × K), $N = 1.6 \times 10^{-9}$ mol/cm^2, $V = 0.126$ cm^3, $F = 0.33$ cm^3/s. The external values of parameters were also taken from Ref. [20], namely $P^0_{CH_4} = 621$ Torr, $P^0_{O_2} = 109$ Torr and $T^0 = 618$ K.

The numerical solutions of system (5) demonstrate that the variables θ_{CHx} $(x = 1, 2, 3)$ can be considered as the fast variables both in transient and in self-oscillating dynamic regimes. Thus one can simplify the system (5) replacing the dynamic values of variables $\theta_{CHx}(t)$ with their 'quasi-stationary' values $C\bar{H}_x(t)$. To obtain these values one has to set the right hand sides of the first three equations in the system (5) equal to zero and to solve the resulting algebraic equations, assuming that $\theta_* \neq 0$. Since at the chosen parameters $k_{1.2} = k_{1.3} = k_{1.4} = k_d \gg 3k_1 P^0_{CH_4}$, CH_x species can be totally eliminated from the model assuming the following expression for θ_*:

$$\theta_* = 1 - \theta_O - \theta_C - \theta_H \tag{6}$$

In the reduced model the equations for the variation of H and C atoms can be written as follows:

$$\begin{aligned} \theta_H' &= 4r_1 - 2r_{-1} - 2r_4 - 2r_8, \\ \theta_C' &= r_1 - r_3 - r_6, \end{aligned} \tag{7}$$

The reduced model, describing methane oxidation over one single foil with a surface area S_0 can be re-written in the following concise form:

$$\begin{aligned} x' &= F(x, p) \\ p' &= G_\tau(p) - G(x, p) \end{aligned} \tag{8}$$

where x is the vector of variables $\{\theta_H, \theta_c, \theta_O, \theta_{Cv}, \theta_{Ov}, T\}$ determining the state of the catalyst and p is the vector of partial pressures $\{P_{CH_4}, P_{O_2}\}$.

The model, describing the dynamic behaviour of methane oxidation over two equal Pd foils in a CSTR can be formulated using the same definitions as follows:

$$\begin{aligned} x_1' &= F(x_1, p) \\ x_2' &= F(x_2, p) \\ p' &= G_\tau(p) - G(x_1, p) - G(x_2, p) \end{aligned} \tag{9}$$

Here two oscillators are coupled via the balance equations corresponding to the CSTR reactor.

The bifurcation diagrams of a stationary catalyst temperature T in dependence of an inlet oxygen pressure P^0_{O2} for models (8) and (9) are shown in Fig. 6a, b.

Only two bifurcation points can be distinguished in the bifurcation diagram for the model (8): the turning point t_I at P^0_{O2} = 6.866 kPa and the Andronov-Hopf bifurcation point h_I at P^0_{O2} = 22.965 kPa. Two unstable steady states (indicated by the dashed line) occur to the right of the point t_I, while only one stable stationary solution (marked with the solid line) exists to the left of the point t_I. The region of oscillations is marked in grey. It locates around the lower unstable steady state and is bounded by the Andronov-Hopf bifurcation h_I to the right and by the homoclinic bifurcation to the left. The nonlinear feedback mechanism, which gives rise to the oscillations is carried out by the multiplier $\exp(-\delta\theta_{Ov})$ in stage 2. The physical meaning of this multiplier is the dependence of the oxygen adsorption rate on the concentration of the subsurface oxygen, detected experimentally in Ref. [21].

In contrast the bifurcation diagram for the model (9) contains two parts of qualitatively different stationary solutions (see Fig. 6b). One basic part marked in black coincides with the stationary solutions of one Pd foil having a double area S = 2S$_0$. It contains the symmetric stationary solution $y_s = \{x, x, p\}$, where $\{x,p\}$ is the stationary solution of the model (8) for one foil with a double area S. Here the

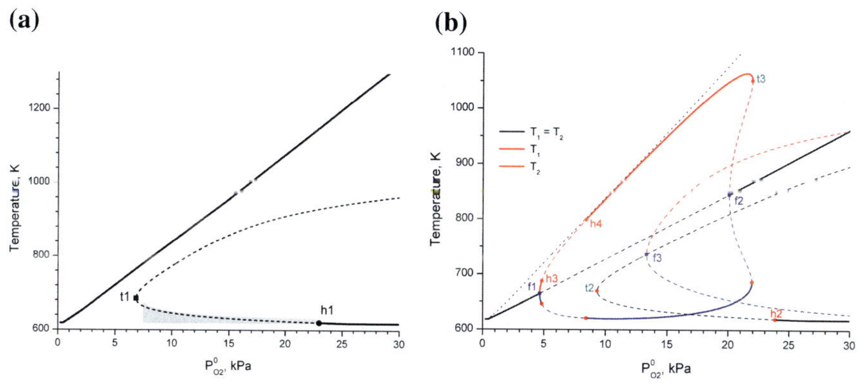

Fig. 6 Bifurcation diagrams of stationary catalyst temperature with $P^0_{O_2}$ as a bifurcation parameter. **a** Model (8): *solid line* stable solutions; *dashed line* unstable solutions. h_1 denotes the Andronov-Hopf bifurcation, t_1 indicates the turning point. The region of oscillations is marked in *grey*. **b** Model (9): T_1 and T_2—temperatures of both foils. *Solid line* stable solutions; *dashed line* unstable solutions. *h* denotes Andronov-Hopf bifurcations, *t* indicates turning points, while *f* marks the pitch-fork bifurcations

temperatures of two foils coincide. The second part contains the asymmetric solution $y_a = \{x_+, x_-, p\}$ together with its 'mirror' asymmetric solution $\{x_-, x_+, p\}$. The symbols x_+ and x_- denote the states of high and low temperatures correspondingly. As can be seen from Fig. 6b the asymmetric solution corresponds to the situation when the catalyst temperature of two similar foils is significantly different indicating that one of the foils becomes more active than the other. Moreover with the P^0_{O2} increase the high temperature branch of the asymmetric state approaches the stationary temperature of the model (8) (thin dotted line in Fig. 6b). This means that the high activity foil behaves as if it is alone in the reactor and in a dual catalyst only one foil is active in methane oxidation.

There are two regions of the oscillatory behaviour in the model (9): one associated with the asymmetric solution and the other with the symmetric solution. The first type of oscillations is shown in Fig. 7.

It occurs in the region between h_3 and h_4 points. Here only "out of phase" oscillations can be detected. Figure 7 demonstrates that during antiphase oscillations one of the foils spends the most time in a high active state, which corresponds to the steady state temperature obtained in the model (8) for a one foil with the area $S = S_0$. The other foil at this time is totally non-active. It becomes active only when the activity of the first foil begins to decrease demonstrating two opposite balanced processes.

The second type of oscillations is related with the symmetric solution. It occurs in the region between t_2 and h_2 similar to the oscillations obtained in the model (8) for a one foil. The properties of the periodic oscillations change with the variation of P^0_{O2} . At low initial oxygen pressure "out of phase" oscillations occur, while at

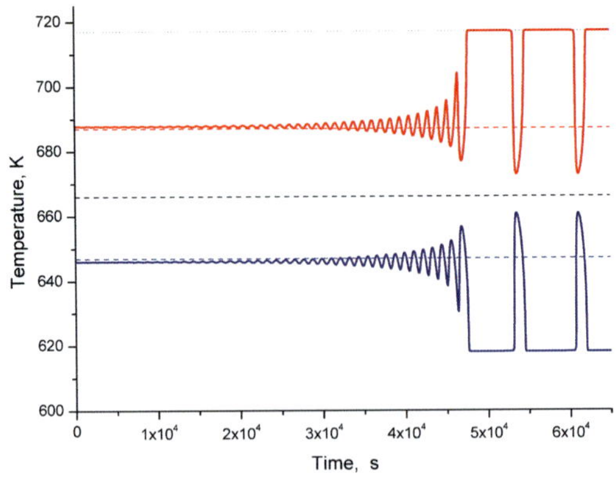

Fig. 7 The development of temperature oscillations of both foils T_1 and T_2 at $P^0_{O_2}$ = 4.825 kPa. *Thick dashed line* indicates unstable symmetric stationary temperature and *thin dashed lines* indicate asymmetric stationary temperature in the model (9). *Dotted line*—the stable stationary temperature in the model (8). (Reprinted from Ref. [19])

high P^0_{O2} the "in-phase" synchronization is detected. Figure 8 demonstrates the examples of 'out of phase' and 'in phase' oscillations for various values of P^0_{O2}.

The results of simulations demonstrate that the division of the Pd foil in two identical parts can lead not to 'in-phase' synchronized oscillations, as was always detected earlier for equal oscillators [16, 17], but to the appearance of "out of

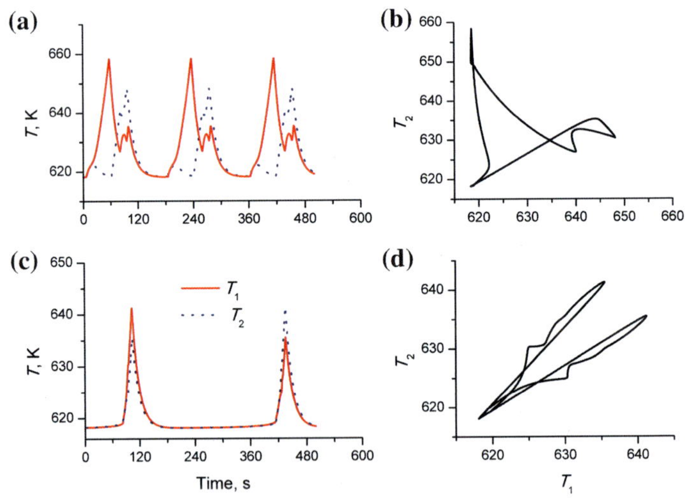

Fig. 8 Time series of T_1 (in *blue*) and T_2 (in *red*) at various initial oxygen pressure: **a** $P^0_{O_2}$ = 8.5 kPa and **b** $P^0_{O_2}$ = 22.6 kPa. The projection of the phase trajectory on the T_1–T_2 parameter plane. **c** $P^0_{O_2}$ = 8.5 kPa and **d** $P^0_{O_2}$ = 22.6 kPa. (Reprinted from Ref. [19])

phase" reaction rate oscillations over two equal separated parts. Moreover the inhomogeneous stable steady state (one foil with high and the other with a low activity) can come out after cutting of the Pd foil in two identical parts. Under some conditions the activity of the dual foil coincides with the activity of a single foil. It happens, when one of the foils behaves as if it is alone in the reactor while the second foil is not active at all. The same situation happens periodically during "antiphase" oscillations shown in Fig. 7 when one of the foils is in an active state, while the second foil is completely inactive.

The obtained results demonstrate that there is no any linear dependence of the catalytic activity upon the catalyst surface and the number of active centers in contrast to the Boreskov theory of constant specific catalytic activity in catalysts of constant chemical composition [22]. Due to the nonlinearity of the system and the ability to form spatial structures two Pd foils can produce the same catalytic activity as one Pd foil.

4 Conclusions

The above examples show how the nonlinear phenomena influence upon the traditional concepts of catalysis. Contrary to the principle of superposition the reaction rate is not necessarily increased with the number of active centers and due to the formation of some kind of a spatial structure some part of the catalyst may be non-active under the well mixed homogeneous conditions. The results presented in Sect. 3 demonstrate that under certain conditions half of the bigger foil can be removed from the reactor without any loss in the degree of conversion.

The synergetic effect in heterogeneous catalysis also originates in violation of the principle of superposition. The results given in Sect. 2 show that the surface coupling of different parts of the catalyst may lead not only to the improvement of the catalyst activity, but also to the appearance of the new dynamic behavior. They demonstrate how two nonlinear bistable systems can produce an oscillatory behavior and reveal the new type of kinetic oscillations that originate due to the interplay of surface diffusion and reaction rates on various parts of a catalyst.

Acknowledgements We thank Dr. Natalya Semendyaeva and Dr. Tatjana Lysak for the help in the bifurcation analysis of models (5) and (9). M.M.S and N.V.P. acknowledge the financial support from the Russian Science Foundation (grant N 17-13-01057).

References

1. R. Imbihl, G. Ertl, Oscillatory kinetics in heterogeneous catalyses. Chem. Rev. **95**, 697–733 (1995)
2. M.M. Slinko, N. Jaeger, *Oscillatory heterogeneous catalytic systems, studies in surface science and catalysis*, vol. 86 (Elsevier, Amsterdam, 1994)

3. H.H. Rotermund, W. Engel, M. Kordesch, G. Ertl, Imaging of spatio-temporal pattern evoluation during carbon monoxide oxidation on platinum. Nature **343**, 355–357 (1990)
4. F. Schüth, B.E. Henry, L.D. Schmidt, Oscillatory reactions in heterogeneous catalysis. Adv. Catal. **39**, 51–127 (1993)
5. V.P. Zhdanov, B. Kasemo, Simulations of the reaction kinetics on nanometer supported catalyst particles. Surf. Sci. Rep. **39**, 25–104 (2000)
6. V. Gorodetskii, J. Lauterbach, H.-H. Rotermund, J.H. Block, G. Ertl, Coupling between adjacent crystal planes in heterogeneous catalysis by propagating reaction diffusion waves. Nature **370**, 276–279 (1994)
7. G. Ertl, Oscillatory catalytic reactions at single-crystal surfaces. Adv. Catal. **37**, 213–277 (1990)
8. R. Imbihl, Oscillatory reactions on single crystal surfaces. Progr. Surf. Sci. **44**, 185–343 (1993)
9. M.G. Slinko, V.I. Bykov, G.S. Yablonskii, T.A. Akramov, Multiplicity of the steady states in heterogeneous catalytic reactions. Dokl. Akad. Nauk SSSR **226**, 876–879 (1976)
10. A.V. Kalinkin, A.V. Pashis, V.I. Bukhtiyarov, CO oxidation over the Pt-Rh system, 3. Reaction on a heterophase surface. React. Kinet. Catal. Lett. **78**, 121–127 (2003)
11. V. Fiorin, D. Borthwick, D.A. King, Microcalorimetry of O_2 and NO on flat and stepped platinum surfaces. Surf. Sci. **603**, 1360–1364 (2009)
12. T.W. Root, L.D. Schmidt, G.B. Fisher, Adsorption and reaction of nitric oxide and oxygen on Rh(111). Surf. Sci. **134**, 30–45 (1983)
13. A.G. Makeev, N.L. Semendyaeva, M.M. Slinko, Synergetic effect and oscillatory behavior of CO oxidation over a bimetallic composite catalyst. Chem. Eng. J. **282**, 3–10 (2015)
14. J. Lauterbach, H.H. Rotermund, Gas-phase coupling in the CO oxidation reaction on polycrystalline platinum. Catal. Lett. **27**, 27–32 (1994)
15. P.K. Tsai, M.B. Maple, R.K. Herz, Coupled catalytic oscillators: CO oxidation over polycrystalline Pt. J. Catal. **113**, 453–465 (1988)
16. M.M. Slinko, E.S. Kurkina, M.A. Liaw, N.J. Jaeger, Mathematical modelling of chaos, synchronisation and phase death phenomena during CO oxidation over Pd zeolite catalysts. J. Chem. Phys. **111**, 8105–8114 (1999)
17. V.Yu. Bychkov, YuP Tyulenin, M.M. Slinko, V.N. Korchak, Nonlinear behavior during methane and ethane oxidation over Ni, Co and Pd catalysts. Surf. Sci. **603**, 1680–1689 (2009)
18. V.Yu. Bychkov, YuP Tyulenin, M.M. Slinko, V.N. Korchak, Synchronization of local oscillators in oxidation reactions of C_1–C_4 hydrocarbons over metal catalysts. Kin. Catal. **55**, 97–102 (2014)
19. N.V. Peskov, M.M. Slinko, VYu. Bychkov, V.N. Korchak, Mathematical modelling of oscillatory behaviour during methane oxidation over palladium catalysts. Chem. Eng. Sci. **84**, 684–694 (2012)
20. V.Yu. Bychkov, YuP Tyulenin, M.M. Slinko, D.P. Shashkin, V.N. Korchak, The study of the oscillatory behavior during methane oxidation over Pd catalysts. J. Catal. **267**, 181–187 (2009)
21. S. Ladas, R. Imbihl, G. Ertl, Kinetic oscillations during the catalytic CO oxidation on Pd (110): the role of subsurface oxygen. Surf. Sci. **219**, 88–106 (1989)
22. G.K. Boreskov, Activité catalytique des métaux en rapport avec la réaction d'oxydation de l'hydrogène. J. de Chimie Physique **51**, 759–768 (1954)

Catalytic Oxidation of CO—A Striking Example of Synergetics

Experimental Studies of Coupled Zeolite Catalyst Wafers in a Tube Reactor

Peter J. Plath and Carsten Ballandis

Abstract The dynamics of coupled catalyst wafers in a tube reactor was studied and traced by examining the oxidation of CO on a supported palladium zeolite catalyst under normal pressure in a continuous flow of reactants. An X-type zeolite loaded with 0.05% by weight palladium was used as catalyst. The overall conversion rate exhibits a dynamic behavior with self-affine patterns of excursions to smaller conversion rates on a time scale in the order of some seconds. Similar patterns were observed for the temperature time series of the catalyst wafers. The influence of the flow rate on catalyst wafer coupling was studied. Increasing the flow rate causes decoupling of the catalyst wafers, increased frequency of maximum excursions, increasing rate of smaller excursions, increasing pattern complexity and a decrease in maximum conversion (baseline).

1 Introduction

Palladium-catalyzed CO-oxidation on support catalysts has long been one of the famous examples for studying oscillating phenomena in heterogeneous catalytic systems [1–6]. In most cases, the reaction exhibits very strange time series which show self-similar patterns of breakdowns in conversion. However, there are also conditions in which regular oscillations can be found [7, 8].

Cellular automata models have been developed to describe self-affine temporal patterns qualitatively [9–13]. These models are based on the local interaction of elementary reactors that show only two states of reaction behavior: active or passive, depending on the reduced or oxidized state of the palladium (Pd) particles of the catalyst. In the automaton model of Liauw et al. [14], the phase transition between Pd and PdO was modelled explicitly for the first time. In said model, there

P.J. Plath (✉) · C. Ballandis
Universität Bremen, Physikalische Chemie, 28359 Bremen, Germany
e-mail: peter_plath@t-online.de

P.J. Plath
Fritz-Haber Institute, Physical Chemistry, Faradayweg 4, 14195 Berlin, Germany

S.C. Müller et al. (eds.), *Complexity and Synergetics*,
https://doi.org/10.1007/978-3-319-64334-2_8

is no spatial interaction of the elementary reaction units. Instead, Liauw took into consideration only the global coupling via the gas phase.

Based on experimental results, Svensson et al. [7] and Slin'ko et al. [8] developed a reaction-diffusion model in order to describe the regular oscillations observed [15]. Both classes of models describe very different experimental situations. It is not the mathematical tool used for these models that accounts for the differences, but the physico-chemical background on which the models are based. The reason for this is the experimental situation involving pre-treatment of the catalyst, which provides quite different results. We decided to focus on the chaotic "stationary" state of the catalytic system, which starts in the reduced state and which is structurally stable across a wide range of flow rates and catalyst particle sizes.

Our model is therefore confined to these specific experimental conditions. Our aim was to describe qualitatively and even semi-quantitatively these chaotic dynamics of CO oxidation catalysed by Pd supported on an X-type zeolite or Al_2O_3. The qualitative success of the discrete models [3, 4, 7, 9, 10, 16] encouraged us to enhance these models

(a) by introducing different sizes of the "elementary palladium reactors" and
(b) by including the non-isothermal behavior of the strong exothermic CO oxidation which has already been used [16].

There is a particle size distribution among the single Pd crystals in the zeolite crystal framework. For a given number of reactants which touch the surface in a given unit of time, there exists a very large set of particles which behave in just the same way, since they have almost the same size and surface area. All these Pd particles thus become synchronized if they can be regarded as only one "elementary palladium reactor". This is true only if the accessibility of the particle surface for the reactants is not restricted by diffusion.

Our aim is to describe the experimentally observed temporal pattern formation obtained solely by coupling these elementary palladium reactors via the gas phase. In this way, we are following the advice of Hermann Haken: *"Synergetics deals with systems composed of many subsystems of quite different natures,..." "In particular, our attention will be focused on those situations in which these structures arise in a self-organized fashion, ..."* [17]. However, in contrast to the usual chemical procedure for describing heterogeneously catalyzed reactions, our interacting elementary units are not molecules and surface sites, but each consist of a very large set of commonly reacting Pd particles of similar sizes which we take as a new chemical entity—the "elementary palladium reactor". A similar ansatz for new reacting catalytic entities—but coupled locally—has previously been described in order to simulate the catalytic behavior of the Pd-catalyzed oxidation of methanol [10].

Given that the production of CO_2 is an exothermic reaction all particles of almost the same size produce an equivalent amount of heat within the same time. This will cause an enormous increase in particle temperature, because the particles are very small. Consequently, all the reaction properties will be synchronized, as well. By choosing enough elementary reactors and a proper "particle size distribution" close to the reality of the Pd particles in the catalyst support, one should be able to describe the experimental results even quantitatively in a convincing way.

In this new work, our aim was to leave the approximately two-dimensional system comprising only one catalyst wafer by introducing a second wafer in order to model a fixed-bed reactor in a tube by means of two differential reactors. The reactants flew through these catalyst wafers one after the other. The coupling of the catalyst wafers was observed for different flow rates and different distances between the catalyst wafers. The results were simulated by using an extended version of the model described earlier [16]. The results of the experimental studies are presented in Sect. 1, while the results of the theoretical studies are described briefly in Sect. 2. Finally, our conclusions and remarks are provided in Sect. 3.

2 Results of the Experimental Studies

2.1 Experimental System and Procedure

The catalyst chosen for the oxidation of CO was an X-type zeolite, used as a matrix for 0.05% by weight highly dispersed Pd. The concentration of Pd in the catalyst is thus only one tenth of the Pd concentration used in previous work [16]. The zeolite was loaded by ion exchange with an aqueous solution of $[Pd(NH_3)_4]Cl_2$. The Pd ions were subsequently reduced under hydrogen flow at 623 K. Under these conditions, Pd particles with an average size of 4 nm can be obtained. Preparation of the catalyst was part of our experimental work. The details of preparation and characterisation are described by Jaeger et al. [2], Svensson et al. [7] and Slin'ko et al. [8]. Oxidation occurred in a continuous flow reactor under non-isothermal conditions.

A new reactor was designed specifically for these experiments. Some important changes were made relative to the reactor which was used in previous work [16]. The new reactor was designed such that a second catalyst layer could be added to the reactor. To this end, several glass tripods could be inserted into the reactor from the top. These tripods were placed on the catalyst wafer, which was positioned on a permanently installed glass frit, because the catalyst wafers had almost the same diameter as the reactor. On top of the tripod was another glass frit on which a second catalyst wafer could be positioned. The length of the legs of the different tripods could be varied from 1 to 10 cm, so it was possible to change the distance between the two catalyst wafers. In order to measure the dynamic behavior of each catalyst layer, three NiCr-Ni thermocouples were positioned in the reactor. Silver plates (diameter 3 mm; weight 14.8 mg and 16.0 mg for wafer 2, and one of 17.8 mg for wafer 1) were soldered to the end of each NiCr-Ni thermocouple. The silver plates were used to enhance local heat conductivity to the catalyst wafers. Two thermocouples were positioned in the lower glass frit. These two thermocouples were inserted into the reactor via a hole in the glass reactor and then through two holes in the lower glass frit. Above the lower glass frit, the silver plates were soldered to each thermocouple. The hole in the reactor for the thermocouples was sealed with cement.

The present paper discusses only the influence of the gas flow rate on the dynamics of CO conversion. The distance between the two catalyst wafers was

5 cm. The CO inlet concentration was 1.23% by volume and the O_2 inlet concentration 13.5% by volume. There was therefore a large excess of oxygen.

The reactor was heated up in a tube oven, the temperature of which was set to 473 K (200 °C) with a precision of about ±0.3 K, and measured with a NiCr-Ni thermocouple in the oven and in the reactor. The gas mixture first flowed through a preheating coil which had been placed in the oven. 23.4 mg (for wafer 1) and 24.3 mg (for wafer 2) of the catalyst powder were pressed into wafers with a diameter of 13 mm. The height of the catalyst wafers ranged from 0.1 to 0.2 mm. These catalyst wafers were then placed on the glass frits in a glass reactor. The flow rate of the gas mixture was varied from 40 to 200 ml/min.

Gases supplied for the reaction comprised a mixture of CO and N_2 (10.1% by volume CO in N_2, 99.97% pure, Messer-Griesheim), O_2 (99.995% pure, Messer-Griesheim) and N_2 (99.996% pure, Messer-Griesheim) under atmospheric pressure. The flow rates of the feed were controlled by thermal mass flow meters (±0.1% precision, HI-TEC Series F-100/200). The mass flow meters were connected to a controller (HI-TEC E-5500). The geometric volume of the reactor was 9 ml for a distance between the catalyst wafers of 5 cm. The residence time was therefore 3.6 s for a flow rate of 150 ml/min.

The CO and CO_2 concentrations were measured at the outlet of the reactor by non-dispersive infrared absorption spectroscopy (URAS 3G, Hartmann and Braun). In addition to the CO and CO_2 concentrations, the temperatures of the catalyst wafers were continuously monitored by the three thermocouples which were soldered to the silver plates touching the catalyst wafers. The voltages of the thermocouples were reduced by counter voltage devices (Novocontrol, Hundsangen), and were each amplified by a factor of 5000 times using two measurement amplifiers (Linseis, T-MV-100, one with factor 100 and one with factor 50). The concentrations of CO and CO_2 and the modified voltages of the thermocouples were recorded on a computer (Intel Pentium MMX, 90 MHz with a 4.3 GB Quantum Fireball hard disk) using a 12 bit A/D transformer card (P 1000, BMC Systeme GmbH) and the "NextView Version 2.5" software (BMC Systeme GmbH). Due to these modifications, the output voltages of the small range of temperature oscillations measured at the catalyst wafers (about 1 or 2 K) were amplified to 3 V. Even temperature changes of about 0.01 K could thus be recorded with precision.

2.2 Experimental Results

By performing the catalytic oxidation in the new reactor (Fig. 1) with only one catalyst wafer, self-affine time series are obtained (Fig. 2), as was the case with the old reactor [16]. These series do not depend on the position of the wafer in the reactor.

Figure 3 shows the correlation between the temperature events on the wafers (wafer 1 and wafer 2) and the CO concentration in the outlet. The delay of about 23 s between the temperature breakdown-events at the catalyst wafers and the

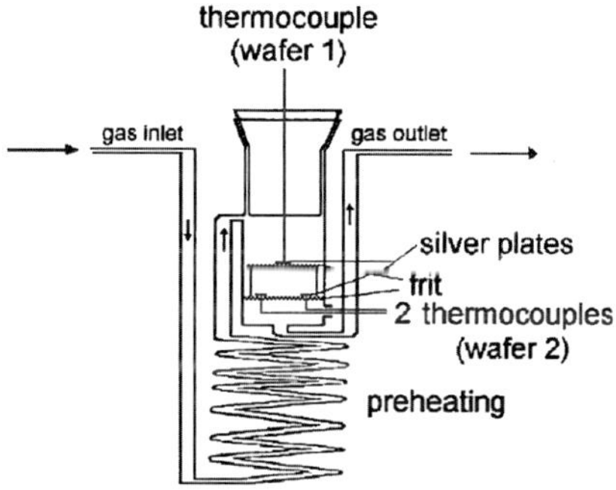

Fig. 1 Flow-reactor with two catalyst wafers (wafer 1 and wafer 2) placed on two frits

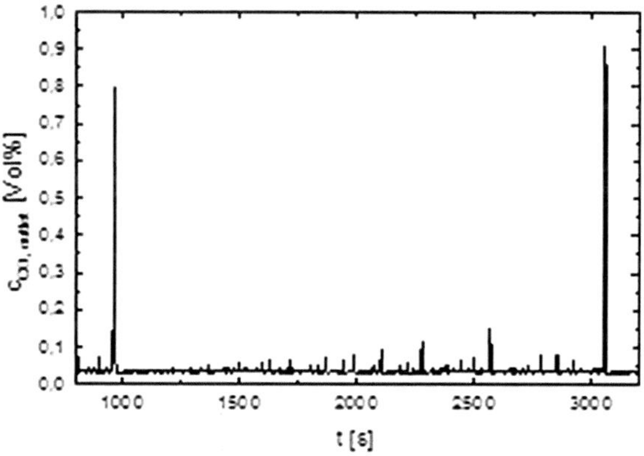

Fig. 2 Self-affine time series for only one catalyst wafer in the reactor at a flow rate of 100 ml/min

observed increase in CO concentration are based on the time required by the gas flow from the reactor to the URAS detector.

Each peak of the CO concentration is correlated with a temperature decrease of at least one of the catalyst wafers. The numbered peaks show exemplarily the correlation of temperature and CO conversion. Same numbers in Fig. 3a, b belong to the same events. The events (3) and (4) are not coupled although they are temporally close. The detailed examination of the time series has shown that the delay times between these two events are too long in order to result from a coupling. The same applies to the events (1a) and (1b) which are temporally close at random.

The self-affine structures of the time series are again evident. However, the self-affine structure is not as well developed as in the case of only one catalyst wafer (Fig. 2), since the reactions on the first catalyst wafer influence the reactions on the second. Nevertheless, each breakdown of the conversion—this means each peak in the CO outlet concentration—can be correlated to a very sharp decrease in the temperature of at least one of the catalyst wafers. Even the excursion magnitudes of the temperature and concentration events can be correlated. Depending on the flow rate, decoupling or different kinds of coupling of the catalyst wafers were found. These different behaviors are described in the following.

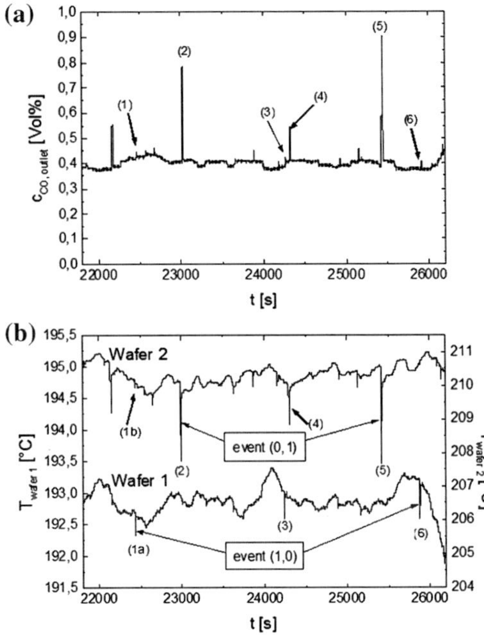

Fig. 3 Measured time series for two catalyst wafers spaced 5 cm apart, for a flow rate of 100 ml/min; **a** of the CO concentration in the outlet and **b** of the temperature of the two catalyst wafers (wafer 1 and wafer 2)

2.2.1 Decoupling of Catalytic Wafers

There are also events of temperature decrease which appear in one of the catalyst wafers only, either in wafer 1 (event (1, 0)) or wafer 2 (event (0, 1)) (see Fig. 3), even though wafer 1 is exposed first to the flow of gaseous reactants. This behavior occurs at higher flow rates, especially, for example at 100 ml/min. This means that both catalyst wafers start to react independently.

Very surprisingly, the breakdown in conversion at the first catalyst wafer (event (1, 0)) does not seem to have any influence at all on the reaction at the second catalyst wafer. The reason is the following: the Pd particles of the second catalyst wafer are in a state that is far from the phase transition function (Fig. 5, point A). There may be a slight increase in the mean temperature of wafer 2, but no set of Pd particles oversteps this function. Therefore, no breakdown in conversion at the second catalyst wafer can be observed.

2.2.2 "In-Phase" Phase Transitions

If a breakdown in conversion occurs at the first catalyst wafer (wafer 1), a sudden exponential decrease in its temperature can be observed. At medium or high flow rates (60 ml/min and higher), a breakdown in conversion at the first wafer is

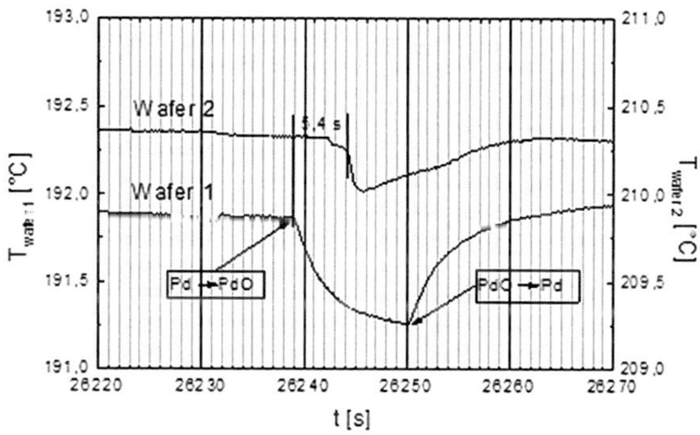

Fig. 4 "In-phase" phase transition: measured time series for two catalyst wafers spaced 5 cm apart, for a flow rate of 100 ml/min

sometimes followed by a breakdown in conversion at the second catalyst wafer, as well (Fig. 4). The delay in the breakdowns occurs over the time needed for the CO peak from the breakdown of the first catalyst wafer to reach the second wafer. Assuming plugged flow behavior; the said time can be calculated for a flow rate of 100 ml/min:

$$\frac{9ml\ reactor\ volume}{100\,ml/min} \cdot 60 = 5.4\,\mathrm{s}$$

In Fig. 4, it can be seen that exactly this calculated time delay can be found in the measured time series.

The breakdown in conversion is caused by the phase transition of some or all Pd particles in the catalyst. As has been demonstrated earlier [16], a breakdown can be described as a phase transition from the catalytically active Pd to the inactive PdO of a set of Pd particles with similar diameter, and subsequently a very fast cascade of phase transitions of all sets of smaller Pd particles. This phase transition from an active to an inactive state of the catalytic Pd particles leads immediately to an exponential decrease in the temperature of the first catalyst wafer (see Fig. 4).

Due to the breakdown in CO conversion, the remaining CO which is not converted to CO_2 at the first catalyst wafer will increase the CO concentration at the second wafer. If the catalytic Pd particles in the second wafer are in an active state close to the inactive state—for example if their degree of oxidation is high enough—the increase in the CO concentration in the feed will also shift these critical Pd particles to the inactive PdO state (see Fig. 5, Point B). As a consequence, we can observe an exponential decrease in the temperature in the second wafer as well. Both catalytic wafers are cooperating "in-phase".

Increase of CO concentration due to conversion breakdown at wafer 1 can produce a breakdown at wafer 2 (point B; "in-phase" coupling) or not (point A; decoupling), if the Pd-particles of wafer 2 are in a state far from the phase transition function.

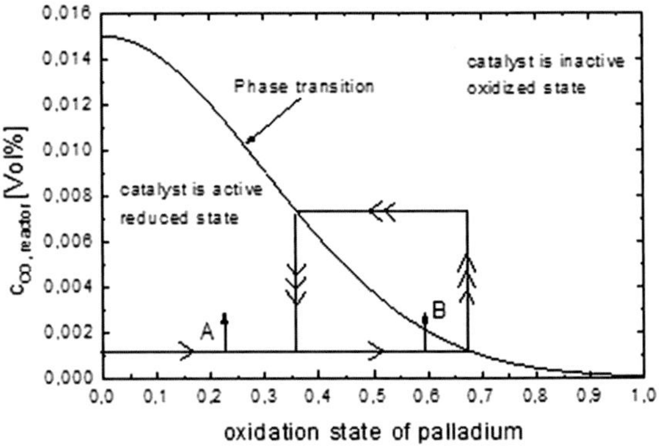

Fig. 5 Schematic representation of the phase transition function of the Pd-particles

2.2.3 Compensation of Conversion between Different Catalyst Wafers

At low flow rates, a remarkable effect occurs, as evident in Fig. 6. We call this the "compensation effect". A breakdown in conversion at the first catalyst wafer (wafer 1) leads to an excess of CO in the feed downstream from that catalyst. When this pulse of increased CO concentration passes the second catalyst wafer, the over-supply of CO leads to a strong increase in the absolute rate of CO to CO_2 conversion compared to the CO supply which normally reaches the second catalyst wafer. This results in sharp temperature peaks in the time series of the latter catalyst (wafer 2). Watching the total CO concentration at the outlet of the reactor (see Fig. 6b), the lack of conversion at the first catalyst wafer is somehow compensated by a correlated excess in conversion on the second catalyst wafer.

(a) Measured temperature time series of the two catalyst wafers. The events are marked where the conversion breakdowns of wafer 1 are compensated at wafer 2.
(b) Measured time series of the CO outlet concentration. Although the catalyst wafer 1 stops CO conversion, a large amount of the 1.23 vol.% CO inlet concentrations is converted at the catalyst wafer 2 due to the "compensation effect".

Although catalyst wafer 1 stops CO conversion, the CO outlet concentration remains in a range between 0.5 and 0.8% by volume. This means that a large amount of the 1.23% by volume CO inlet concentration is still converted at catalyst wafer 2. Three occurrences of this compensation effect are marked in Fig. 6a.

It is of interest to gain deeper insight into the chemical dynamics of this compensation effect. Figure 7 shows a temporal magnification of the experimental temperature time series. The points of ignition and extinction on both catalyst wafers are marked.

As shown by previous results [16], the breakdown in conversion at the first catalyst wafer is due to a common phase transition of a set of Pd particles to PdO particles. This leads immediately to an exponential decrease in the temperature of

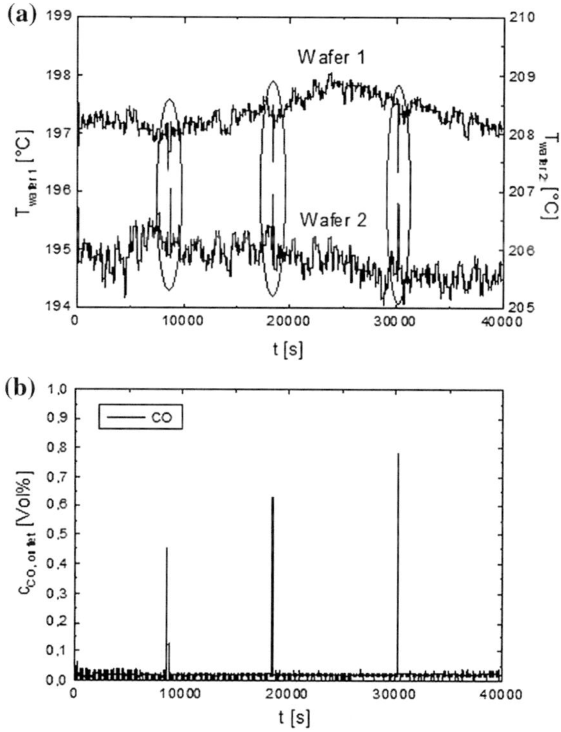

Fig. 6 Experimental time series for two catalyst wafers spaced 5 cm apart, for a flow rate of 40 ml/min

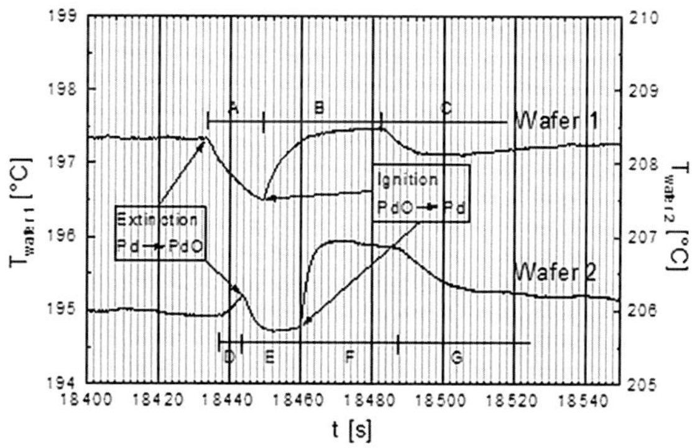

Fig. 7 Magnified time series for two catalyst wafers spaced 5 cm apart, for a flow rate of 40 ml/min (see Fig. 6, second marked event)

this catalyst wafer (see Fig. 7, area A). The effective CO concentration increases in the vicinity of the catalyst, and this set of PdO particles becomes reduced and reactivated due to the reverse phase transition from PdO to Pd (see Fig. 5). The conversion rate and its correlated temperature rise again exponentially in conjunction with this phase transition. The baseline temperature of wafer 1 is temporarily higher than before (see Fig. 7, area B). Finally, the baseline temperature of wafer 1 returns to the normal baseline (see Fig. 7, area C).

The phase transitions are indicated by the points of discontinuity in the derivation of the temperature time series. Despite the time shift due to the flow rate and the volume between the two catalyst wafers, a sudden increase occurs in the temperature and the conversion rate on the second catalyst, when the CO pulse reaches wafer 2 (see Fig. 7, area D). Wafer 2 then switches to the inactive state. The temperature increase accelerates this phase transition from Pd to PdO (see Fig. 7, area E).

When wafer 2 switches back to the active state, its temperature baseline is temporarily much higher than before (see Fig. 7, area F). This can be explained by the excess of CO due to the breakdown in conversion at wafer 1. Finally, the baseline for wafer 2 returns to the normal value also (see Fig. 7, area G).

3 A Short Commentary on the Theoretical Studies

The automaton model for calculating the results for only one catalyst wafer is taken as a basis for the extended model for two catalyst wafers. This basic model has been published in detail [16]. In the following, the basic principles of the one-layer-model will be sketched briefly. For better understanding, some of the key iterative equations of the basic model are provided below. It should be emphasized that we are modelling the reaction system by use of a coupled "iterative function system" with discrete time, not taking into account any spatial dependencies of the reaction.

Using the terms of H. Haken's concept of synergetics the CO-concentration takes over the function of an "order parameter" of the reaction.

For example, the CO concentration in the reactor at the next time step $c_{co}(t{+}1)$ is calculated by:

$$c_{CO}(t+1) = \underbrace{c_{CO}(t)\left(1 - c'_{conv}\right)}_{reaction} + \underbrace{a_f c^0_{CO}}_{gas\ inlet} - \underbrace{a_f c_{CO}(t)\left(1 - c'_{conv}(t)\right)}_{gas\ outlet}$$

Here, "a_f" is the "exchange factor" defined by: $a_f = 1 - e^{-\frac{v}{vol}}$, where "$v$" is the flow velocity, and "*vol*" has the meaning of the volume of the reaction zone, i.e. the volume of the very thin catalyst layer.

Calculations have shown that for the conversion rate the following approximation can be used:

$$c_{conv}(t) = \begin{cases} 1, & if\ active \\ 0, & if\ inactive \end{cases}$$

One of the really new ideas of the basic model is to introduce the temperature of each elementary palladium reactor. The lowest possible temperature of the particles is, of course, the oven temperature. To calculate the temperature of each elementary reactor, the heat production H_{prod} due to reaction heat and the heat loss H_{loss} due to the gas flow $\dot{V}$ are taken into account. Heat production and heat loss are added as the quantity of heat $Q(t)$ for each elementary reactor. The temperature T is obtained from the quantity of heat $Q(t)$, which is the second "order parameter", since the temperature $T(t)$ can simply be derived linearly from $Q(t)$. The quantity of heat for the next time step $Q(t+1)$ is calculated for each model Pd particle:

$$
\begin{aligned}
Q(t+1) &= Q(t) + H_{prod}(t) - H_{loss}(t) \\
H_{prod}(t) &= h_{prod} v_{flow} c_{conv}(t) O_{surface} \\
H_{loss}(t) &= (Q(t) - Q_0)\left(1 - e^{\frac{h_{loss} - O_{surface} t_{step}}{t_{step}}}\right)
\end{aligned}
$$

By including the oxidation and reduction velocities constants k_{ox} and k_{red}, the degree of oxidation for the next time step G(t + 1) can be calculated for each elementary palladium reactor. The functions with form $\left(1 - e^{-k}\right)$ cause the degree of conversion not to cross the value below 0 or above 1;

$$
\begin{aligned}
&\textit{if active}: G_{ox}(t+1) = G_{ox}(t) + (1 - G_{ox}(t))\left(1 - e^{-k_{ox}}\right) \\
&\textit{if inactive}: G_{ox}(t+1) = G_{ox}(t) - G_{ox}(t)\left(1 - e^{-k_{red}}\right)
\end{aligned}
$$

This degree of oxidation is the third "order parameter" of our system. Since we know that the Pd particles undergo strong variations of the temperature because of the heat production and heat loss, the reaction velocities are no longer constants:

$$
\begin{aligned}
k_{ox} &= A_{ox} e^{\frac{E_{ox}}{RT(t)}} \\
k_{red} &= A_{red} e^{\frac{E_{red}}{RT(t)}}
\end{aligned}
$$

The structure of two-wafer model [18] is documented in Fig. 8. One of the crucial differences to the one wafer model is the space between the wafers. A volume, which acts as a storage for the reactants and products is introduced between catalyst layer 1 and catalyst layer 2. This allows us to simulate different spacings between the catalyst wafers.

This storage space is assumed to be a well-mixed zone. It is a simplification of the real reactor, where the behavior of this space seems to behave more as a plugged flow zone.

To illustrate the capability of the enhanced two-wafer model, we demonstrate this briefly in one example. In Sect. 2.2.2, it was reported that an "in-phase" phase transition can occur. Figure 9 shows that the two wafer model reflects this striking behavior.

We calculated temperature time series of two catalyst wafers spaced 5 cm apart and at a flow rate of 200 ml/min. The breakdown of CO conversion at wafer 1

Fig. 8 Structure of the automaton with two wafers

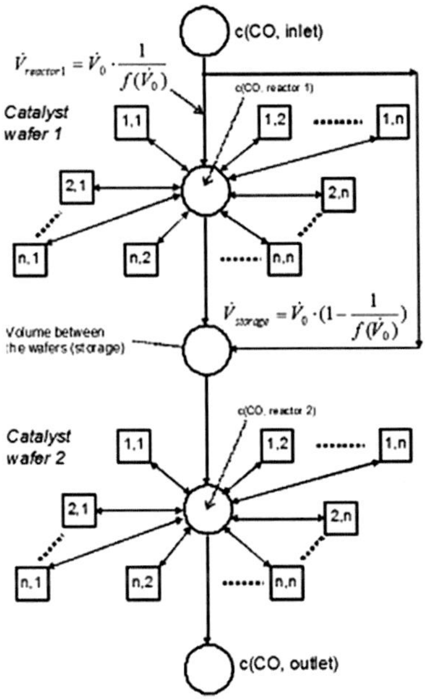

Fig. 9 "In-phase" phase transition calculated by the two-wafer model

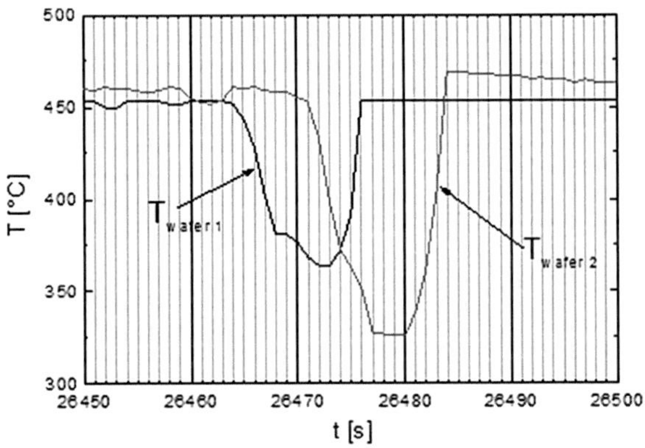

causes a breakdown of CO conversion at wafer 2. Both catalyst wafers are coupled and act in the same way which amplifies the CO decrease at the reactor outlet.

It is remarkable that the seven-second delay between the conversion breakdowns at wafer 1 and wafer 2 is very close to the measured time delay of 5.4 s (see Sect. 2.2.2), although a storage with a continuously stirred tank reactor (CSTR) behavior was used as approximation for the spacing between the catalyst wafers.

Catalyst wafers 1 and 2 are coupled and act in the same way. This causes the decrease in CO concentration at the reactor outlet to be amplified. The catalyst wafers act "in-phase".

4 Conclusion and Remarks

Looking at only one catalyst wafer, one sees that its width is much larger than its thickness. Thus, previous measurements [16] were performed in an approximately two-dimensional system. In the present work, the successive arrangement of two catalyst wafers in the reactor—both acting as differential reactors—was the first step-in proceeding from a two-dimensional to a three-dimensional fixed-bed reactor system. The task was to measure and to simulate the possible kinds of coupling between these catalyst wafers. As described in the previous work [16], the oxidation of CO on Pd support catalysts produces complicated patterns of CO conversion breakdowns. In our model system, the coupling of the Pd particles is assumed to occur via the gas phase only. Therefore, it was correctly anticipated that the patterns of CO conversion breakdowns at the first catalyst wafer can affect the patterns of the CO conversion breakdowns at the second catalyst wafer.

In this work, the influence of the flow rate of the educts gases on the dynamics of the patterns and of the coupling of the catalyst wafers was observed at a constant oven temperature of 473 K (200 °C). Three kinds of coupling between the catalyst wafers were identified:

- "Decoupling of the catalyst wafers". Breakdowns in CO conversion were observed independently at both wafers. A conversion breakdown at one catalyst wafer was not caused by a breakdown at the other wafer. This behavior was observed at higher flow rates ranging from 60 to 100 ml/min.
- "In-phase phase-transition". Discontinuation of CO conversion at the first catalyst wafer could affect a conversion breakdown at the second wafer after a specific time delay. Reverse coupling from the second to the first catalyst wafer was not observed. This behavior occurred at a higher flow rate in the range from 60 to 100 ml/min, also.
- "Compensation of conversion". Excess CO resulting from a conversion breakdown at the first wafer could be partially compensated with respect to the overall conversion by an increase in the activity of the second wafer. This kind of coupling only occurs at relatively low flow rates (in a range from 30 to 50 ml/min) and produces a complicated pattern.

By upgrading the one-layer model [16] to a two-layer system (in preparation), it was possible to simulate the main effects of the real measurements qualitatively in detail. The two-layer model will be published elsewhere in the near future. This proves once again the robustness of the previous one-layer model. Even the changes made to the real system by adding a second catalyst wafer could be simulated very well simply by duplicating the one-layer computer model.

It could be shown clearly that the automaton model of the coupled iterative functions describes perfectly the observed macroscopic temporal pattern formation during catalytic CO oxidation. We have introduced a new kind of chemical entity which we now call "elementary Pd reactors" (in the former work [16] we used the expression "model Pd particle") in order to describe in every detail the highly complicated temporal patterns of the linearly coupled two-layer system of catalytic differential reactors in a tube.

These new chemical entities are neither electrons, or molecules, nor concentrations. One can understand them as mesoscopic ensembles of nanoscale particles of single Pd crystals [18] acting in synchrony to create macroscopic temporal patterns. We thus conclude that the synergetic approach of Haken [17] has been confirmed in an extraordinary manner.

References

1. N.I. Jaeger, K. Möller, P.J. Plath, Z. Naturf. **36a**, 1012–1015 (1981)
2. N.I. Jaeger, K. Möller, P.J. Plath, Chem. Soc. Faraday Trans. **I**(82), 3315–3330 (1986)
3. H. Beusch, P. Fieguth, E. Wicke, Chemie-Ing.-Techn. **44**, 445–451 (1972)
4. A.K. Galvey, P. Gray, J.F. Griffiths, St.M. Hasko, Nature **313**, 668–671 (1985)
5. L.F. Razón, S.M. Chang, R.A. Schmitz, Chem. Eng. Sci. **41**, 1561–1576 (1986)
6. K. Möller, K. Wetzel, M. Eiswirth, G. Ertl, J. Chem. Phys. **85**, 5328–5336 (1986)
7. P. Svensson, N.I. Jaeger, P.J. Plath, J. Phys. Chem. **92**, 1882–1888 (1988)
8. M.M. Slin'ko, N.I. Jaeger, P. Svensson, J. Catal. **118**, 349–359 (1989)
9. P.J. Plath, K. Möller, N.I. Jaeger, J. Chem. Soc. Faraday Trans. **I**(84), 1751–1771 (1988)
10. P.J. Plath, H. Prüfer, Z. Phys. Chemie, Leipzig **268**, 235–249 (1987)
11. P.J. Plath, in *Optimal Structures in Heterogeneous Reaction Systems*, Springer Series in Synergetics, ed. by P.J. Plath (Springer, Berlin, 1989) pp. 1–25
12. A.W.M. Dress, M. Gerhardt, N.I. Jaeger, P.J. Plath, H. Schuster, *Temporal Order,* ed. by L. Rensing (Springer Ser. in Synergetics Vol. 29, 1985), pp. 67–74
13. N.I. Jaeger, K. Möller, P.J. Plath, Ber. Bunsenges. Phys. Chem. **89**, 633–637 (1985)
14. M.A. Liauw, P.J. Plath, N.I. Jaeger, J. Chem. Phys. **104**, 6375–6386 (1996)
15. M.M. Slin'ko, A.A. Ukharskii, N.V. Peskov, N.I. Jaeger, Catal. Today **70**, 341–357 (2001)
16. C. Ballandis, P.J. Plath, J. Non-Equilib. Thermodyn. **25**, 301–324 (2000)
17. H. Haken, *Advanced synergetics* (Springer, Berlin, 1987), p. 1
18. A. Kleine, P.L. Ryder, N. Jaeger, G. Schulz-Ekloff, J. Chem. Soc. Faraday Trans. **I**(82), 205–212 (1986)

Understanding Earth: The Self-organization Concept and Its Geological Significance; on the Example of Liesegang-Structures and Electric Fields

Sabine Dietrich and Karl-Heinz Jacob

Abstract The existence of ordered structures in the lithosphere is a well-known and yet remarkable matter of fact. They appear as banded formations, layered and folded structures, diapirs, or cockade ores and in numerous other shapes and forms. Their size ranges from less than one centimeter up to kilometers. In many cases ordered structures develop in self-organized systems, making this concept a useful tool for better understanding complex and diverse processes in geology. In essence, self-organized systems are autonomously shaped units resulting from their inner determination under the influence of environmental conditions. It is important to recognize ordered structures and interpret them as self-organized with respect to their (external) environment *and* their (inner) components and properties in order to understand their genesis. Conversely, self-organized geological systems that have ordered structures contain valuable information about their genesis that is preserved within the structure. Because geological objects typically preserve only one picture for a very long time-frame, it is important to evaluate the ordered structures from the perspective of self-organization to gain access to the information. Reading this information enables us to learn more about Earth's history and future, and to use it toward the sustainable management of global resources. For this reason, applying the self-organization concept to geology is also important for the society as a whole.

S. Dietrich (✉) · K.-H. Jacob
Department of Earth Sciences, Technical University, Berlin, Germany
e-mail: sabine.dietrich@alumni.tu-berlin.de

K.-H. Jacob
e-mail: jacob@bg.tu-berlin.de

S.C. Müller et al. (eds.), *Complexity and Synergetics*,
https://doi.org/10.1007/978-3-319-64334-2_9

1 Introduction

Geological structures come in numerous formations and sizes, ranging from sub-millimetre structures to kilometre-long mountain ranges. Particularly striking are, for instance, geological folds, mineral deposits and banded ore bodies, which are formed by complex and diverse processes. In many cases, and especially in the case of ordered structures, they show the typical features of a self-organized genesis.

Today, geoscience generally recognizes ordered structures relating to the formation of rocks as a way of describing the shared boundaries of the different components, namely parts of rocks and their last displacement. According to Murawski [1], *the spatial position of a structure's building block marks its last movement*. Press and Siever [2] state the same in different words. According to them, the "displacement" of a structure depends directly and solely on external parameters.

However, for ordered structures of rocks, as shown in Fig. 1, the predominant hypotheses about their genesis are not an adequate fit and do not provide a sound explanation of their genesis. Considering geological systems as complex dynamic systems and not as systems in thermodynamical equilibrium, however, leapfrogs the current prevailing geoscientific concept and offers additional differentiated explanations.

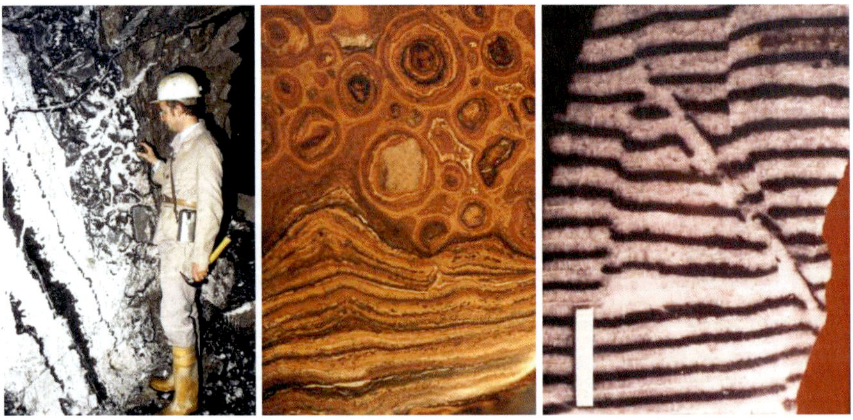

Fig. 1 Examples of geological structures for which the prevailing geo-scientific models for structure forming are disputable and which should be interpreted using the self-organization concept. (*Left*) Cockades structure of a lead-zinc ore-vein, Harz Mountains, Germany; (*Center*) Karoo sediment, RSA; (*Right*) Zebra fluorspar, Almeria, Spain. *Photographs* K.-H. Jacob

2 Some Principles in the Study of Self-organized Systems Featuring Ordered Structures

According to Ebeling [3], *a structure is the space and time dependent distribution of components as well as properties of a system*. This is a promising approach to understanding that "control parameters," such as energy or material supply, may not only act on a system as a whole but also influence it at the level of its components and its properties. The latter in turn interact as a network and keep adapting the relations to each other, which in turn keep changing under the influence of the control parameters. Due to the inner determination of the system and under certain conditions, these interactions and the relations between the components and properties produce, meaning self-organize, autonomously shaped units. The emerging units, for example, rock structures, display properties of a new quality. In the sense of thermodynamics, a phase transition takes place. If there is a feedback mechanism, the transition may repeat itself leading to the repetition of time- and space- dependent patterns or ordered structures. The macroscopic parameter that is used to describe the different phases of the system's components and properties is called "order parameter." The identification of the appropriate order parameter is not a trivial issue and requires a deep understanding of the nature of the system. Because of its nonlinear coupling to the control parameters, the order parameters represent the macroscopic descriptors for self-organized units, Haken [4]. In self-organized systems the dynamics of the order parameters have lower dimension than the dynamics of the control parameters. Hence, macroscopic ordered structures seen from the perspective of a self-organized system, and using the appropriate order parameter, are easier to describe, and provide a useful tool for understanding the nature and origin of geological structures.

A list of geochemical self-organization phenomena was already created in 1987 by Ortoleva [5], who also did extensive basic research into this topic. Nobel laureate Prigogine [6] said that applying ideas of non-equilibrium physics to geology will have a profound influence on the interpretation of numerous geological depositions *"Self-organization is, how nature really works!"*

3 Liesegang-Bands—An Example for Self-organized Band Formation Due to Nonlinear Coupling of Reaction-Transport-Processes

Since the middle of the nineteen 19th century we know from physical chemistry that combining at least two continuous processes under certain conditions can lead to rhythmic structures—although external rhythms in time or space do not apply. In Liesegang-experiments, the diffusion of aqueous solutions into a homogeneous matrix containing another substance creates a salt of lower solubility through chemical reaction. It does not precipitate homogeneously but in distinct lines within

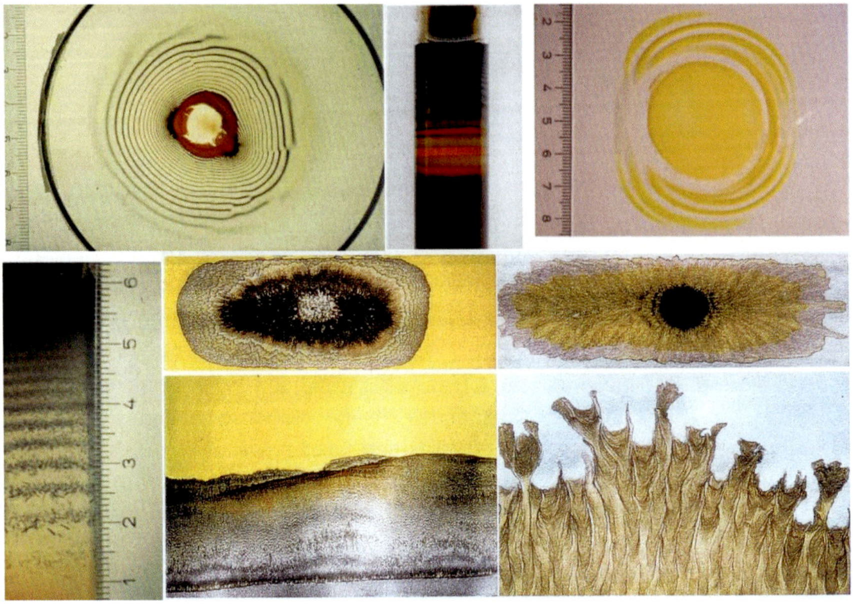

Fig. 2 Examples of Liesegang and Runge experiments. (*Top*) Liesegang rings in a 1 mm thick gelatin layer around a droplet from silver nitrate solution; Liesegang bands in a gelatin matrix that is several centimetres thick topped with silver nitrate solution; Liesegang rings from PbJ_2; (*Bottom left*) Liesegang bands in silica gel matrix; (*Bottom: centre* and *right*) Runge patterns on blotting paper: using a mixture of manganese and potassium sulphate solution (*Center*), and manganese and copper sulphate solution (*Right*)—spreading the solution from a droplet onto the blotting paper (*Center*), and rising into the blotting paper by dipping it into the solution (*Bottom*). *Photographs* S. Dietrich

certain distances, the so called A-lines (Liesegang [7]). The findings from physical-chemistry had already been transferred to geoscience at that time, although the interpretation of the ordered structures remained phenomenological (Liesegang [8] and Watanabe [9]).

How the evolving types of structures for the different substances and matrices depend on the inner determination, meaning the interactions and relations of the system's components and its properties, is illustrated by a selection of examples, Fig. 2.

First, different substances may form the same type of ordered structure. Their characteristic appearance, however, varies depending on the inner relations and specific properties of the different substances themselves (Fig. 2 *Top*: *left and right*). Second, it illustrates the influence of parameters such as the particle growth rate or surface tension depending on the surrounding matrix. The same substances at same concentration allow amorphous or crystal-loaded bands to grow in gelatin as opposed to in a silica matrix (Fig. 2 *Top: center; bottom: left*). Third, in addition to the reacting substances, the type of transport mechanism influences the shape of ordered structures. Classical Liesegang experiments are carried out in gel matrices with diffusion-controlled transport. In blotting paper, the same substances migrate mainly by capillary transport (Runge [10]). Remarkably, Runge gave his book on this topic the title "The formative tendency of substances illustrated by autonomously developed images" (Self-published, 1855).

Scientific interpretation of the structures has progressed since Liesegang's finding (Kuhnert and Niedersen [11]). Two main models have been established: the Ostwald-Prager (super-saturation) model and the Ostwald ripening model, also known as competitive particle growth model or Turing instability. Müller and Ross [12] analyzed the findings of a selection of experiments with respect to the inner determination of the components and evaluated the theoretical evidence of the models. They state: *We found extensive experimental evidence for the second model for a large range of initial concentrations of electrolytes as well as theoretical evidence, and we find the experimental evidence presented for the validity of the first model alone to be insufficient.* They also point out that there is an indication of colloid forming prior to banding. The colloid will later undergo ripening processes to form the rings. Depending on the concentration of the substances, which lead to weaker or stronger concentration gradients around a growing particle, the resulting structures will show different shapes. Bands develop if the gradient exceeds a certain value. At weak potentials, clusters are established or homogenous distribution takes place. Furthermore, Chopard et al. [13] analysed three models, namely the coagulation model, the competitive particle growth model, and super-saturation model. They state that none of these models can account for all experimental observations. They *believe, … that super-saturation mechanism based on Ostwald's ideas plays a crucial role in band formation.* By using a cellular automata model, they obtained a phase diagram showing the different possible patterns and phases (amorphous, crystals, dendrites, none of them), described as a function of critical super-saturation value and growth rate. Every type of the patterns can also be observed in the lithosphere.

Reseach into the formation of fragmented periodic bands revealed that the models are also suited to describe the evolution of lateral instabilities of Liesegang bands, the so-called bifurcations (Krug et al. [14, 15], Polezhaev and Müller [16], Krug and Bandtstädter [17]). This work provides the theoretical and experimental proof that fissure-containing structures are not necessarily the result of mechanical fractioning, Fig. 3.

Slight changes in the components or properties of the system or their interactions cause ordered structures to develop with bifurcation-like shapes, fissures or spirals. It is necessary to emphasize that the irregularities are explicitly caused by the complex inner determination of the system itself and not by any primary fissures, faults, mechanical influences or displacements.

Liesegang band-like and other ordered structures in the lithosphere can reach huge extensions. Especially mineral deposits and ores in the Earth's crust represent significant accumulations of usable minerals that occur in dimensions and by concentrations that make them worth mining. It is worth noting that the majority of the mineral deposits and ore bodies exhibit ordered structures.

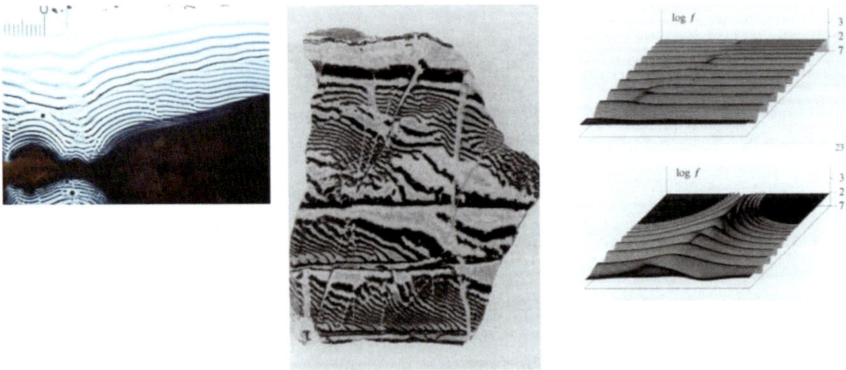

Fig. 3 (*Left*) Liesegang-bands around a larger irregularly shaped drop of silver nitrate. The anastomoses due to its irregular shape. (*Center*) Zebra mineralization from Bakal siderite deposit Ural, Russia. The banded patterns repetitively consist of pure siderite (*white*) and soiled siderite containing a small amount of clay minerals and organic impurities (*dark*). Maximum scale of sample is 15 cm. (*Right*) Numerical simulations of Liesegang bands based on super-saturation model for a silver chromate system (Krug et al. [15]). *Photographs/Images* S. Dietrich, R. Ellmis and H.-J. Krug

4 Electric Fields—An Example for Permanent Energy or Material Supply and Structure Forming Forces in the Environment

Sometimes lucky coincidence leads down a fruitful path. While solving a problem with damp walls in his home, K.-H. Jacob used a professional electro-kinetic masonry dehumification system to remove the moisture. He placed electrodes made from simple sheet iron into the walls and applied the prescribed voltage of just 1.0 V. The walls dryed as desired, but in addition, various new salt minerals grew between and near the electrodes. The new mineral formations featured ordered structures like the ore deposits K.-H. Jacob was familiar with from his experience as a mining engineer. However, in contrast to the explanations provided by state-of-the-art genesis models in geoscience, there was no sequential input of mass or change in energy supply during the wall's entire drying procedure. Curious about this, he performed cursory experiments, Fig. 4. In all of the experiments, colored redox horizons developed, shaped just as natural mineral fabrics (Jacob [18]).

These results are reproducible and indicate *the great importance of electricity for geology* as already postulated by the English scientist Priestley in 1772 [19]. During 18th and 19th century, frequent attempts were made to demonstrate the importance of electricity for geological processes and structure forming, but unfortunately proved unsuccessful. Consequently, all interest in electricity as a driving force for the genesis of mineral structures was lost.

During more systematic experiments at the Technical University Berlin during the 1990th, Klein [20], revealed that applying a weak electric field resulted in the

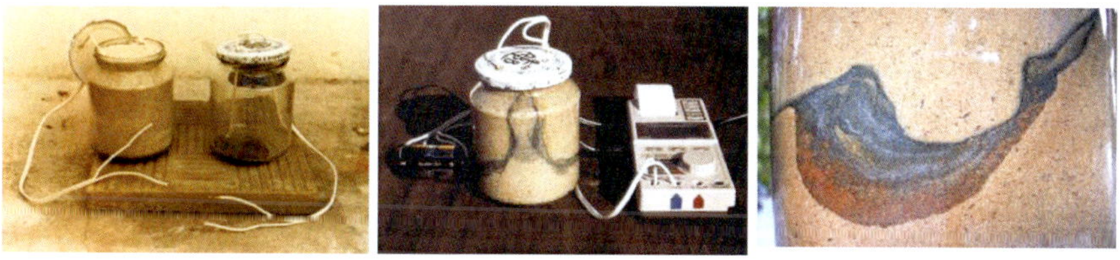

Fig. 4 (*Left*) Glasses with electrodes in place and already filled with moist sand, (*Center*) Visible effect of electric fields applied by a common battery, (*Right*) Development of colored redox horizon caused by electric energy input. *Photographs* K.-H. Jacob

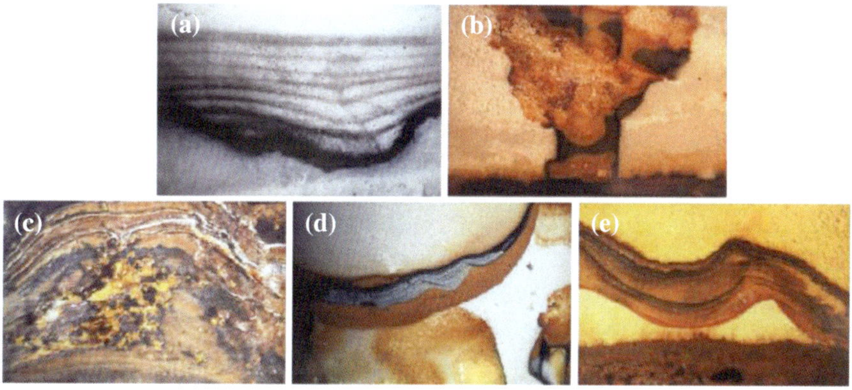

Fig. 5 Mineral fabrics, self-organized due to electrically induced non-equilibrium in laboratory experiments: **a** rhythmic bands, **b** vertically "displaced" horizons, **c** banded minerals, **d** undulatory, fold-like structures, **e** discordant banding. *Photographs* K.-H. Jacob

formation of "artificial structures" that showed significant phenomenological similarity to natural ones, see Figs. 5 and 6, and demonstrably can produce artificial "iron ores", see Figs. 7 and 8.

Noteworthy is that a characteristic structure/pattern evolves for every experiment that is performed with the same setup. Their individual features, however, vary as they are influenced by fluctuations in the same basic starting conditions. Some representative structures are shown in Fig. 5. Micro-diapirs, rosettes (cockades) of different dimensions, fractals such as dendrites or cauliflower structures, and patchy and spotty patterns similar to Ostwald ripening patterns, also develop. They form without any displacement of rocks and without any flow of fluids.

In contrast, the prevailing geoscientific consensus is that interactions between mineral systems and their environment are based on energy fluxes or material displacements caused by gravitation, temperature and pressure, migrating (hydrothermal) solutions or magma.

But gradients in temperature, pressure, chemical concentrations, and solubility—to name but a few—may result in more than just processes such as heat conduction, displacements, and diffusion. At the same time, they can also generate corresponding

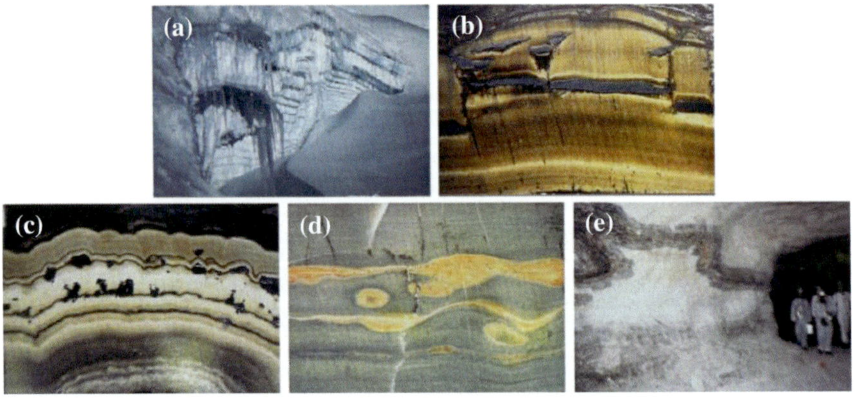

Fig. 6 Natural mineral structures: **a** snow sediments, **b** metamorphic "tiger eye" with vertically staggered hematite-horizon from Banded Iron Formation (BIF), Postmasburg district, RSA, **c** Lead-Zinc ore (Schalenblende) from Kalusz, Poland, **d** layered schist from Harz Mountains, Germany, **e** folded anhydrite layer in a German salt mine within the Zechstein horizon. *Photographs* K.-H. Jacob

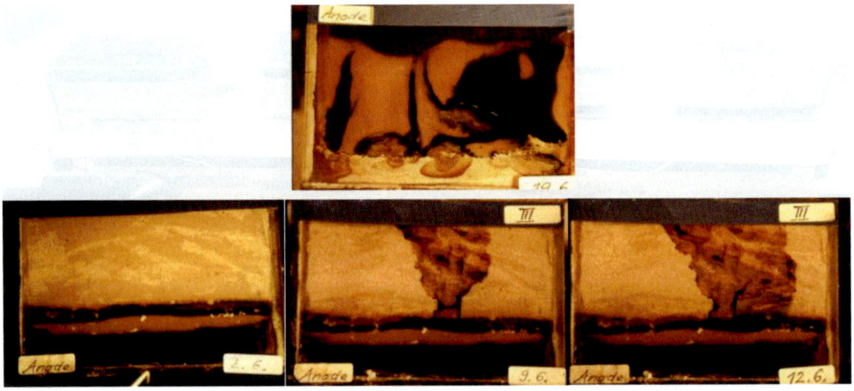

Fig. 7 Laboratory experiments with horizontally positioned electrodes: (*Top*) Anode (positive electrode) situated at the *top*: Iron ions move downwards with gravity. Irregular structures evolve. (*Bottom*) Anode situated at the *bottom*: Iron ions move in the opposite direction to gravity. Uniform structures that look like coal seams develop. Unexpectedly, the same experiment 7 days later shows a "broken structure". Three days later, these structures widen. *Photographs* K.-H. Jacob

currents and fields. For instance, electric fields can originate from mechanisms such as the well-known Thompson and Seebeck effects, Fig. 9. Klein [20] investigated processes in the lithosphere by inducing electric potentials and grouped them into six categories: electrochemical potentials (Eh/pH) as for redox reactions, tensor-induced potentials (by pressure/temperature gradients), potentials as a function of phase transformation (phase boundary), potentials produced by induction

Fig. 8 Laboratory experiment with vertically positioned electrodes, so-called “vein experiment”. (*Top left*) Full image of the 20 cm × 20 cm × 15 cm test container; (*Right*) Enlarged section of the symmetrically evolved veinlet; (*Bottom left*) Ore microscopy analysis Step I: shows quartz grains surrounded by magnetite- like cockade ore; (*Center left*) Step II: square section of 20 mm: the quartz grains dissolved and banded magnetite deposited instead. It looks like a “hydrothermal ore-vein”. *Photographs* K.-H. Jacob

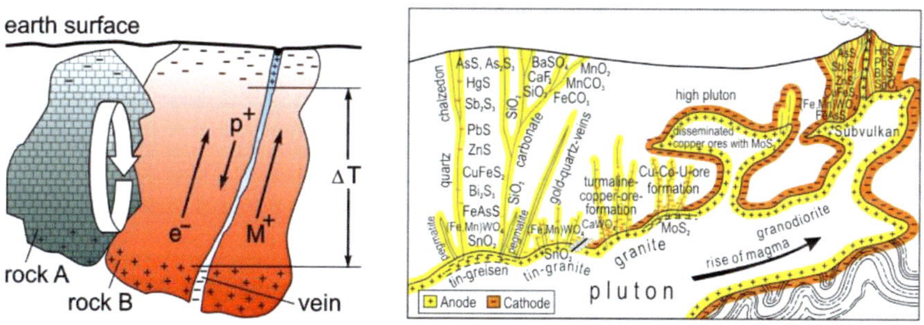

Fig. 9 (*Left*) Seebeck (1826): Electricity, generated at contact zones of different materials (metallic/ionic conductors) exposed to different temperatures. Thomson (1850): Electric currents, generated within metals exposed to a temperature gradient. They cause concentration exchange along contact zones of different materials (minerals) or along temperature gradients within a material; e.g., veins, horizons and filled gaps, fractures and even homogeneous bulk material. (*Right*) Borchert-Model [30] for ore deposits extended with thermoelectric chemical (TEC) processes according to Oesterle [31]. *Graphics* W. Klein, O. Oesterle after Borchert

(polarization), electro kinetic potentials (electrophoresis/electro-osmosis), and electro-potentials produced by radioactivity.

According to the standard Borchert-Model, large quantities of hydrothermal fluids are needed to explain the genesis of ore deposits. However, these quantities are not detectable in-situ. In contrast, the ion diffusion-based TEC process does not need any fluid flows to transfer mass. Instead, thermal gradients, which are typically available in particular at granite intrusions, not only lead to thermo-diffusion, they also generate corresponding electric fields, which are responsible for the ion-and charge-diffusion-based transfer. This new approach and type of research is in an early stage of development and needs to be continued.

Laboratory experiments demonstrate that at least in moist natural systems, electrical fields are able to generate transport processes and phase transitions. They cause mobilization and accumulation of substances as well as the formation of new minerals featuring ordered structures, and may also be responsible for the development of geological structures, and most likely responsible for many ore-bearing structures too (Jacob et al. [21–23]).

The results of the experiments clearly show the conditions under which typical "natural" structures of minerals and rocks form by applying continuous field forces, and without changing mechanical inputs. We believe that this contributes to the acceptance of the self-organization principles also in geology.

5 Recent Mineral Formation—An Example for Self-organized Layering in Geological Structures

To humans, geological objects typically appear to be in steady state. They are "carved in stone" so long as there are no volcano eruptions, earthquakes, floods or intensive desertification or erosion. Other, less spectacular but permanently acting processes, may be ignored unintentionally and the significance of their contribution to geological systems overlooked.

We took the opportunity of evaluating the development of a recently structured iron-manganese mud to identify the layering processes in-situ. The object of the investigation is located in the abandoned but accessible underground mine "Lautenthals Glück" in the Harz Mountains, Germany. Column-probing the material clearly displayed visible changes of red-brownish and black layers up to 4–5 cm on top, but getting more fine-banded at the bottom. During a period of 2 years, we observed the 20 cm thick mud that had been accumulating for 40 years. We took monthly probes, measured key parameters hourly (automated), performed time-series and substance analyses and used the results to derive a genesis-model.

In order to proof the postulated model, the recent early-diagenetic mineral accumulation was simulated numerically. The simulation basically combines two approaches: first, the mud accumulation is described time-dependent with changing manganese content according to the monthly amount of rain by some kind of super-saturation model. The second approach consists of ripening the components within the continuously growing mud, the so-called Ostwald ripening. It is described by using the competitive particle growth model together with the scaling procedure according to Feeney et al. [24]. The results of the simulation are shown in Fig. 10 (*Right*), (Dietrich [25]). Clearly visible is the increasing impact of the

Fig. 10 (*Left and center*) Accumulated mud in the abandoned mine in Lautenthal. This "natural laboratory" is situated ca. 200 meters underground, completely decoupled from daily changing temperatures sun radiation. However, permanent coupling to the environment is provided by the inflow of water that has migrated through the covering mountain. (*Center*) Taking time series of hourly measured key parameters; (*Right*: from right to left) monthly sums of the rain, numerically calculated mud profile without Ostwald ripening, Column-probe: ca. 20 cm, numerically calculated mud profile including Ostwald ripening. *Photographs/Images* K.-H. Jacob, S. Dietrich

secondary banding, which becomes more and more dominant with the growing age of the material, i.e. with depth of the profile.

The long-term observation and analysis of characterizing chemical and physical parameters for the recent mineralization led to the result that numerous complex processes are contributing to the formation of the banded iron-manganese mud. It also revealed that these processes are linked to each other both linearly and non-linearly. The ripening of low crystallized colloidal, and therefore thermo-dynamically unstable hydrated iron and manganese oxides, causes a rearrangement of a primary banding toward a secondary fine banding of the mud. Primary banding is caused by varying oxygen content and therefore changing Eh conditions and electrical potential gradients of the system. They are externally controlled by the amount of rain, i.e., water throughput, and are responsible for the amount of manganese precipitating together with the iron-containing compounds. However, the accumulation is superimposed by processes such as Ostwald ripening, redox reactions and microbial activity, providing nonlinear coupling to parameters such as the size of particles, surface tension, chemical potentials, ion concentrations and others. They create self-organized secondary bands, which become predominat with the growing age of the older mud partitions, i.e., along with depth, while the mud keeps accumulating. At the same time, the mud shows the primary banding in the freshly deposited partitions. The bands are already clearly visible within a few decades of mud accumulation, a very short time span from geological point of view, and it is forming at constant temperature and normal pressure.

It must be emphasized that structures such as those evaluated in the mud are very common in the lithosphere and are described, for instance, by Cornu [26] and Landmessser [27]. They appear as hydrogels. Already in 1909, Cornu stated that hydrogels are widespread and found in the oxidation zone of nearly every ore body in the Earth's crust. The results of the long-term evaluation of a recent iron-manganese mud indicate how the self-organization concept leads to a new understanding of the layered geological structure and therefore its significance for geology. More systematic research is needed to improve the understanding of the genesis of the various mineral systems and self-organized ordered structures in the lithosphere.

6 Summary

After a short introduction to the self-organization concept and its relevance to structure-forming processes in general, we looked at Liesegang bands as one example of nonlinear coupling of reaction-transport processes able to produce macroscopic ordered structures. We showed how the inner determination of the system influences the evolving structures and how slightly changing parameters can cause spontaneous growth of discontinuities even in homogeneous bulk systems. We then looked at an example of permanent energy and material supply, the electric field, which couples a system to the environment and being able to causes phase transitions.

Finally, the results of the long-term evaluation of a recent banded iron-manganese mud show that electric processes also contribute primarily to the

Fig. 11 Examples of ordered structures in ores and minerals whose components might have had hydrogel phase before or might have developed because of the contrast in competency. (*Left*) Banded Iron Formation (BIF) from Hamersley Range, Western Australia, (*Right*) Boudinage structure from Wadi ad Dil, Oman. *Photographs* K.-H. Jacob, S. Dietrich

accumulation of the mud and secondarily, combined with processes that are responsible for the formation of Liesegang bands, lead to the formation of the repetitive macroscopic structures that can be observed. The examined early-diagenetic iron-manganese system consists of hydrated iron and manganese oxides in the colloid phase. These hydrogels are wide spread in the mineral world. For almost every hydrogel, i.e., thermodynamically energy rich phase, there is a comparable crystalline ore body, i.e., the dehydrated and thermodynamically more stable phase. We showed that a demonstrable accumulation of material featuring macroscopic ordered structures—namely bands—grows from a considerably low-mineralized aqueous material source within a geologically short time span under undisturbed environmental conditions.

Fig. 12 The first 5 pictures were taken at a landfill only six years after the sludge had been filled in that place and stayed undisturbed. The structure shows remarkable similarity to banded "metamorphic rocks". Striking is not only the rhythmic banding, but also the cleavage, which is characteristic for schistose rocks. The picture in the bottom left corner shows a special type of soil: the *banded pararendzina* (Rid [32]). Structures such as these occur relatively often in nature and could be explained by means of self-organization (Radebold [33]). *Photographs* K.-H. Jacob

In the end these structures turned out to be independent from externally imposed cycles and very common in the Earth's crust.

Consequently, the concept of self-organization may enable us to answer hitherto unanswered questions about the formation of ordered structures in ores, rocks and minerals, for example, structures such as shown in Figs. 11 and 12.

We believe that our findings have fundamental relevance for the interpretation of various geological structures. The acceptance of self-organization processes occurring in the Earth's crust might lead to a new perspective in Earth science [28, 29].

It is important to recognize ordered structures and interpret them as self-organized with respect to their (external) environment and their (inner) components and properties in order to understand their genesis. Conversely, self-organized geological systems that have ordered structures contain valuable information about their genesis that is preserved within the structure. Because geological objects typically preserve only one picture for a very long time-frame, it is important to evaluate the ordered structures from the perspective of self-organization to gain access to the information. Reading this information enables us to learn more about Earth's history and future, and to use it toward the sustainable management of global resources. For this reason, applying the self-organization concept to geology is also important for the society as a whole.

However, more systematic research is needed to support our theses.

Acknowledgements Results of this research and new discoveries are based on several years of cooperation with W. Borges, W. Driehaus, R. Ellmies, P. Füssel, W. Klein, H.-J. Krug, M. Markgraf, J. Niemeyer, O. Oesterle, R. Radebold, W. Radebold, E. Schöll, G. Siems, J. Thieme, W. Zimmerle, and others. Some of the images, figures and/or photographs used in this article were used in prior publications as stated in the list of references. We would like to express our gratitude to P. Möller and M. Wipki at GFZ Potsdam for their constructive feedback and D. McCartney for her revision of this text.

References

1. H. Murawski, *Geologisches Wörterbuch*, 9th edn. (Enke, Stuttgart, 1992)
2. F. Press, R. Siever, *Understanding Earth* (W.H. Freeman and Company, New York, 1994)
3. W. Ebeling, *Strukturbildung bei irreversiblen Prozessen* (Math. Naturwiss, Bibliothek (Teubner, Leipzig), 1976)
4. H. Haken, *Synergetics—An Introduction* (Springer, Berlin, 1978)
5. P. Ortoleva, E. Merino, C. Moore, J. Chadam, Geochemical self-organization I: reaction-transport feedbacks and modeling approach. Am. J. Sci. **287**, 979–1007 (1987), P. Ortoleva, *Geochemical Self-organization.* Oxf. Monogr. Geol. Geophys. **23**, Oxford, 411 (1994)
6. G. Nicolis, I. Prigogine, *Die Erforschung des Komplexen* (Pieper, Zürich, 1987)
7. R.E. Liesegang, A-Linien, Liesegangs photographisches Archiv **21**, 321–326 (1896)
8. R.E. Liesegang, *Geologische Diffusionen* (Theodor Steinkopff, Dresden-Leipzig, 1913)
9. M. Watanabe, Zonal precipitation of ores from a mixed solution. Econ. Geol. **19**, 497–503 (1924)
10. F.F. Runge, *Zur Farben-Chemie. Musterbilder für Freunde des Schönen* (Mittler & Sohn, Berlin, 1850)

11. L. Kuhnert, U. Niedersen, *Selbstorganisation chemischer Strukturen.* Ostwalds Klassiker der exakten Wissenschaften **273** (Geest und Portig, Leipzig, 1987)
12. S.C. Müller, J. Ross, Spatial structure formation in precipitation reactions. J. Phys. Chem. A **107**, 7997–8008 (2003)
13. B. Chopard, P. Luthi, M. Droz, Reaction-diffusion cellular automata model for the formation of liesegang patterns. Phys. Rev. Lett. **72**, 1384–1387 (1994)
14. H.-J. Krug, S. Dietrich, K. H. Jacob, The formation and fragmentation of periodic bands through precipitation and ostwald ripening, in *Fractals and dynamic systems in geosciences*, ed. by J.H. Kruhl (Springer, Berlin, 1994), pp. 269–289
15. H.-J. Krug, H. Brandtstädter, K.-H. Jacob, Morphological instabilities in pattern formation by precipitation and crystallization processes. Geol. Rundsch. **85**, 19–28 (1996)
16. A.A. Polezhaev, S.C. Müller, Complexity of precipitation patterns: comparison of simulation with experiment. Chaos **4**, 631–636 (1994)
17. H.-J. Krug, H. Brandtstädter, Morphological characteristics of liesegang rings and their simulations. J. Phys. Chem. A **103**, 7811–7820 (1999)
18. K.-H. Jacob, Künstliche Bänderungen, Wissenschaftsmagazin der TU Berlin, 75–78 (1988)
19. J. Priestley, *History and present state of electricity. Geschichte und gegenwärtiger Zustand der Elektricität nebst eigenthümlichen Versuchen* (G.A. Lange, Berlin und Stralsund, 1772)
20. W. Klein, *Dissipative Gefügebildung durch elektrische Felder*, Dissertation TU Berlin (1990)
21. K.-H. Jacob, H.-J. Krug, S. Dietrich, Lagerstättenbildung durch Energiepotentiale in der Lithosphäre, Erzmetall **45**, 505–513 (Weilheim 1992)
22. K.-H. Jacob, W. Zimmerle, Some diagenetic phenomena in carboniferous sedimentary rocks seen as rhythmic structures produced by energy dissipation in open systems. Zbl. Geol. Paläont. Tl. **1**, 437–459 (Stuttgart 1993)
23. K.-H. Jacob, S. Dietrich, H.-J. Krug, Selforganization of mineral fabrics, in *Fractals and dynamic systems in geosciences*, ed. by J.H. Kruhl (Springer, Berlin, 1994), pp. 259–268
24. R. Feeney, S.L. Schmidt, P. Strickholm, J. Chadam, P. Ortoleva, Periodic precipitation and coarsening waves: applications of the competitive particle growth model. J. Chem. Phys. **73**, 1293–1311 (1983)
25. S. Dietrich, *Strukturierung Eisen-Mangan-haltiger Schlämme durch komplexe dynamische Prozesse. Ein Beitrag zur frühdiagenetischen Bildung von Mineralgefügen durch Selbstorganisation*, Berliner geowiss. Abh. **180**, 86, (Dissertation TU Berlin, 1996); S. Dietrich, K.-H. Jacob, Self-organization of recent rhythmic iron-manganese precipitations in underground—mines in the Harz mountains, in *Proceedings of the 2nd International Symposium of Fractals and Dynamic Systems in Geosciences, Frankfurt-Gelnhausen, 4–7 April 1995*. Geol. Rundschau **85**, 29–37 (1996)
26. F. Cornu, Die Bedeutung der Hydrogele im Mineralreich. Z. Chem. Ind. Kolloid **4**, 15–18 (1909)
27. M. Landmesser, Selbstorganisation und Achatgenese. Wissenschaftsgeschichte, Problemfacetten und Resultate der neueren Forschung, in *Selbstorganisation*, Jahrbuch für Komplexität in den Natur- und Sozialwissenschaften, ed. by H.-J. Krug, J.H. Kruhl (Duncker & Humblodt, Berlin, 2001)
28. H.J. Walther, K.v. Gehlen, G. Haditsch, H.J. Maus, *Lagerstättenkundliches Wörterbuch*, GDMB, 688 (1999)
29. K.-H. Jacob, Über Selbstorganisation und ihre Bedeutung für die Geologie. About self-organization and its Importance in Geology, Z. Geol. Wiss., Berlin **38,** 295–310 (2010)
30. H. Borchert, *Lehrblätter zur Geochemie und Lagerstättenkunde* (Glückauf Verlag, Essen, 1979)
31. O. Oesterle, K.-H. Jacob, Über Lagerstättenbildung durch elektrische Felder, Zeitschrift der Förderer des Bergbaus und des Hüttenwesens an der TU Berlin, 21–29 (1994)
32. H. Rid, *Das Buch vom Boden* (Eugen Ulmer, 1984)
33. W. Radebold, *Chemoelektrische Vorgänge in rezenten oberflächennahen Systemen - eine wesentliche Ursache für die Neubildung von Mineralstrukturen* (Dissertation TU Berlin, 2004)

Pattern Formation in Microemulsions Affected by Electric Fields

Patricia Dähmlow and Stefan C. Müller

Abstract In living nature and biological morphogenesis, Turing's mechanism plays an important role. Accordingly, patterns can be found on animal skins or in chemical reactive systems. The formation of these structures is governed by gradients of chemical reactants and ions and thus, of electric fields. Here, an electric field is applied to a chemical compartmentalized reaction (i.e., a water-in-oil microemulsion), in which Turing patterns may form. In this system, percolation can occur when the volume of water is large compared to that of the oil. Thus, water droplets generate a network of water channels. Due to the presence of ions, this formation can be manipulated by an electric field. Turing patterns show a spatial drift, caused by the electric field. The strength of the field resolves the resulting drift velocity of the patterns. Additionally, a reorientation of the patterns is induced by a gradient generated by the electric field.

1 Introduction

Patterns in time and space occur in the inamimate nature and in living systems. They may be periodic and regular or they develop irregular and complex features [1, 2]. There are numerous examples in many scientific disciplines dealing with the properties of such systems, as described in Chap. 2 of this book. And either well a great number of scientists consider the phenomenon of complexity, as exemplified in many other chapters.

In this contribution, we are concerned with a "classical" example in liquid chemistry—the Belousov-Zhabotinsky (BZ) reaction, which shows, under appropri-

P. Dähmlow (✉) · S.C. Müller
Institute of Experimental Physics, Otto-von-Guericke University Magdeburg, Universitätsplatz 2, 39106 Magdeburg, Germany
e-mail: patricia.daehmlow@ovgu.de

S.C. Müller
e-mail: stefan.mueller@ovgu.de

S.C. MÃijller et al. (eds.), *Complexity and Synergetics*,
https://doi.org/10.1007/978-3-319-64334-2_10

ate conditions, regular and irregular oscillations in homogeneous aqueous solution. If an excitable regime is prepared, propagating wave fronts with circular, spiral-shaped or more complex geometry are formed in the reaction [3, 4]. In this classical case, the prepared solution initiates the oxidation of an organic substrate (usually malonic acid) by bromate in an acidified aqueous medium in the presence of a metal ion catalyst (e.g., ferroin). In this form many salient investigations have been made, rendering the BZ reaction a standard system for spatio-temporal self-organization. Later on, significant modifications have been introduced by replacing, for instance, the substrate malonic acid by CHD (1,4-cyclohexanedione), a substance which does not produce the normally emerging CO_2 bubbles [5, 6]. Other modifications were concerned with the choice of the medium in which the reaction proceeds. Several hydrogels can be used, in which disturbances like hydrodynamic flows due to evaporation or chemical fronts are avoided [7]. Furthermore, such gels offer opportunities to feed substrates to the reactive system in a controlled way.

Recently the usefulness of microemulsions (ME) has been detected [8]. If the BZ-solution is embedded in a reverse ME and experiments are performed below the percolation threshold (see below), the activator of the autocatalytic reaction is trapped in water droplets (where the BZ reaction takes place) and diffuses with their velocity. The non-polar inhibitor, on the other hand, is able to diffuse in the oil phase, and has therefore a higher diffusion coefficient. These conditions produce conditions, where Turing or Turing-type patterns may form, a most attractive application of the properties of the BZ reaction.

Beyond these chemical means to influence and control the pattern-forming system, there are physical parameters that prove to be efficient in their control function. These are light effects in photosensitive versions of the BZ reaction [9], and electric fields which act on the many charged variables in the BZ solution [10].

In this work we investigate the BZ reaction under chemically and physically specific conditions: we choose a microemulsion for controlling the diffusion coefficients of activator and inhibitor variables. Our system will be subjected to an electrical field, thus making observations of the pattern dynamics under an external forcing possible. The observed behavior will be compared to previous electric field effects, as found in the reaction under "classical" conditions.

2 Experimental

Experiments are performed preparing two non-reactive microemulsions with the same molar ratio between water and the anionic surfactant AOT ω ($\omega = [H_2SO_4]$ $[AOT]^{-1}$, where $[H_2O]$ and $[AOT]$ are the molar concentrations of H_2O and AOT, respectively) and the same droplet fraction of the dispersed phase φ_d ($\varphi_d = (V_{AOT} + V_{H_2O})(V_{AOT} + V_{H_2O} + V_{oil})^{-1}$, where V_i are the volumes of AOT, water and oil). The droplet fraction indicates the ratio of droplets compared to the whole system.

The first microemulsion (ME1) consists of AOT (*Sigma*), water, malonic acid (*Merck*) and sulfuric acid (*Fluka*). The second microemulsion (ME2) consists of

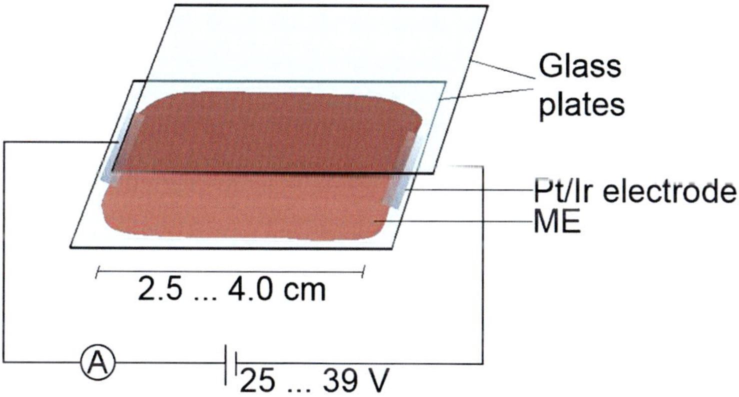

Fig. 1 Reactor for observing pattern formation in microemulsions subjected to an electric field. The microemulsion is sandwiched between two Plexiglas plates. The platinum/iridium electrodes play the role of a spacer and are connected to a voltage source. The electric current is measured with an ammeter. The distance between the electrodes can be varied

AOT, water, sodium bromate (*Merck*) and ferroin. Both solutions appear milky after all components are added. Thus, they must be stirred for several minutes until they appear homogeneously. The BZ reaction does not start until both microemulsions are brought together. They are mixed with a ratio of 1:1, and a certain volume of the solvent (n-octane; *Arcos Organics*) is added, depending on the requested droplet fraction φ_d, which is chosen below and above the percolation transition with values of 0.48 and 0.52.

The reactive microemulsion is sandwiched between to Plexiglas plates. Platinum/iridium (90/10 wt%) electrodes (*Alfa Aesar*) with a thickness of 100 μm (length 2.5 cm) are used, additionally taking the role of a spacer (Fig. 1). They are immersed directly into the microemulsion, since they do not interfere with the reaction. No electrolysis can be observed on the electrodes. Differently sized cover plates allow a variation of the distance between the electrodes between 2.5 and 4 cm in 0.5 cm steps. Measurements with direct current (DC) are performed as follows: The electrodes are connected to a voltage source (Voltcraft DIGI 40) and an ammeter (Prächitronic MV 40) as in Fig. 1. When the patterns are fully developed, the electric field is switched on. Simultaneously, the current is measured with the ammeter. The electric field strength is changed between 3.0 to 15.7 V cm^{-1}. Patterns forming in the reactor are observed under a stereo microscope (Zeiss GSZ 2T) equipped with a CCD camera (AVT Stingray F-046B), which is connected to a computer. The used frame rate is 0.495 Hz. The illumination of the patterns is realized by a self-made LED panel with 100 white LEDs (maximum intensity at ≈455 nm) through an interference filter (480 nm, bandwidth 10 nm). This wavelength is chosen, since the absorption spectrum of ferroin and ferriin has a maximum difference around 480 nm, which yields a good contrast.

3 Results

In this section, the effect of homogeneous electric fields on stationary Turing patterns is investigated. These patterns are well suited to study their drift induced by an electric field. In the BZ reaction, many charge carriers are present, such as Br^-, H^+, BrO_3^-. Even the catalyst ferroin has positive charges. Depending on its state the charges change between Fe^{2+} (ferroin, red color) and Fe^{3+} (ferriin, blue color). Additionally to all these charges, embedding the BZ reaction in a microemulsion leads to further charges, since the surfactant AOT has a SO_3^--and a Na^+-group located on the head group.

This large number of charge carriers causes the patterns, formed in the BZ reaction, to drift. It is known from the aqueous BZ reaction, that the velocity of a propagating wave front is reduced, when an electric field is applied with a polarity counteracting the overall motion of the bromide ion, the negatively charged inhibitor [10]. Thus, a drift of stationary Turing patterns caused by the electric field can be expected in this work.

The recipe for all measurements shown in this chapter is: [MA] = 0.259 M, [$NaBrO_3$] = 0.155 M, [H_2SO_4] = 0.207 M and [ferroin] = 1.74 mM, $\omega = 18.03$. The

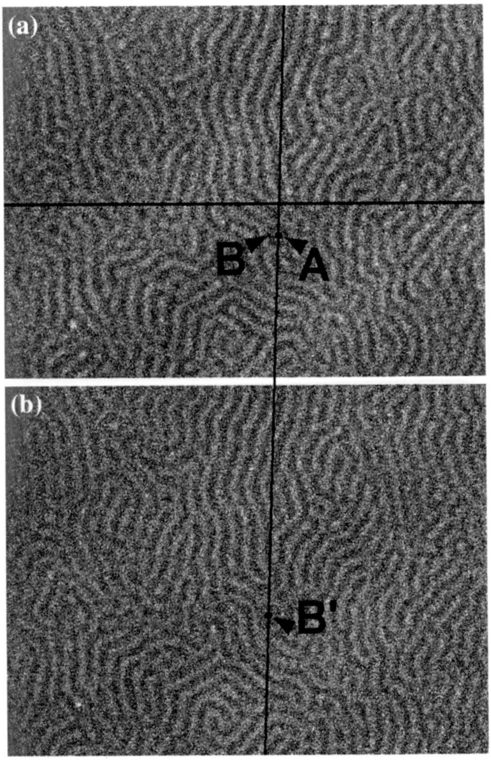

Fig. 2 Two snapshots of drifting Turing patterns at $E = 8.7\,\text{V cm}^{-1}$ with $\varphi_d = 0.48$ at different instants of time. A time-space plot is extracted along the *horizontal line* in **a** (Fig. 3). Patterns are drifting towards the positive electrode, which is located at the left side of the image. **a** Label "A" marks the position of a bright spot in the image at $t = 55.3$ min. **b** Here, label B' marks the identical spot as "A" at a later instant and thus, at another position. The projection of B' is marked with "B" in image **a** at $t = 63.7$ min. Labels B' and "A" are connected through the inclined line. Image size: $4.2 \times 3.1\,\text{mm}^2$. Reprinted from [13], with the permission of AIP Publishing

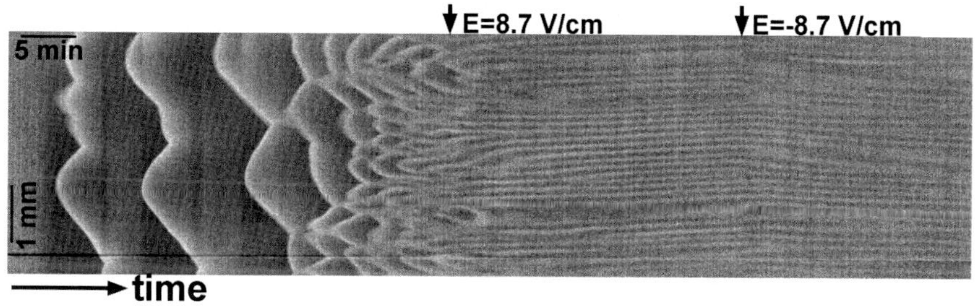

Fig. 3 Time-space plot of drifting Turing patterns with $\varphi_d = 0.48$. An electric field is applied when Turing patterns are fully developed (*left arrow*), causing a drift towards the positive electrode. After 27 min, the polarity is changed (*right arrow*) and the drift reverses its direction immediately. Reprinted from [13], with the permission of AIP Publishing

droplet fraction is chosen between 0.48 and 0.52. Both values are close to the percolation threshold ($\varphi_d \approx 0.5$) [11]. However, in both cases Turing patterns can be observed. Although the value of 0.52 lies above the percolation transition, Turing patterns can still be observed [12]. This indicates, that percolation is not completely reached in this system. It can be assumed, that already some lager droplet clusters have been formed, but not an infinite cluster. Otherwise, Turing patterns cannot be observed.

A homogeneous electric field (direct current) is applied to the patterns in the range of 3.0 and 15.7 V cm^{-1}. Turing patterns are found to evolve a drift under the influence of the electric field, which is constant for a certain value of the electric field strength E (compare Figs. 2 and 3). In Fig. 2 fully developed labyrinth-like Turing patterns are shown, which move due to the electric field. A large number of negatively charged bromide ions, which are able to diffuse into the oil phase, force the pattern to drift towards the left side of the image. Here, the positive electrode is located. Labels "A" and B' mark the same spot with a time delay of 8.4 min. The inclined line between both labels clarifies the distance, the patterns traveled within the time interval of 8.4 min. Label "B" presents the projection of B' at the earlier instant.

In Fig. 3 the time-space plot extracted from Fig. 2 is depicted. The arrows on the top of the image indicate, when the electric field is applied (left arrow) and when the polarity of the field is changed (right arrow). The patterns react immediately on the polarity change and reverse their drift direction. The overall movement of the patterns is fully reversed within a minute. In addition to the intensity of the electric field, the current is measured. Before turning on the field, the current is zero. Immediately after the application of the gradient, the current increases rapidly until it reaches a maximum value and decreases slowly up to a final value (Fig. 4a). The maximum value, which is reached by the current lies between 250 and 400 μA and the final value around 5 or 60 μA, which remains almost constant for the rest of the experiment. In Fig. 4a the evolution of the current is shown exemplarily for a droplet fraction of 0.52. Here, the current increases up to 270 μA within 3 min (the arrow marks the instant, when the current is turned on). Afterwards, the current decreases

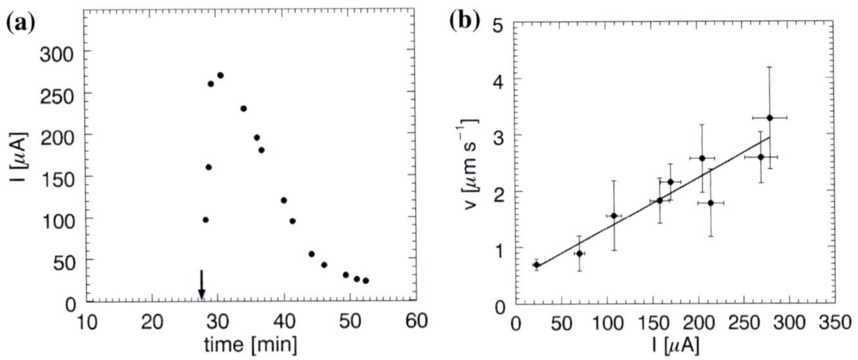

Fig. 4 **a** Course of current during a measurement for $\varphi_d = 0.52$. After switching on the constant electric field (*arrow*), the current increases quickly and subsequently decreases much slower, until reaching a final value, which remains constant for the remaining time of the experiment. **b** Drift velocity of Turing patterns versus the electric current. The velocity is measured in a time interval between 100 and 600 s during the period of current decrease (declining part in (a)). A linear fit has been computed with a correlation coefficient of 0.934. Reprinted from [13], with the permission of AIP Publishing

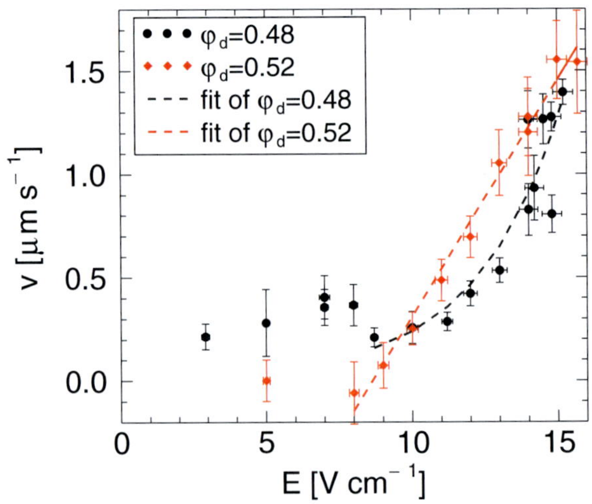

Fig. 5 Drift velocity v of Turing patterns versus the applied field strength E. Below the percolation threshold (*black circles*), a nonlinear (presumably exponential) increase of the velocity with increasing field strength is observed (correlation coefficient = 0.8382). Above the threshold (*red diamonds*), a linear behavior is observed (correlation coefficient = 0.9874). The dashed lines are the corresponding fitted curves. Note, that each data point represents an individual experiment. Reprinted from [13], with the permission of AIP Publishing

and remains around 23 μA. The maximum value of the current above the percolation transition ($\varphi_d > 0.5$) and the final value are typically higher than below the transition.

The slow decrease of the current can be used to calculate the drift velocity of the patterns, depending on the current. The value of the velocity is calculated from the time-space plot. Here, a linear part of the inclined lines is needed to estimate the velocity. In Fig. 4b the velocity is calculated during the decreasing tail of (a), which results in a linear behavior between current and speed. The continuous decrease of the current makes the estimation of the drift velocity challenging, since it requires a linear part of a line in the corresponding time-space plot. Thus, the quality of the time-space plot determines the estimation of the pattern's drift velocity.

As seen from the current measurement (i.e., from the values of the maximum and final value), the drift of patterns depends on whether the experiments are performed below or above the percolation threshold. Thus, experiments with two different droplet fractions near the threshold are performed (i.e., at 0.48 and 0.52). In Fig. 5 the course of the pattern drift velocity versus the electric field strength E is shown for $\varphi_d = 0.48$ and 0.52. Above the percolation transition, a linear increase of the drift velocity of Turing patterns with increasing electric field strength is found (red diamonds in Fig. 5). Below the percolation transition, the course of the drift velocity is nonlinear, presumably exponential (black circles).

An exponential fit of the data for the drift velocity at $\varphi_d = 0.48$ is performed for $E > 8.7\,\mathrm{V\,cm^{-1}}$. At lower values the velocity is scattered in the range between 0.2 and 0.4 $\mu\mathrm{m\,s^{-1}}$. Thus, the errors are too large to include them in the exponential fit. A statistical test [14] is performed to examine, whether a linear or an exponential fit is the better choice for the data points below the percolation transition. The correlation coefficient of a linear fit is found to be smaller than that of an exponential fit on a significance level of 99% (see Ref. [13]).

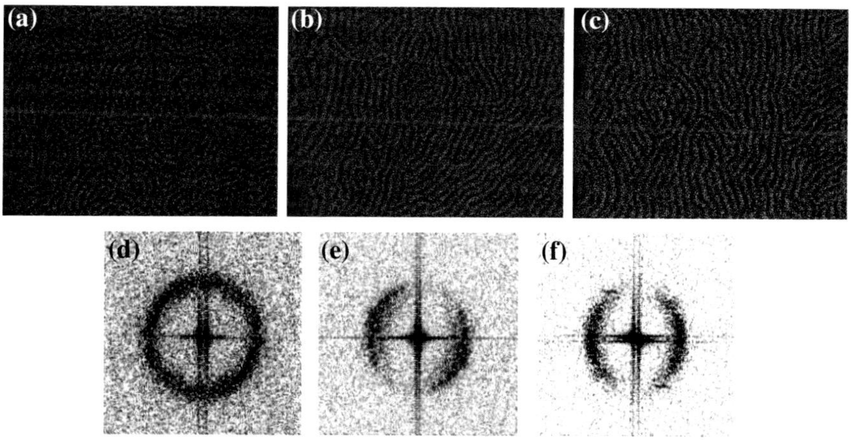

Fig. 6 Alignment of Turing patterns at different field strengths for $\varphi_d = 0.52$. The cathode is located on the left side and the field lines are oriented horizontally. Turing patterns **a** are randomly oriented at 5 $\mathrm{V\,cm^{-1}}$. **b** A mean field strength (12 $\mathrm{V\,cm^{-1}}$) yields an almost parallel alignment of the line with a certain angle to the field vector. In **c** the line patterns is oriented almost perpendicular to the field lines at 15.7 $\mathrm{V\,cm^{-1}}$. Image size: $5.2 \times 3.8\,\mathrm{mm^2}$. **d**–**f** Amplitude spectrum of the Turing patterns **a**–**c** to illustrate the reorientation of the patterns. Reprinted from [13], with the permission of AIP Publishing

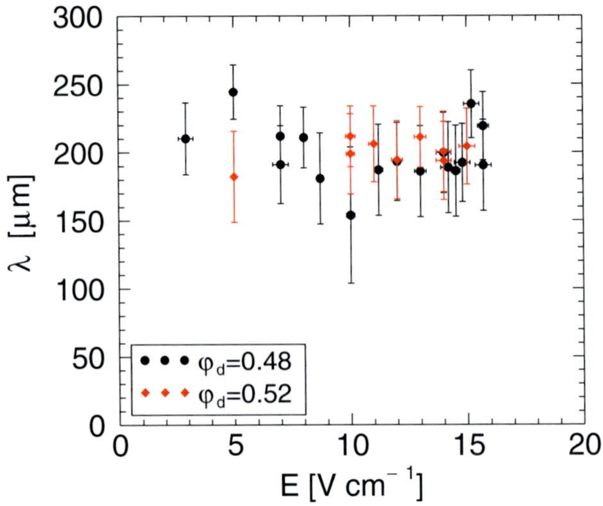

Fig. 7 Characteristic wavelength of Turing patterns as a function of the electric field strength with different droplet fractions. Below (*black circles*) and above the percolation transition (*red diamonds*) the wavelength of Turing patterns is almost equal. Reprinted from [13], with the permission of AIP Publishing

Further investigation of the patterns themselves has revealed, that the patterns reorient under the influence of the electric field (Fig. 6). The electric field lines are parallel to each other and lie horizontally. Depending on the intensity of the electric field, labyrinth-like Turing patterns show either a random orientation or an ordered configuration, where the lines of the patterns are almost parallel to each other. A weak field up to 5 V cm^{-1} yields the mentioned random orientation as depicted in Fig. 6a. With an intermediate intensity (5–12 V m^{-1}), the lines of the labyrinth structures become almost parallel to each other, but are inclined with respect to the horizontal field lines (Fig. 6b). With increasing field strength (about 14 V cm^{-1}), the parallel alignment of the structures becomes more pronounced and they orient perpendicularly to the field lines (Fig. 6c). To illustrate the reorientation, a spatial Fourier transform of the patterns (a–c) is computed to obtain the amplitude spectrum of the original images (Fig. 6d–f).

A low field strength shows a well-defined homogeneous wavelength distribution around the point of origin (Fig. 6d). This means, that in all directions in space an equal wavelength can be found. At a medium field strength, the distribution loses its spatial homogeneity. It shows a gap in the upper right (and lower left) corner (Fig. 6e). The position of the gap corresponds to the angle between the lines of the Turing patterns and the field vector. With further increase of the field strength the gap rotates, such that it is located at the top (and bottom) of the wavelength distribution (Fig. 6f). When this alignment is reached, the reorientation of the Turing patterns is completed. The reorientation occurs in a continuous fashion and can be found above and below the percolation threshold. The characteristic wavelength of Turing patterns evolving under conditions of different field strengths is shown in Fig. 7.

Here, the wavelength above and below the percolation transition is estimated from the spatial Fourier transform. For both droplet fractions ($\varphi_d = 0.48$ and 0.52), the wavelength lies around 200 μm and is almost the same for both values of φ_d over the whole used range of electric field intensity. Even the reorientation of the patterns does not affect the values of the wavelength.

4 Discussion

The effect of an electric field on complex structures, which can be found in pattern-forming chemical or biological systems, belongs to the strongest methods to manipulate the inherent dynamics of such systems. They contain many electrically charged components which are known to govern the spatio-temporal organization. One can imagine, that an electric field dominates the formation process of patterns, as long as the activating and inhibiting species of such a self-organizing chemical reaction carry an electric charge. This is true for the BZ reaction, where the negatively charged bromide ion plays the role of the inhibitor, dominating the dynamics to a large extent. Manipulating the dynamics of the latter by an electric force, change the interior dynamics of the system and, e.g., a spatial drift of macroscopic patterns can be observed [15].

Modeling the role of an electrical or chemical gradient in the BZ reaction, a realistic reaction scheme of the reaction is taken, for instance in the Oregonator model [4, 16], and an additional equation for the gradient is added [17]. Here, the drift velocity of patterns depends linearly on the field strength or on the electric current [18, 19], which is also true for other reactions, such as the chlorine dioxide-iodine-malonic acid reaction [20]. To go a step further, the BZ reaction can be incorporated in a reverse emulsion (i.e. a water-in-oil emulsion), which enables us to obverse Turing patterns [8]. Within the water droplets, the BZ reaction proceeds yielding a short range interaction of the activator (which is trapped within the droplets). A long range interaction within this reaction is generated by the inhibitor species, diffusing within the oil phase. Our findings in the investigation of the BZ reaction incorporated in a reverse emulsion under external electrical forcing show a linear or exponential drift of Turing patterns, depending on the initial droplet fraction.

The measured current in the microemulsion changes during any specific experiment, which can be explained by the large number of free charge carriers, such as Br^- ions (located in the oil phase), SO_3^- and Na^+ (ions from the surfactant, located in the water phase). When the electric field is turned on, the ions in the oil phase (Br^-) propagate towards the positive electrode, whereas the ions in the water phase propagate with almost the velocity of a droplet, which is much slower than the velocity of the free ions. If the percolation threshold is reached, the ions in the water phase can move faster, since they can diffuse through the forming water channels. Thus, the final value of the current is higher above the percolation transition than below.

Even the course of the pattern drift velocity depends on the percolation process (Fig. 5). Above the percolation transition, a linear increase in drift velocity with increasing field strength is observed, as in the aqueous BZ system [18, 19]. Below the threshold, a more rapid increase of the velocity with increasing field strength is observed from 8.7 to 15.7 V cm^{-1}. This increase is primarily nonlinear, justifying an empirical exponential fit in the measured range. With higher field strengths, this exponential function might become linear, when the system is percolated due to the electric field. However, there is no appropriate theory yet to explain such an exponential behavior.

Structural changes in the mircoemulsion, induced by the electric field, appear to be responsible for the observed nonlinear velocity increase below the percolation transition. Due to the action of the field, the nanodroplets become more elongated and polarized (mediated by dissociation into SO_3^- and Na^+) [21, 22], thus forming droplet clusters. In these clusters, the monolayer of AOT molecules opens up and water molecules can diffuse from one droplet within the cluster into an adjacent one [23]. As a result, the conductivity of the microemulsion increases [22]. This might be a case of the so-called electric-field induced percolation. This effect may also be the reason for the reorientation of the patterns in Fig. 6.

Moreover, in other systems, like in the polyacrylamide methyleneblue-sulfide-oxygen reaction (PA-MBO), the orientation of Turing patterns is found to change with the strength of the electric field. In this system, a weak electric field (up to 5 V cm^{-1}) leads to a parallel orientation to the field, whereas fields above 10 V cm^{-1} produce patterns, which are oriented perpendicular to the field vector [24].

The results presented here emphasize the complexity of the BZ-AOT system, as suggested by the assumed exponential increase of the drift velocity. This should motivate a more detailed investigation of the physical properties of pattern forming microemulsions by further experimental and theoretical efforts.

5 Conclusion

The reactive system under investigation in this article follows in a broader sense properties that are characteristic for the large class of activator-inhibitor interactions. As long as the reactive species diffuse in water or in an appropriate hydrogel, one frequently finds homogeneous oscillations or spatio-temporal patterns, such as propagating wave fronts (circular or spiral-shaped) and Turing patterns (dots or stripes), some times summarized as "classical" self-organized reactions [1, 4].

Whereas structures in these reactions exhibit relatively simple geometrical shapes (although being mechanistically quite involved), the addition of certain ingredients can lead to substantial changes in evolution and pattern formation, mostly to much more complex ones. One way would be to use a compartmentalized system, supporting different transport properties of activator and inhibitor, as done in this work (and following to some extent the procedures developed in Refs. [8, 11]). This supports the formation of, for instance, Turing-like structures not known from the simpler "classical" preparations. Another way to interfere with these active reactions is based on the idea that practically all the relevant chemical variables carry an electrical charge. Therefore applying an electrical field of a given strength, direction and polarity of the field can be expected to have a clear influence on the pattern formation and its dynamics [13]. This external forcing has been investigated in this work, as it acts on a BZ system in a microemulsion.

A major result is that under such conditions the complexity of the nonlinear system increases in a pronounced way. Detailed mechanistic studies about the experimental observations have not been presented yet. The authors hope for progress in this field in the framework of powerful analytical and numerical techniques, available by modern concepts, as for instance the synergetic approach.

References

1. J.D. Murray, *Mathematical Biology. An Introduction*. Interdisciplinary Applied Mathematics (Springer, New York, 2002)
2. H. Meinhardt, *The Algorithmic Beauty of Sea Shells* (Springer, New York, 1995)
3. G. Nicolis, I. Prigogine, *Self-organization in Nonequilibrium Systems: from Dissipative Structures to Order through Fluctuations* (Wiley, New York, 1977)
4. R. Kapral, K. Showalter, *Chemical Waves and Patterns* (Kluwer, Dordrecht, 1995)
5. K. Kurin-Csörgei, I. Szalai, E. Körös, The 1,4-cyclohexanedione-bromate-acid oscillatory system. II. Chemical waves. React. Kinet. Catal. Lett. **54**, 217–224 (1995)
6. K. Kurin-Csörgei, A.M. Zhabotinsky, M. Orbán, I.R. Epstein, Bromate-1,4-cyclohexanedione-ferroin gas-free oscillating reaction. 1. Basic features and crossing wave patterns in a reaction-diffusion system without gel. J. Phys. Chem. **100**, 5393–5397 (1996)
7. T. Yamaguchi, L. Kuhnert, Z. Nagy-Ungvarai, S.C. Müller, B. Hess, Gel systems for the Belousov-Zhabotinskii reaction. J. Phys. Chem. **95**, 5831–5837 (1991)
8. V.K. Vanag, I.R. Epstein, Pattern formation in a tunable medium: the Belousov-Zhabotinsky reaction in an aerosol OT microemulsion. Phys. Rev. Lett. **87**, 228301 (2001)
9. O. Steinbock, V. Zykov, S.C. Müller, Control of spiral-wave dynamics in active media by periodic modulation of excitability. Nature **366**, 322–324 (1993)
10. H. Ševčíková, M. Marek, S.C. Müller, The reversal and splitting of waves in an excitable medium caused by an electrical field. Science **257**, 951–954 (1992)
11. V.K. Vanag, I.R. Epstein, Patterns of nanodroplets: the Belousov-Zhabotinsky-aerosol OT-microemulsion system, in *Self-Organized Morphology in Nanostructured Materials*, ed. by K. Al-Shamery, J. Parisi, Springer Series, in Materials Science, Vol. 99, (Springer, Berlin, 2008), pp. 89–113
12. J. Carballido-Landeira, P. Taboada, A.P. Muñuzuri, Effect of electric field on Turing patterns in a microemulsion. Soft Matter **8**, 2945–2949 (2012)
13. P. Dähmlow, S.C. Müller, Nonlinear effects of electric fields in the Belousov-Zhabotinsky reaction dissolved in a microemulsion. Chaos **25**, 043117 (2015)
14. Student, The probable error of a mean. Biometrika **6**, 1–25 (1908)
15. A.F. Münster, Simulation of stationary chemical patterns and waves in ionic reactions. Discret. Cont. Dyn. Ser. B **2**, 35–46 (2002)
16. R.J. Field, Oregonator, Scholarpedia **2**, 1386 (2007). revision No.91613
17. D. Kupitz, S. Alonso, M. Bär, M.J.B. Hauser, Surfactant-induced gradients in the three-dimensional Belousov-Zhabotinsky reaction. Phys. Rev. E **84**, 056210 (2011)
18. O. Steinbock, J. Schütze, S.C. Müller, Electric-field-induced drift and deformation of spiral waves in an excitable medium. Phys. Rev. Lett. **68**, 248–251 (1992)
19. K.I. Agladze, P. De Kepper, Influence of electric field on rotating spiral waves in the Belousov-Zhabotinskii reaction. J. Phys. Chem. **96**, 5239–5242 (1992)

20. B. Schmidt, P. De Kepper, S.C. Müller, Destabilization of Turing structures by electric fields. Phys. Rev. Lett. **90**, 118302 (2003)
21. O.D. Bedford, G. Ilgenfritz, Electric field effects in AOT w/o microemulsions: field-induced percolation and dynamics of structure changes. Prog. Colloid Polym. Sci. **278**, 692–696 (2000)
22. H.-F. Eicke, M. Borkovec, B. Das-Gupta, Conductivity of water-in-oil microemulsions: a quantitative charge fluctuation model. J. Phys. Chem. **93**, 314–317 (1989)
23. Y. Feldman, N. Kozlovich, I. Nir, N. Garti, V. Archipov, Z. Idiyatullin, Y. Zuev, V. Fedotov, Mechanism of transport of charge carriers in the sodium bis(2-ethylhexyl) sulfosuccinate-water-decane microemulsion near the percolation temperature threshold. J. Phys. Chem. **100**, 3745–3748 (1996)
24. A.F. Münster, M. Watzl, F.W. Schneider, Two-dimensional Turing-like patterns in the PA-MBO-System and effects of an electric field. Phys. Scr. **1996**, 58–62 (1996)

Unpinning of Spiral Waves

Jiraporn Luengviriya, Malee Sutthiopad, Metinee Phantu, Porramain Porjai, Stefan C. Müller and Chaiya Luengviriya

Abstract Spiral waves are propagating self-organized structures commonly found in excitable media. Spiral waves of electrical excitation in cardiac systems connect to some arrhythmias, such as tachycardia and fibrillations, potentially leading to sudden cardiac death so that they should be eliminated. Such waves may drift and eventually annihilate at the boundary. However, they can be stabilized, when they are pinned to obstacles, that are weakly excitable or unexcitable regions in the medium. Recently, we used the Belousov-Zhabotinsky solutions, the well-known excitable chemical systems, to study the propagation of spiral waves pinned to obstacles and applied electrical forcing to unpin them in different situations of obstacle size and excitability. We employed simulations with the Oregonator model, a realistic scheme for the Belousov-Zhabotinsky reaction, to confirm the experimental findings as well as to reveal the detailed motions of the spiral waves under some specific conditions that are difficult to be realized in the experiments.

J. Luengviriya
Department of Industrial Physics and Medical Instrumentation,
King Mongkut's University of Technology North Bangkok,
1518 Pibulsongkram Road, Bangkok 10800, Thailand

J. Luengviriya
Lasers and Optics Research Center, King Mongkut's University of Technology
North Bangkok, 1518 Pibulsongkram Road, Bangkok 10800, Thailand

M. Sutthiopad · M. Phantu · P. Porjai · C. Luengviriya (✉)
Department of Physics, Kasetsart University, 50 Phaholyothin Road,
Jatujak 10900, Bangkok, Thailand
e-mail: fscicyl@ku.ac.th

S.C. Müller
Institute of Experimental Physics, Otto-Von-Guericke University Magdeburg,
Universitätsplatz 2, 39106 Magdeburg, Germany

S.C. Müller et al. (eds.), *Complexity and Synergetics*,
https://doi.org/10.1007/978-3-319-64334-2_11

1 Introduction

Self-organized rotating spiral waves are commonly observed in many excitable media [1–5]. These waves also have important impacts on human health, since rotating spiral patterns of electrical excitation in the heart and their instabilities concern cardiac tachycardia and life-threatening fibrillations [6, 7]. The rotation of spiral waves may cease and subsequently disappear, when the spiral tip is induced by some gradient to drift until it hits the boundary of the medium. However, spiral waves in cardiac tissues are often stabilized by being pinned to weakly excitable or unexcitable obstacles like veins and scars [3]. Therefore, such obstacles in excitable media act as helpers of the spiral waves by enlarging the lifetime of the waves.

Experimental investigations of spiral waves are mostly performed using an excitable chemical medium, namely the Belousov-Zhabotinsky (BZ) reaction, because the reaction preparation and the wave observation are convenient. Experiments on pinned spiral waves in a thin layer of the photosensitive ruthenium-catalyzed BZ reaction [8] have demonstrated that wave period, wavelength and velocity increase with the diameter of a circular unexcitable obstacle created by a laser spot. Elimination of pinned spiral waves in cardiac tissue cultures can be done by using a high-frequency train of electrical stimuli to induce unpinning and drift of the waves. The necessary frequency increases with the obstacle diameter [9–12]. Such an application of a wave train has been demonstrated in the BZ reaction, where small oil droplets were taken as obstacles [11].

Applying an electric field can induce a linear drift of the rotation center of free spiral waves in the BZ reaction, since the electrical current causes advective motions of key ionic species in the reaction. The drift speed and angle increase with the magnitude of the electric field [13–15]. In 3D media, the applied field causes closed-loop filaments of free scroll rings to reorient, until the unit vector normal to the plane of the filament ring is anti-parallel to the field direction [16]. It has been demonstrated that such electric field application results in an unpinning of a scroll ring from a pair of unexcitable spheres [17].

In this chapter, we review our recent investigations [18–20] on the dynamics of pinned spiral waves and their unpinning in BZ media under electrical forcing to elucidate the effect of the size and shape of the obstacles and the excitability of the medium. The experiments were conducted in uniform thin layers of the BZ reaction [21], and well-defined objects made from chemically inert plastic were used as unexcitable obstacles. We performed simulations using the Oregonator model [22, 23] in close correspondence with the experimental results.

2 Experimental

The BZ reaction is prepared with the following initial concentrations: $[NaBrO_3] =$ 50 mM, [MA] = 50 mM, [ferroin] = 0.625 mM, and $[H_2SO_4]$ 160 or 200 mM in the study of the influence of obstacle size and shape. The effect of excitability is investigated using the reaction with [H_2SO_4] varied between 100 and 280 mM. A high [H_2SO_4] corresponds to a stronger excitability. To prevent any hydrodynamic perturbation, the reaction is embedded in a 1.0% wt/wt agarose gel. Various circles or rectangles made from chemically inert plastic are used as obstacles. In each experiment, an obstacle is attached to a transparent flat reactor with a size of 100 × 100 × 1.0 mm^3, before the BZ solution is filled into the reactor. A spiral wave pinned to the obstacle is initiated using a two-layer method [18].

To study the unpinning of the spiral wave, a constant electrical current is applied via two electrolytic compartments (size of each 25 × 100 × 2.0 mm^3), which are attached to the left and the right boundaries of the main reactor. The density of applied electrical current is stepwise increased every few spiral rotations until the spiral wave is detached from the obstacle [18, 19]. This critical value of the current density is defined as J_{unpin}. The influence of excitability on the termination of free spiral waves is studied by forcing the spiral tip to hit the bottom of the reactor, and the critical current density is called J_{term} [19]. In order to control the temperature, the reactor is placed vertically in a transparent thermostating bath, which is located between a white light source and a color CCD camera to record the images.

Examples of pinned spiral waves in the BZ reaction with different concentrations of sulfuric acid or obstacle sizes and shapes are shown in Fig. 1. For a given circular obstacle, the wavelength increases while [H_2SO_4] decreases (Fig. 1a and b). For the same concentration, the wavelength increases with the circle diameter (Fig. 1b and c). In the case of obstacles with the same area size, the wavelength becomes larger with the circumference (Fig. 1c and d). Our detailed investigation of propagating spiral waves pinned to circular and rectangular obstacles shows as common features that the wavelength, wave period, and velocity of pinned spiral waves increase with the circumference, regardless of the obstacle area [20].

Typical spiral structures unpinned from the obstacle due to the action of an applied electrical current are illustrated in Fig. 2a–d. For a small current density

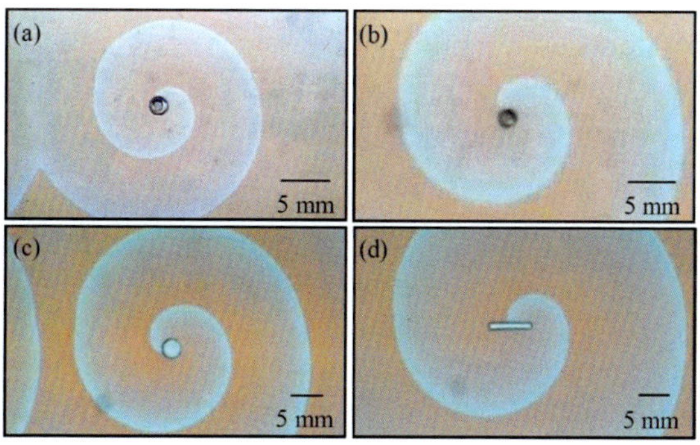

Fig. 1 Pinned spiral waves in the BZ reaction. Concentrations of sulfuric acid are **a** 200 mM and **b–d** 160 mM. Obstacles are circles with a diameter of **a** 1.5 mm, **b** 1.5 mm, and **c** 2.8 mm and **d** a rectangle with dimensions 6.5 mm × 0.9 mm. The wavelengths in **a–d** are 7.25 mm, 11.5 mm, 14.3 mm, and 20.3 mm, respectively [20]

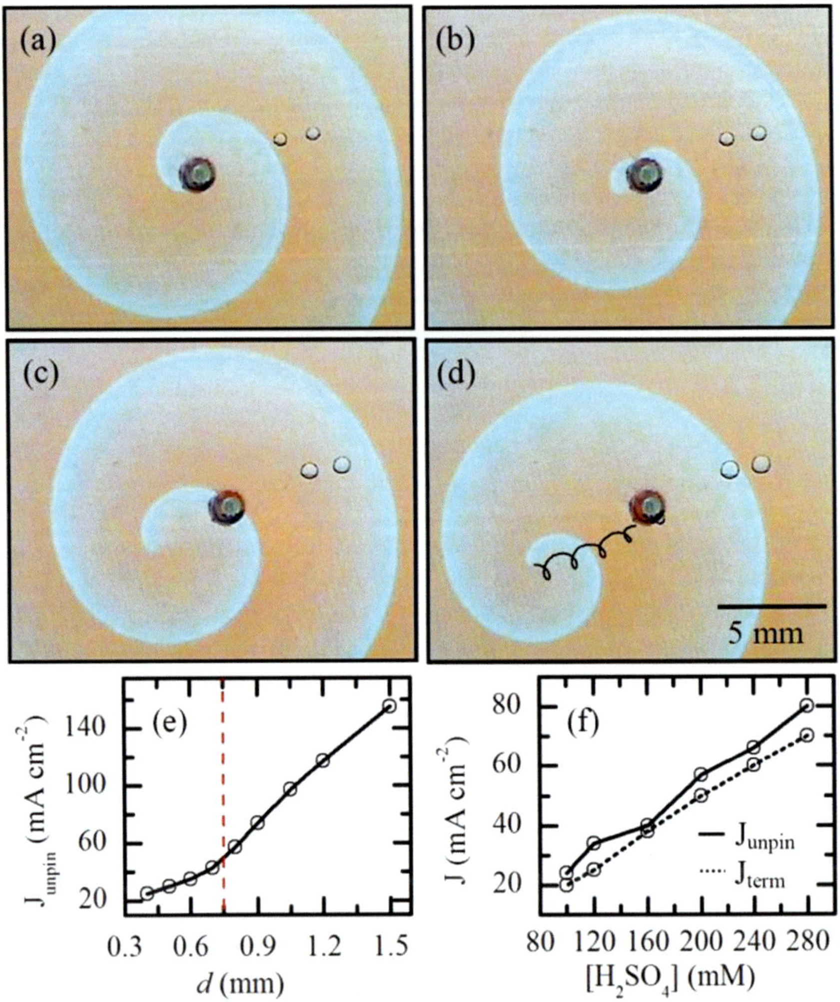

Fig. 2 Unpinning of spiral waves in the BZ reaction. **a–d** Time sequence of images of a spiral wave unpinned from a circular obstacle by electrical current. The *curling line* indicates the tip trajectory of the spiral wave leaving the obstacle [18, 19]. **e** J_{unpin} increases with the obstacle diameter. The *dashed line* indicates the core diameter of 0.8 mm of a free spiral wave [18]. **e** Both J_{unpin} and J_{term} increase with the concentration of sulfuric acid [19]

$J < J_{unpin}$, the spiral wave remains attached to the obstacle and the wave has an anisotropic distorted form (Fig. 2a). This results from the fact that the electrical current accelerates/decelerates the front propagating towards/away from the positive electrode [15]. When J reaches J_{unpin}, the spiral wave is detached from the obstacle (Fig. 2b). Under a constant electrical current, the unpinned spiral tip moves towards the positive electrode with an angle and the front structure changes with

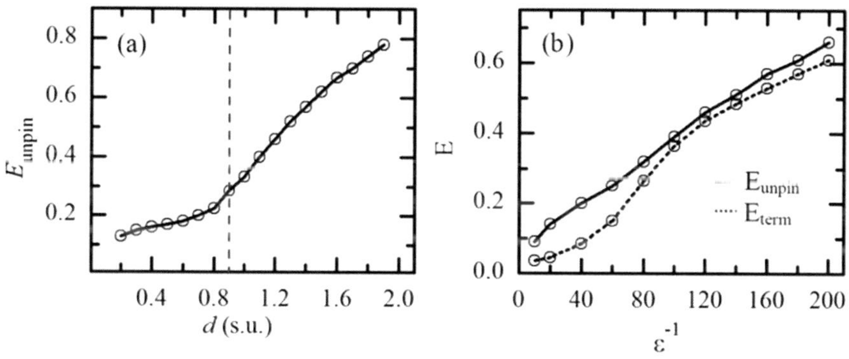

Fig. 3 Unpinning of spiral waves in the Oregonator model. **a** E_{unpin} increases with the obstacle diameter *d*. The *dashed line* indicates the spiral core diameter of 0.9 s.u. [18]. **h** Both E_{unpin} and E_{term} increase with the value of the parameter ε^{-1} [19]

time (Fig. 2b and c). Finally, the deformed wave structure (Fig. 3d) is similar to a drifting spiral wave under electrical forcing in the absence of obstacles [13].

The effect of the diameter of circular obstacles on the unpinning is shown in Fig. 2e. The experiments have been done with the BZ solution with $[H_2SO_4]$ = 200 mM and at a temperature of 24 ± 1°C [18]. In the absence of the obstacle, the tip of a free spiral wave rotates around the spiral core, which is a small circle with a diameter of 0.75 mm. When the obstacle diameter *d* increases from 0.4 to 1.5 mm, the required current density J_{unpin} increases monotonously. Linear fits for two ranges of the obstacle diameter show that the slope $\Delta J_{unpin}/\Delta d = 0.590 \pm 0.052$ and 1.389 ± 0.048 A cm^{-3} for $d < 0.75$ mm and $d > 0.75$ mm, respectively. This means that J_{unpin} increases with a much higher rate for the large obstacles in comparison with the small ones. The results show that it is more difficult to release spiral waves pinned to larger circles, especially when the obstacle diameter is larger than that of the free spiral core.

The influence of the medium excitability on the unpinning as well as the termination of spiral waves at the boundary is shown in Fig. 2f. The same circular obstacle with a diameter of 0.8 mm was used in the experiments, and the temperature was set at 24 ± 1°C [19]. The results show that the critical density of the applied electrical current for unpinning J_{unpin} increases with the concentration of H_2SO_4. This is due to the fact that the intrinsic properties of the spiral waves in the absence of the obstacle depend on the excitability. The spiral core diameter d_S of free spiral waves decreases from 1.20 to 0.45 mm when $[H_2SO_4]$ in the BZ solutions is increased from 100 to 280 mM. Thus, d_S ranges from a value larger to smaller than the obstacle diameter ($d = 0.8$ mm). Therefore, the smaller the free spiral core, the more difficult to unpin the spiral wave from a given obstacle. The excitability also affects the termination of free spiral waves. Under an applied electrical current, the spiral tip is forced to drift towards the bottom of the reactor, and subsequently it drifts parallel to the boundary with a distance of about 1 mm. The tip is forced to drift closer to the bottom and finally hits the boundary, when the current density reaches the critical value J_{term}, so that the rotation of the spiral wave is terminated. J_{term} also increases with [H_2SO_4] as in Fig. 2f. Therefore, it is more difficult to release a spiral wave pinned to a given obstacle, and either well to terminate a free spiral wave at the boundary in a medium with higher excitability.

3 Simulations

Simulations are performed using the two-variable Oregonator model to describe the dynamics of two variables u and v. The additional advection terms for u and v account for the electric field E applied in the x direction [18, 19]:

$$\begin{aligned}\frac{\partial u}{\partial t} &= \frac{1}{\varepsilon}\left(u - u^2 - fv\frac{u-q}{u+q}\right) + D_u\nabla^2 u - M_u E\frac{\partial u}{\partial x},\\ \frac{\partial v}{\partial t} &= u - v + D_v\nabla^2 v - M_v E\frac{\partial v}{\partial x}.\end{aligned} \tag{1}$$

The parameters are chosen as follows: $q = 0.002$, $f = 1.4$, diffusion coefficients $D_u = 1.0$ and $D_v = 0.6$, ionic mobilities $M_u = -1.0$ and $M_v = 2.0$, and $\varepsilon^{-1} = 100$ in the study of the influence of obstacle size and shape. The effect of excitability is investigated by modulating the parameter ε^{-1} between 20 and 200. A high ε^{-1} corresponds to a stronger excitability.

We use an explicit Euler method with a 9-point approximation of the two-dimensional Laplacian operator and a centered-space approximation of the gradient term in a discrete system with a uniform grid space Δx between 0.025 and 0.1 system unit (s.u.) and a time step $\Delta t \leq (3/8)(\Delta x)^2$ time unit (t.u.) as required for numerical stability [24]. A circular or rectangular area in the middle of the system is defined as the unexcitable obstacle so that the system and the obstacle have no-flux boundaries described in detail in [19].

A spiral wave is initiated by triggering a planar wave ($u = 1.0$ and $v = 0$). When the propagating wave is located at the middle, a half of the medium is reset to an excitable state ($u = 0$ and $v = 0$) leading to wave front end, which subsequently curls to form a rotating spiral wave pinned to the obstacle or a free spiral wave in the absence of the obstacle. To unpin the spiral wave, a constant electric field E is applied. E is increased with a fine step $\Delta E = 0.005$ every few spiral rotations until the spiral wave is detached from the obstacle. The critical value of the electric field E is called E_{unpin}. Finally, the influence of excitability on the termination of free spiral waves is studied by forcing the spiral tip to hit the bottom and the critical value of E is called E_{term}.

How the size of circular obstacles influences the unpinning was investigated in a system with a given $\varepsilon^{-1} = 100$ [18]. At this parameter value, the spiral tip rotates around the spiral core with a diameter of 0.9 s.u. in the absence of the obstacle. As shown in Fig. 3a, the critical value of the applied field E_{unpin} increases with the obstacle diameters d. The growth rate $\Delta E_{unpin}/\Delta d$ is about 0.145 ± 0.011 s.u.$^{-1}$ and 0.501 ± 0.016 s.u.$^{-1}$ for $d < 0.9$ s.u. and $d > 0.9$ s.u., respectively. Thus

E_{unpin} increases with a much higher rate for obstacles larger than the free spiral core as found in the experiments in Fig. 2e.

The effect of the excitability on the unpinning as well as the termination of spiral waves at the boundary of the simulated system was investigated for the parameter values $\varepsilon^{-1} = 20$–200 [19]. When ε^{-1} is increased in this range, the core diameter d_S of free spiral waves decreases from 1.65 to 0.70 s.u. For a given circular obstacle with a diameter of 1.0 s.u., E_{unpin} increases with the value of ε^{-1}, i.e., the excitability, as shown in Fig. 3b. The excitability has the same effect on the termination of free spiral waves. When the electric field reaches the critical value E_{term}, the spiral wave is forced to terminating at the boundary. Therefore, the simulations show the same trend as found in the experiments (Fig. 2f).

Besides the confirmation of the experimental results, we can use the simulations to reveal the details of the pinning and unpinning phenomena, since the parameter adjustment and the analysis can be done precisely. As in Fig. 4a, a small circular obstacle (0.5 s.u. in diameter) is added into the larger spiral core (0.9 s.u. in diameter). Even though the center of the obstacle is located exactly at the middle of the spiral core at the beginning, the motion of the spiral tip is affected by the smaller obstacle leading to a temporary pinning—the spiral tip is alternately attached to and detached from the obstacle in the absence of an electric field, e.g., as in Fig. 4b and c [18].

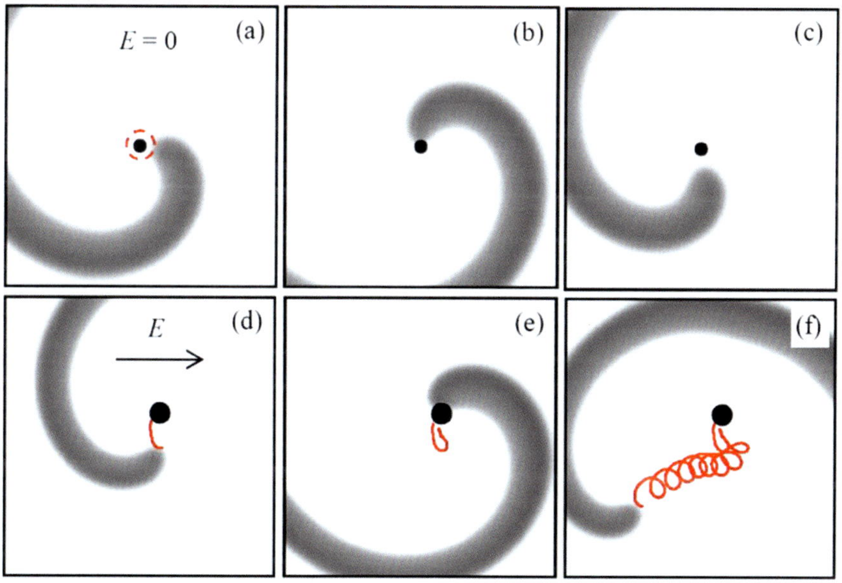

Fig. 4 Detailed motion of spiral waves in the Oregonator model. **a–c** A spiral wave is alternately attached to and detached from a small obstacle (0.5 s.u. in diameter) in the absence of an electric field (E = 0) [18]. The *dashed circle* indicates the spiral core with 0.9 s.u. in diameter. **d–f** Sequential images of a spiral wave unpinned from an obstacle (1.0 s.u. in diameter) by the electric field E_{unpin}. The lines represent the trajectory of the spiral tip [19]

The fine adjustment of the applied electric field allows us to observe a temporary unpinning as in Figs. 4d–f [19], which was not found in the experiments. When the electric field reaches the critical value E_{unpin}, the tip of the pinned spiral wave leaves the obstacle (Fig. 4d) but then it approaches the obstacle again (Fig. 4e). The forward—backward motion of the spiral tip occurs for a few rotations before the tip permanently moves away from the obstacle (Fig. 4f).

4 Discussion

Experiments using the BZ reaction and corroborated by simulations with the Oregonator model were used in our studies of pinned spiral waves and their unpinning by electrical forcing. The propagation of pinned waves depends on the excitability of the medium as well as the obstacle size that is more influential by the obstacle circumference than by the area, at least in cases of circles and rectangles.

Under electrical forcing, the unpinning of spiral waves from circular obstacles occurs when the strength of the forcing reaches a critical value that increases with the obstacle diameter. Furthermore, the electrically forced unpinning becomes a tough endeavor, when the obstacle diameter larger than the free spiral core. Our findings are consistent with earlier reports [11, 25, 26], where the unpinning occurs only when the frequency of the applied wave train is higher than a critical value that increases with the obstacle diameter. However, the highest frequency of waves is limited by the refractory period of the excitable medium so that the unpinning is impossible when the obstacle is very large [11, 25].

The electrically forced unpinning of spiral waves is also affected by the excitability of the medium: the critical value of the electrical forcing increases with the excitability. These results agree well with a numerical study [26] which shows that one can improve the success of unpinning of spiral waves induced by a train of stimuli by reducing the excitability of the medium which results in an enlargement of the spiral core. Other experimental studies using cardiac specimens have shown that unpinning of spiral waves occurred after application of anti-arrhythmic agents, which reduced the excitability [27, 28].

Acknowledgements We thank the Faculty of Science, the Research and Development Institute (KURDI), the Center for Advanced Studies of Industrial Technology, and the Graduate School, Kasetsart University, and the Office of the Higher Education Commission and King Mongkut's University of Technology North Bangkok (contract no. KMUTNB-NRU-58-03) for financial support.

References

1. S. Nettesheim, A. von Oertzen, H.H. Rotermund, G. Ertl, Reaction diffusion patterns in the catalytic CO-oxidation on Pt(110), front propagation and spiral waves. J. Chem. Phys. **98**, 9977–9985 (1993)

2. F. Siegert, C.J. Weijer, Digital image processing of optical density wave propagation in *Dictyostelium discoideum*. J. Cell Sci. **93**, 325–335 (1989)
3. J.M. Davidenko, A.M. Pertsov, R. Salomonz, W. Baxter, J. Jalife, Stationary and drifting spiral waves of excitation in isolated cardiac muscle. Nature **335**, 349–351 (1992)
4. A.T. Winfree, Spiral waves of chemical activity. Science **175**, 634–636 (1972)
5. A.T. Winfree, Scroll-shaped waves of chemical activity in three dimensions. Science **181**, 937 939 (1973)
6. A.T. Winfree, Electrical turbulence in three-dimensional heart muscle. Science **266**, 1003–1006 (1994)
7. E.M. Cherry, F.H. Fenton, Visualization of spiral and scroll waves in simulated and experimental cardiac tissue. New J. Phys. **10**, 125016 (2008)
8. O. Steinbock, S.C. Müller, Chemical spiral rotation is controlled by light-induced artificial cores. Phys. A **188**, 61 (1992)
9. K. Agladze, M.W. Kay, V. Krinsky, N. Sarvazyan, Interaction between spiral and paced waves in cardiac tissue. Am. J. Physiol. Heart Circ. Physiol. **293**, H503–H513 (2007)
10. A. Isomura, M. Hörning, K. Agladze, K. Yoshikawa, Eliminating spiral waves pinned to an anatomical obstacle in cardiac myocytes by high-frequency stimuli. Phys. Rev. E **78**, 066216 (2008)
11. M. Tanaka, A. Isomura, M. Hörning, H. Kitahata, K. Agladze, K. Yoshikawa, Unpinning of a spiral wave anchored around a circular obstacle by an external wave train: common aspects of a chemical reaction and cardiomyocyte tissue. Chaos **19**, 043114 (2009)
12. A. Pumir, S. Sinha, S. Sridhar, M. Argentina, M. Hörning, S. Filippi, C. Cherubini, S. Luther, V. Krinsky, Wave-train-induced termination of weakly anchored vortices in excitable media. Phys. Rev. E **81**, 010901(R) (2010)
13. O. Steinbock, J. Schütze, S.C. Müller, Electric-field-induced drift and deformation of spiral waves in an excitable medium. Phys. Rev. Lett. **68**, 248–251 (1992)
14. K.I. Agladze, P. De Kepper, Influence of electric field on rotating spiral waves in the Belousov-Zhabotinsky reaction. J. Phys. Chem. **96**, 5239–5242 (1992)
15. A.P. Muñuzuri, V.A. Davydov, V. Pérez-Muñuzuri, M. Gómez-Gesteira, V. Pérez-Villar, General properties of the electric-field-induced vortex drift in excitable media. Chaos, Solitons Fractals **7**, 585–595 (1996)
16. C. Luengviriya, S.C. Müller, M.J.B. Hauser, Reorientation of scroll rings in an advective field. Phys. Rev. E **77**, 015201 (2008)
17. Z.A. Jiménez, Z. Zhang, O. Steinbock, Electric-field-controlled unpinning of scroll waves. Phys. Rev. E **88**, 052918 (2013)
18. M. Sutthiopad, J. Luengviriya, P. Porjai, B. Tomapatanaget, S.C. Müller, C. Luengviriya, Unpinning of spiral waves by electrical forcing in excitable chemical media. Phys. Rev. E **89**, 052902 (2014)
19. J. Luengviriya, M. Sutthiopad, M. Phantu, P. Porjai, J. Kanchanawarin, S.C. Müller, C. Luengviriya, Influence of excitability on unpinning and termination of spiral waves. Phys. Rev. E **90**, 052919 (2014)
20. M. Sutthiopad, J. Luengviriya, P. Porjai, M. Phantu, J. Kanchanawarin, S.C. Müller, C. Luengviriya, Propagation of spiral waves pinned to circular and rectangular obstacles. Phys. Rev. E **91**, 052912 (2015)
21. C. Luengviriya, U. Storb, M.J.B. Hauser, S.C. Müller, An elegant method to study an isolated spiral wave in a thin layer of a batch Belousov-Zhabotinsky reaction under oxygen-free conditions. Phys. Chem. Chem. Phys. **8**, 1425–1429 (2006)
22. R.J. Field, R.M. Noyes, Oscillations in chemical systems. IV. Limit cycle behavior in a model of a real chemical reaction. J. Chem. Phys. **60**, 1877–1884 (1974)
23. W. Jahnke, A.T. Winfree, A survey of spiral-wave behaviors in the Oregonator model. Int. J. Bif. Chaos **1**, 445–466 (1991)
24. M. Dowle, R.M. Mantel, D. Barkley, Fast simulations of waves in three-dimensional excitable media. Int. J. Bif. Chaos **7**, 2529–2545 (1997)

25. Y.-Q. Fu, H. Zhang, Z. Cao, B. Zheng, G. Hu, Removal of a pinned spiral by generating target waves with a localized stimulus. Phys. Rev. E **72**, 046206 (2005)
26. A. Pumir, S. Sinha, S. Sridhar, M. Argentina, M. Hörning, S. Filippi, C. Cherubini, S. Luther, V. Krinsky, Wave-train-induced termination of weakly anchored vortices in excitable media. Phys. Rev. E **81**, 010901(R) (2010)
27. Z.Y. Lim, B. Maskara, F. Aguel, R. Emokpae, L. Tung, Spiral wave attachment to millimeter-sized obstacles. Circulation **114**, 2113–2121 (2006)
28. C. Cabo, A.M. Pertsov, J.M. Davidenko, W.T. Baxter, R.A. Gray, J. Jalife, Vortex shedding as a precursor of turbulent electrical activity in cardiac muscle. Biophysical J. **70**, 1105–1111 (1996)

Battery—Determination and Forecast via Synergetics

Ernst-Christoph Haß, Katharina Knicker, Uwe Sydow, Matthias Schulz and Peter J. Plath

Abstract A rechargeable battery can be considered as a complex system which is self-organizing under charging, discharging and relaxation. Depending of the type of battery and the kind of external influences, the battery is stressed in many different ways. This leads to an alteration of the internal self-organization processes which are indicative for the present internal state and the future development of the battery. To get detailed information about the State-Of-Charge (SOC) and the State-Of-Health (SOH) of such electrochemical systems we carried out on the one hand *internal* spatio-temporal measurements of the half-cell potentials within lead-acid batteries. By this, significant structure formation can be observed and is represented using several methods of synergetics. On the other hand, we performed *external* short-time measurements of the battery's voltage with differently triggered charging and discharging currents. The response curves of repeated excitations are characteristic for the dynamic behavior of the electrochemical system providing information on the battery's aging state. The fits of these curves are comprehended in a small set of significant parameters.

E.-C. Haß (✉) · K. Knicker
Institut für Angewandte und Physikalische Chemie, Arbeitsgruppe Synergetik,
Universität Bremen, Bibliothekstraße NW2, 28359 Bremen, Germany
e-mail: echass@uni-bremen.de

U. Sydow · M. Schulz
Fraunhofer-Institut für Keramische Technologien und Systeme IKTS,
Stationäre Energiespeicher, Michael-Faradaystraße 1, 07629 Hermsdorf, Germany
e-mail: uwe.sydow@ikts.fraunhofer.de

P.J. Plath
Fritz-Haber-Institut der Max-Planck-Gesellschaft, Faradayweg 4-6,
14195 Berlin-Dahlem, Germany
e-mail: peter_plath@t-online.de

S.C. Müller et al. (eds.), *Complexity and Synergetics*,
https://doi.org/10.1007/978-3-319-64334-2_12

1 Introduction

Already in the beginning of this century it was forecasted that the overall growth of world consumption of energy will increase by around 40% over the next 20 years, and concluded that energy storage will be a key technology for global energy sustainability [1]. At present days, major efforts are done to improve the utilization of renewable energy resources for electro mobility as well as for stationary energy storage [2, 3]. Also in battery-powered medical devices, many unresolved problems exist such as potential battery loss and failure at frequent use [4]. Some crucially important unresolved problems are non-sufficient lifetime of the batteries, lack of reliability at on-line remote measurements of the remaining capacity and prediction of the future performance of the batteries. Therefore, a reliable prediction of the actual state (State-Of-Charge, SOC, and State-Of-Health, SOH) of an accumulator is highly desirable [5, 6] (and references cited therein).

In the first part of our work, local potential measurements in lead-acid batteries are realized in order to find reliable quantities for State-Of-Charge and State-Of-Health evaluations. It is known that in lead-acid batteries as an example for strongly coupled non-linear electrochemical systems, spatial inhomogeneities can be found like acid stratification due to inhomogeneous current density distribution. To detect such inhomogeneities in situ, we carried out spatially resolved potential measurements between the plates of a cell stack under charging/discharging conditions. The resulting spatially resolved potential curves from inside the cell stack give additional information about the driven electro-chemical process of charging and discharging. This information cannot be obtained by potential and current monitoring at the outside poles of the battery. Nevertheless, such internal information might be suitable for developing a reliable prognosis of the battery's service time.

In the second part, we imitated the influence of the outside world on the battery in order to achieve a forecast of the capability of the battery without knowing what happens internally. Clearly, one cannot simulate the total history of an individual battery, but it is possible to investigate the responsiveness of the battery on specific complex external signals. The evaluation of the battery's answers using methods of synergetics [6] enables forecasting its lifetime and capability without having an insight on the internal processes. Although we concentrated our experiments mainly to lead-acid batteries, this method is not restricted to this type of accumulators.

2 Local Potential Measurements with Auxiliary Lead Electrodes

2.1 Experimental Setup

In order to measure local potentials, small auxiliary lead electrodes have been developed and were placed inside the lead-acid battery. The principal setup for these measurements is shown in Fig. 1.

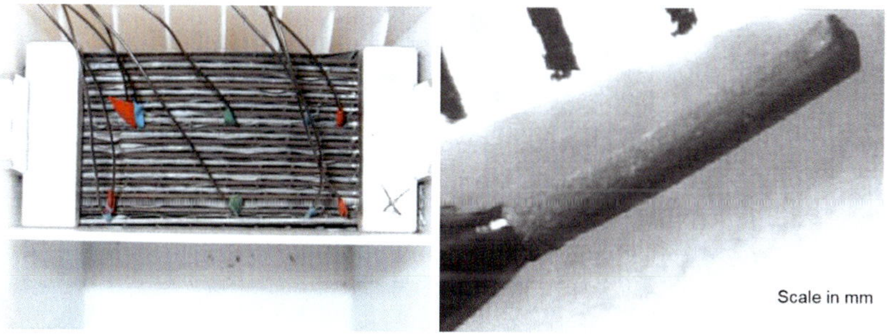

Fig. 1 Setup of a lead-acid battery for local potential measurements (*left*) and an enlarged single auxiliary lead sensor as reference electrode (*right*)

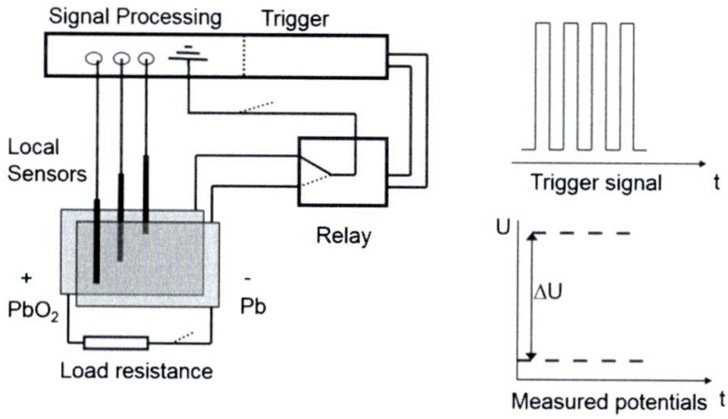

Fig. 2 Electrical setup for local potential measurements between internal probes and the battery plates. The auxiliary lead electrodes (local sensors) are periodically switched between the positive and the negative plate with a triggered relay. The frequency of the relay was set to 0.2 Hz, the sampling rate was 2 Hz and the input impedance of the analog/digital channels (signal input) was >1 GΩ

Lead electrodes were prepared by soldering a small piece of pure lead wire (Alfa Aesar puratronic 99.998%, 0.5 mm ∅) to the end of an insulated copper wire (0.5 mm ∅). The contact between the lead and the insulated copper wire was coated to achieve acid resistance. Only the tip was left uncoated to enable local potential measurements.

In order to get information about the local potentials between the auxiliary electrodes (local sensors, probes) and both the negative and positive battery plates, the electrical setup was chosen in such a way that the probes could be switched by a relay periodically to either the positive or the negative cell plate (see Fig. 2).

In order to get local potentials at different horizontal and vertical positions as well as from positions between different plates in a stack, an arrangement of auxiliary electrodes was chosen as shown in Fig. 3 (see also Fig. 1).

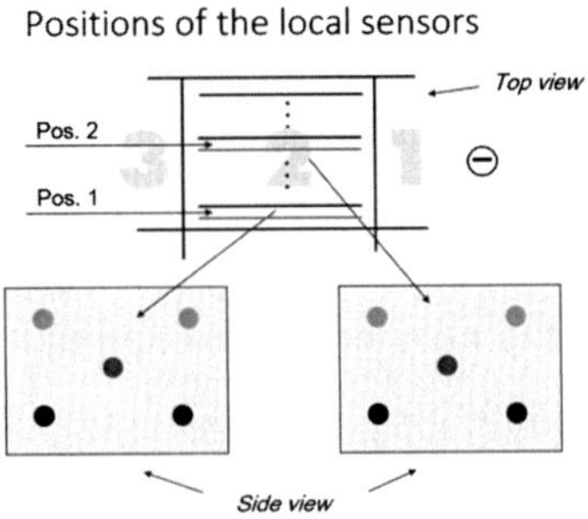

Fig. 3 Arrangement of local sensors in the cell stack. Each five probes are placed in the same way between two plates (Pos. 1 and Pos. 2). Two of them are located *left* and *right* at *top* (*light gray*), one is placed in the *center* (*gray*) and the remaining two are situated at *bottom* (*black*). This takes account of the fact that the electrolytic density varies from *top* to *bottom* due to increasing strength of sulfuric acid

2.2 Industry-Like Long-Term Measurements—Aging of a Battery

In industry, long term stress tests are carried out to get information about aging and lifetime of the battery. These tests consist of mutually strong load for a short time with subsequent moderate recovery and periodically repeated discharging/charging cycles (up to 100 times) with constant currents. Each of the short time stress tests leaves traces of aging in the battery which should be detectable. The repetition of about 100 discharging/charging cycles corresponds to an aging of approximately one year.

A typical example is the so-called *Großglockner test* where the downhill drive of a car with strong braking and running with weak load is simulated. The measured voltages at the poles of the battery for such a test is shown in Fig. 4.

After about 3 month of such tests the battery operation breaks down. Figure 5 illustrates exemplarily a damaged lead plate of a 'dead' battery.

In the case of lead-acid batteries the mechanism of deterioration of the electro-chemical system is explained mainly by acid stratification due to chemical reactions shown in Fig. 6.

During charging, the acid density of H_2SO_4 between the plates increases and thus pure acid moves down due to gravity. Conversely, during discharging, the acid density decreases. As a result of these processes, a density gradient evolves with low density at the top and high density at the bottom. This means that during the lifetime of a battery the internal inhomogeneities are growing, hardly noticeable in the beginning and over a long time, but almost suddenly at the end of the battery's life leading to its breakdown.

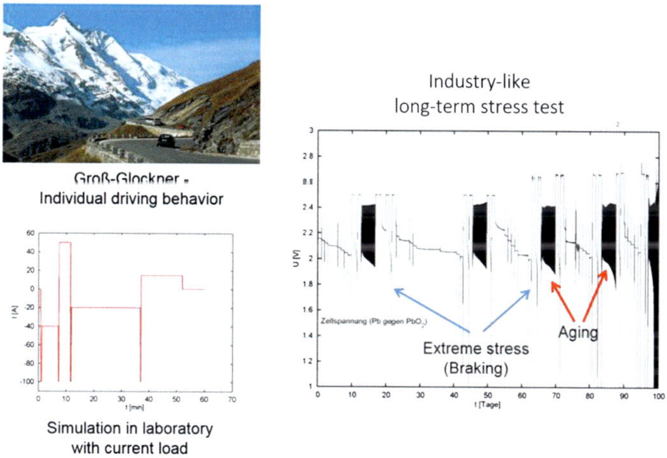

Fig. 4 Long term stress test of a battery with alternating simulation of downhill driving and 100 periodically repeated discharging/charging cycles. The small peaks are due to extreme stress (sharp braking). After about two thirds of battery lifetime, one can observe a beginning structural change of discharging/charging cycles due to aging; after about 100 days the battery is 'dead'

Fig. 5 Lead plate of a 'dead' battery cell after about 3 month of stress tests. The *light gray* area at the bottom of the plate is caused by $PbSO_4$ particles which crystallized out at higher concentrations of sulfuric acid

Negative Electrode:

$$Pb \longrightarrow Pb^{2+} + 2e^-$$

$$Pb^{2+} + SO_4^{2-} \longrightarrow PbSO_4 \qquad E^0 = -0.356V$$

Positive Electrode :

$$PbO_2 + 4H^+ + 2e^- \longrightarrow Pb^{2+} + 2\,H_2O$$

$$Pb^{2+} + SO_4^{2-} \longrightarrow PbSO_4 \qquad E^0 = 1.685V$$

Combined Reaction:

$$Pb + PbO_2 + 2\,H_2SO_4 \longrightarrow 2\,PbSO_4 + 2\,H_2O$$

$$E^0 = 2.041V$$

Fig. 6 Basic chemical reactions in lead-acid batteries

2.3 Results of Long-Term Local Cell Voltage Measurements

Starting from the observations mentioned above, one can consider the battery as a complex dynamical system with pronounced self-organization under charging, discharging and relaxation conditions and thus may raise the question whether its State-Of-Health (SOH) is detectable by short time investigations long before it breaks down. One method we have chosen consists of temporarily and spatially resolved measurements of local voltages using the experimental setup described in Figs. 1, 2 and 3.

Figure 7 shows the time series of local potentials recorded for sensors at top right and bottom right positions inside the battery cell during the industrial stress

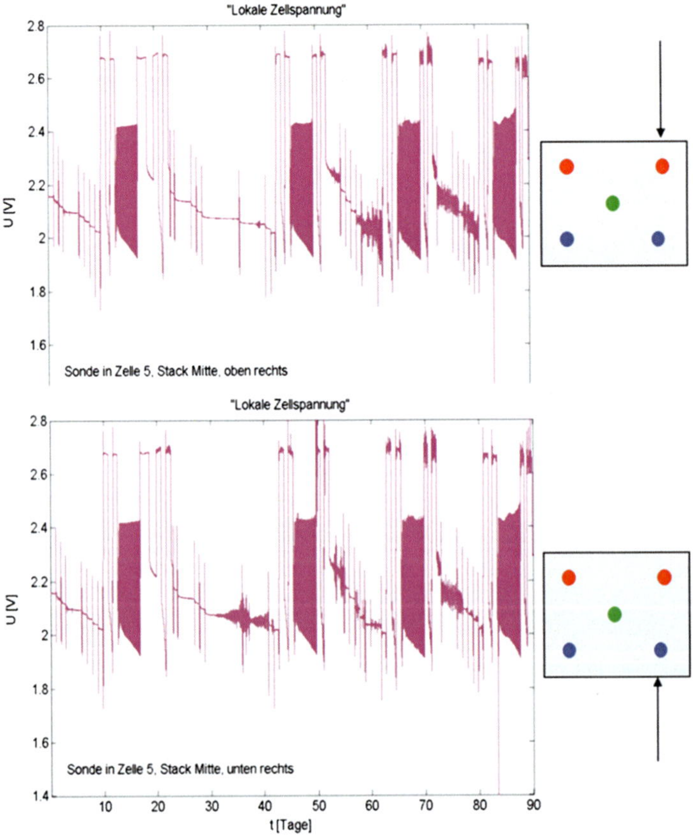

Fig. 7 Time series of local potentials from industrial stress tests The upper diagram is recorded at *top right* sensor position and the lower diagram at *bottom right* position. The latter shows already after about 35 days pronounced potential fluctuations during strong load indicating a higher discharge rate in the lower cell position than in the upper one. This may be explained by higher concentration of sulfuric acid and beginning $PbSO_4$ crystallization

test discussed in the previous section. A comparison with the time series of voltages measured at the poles (Fig. 4) indicates already significant differences of the breaking behavior in the middle of the battery's lifetime (after about 35 days and even more pronounced between 50 and 60 days).

More information about the health state of the battery can by gained if one compares local cell voltages. For this purpose, we plotted the differences between two local cell potentials against time. In Fig. 8, this is exemplarily demonstrated for two vertically positioned and for two horizontally positioned probes, respectively.

In the first case, a beat like pattern is visible in higher resolution during the first period of repetition cycles (day 12.5–16.5) of a healthy battery reflecting the almost regular periodic charging/discharging process. In the latter case, a complex pattern can be seen during stress tests (day 25–32), if the battery is slightly more aged. Both examples demonstrate that local dynamics can be quite different in various regions between the plates of the electro-chemical cell and thus local measurements can give more information than recording the integral cell voltage.

Another method to obtain information of space-time resolved measurements of dynamical systems which is frequently used in synergetics is the phase space representation of two or three time series. To give an example, Fig. 9 shows four time series of local potentials corresponding to four partial systems measured over a period of about half a week during cycling (30 min discharging, 40 min charging, both with about 53.9 A). The four auxiliary electrodes are placed at the four upper positions of the battery cell, two each between different electrode plates (see Fig. 3). Only the (half-) potentials against the positive PbO_2 plate are shown in Fig. 9.

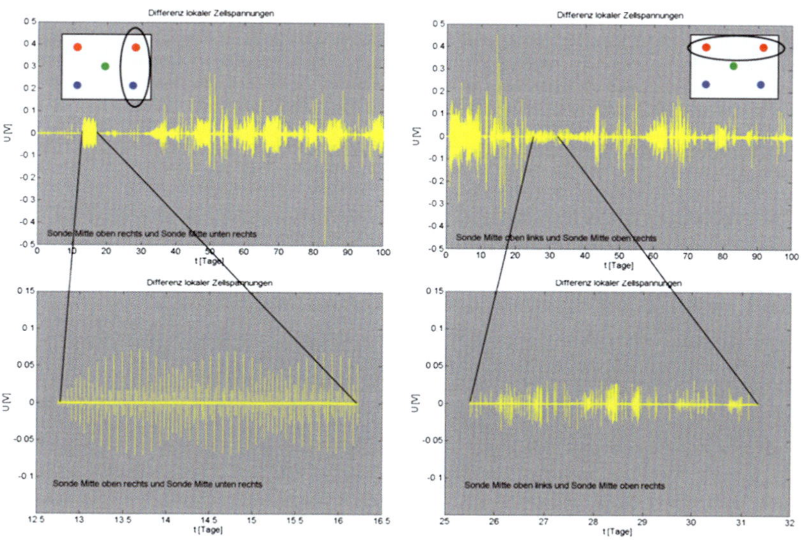

Fig. 8 Time series of differences of local cell voltage measurements between two vertically positioned probes (*left*) and between two horizontally positioned probes (*right*). In both cases, the upper image corresponds to the full time scale of aging and the lower one is a significant section of a few days

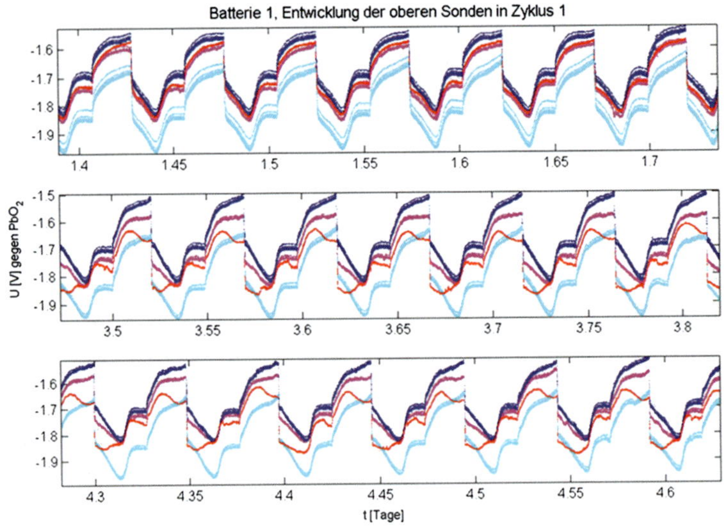

Fig. 9 Time series of the local (half-) potentials measured with four auxiliary electrodes in upper position versus the positive mass PbO_2. The three plot segments which are arranged under each other correspond to cycling (30 min discharging, 40 min charging, both with about 53.9 A) in three different time intervals

A phase space representation in three dimensions of the time series of the potential responses, $U_1(t)$, $U_2(t)$ and $U_3(t)$, corresponding to three of these upper probes is given in Fig. 10. The attractors match three different time intervals: 1.4–1.7 days (*blue*), 2.5–3.8 days (*yellow*) and 4.3–4.8 days (*red*). The points at the highest potential values reflect the situation of maximum charge and those at lowest potential values the situation of maximum discharge. Figure 10 demonstrates that after a few days different scenarios have been established in different regions of the stack due to short time aging of the battery which are indicative for its State-Of-Health.

Another possibility to gain insight into the internal state of a battery cell is to compare the values of a single auxiliary electrode at different times. This can be achieved by a phase diagram with delay coordinates as shown in Fig. 11. In this example, the two-dimensional delay attractors of local cell voltages obtained by a probe in the upper left position in the middle of a cell stack (see Fig. 3) are depicted. The potentials were compared through charge/discharge cycling by a delay of $\tau = 150$ s in time intervals of about 2 weeks over a period of 1 month.

From Fig. 11 it is obvious, that the ranges of attractors are widening in time. In the beginning of a cycling period, the attractor comprises the smallest area which is indicative for a healthy state of the battery. A few weeks later, the extension of the attractor corresponds to a warning state, and after a very long time (more than a month) of cyclic stressing, the expansiveness of the attractor indicates a critical state.

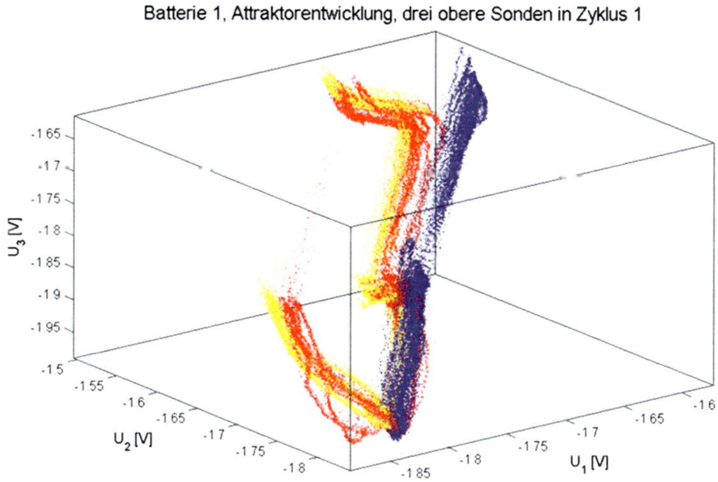

Fig. 10 Phase space representation of the attractors of local (half-) potentials measured with three auxiliary electrodes in upper position versus the positive mass PbO_2 corresponding to three different time intervals (1.4–1.7 days (*blue*), 2.5–3.8 days (*yellow*) and 4.3–4.8 days (*red*)). This figure shows clearly, that the attractors are changing in time due to short time aging of the battery

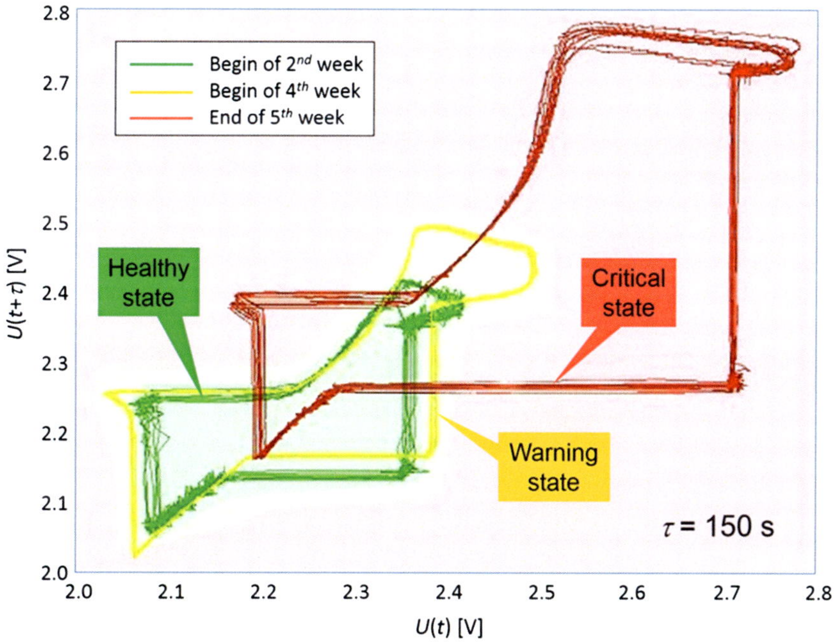

Fig. 11 Delay attractors of local potentials measured by an auxiliary electrode located in the *upper left* position in the middle of a cell stack. The delay time of each attractor is $\tau = 150$ s. The attractors are obtained at begin of the 2nd week, begin of the 4th week, and the end of the 5th week of a cyclic charging/discharging period, respectively

This example demonstrates that one can make a prediction about the development of the battery's health state if one measures the local potential inside the cell with only one auxiliary electrode which is located in a suitable position.

3 Short Time Measurements at the Battery's Poles with Periodic Excitation

Since measurements inside the battery are relatively costly, in particular because of preparation of the auxiliary electrodes and putting them into the accumulator, we tested an additional way to get knowledge about its State-Of-Health and making a prediction about its life-time without having information about its history.

Considering the battery as a dynamical system, it should react to external influences in an individual way. This was investigated by completely unloading the battery below a certain limit superimposed by specific short-time trigger functions. The voltage response to the excitation currents was measured at the poles of the battery. In Fig. 12 the discharge curves of 8 batteries in different aging states are shown which are obtained on the one hand analogously to industrial tests and on the other hand overlaid by asymmetric periodic excitations.

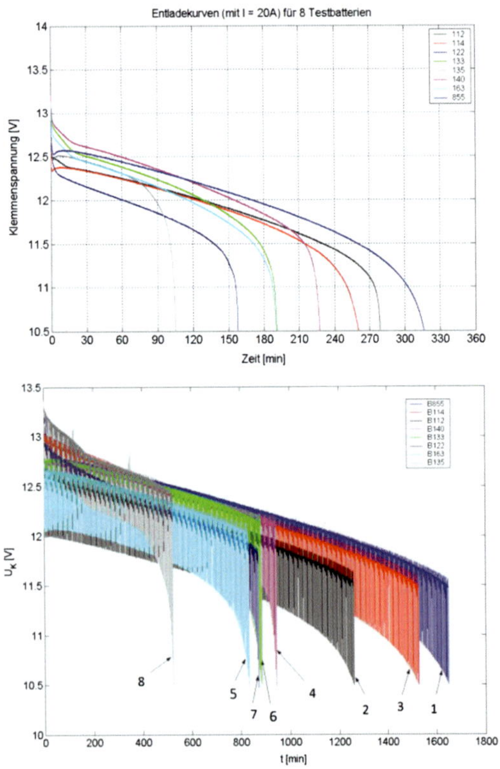

Fig. 12 Voltage response curves for complete unload below a voltage limit of 10.5 V measured on 8 batteries in different aging states. The upper diagram shows the graphs for discharging with a constant current of 20 A analogously to industrial tests. The lower diagram shows the resulting curves if the discharging process is overlaid by an asymmetric periodic excitation

3.1 *Measuring Principle and Evaluation Method*

The asymmetric periodic excitations are performed in such way that in pre-defined time intervals the battery cell is abruptly partially discharged and after a short relaxation time again charged to a certain amount. This process was repeated every 4 min in the actual measurements. The constant discharge and charge currents flowing thereby and the respective time intervals are shown in Fig. 13. The resulting curve of voltage response measured at the battery poles for a single asymmetric excitation period is represented in Fig. 14.

During complete unloading below a certain voltage limit, the shape of the response curve of each section should evolve systematically in subsequent asymmetric excitation periods due to aging of the battery. To get a quantitative measure, we approximated the discharge curves (Sect. 1) of successive periods by the 3-parametric fit function given in Eq. (1):

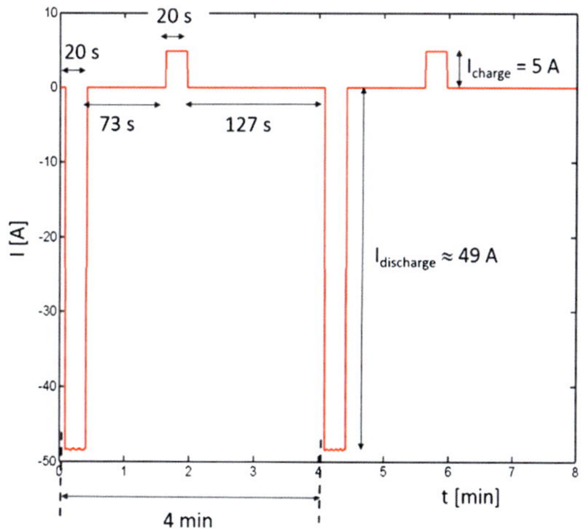

Fig. 13 Current trigger signal for two asymmetric excitation periods of each 4 min: 20 s discharge at ≈ 49 A, 73 s relaxation, 20 s charge at 5 A and 127 s relaxation

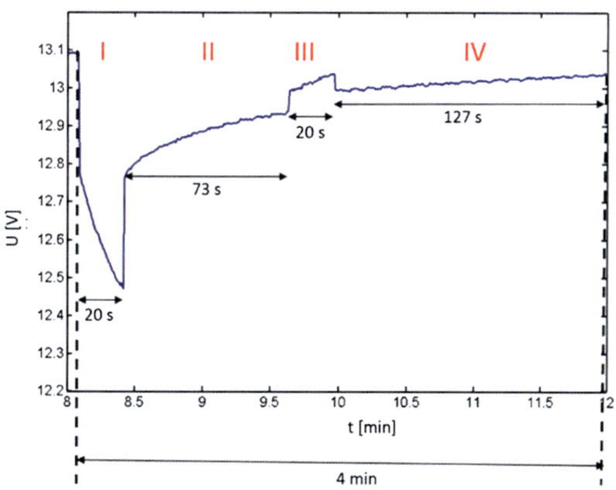

Fig. 14 Single period of voltage response at one asymmetric excitation. Section 1 corresponds to discharging, Sect. 2 to relaxation after discharging, Sect. 3 to charging and Sect. 4 to relaxation after charging

$$U(t) = U_0 \exp(-bt^c). \tag{1}$$

This fit method turned out to be optimal with respect to flexibility and quality by comparing with various other fit approaches.

Starting from these spline functions of single periods, another function $c = f(b)$ was used for further evaluation which combines the single pairs of values b and c obtained by successive splines of discharge cycles and represents a time-independent characteristic of the battery. As fit approach we have chosen the two-parametric spline function (2),

$$c = A/b^B, \tag{2}$$

which proved to be most flexible with an optimal regression coefficient by comparing different fit trials. The resulting parameters A and B can be conceived as specific values for each battery.

3.2 *Results of Discharge Measurements with Asymmetric Periodic Excitations*

As an example, Fig. 15 shows the discharge spline function of voltage response during the third period for a new battery with 100 Ah capacity and a 3D-plot of fitting parameters of all discharging splines from an unload process down to 10.5 V.

The dependency between spline parameters b and c of all periods together with the fitting function according Eq. (2) is visualized in Fig. 16. The fit provides the parameters A and B, i.e. the individual battery characteristic, in good approximation (r^2 = 0.996). Surprisingly, almost the same values for A and B are obtained if one

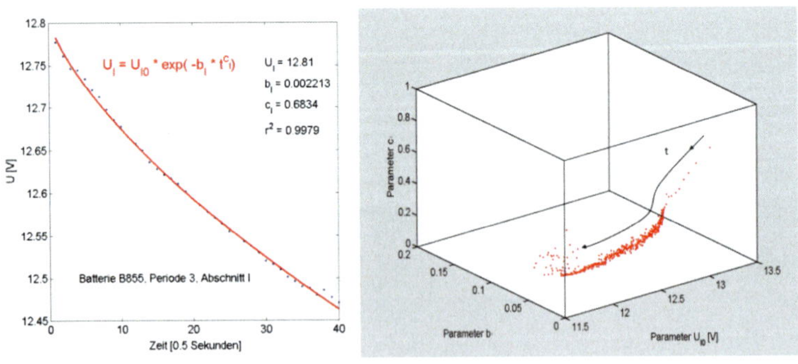

Fig. 15 *Left* Discharge spline function of voltage response. *Right* 3D plot of fitting parameters U_0, b and c of all discharging splines from an unload process down to 10.5 V. The arrow indicates the time development

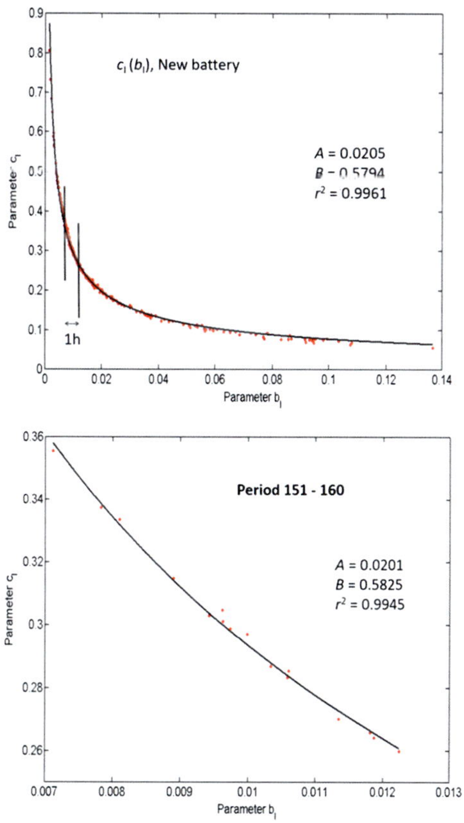

Fig. 16 Dependency between spline parameters c and b as well as their fitting function (Eq. (2)).The upper diagram shows the fit over all discharging periods, whereas the lower diagram covers only the periods of about 1 h. The resulting characteristic values A and B are quite similar, i.e. one only needs a small interval to get the required information about the individual battery characteristic

fits only a small interval (about 1 h) of the spline function $c = f(b)$ (see also Fig. 16).

To summarize our results, the courses of spline parameters b and c depending on the asymmetric excitation periods are represented in Fig. 17 for the 8 batteries mentioned above (see Fig. 12). One can clearly see that the batteries are divided into two groups with distinctly different dynamical behavior, one of them containing 6 and the other containing 2 accumulators. This reflects the fact, that they are produced by two different manufacturers.

In addition, from the length and the structure of the plotted curves one can already conclude on the aging state of the respective battery. For instance, a long and comparatively smooth curve is indicative for a newer battery, and a short and noisy curve, in particular with a maximum of b, suggests an aged one.

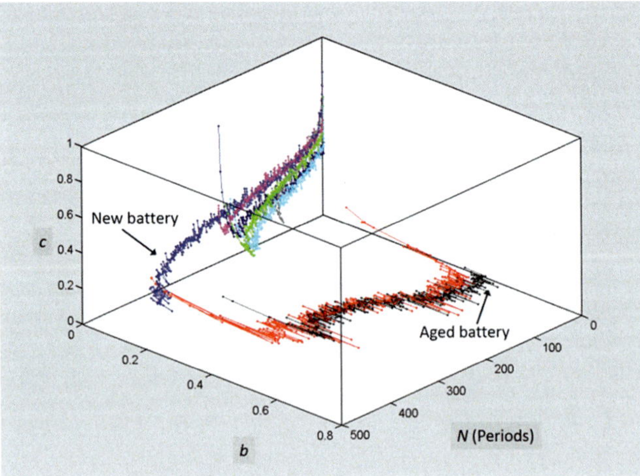

Fig. 17 Three dimensional plot of spline parameters b and c depending on the number of periods for the 8 batteries mentioned above. There are two groups of battery plots due to different manufactures: 6 in the *upper left* and 2 almost in the plane spanned by coordinates *N* and *b*. The *blue* curve corresponds to a new battery, and the *black* one to an aged one short before the end of its lifetime

4 Conclusion and Outlook

Local potential measurements in Pb/PbO_2 cells using special sensors can be carried out to get reliable quantities about the State-Of-Charge (SOC) and the State-Of-Health (SOH). It is generally known that lead-acid batteries are an example of strongly coupled non-linear electrochemical systems in which special inhomogeneities and acid stratification occur due to uneven distribution of current density and hindered diffusion. Such inhomogeneities can be detected by spatio-temporal potential measurements, but it requires a considerable measuring effort and hence this method is relatively costly.

On the other hand, one can assume that many of the different local states and potential distributions are reflected in measured values of the total current and the total voltage at the poles of an accumulator. We are convinced that aging of electrochemical systems, such as Pb/PbO_2 or Li-Ion batteries, can be ascribed to chemical modifications within the electrolyte as well as to self-organized structure formation, in particular in the vicinity of the electrodes.

In the second part of this article we have exemplarily demonstrated that one can get information about aging and the future performance of an electrochemical system by studying the response behavior to small short-term external perturbations. Using asymmetric periodic excitations we obtained individual characteristics of the batteries from fitted spline functions of the voltage response behavior which can be summarized in battery specific parameters *A* and *B*.

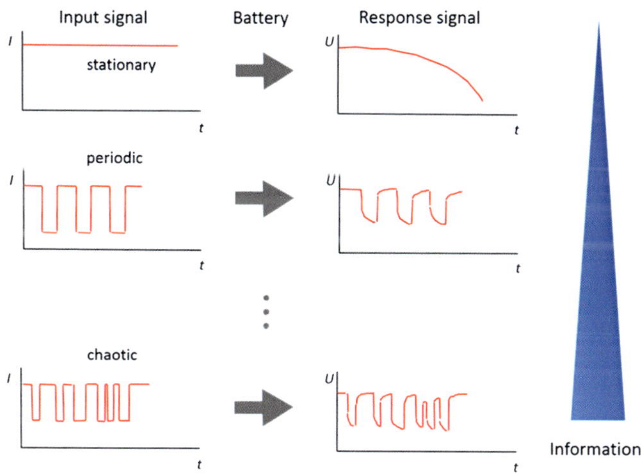

Fig. 18 Increasing complexity of the current input signal raises the information content of the voltage response (*stationary* < *periodic* < *chaotic*)

For future work, it is conjectured that by applying more complex charge and discharge current profiles the information content of the responding voltage signal under the imposed current application can be strongly increased. Figure 18 demonstrates schematically the increase of information using a stationary, periodic or chaotic current input.

For instance, imposing a chaotic current profile will give a chaotic voltage answer of the battery which can be evaluated in frequency space leading to a set of Fourier coefficients which may allow a refined prediction of the battery's state.

References

1. J. Leadbetter, L.G. Swan, Selection of battery technology to support grid-integrated renewable electricity. J. Power Sour. **216**, 376–386 (2012)
2. D. Oertel, *Energiespeicher—Stand und Perspektiven* (2008), http://www.tab-beim-bundestag.de/de/pdf/publikationen/berichte/TAB-Arbeitsbericht-ab123.pdf . Accessed 21 April 2016
3. W.Y. Chang, The state of charge estimating methods for battery: A review. ISRN Appl. Math. 2013, Article ID 953792, 7 p. (2013). https://doi.org/10.1155/2013/953792
4. M. Vockley, All charged up: The many challenges of battery maintenance. Biomed. Instrum. Techn. **48**, 86–96 (2014). https://doi.org/10.2345/0899-8205-48.2.86
5. R. Ottensmeier, E.C. Haß, U. Sydow, M. Buhlert, P.J. Plath, in *Proceedings of the Third International Advanced Automotive Battery Conference. Poster Presentations*, Nice, France, 10–13 June 2003, pp. 57–60
6. U. Sydow, M. Buhlert, E.C. Haß, P.J. Plath, in *Nonlinear Dynamics of Production Systems*, ed. by G. Radons, R. Neugebauer (Wiley, Weinheim, 2004), pp. 593–605

Dynamics and Control of Spiral and Scroll Waves

Sumana Dutta, Nirmali Prabha Das and Dhriti Mahanta

Abstract Excitable systems can sustain different kinds of wave forms like target patterns, two-dimensional spiral waves or their three-dimensional counterparts, the scroll waves. The dynamics of these excitation patterns and their responses to different kinds of internal and external perturbations are being looked into. These waves interact with neighboring vortices, that could lead to either attraction or repulsion and sometimes even their merging. Thermal gradients and electric fields can be used to control the motion of spiral and scroll waves. Scroll waves anchor to unexcitable heterogeneities and external field gradients can be used to unpin them from such obstacles. Our experiments with the Belousov–Zhabotinsky reaction are explained on the basis of numerical simulations using the Barkley model.

1 Introduction

Spiral and scroll wave activities are often encountered in different kinds of excitable media, like slime molds, viz. *Dictyostelium discoideum* [1], the nervous [2], retinal [3] and cardiac tissues [4, 5], catalytic [6] and precipitation chemical processes [7], etc. These waves have a higher frequency and shorter wavelength than the plane waves or target waves in the same medium [8]. Hence, once formed, they take over the entire excitable space, casting aside any target waves that may have been present. In a system such as the heart, this gives rise to arrhythmia, or fast and irregular heartbeats. This means that the heart cannot pump enough blood to the organs [9]. The

S. Dutta (✉) · N.P. Das · D. Mahanta
Indian Institute of Technology Guwahati, Guwahati 781039, India
e-mail: sumana@iitg.ernet.in

N.P. Das
e-mail: das.nirmali5@gmail.com

D. Mahanta
e-mail: mdhriti@gmail.com

S.C. Müller et al. (eds.), *Complexity and Synergetics*,
https://doi.org/10.1007/978-3-319-64334-2_13

situation may further worsen when there is spiral breakup or turbulence, as that is reflected in a medical condition called Fibrillation, which when occurring in the ventricles of the heart (VF) can cause immediate cardiac arrest and death. The understanding of such phenomenon requires a complete knowledge of the dynamics of these wave forms. Further, in order to find measures of cure, one needs to be able to control these waves. Current methods of defibrillation employ a strong and quick electric pulse that aims at erasing all wave forms and depends on the pacemaker cells of the heart to restart the supply of the regular waves. But such strong electric shock can create scars in the cardiac tissues which will act as seeds of future fibrillation. Milder methods of control of spiral and scroll waves are hence sought after. The heart is a complicated biological organ, and experiments on a live heart are neither easy nor cost-effective to be carried out on a regular basis. Another excitable system which shows similar kinds of wave forms is the Belousov–Zhabotinsky (BZ) reaction [10, 11]. This is an oscillatory redox reaction that can sustain spiral and scroll waves in an unstirred medium. Though there are several mismatches between the two systems: the heart is highly heterogeneous and the conduction velocity of the cardiac waves are around 0.2–0.5 $m\,s^{-1}$, whereas the BZ system is mostly homogenous and its waves have a typical velocity in the range of 0.01–0.1 $mm\,s^{-1}$; nevertheless, the exact nature of the waves that they sustain, makes the BZ system one of the simplest laboratory models for the study of spiral and scroll waves.

Through this contribution we try to summarize the results of some experimental research of spiral and scroll waves, specifically in the BZ system. Our main focus is on the methods employed for controlling the dynamics of these waveforms. We explore how vortices interact with each other, anchor to unexcitable obstacles, and how they respond to external stimuli. Simple numerical simulations used to explain the experimental findings are also discussed.

2 Spiral and Scroll Waves in the BZ System

In the last couple of decades, several studies have employed the BZ system to understand spiral and scroll wave dynamics [12]. The BZ system constitutes the oxidation of an organic acid by sodium bromate, in the presence of sulphuric acid. Ferroin as well as some ruthenium or cerium complex can act as a catalyst in the reaction that can show a change of color when stirred. In unstirred layers of BZ solution, the system can sustain two-dimensional (2D) spiral and three-dimensional (3D) scroll waves (Fig. 1). A typical BZ reaction in 2D constitutes a 2 mm layer of agarose gel with the required concentrations of the reagents. A gel is used to rule out convection effects. A 3D BZ system involves two gel layers (of 4 mm each) placed one above the other.

In a typical experiment, the reaction is carried out in a flat reaction chamber (like a petri dish) that is illuminated from below by a white light source and viewed from the top. In recent times, a charge coupled device camera connected to a personal computer is frequently used to take snapshots of the experiment and record the data.

The images are later analyzed by using different computational techniques, and the variation of the intensity studied to make observations on the dynamics of the wave forms. Tomographic techniques are also used by some groups to analyze the 3D scroll waves [13]. This enables one to get a three-dimensional view of the entire scroll as opposed to the top view. The latter gives an intensity profile that is basically the sum of intensities of every layer in the 3D object, more like a projection image.

Numerical studies of spiral and scroll waves are also very beneficial to understand the dynamics of the tips and the filaments. There are several mathematical models that mimic reaction diffusion systems. The simplest and most versatile amongst them is the Barkley model [14]. This is purely a mathematical model of an activator—inhibitor system that supports the formation of spiral and scroll waves (Fig. 2).

$$\frac{\partial u}{\partial t} = \frac{1}{\epsilon}\left\{u(1-u)\left(u - \frac{v+b}{a}\right)\right\} - \nabla J_u, \tag{1}$$

$$\frac{\partial v}{\partial t} = u - v - \nabla J_v. \tag{2}$$

where u and v are the concentrations of the activator and inhibitor respectively, ϵ, a and b are system parameters of the system, J_u and J_v are the fluxes of the ions. In the presence of diffusion, the fluxes are given by, $J_u = -D_u \nabla u$ and $J_v = -D_v \nabla v$, where D_u and D_v are the translational diffusion coefficients of u and v respectively.

The Oregonator model, which has been derived from the simplified kinetics of the BZ system, is also widely used in the theoretical study of these systems.

A two-dimensional spiral wave has a singularity, its tip. The trajectories of spiral tips are very interesting, and several attempts have been made to study them. Oftentimes the tip traces out trajectories that are aesthetically pleasing (Fig. 3a). The trajectory could be circular (Fig. 3b) or a meandering one (Fig. 3c). Meandering trajectories could be epicycloid or hypocycloid in nature [15].

The three-dimensional scroll waves also have a one-dimensional singularity. This is the filament of the scroll (Fig. 3d). The filament is that geometry around which the vortex rotates [16]. The ends of the filament will either close in on itself or touch the boundaries. Filaments are geometrical entities that follow the curvature dependent

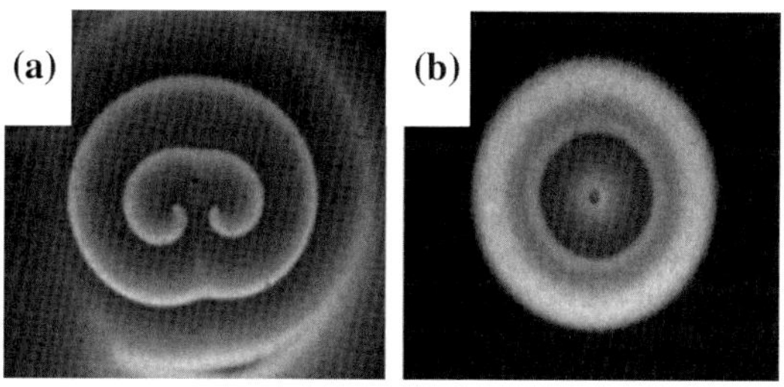

Fig. 1 **a** Spiral and **b** scroll waves in the Ferroin catalyzed BZ system. The initial concentrations of reactants in the 0.8% (w/v) agarose gel layers are, $[NaBrO_3] = 0.04$ M, $[CH_2(COOH)_2] = 0.04$ M, $[H_2SO_4] = 0.16$ M, and $[Fe(o\text{-}phen)_3]SO_4 = 0.5$ mM

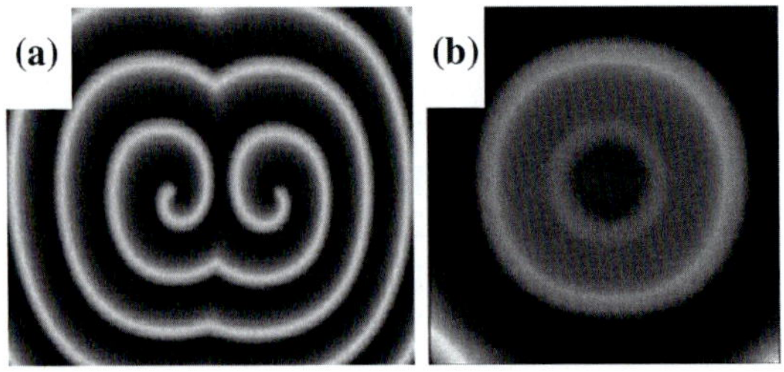

Fig. 2 Numerical simulation of the Barkley model showing formation of a spiral in two dimensions (**a**) and a scroll in three dimensions (**b**). Parameter values taken are $a = 1.1$, $b = 0.84$, and $\epsilon = 0.02$. $D_u = 1.0$, $D_v = 1.0$. Zero flux boundary condition is set for each diffusing species along all the boundaries

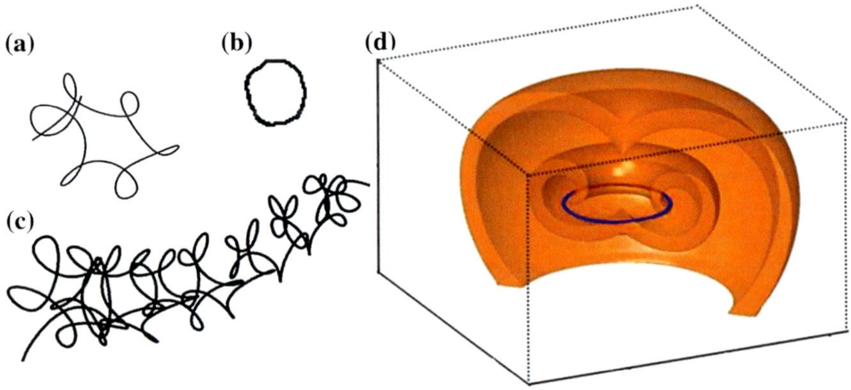

Fig. 3 **a–c** Various tip trajectories of spiral waves in the experimental BZ system. **d** A three-dimensional scroll wave obtained by simulation of the Barkley model. The filament, which is traced in *blue*, is identified at regions where $u = 0.5$ and $v = a/2 - b$. Other parameters same as in Fig. 2

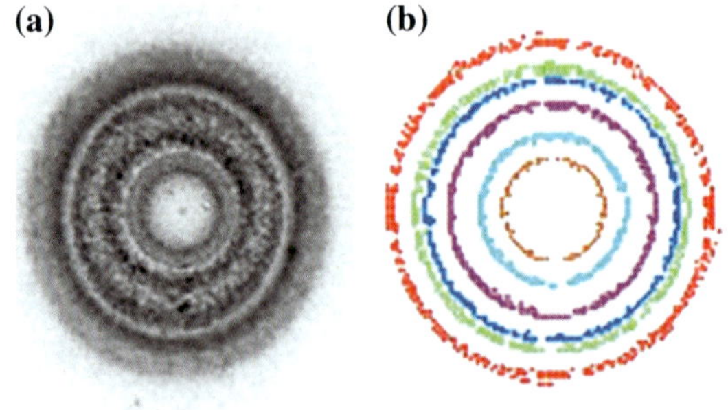

Fig. 4 Time evolution of the filament of a scroll ring. The filament shrinks with time. **a** Filaments from four snapshots of an experiment with the BZ system, superposed on each other. Each snapshot is 15 min apart. **b** Filaments of a simulation involving a scroll ring in the Barkley model

motion, according as

$$\frac{\mathrm{d}\mathbf{s}}{\mathrm{d}t} = \alpha\kappa\hat{\mathbf{N}} + \beta\kappa\hat{\mathbf{B}}, \tag{3}$$

where α is the filament tension, β the translational drift coefficient, κ the local curvature, $\hat{\mathbf{N}}$ the unit normal vector, and $\hat{\mathbf{B}}$ the unit binormal vector of the curve. It has been shown that the filament of a scroll wave in the BZ reaction has positive filament tension $\alpha > 0$, meaning a round scroll ring will always shrink (Fig. 4). Also $\beta \sim 0$ in an appreciable concentration range, choosing which would ascertain that there is no movement in the z−direction. Alternatively, negative filament tension paves the way for interesting situations like spiral wave turbulence.

3 Interaction of Spiral and Scroll Waves

Spiral waves of same frequency and wavelength will coexist side by side in a homogenously excitable system. Spiral waves generally occur in pairs. In a homogeneous environment, the two tips rotate with the same frequency and hence the mutual distance between them does not change with time. But if the excitability of one of the two tips of the spiral pair can be changed, then the faster arm will perturb the slower arm, and the two tips will move away from each other (Fig. 5).

Again a spiral arm may act as a defect when it is forced by a high-frequency wave train that may be originating from a target source or another fast moving spiral (Fig. 6) [17].

Depending on the relative frequencies of the spiral wave and the forcing wave, the direction of drift may vary. Two or three forcing waves can allow the positioning of the spiral at a specific point in space (Fig. 6d).

Scroll waves may interact with each other if they are close enough [18]. Depending on their mutual orientation, two scroll rings can push each other away and rupture on touching the system boundary, or they can reconnect to form a single, large ring (Fig. 7). Reconnection only occurs when the filaments lie within one core length of each other (Fig. 8). The reconnected filament has highly extended lifetimes, which could have serious implications in systems where they occur.

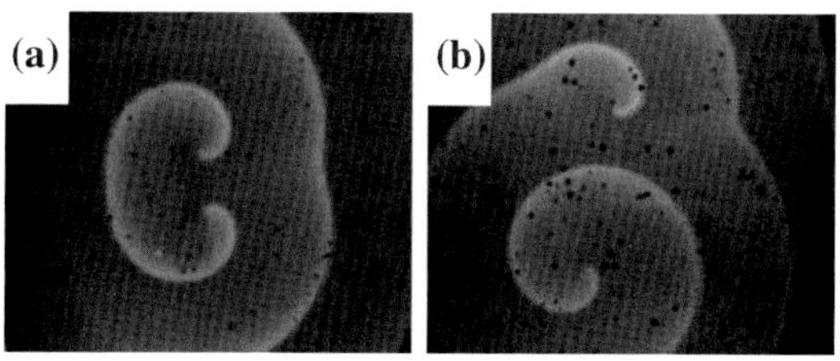

Fig. 5 Faster arm (*lower*) of a spiral pair pushing away the slower arm in experiments involving a thin layer of the BZ system. **a** 15 min and **b** 100 min after initiation of spiral

4 Influence of External Gradients

Reaction-diffusion systems react to external field gradients. Noise, electromagnetic fields and thermal gradients can bring about instability in a stable system [19]. The ions in the BZ system are also expected to be directly affected by the application of such external fields. The flux in these cases get modified due to the presence of the gradients. In the presence of an electric field, the flux for a species u is of the form:

$$J_u = -D_u \nabla u + D_u z_u u E \tag{4}$$

where z_u is the charge on u, and E the electric field or electric potential gradient. The flux in the presence of a thermal gradient takes the form [20]:

$$J_u = -D_u \nabla u - D_{T_u} u(1 - r_u)\nabla T \tag{5}$$

where $D_{T_u} = (D_u S_{T_0})/(1 + k_s u)$, is the thermal diffusion coefficient of species u, r_u is its relative concentration given by, $r_u = u/(u + v)$, and ∇T is the thermal gradient

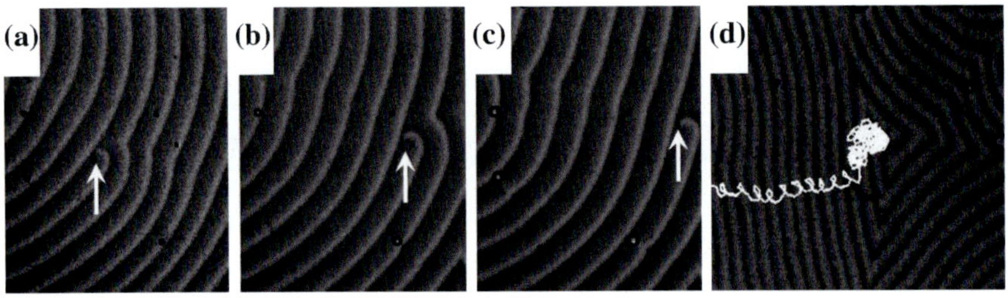

Fig. 6 **a–c** Drift of a spiral defect in a BZ system. The time between subsequent snapshots is 200s. The *arrows* indicate the position of the defect at every instant. **d** Trapping of a defect by three forcing waves. The snapshot is from a later stage of the experiment, after the trapping of the defect. The trajectory of the defect has been superscribed in *white*

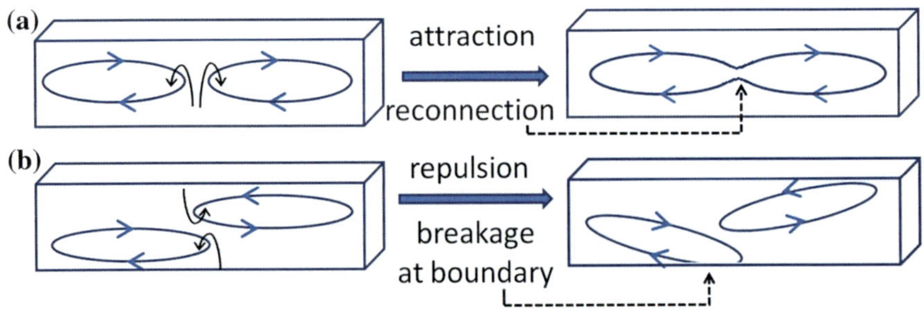

Fig. 7 Schematic representation of filament interaction [18]. Small *black arrows* denote the motion of the constituent spirals around the points of nearest approach of the filaments. **a** Reconnection of vortex filaments. The nearest approaching spirals have opposite sense of rotation. **b** Repulsion of filaments and possible rupture at the boundary. Here the nearest spirals have same sense of rotation

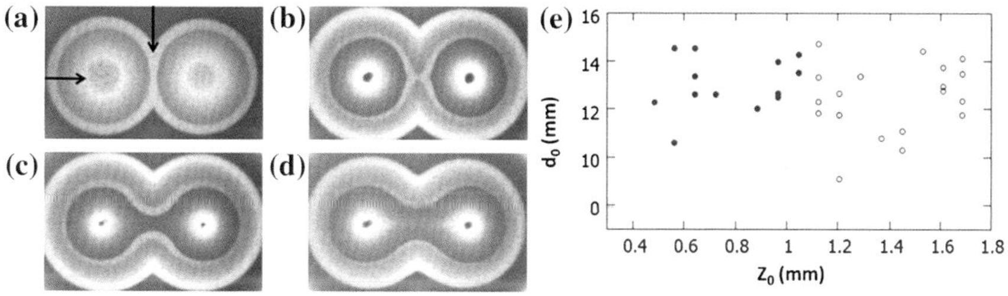

Fig. 8 **a–d** Reconnection of scroll wave filaments in the BZ system. Snapshots of a pair of coplanar scroll waves at **a** 7 min, **b** 21 min, **c** 35 min, and **d** 49 min after scroll-wave initiation. **e** Reconnection as a function of inter-filament distance Z_0, and ring diameter, d_0. Reconnected filaments are marked as *full* (*red*) *circles* and nonreconnected ones as *open* (*blue*) *circles*

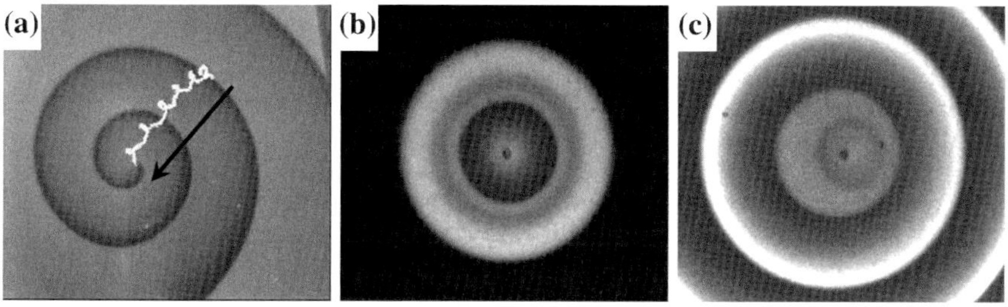

Fig. 9 **a** Movement of a spiral under the influence of an electric field (1.5 V cm^{-1}, positive electrode on the *left*). The snapshot has been taken 10 min after the field has been applied, and the tip trajectory has been superposed on it to show the path the spiral has traced. **b–c** Reorientation of a scroll ring under a thermal gradient ($\nabla T = 1\,°C\ cm^{-1}$, *right* to *left*). The two snapshots have been taken **b** just before and **c** 26 min after applying the gradient

between two sides of the reaction vessel and a constant for a reaction. The flux of species v also gets similarly modified in the presence of external gradients.

Spiral waves meander away from their point of initiation under the influence of electrical and thermal gradients [21] (Fig. 9a). Scroll waves on the other hand try to align themselves perpendicular to the direction of the gradient [22, 23] (Fig. 9b, c). It was found that the reoriented scroll ring later moves in the direction of the hot end (thermal gradient) or the positive electrode (electric potential gradient).

This observation can be further extended to control the position of spiral and scroll waves. A scroll ring can be made to traverse in a particular direction by employing moving or multiple gradients (Fig. 10) [24].

5 Pinning of Scroll Waves to Unexcitable Obstacles

The breakage of a plane wave due to some unexcitable obstacle, in conducive conditions of excitability, can lead to the formation of spiral and scroll waves. Again, such

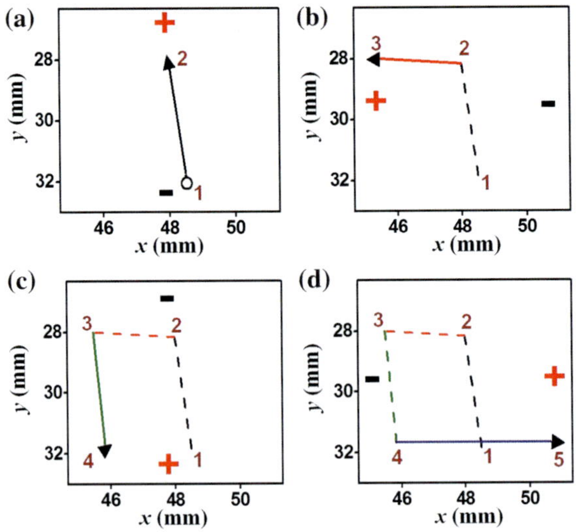

Fig. 10 Trajectory of a scroll ring under a moving electric potential. The field strength is 1.5 V cm^{-1}, and the position of the electrodes are changed after every 1 h. The continuous *bold line* in every *box* shows the trajectory for the current position of the electrodes, while *broken lines* denote past trajectory. *Open circle* in **a** denotes the initial position of the center of the scroll ring and 5 in **d** is the final position of the ring center. For any orientation of the gradient, it is observed that the scroll ring moves in the direction of the positive electrode

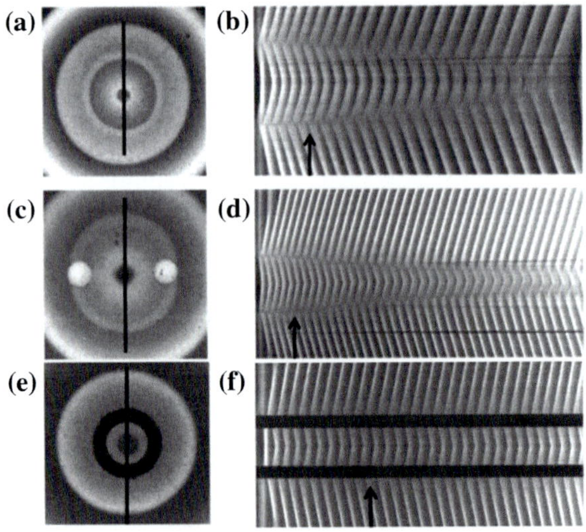

Fig. 11 Pinning of scroll waves. Snapshots (**a**, **c**, **e**) and corresponding time-space plots (**b**, **d**, **f**) of three experiments. **a**, **b** freely collapsing scroll; **c**, **d** vortex pinned to glass beads of 1 mm radius; **e**, **f** scroll ring pinned to a rubber O-ring of diameter 6.0 mm and thickness 2.0 mm. The time-space plots have been constructed by plotting intensity profiles with time, along the straight *black lines* seen in the corresponding snapshot. These plots (time going *left* to *right*) span time intervals of 120 min (**b**), 210 min (**d**) and 183 min (**f**)

unexcitable obstacles can also act as anchors to these wave forms. Pinning is a phenomenon which can be seen in the BZ reaction, where spiral waves and scroll waves get attached to glass beads and rubber rings (Fig. 11). Once pinned, the filament tends to be stabilized and no longer shrinks like a free filament. This increases the lifetime of the scroll rings infinitely [25, 26]. In the cardiac tissue, scroll waves may get pinned to scar tissues and other heterogeneous anomalies. The elongated lifetimes of these pinned vortices will be further detrimental to cardiac health. Hence, along with the control of free scroll waves, the mechanism of pinning and possible ways of unpinning of scroll waves are topics of major concern.

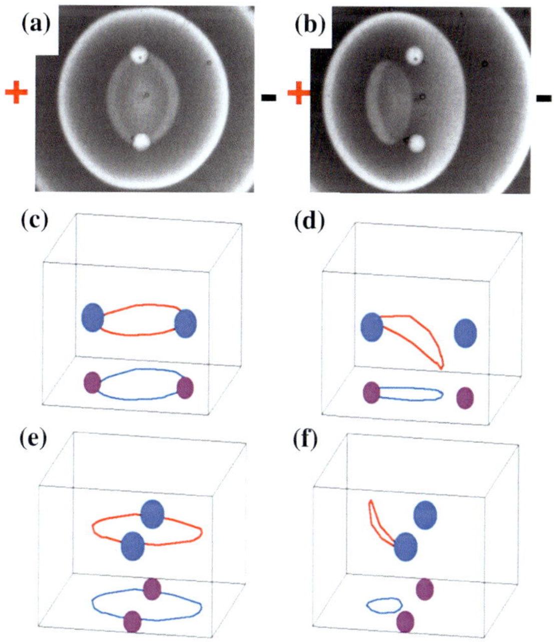

Fig. 12 **a–b** Electric field controlled unpinning of scroll ring in an experimental BZ system. The two snapshots are taken just before (**a**) and 23 min after (**b**) the application of electric field (1.5 V cm^{-1}). **c—f** Filament dynamics of unpinning under a thermal gradient (*right* to *left*) for different orientations of inert obstacles. Results of two sets of simulation for the Barkley model: interbead axis parallel to the gradient vector in (**c**, **d**) and perpendicular in (**e**, **f**). The *curves* (*red*) in the center of the cube are the filaments, the *spheres* (*blue*) are the pinning obstacles, and the *curves* (*blue*) along with the *circles* (*purple*) at the *bottom* of the box are the two-dimensional projections of the filament and the obstacle

6 Unpinning of Scroll Waves by External Gradients

In the earlier sections we have seen that spiral and scroll waves can be controlled by external field gradients. It is hence worthwhile to speculate if such external gradients could be successful in unpinning the waves anchored to inhomogeneous obstacles. Further studies showed that thermal [27] and electric potential [28] gradeints were able to unpin stable scroll rings attached to beads, in the BZ system (Fig. 12).

7 Discussion and Future Direction

The study of 2D spiral waves and 3D scroll waves in the homogeneous BZ system has led to an appreciable understanding of their dynamics. Activity of free and pinned waves, and their modified dynamics under the influence of neighboring sources of excitation waves and external field gradients have been well studied. It has been shown that such external fields can be used to control them and position them in space. Further studies aiming at the annihilation of trapped defects needs to be carried out. However, the real challenge lies in extending our knowledge to the control of such waves in biological systems, like the cardiac tissues. For this, we have to understand better the effect of irregular heterogeneities and other complexities present in such excitable systems. Once we are successful in bridging this gap and a correlation has been established between the waveforms in the BZ system and the heart, it may be possible to design medical gadgets that will pave the way for better treatment of conditions like VT and VF that results from these waves.

Acknowledgements S. Dutta would like to acknowledge Deb Shankar Ray and Oliver Steinbock for introducing her to the astonishing field of nonlinear dynamics. Portions of the work mentioned here were financially supported by the Department of Science and Technology, India (Grant No. SB/S1/PC-19/2012) and the National Science Foundation, USA (Grant No. 0910657).

References

1. P. Devreotes, Dictyostelium discoideum: a model system for cell-cell interactions in development. Science **245**, 1054–1058 (1989)
2. R. Larter, Understanding complexity in biophysical chemistry. J. Phys. Chem. B **107**, 415–429 (2003)
3. N.A. Gorelova, J. Bures, Spiral waves of spreading depression in the isolated chicken retina. J. Neurobiol. **14**, 353–363 (1983)
4. R. Majumder, A.R. Nayak, R. Pandit, Scroll-wave dynamics in human cardiac tissue: lessons from a mathematical model with inhomogeneities and fiber architecture. PLOS one **6**, e18052 (2011). https://doi.org/10.1371/journal.pone.0018052
5. E.M. Cherry, F.H. Fenton, Visualization of spiral and scroll waves in simulated and experimental cardiac tissue. J. Phys. **10**, 125016 (2008)

6. E. Gerhard, Reactions at surfaces: from atoms to complexity. Angew. Chem. **47**, 3524–3535 (2008)
7. F. Haudin, J.H.E. Cartwright, F. Brau, A.D. Wit, Spiral precipitation patterns in confined chemical gardens. Proc. Natl. Acad. Sci. **111**, 17363–17367 (2014)
8. I.R. Epstein, J.A. Pojman, *An Introduction to Nonlinear Chemical Dynamics: Oscillations, Waves, Patterns, and Chaos* (Oxford University Press, New York, 1998)
9. J. Jalife, M. Delmar, J. Anumonwo, O. Berenfeld, J. Kalifa, *Basic Cardiac Electrophysiology for the Clinician*, 2nd edn. (Wiley-Blackwell, Oxford, UK, 2009)
10. A.T. Winfree, Spiral waves of chemical activity. Science **175**, 634–636 (1972)
11. A.T. Winfree, Scroll-shaped waves of chemical activity in three dimensions. Science **181**, 937–939 (1973)
12. R. Kapral, K. Showalter, *Chemical Waves and Patterns* (Kluwer, Dordrecht, 1995)
13. D. Stock, S.C. Müller, Three-dimensional reconstruction of scroll waves in the Belousov–Zhabotinsky reaction using optical tomography. Physica D **96**, 396–403 (1996)
14. D. Barkley, M. Kness, L.S. Tuckerman, Spiral-wave dynamics in a simple model of excitable media: The transition from simple to compound rotation. Phys. Rev. A **42**, 2489–2492 (1990)
15. C. Luengviriya, U. Storb, M.J.B. Hauser, S.C. Müller, An elegant method to study an isolated spiral wave in a thin layer of a batch Belousov–Zhabotinsky reaction under oxygen-free conditions. Phys. Chem. Chem. Phys. **8**, 1425–1429 (2006)
16. J.P. Keener, The dynamics of three dimensional scroll waves in excitable media. Physica D **31**, 269–276 (1988)
17. S. Dutta, O. Steinbock, Spiral defect drift in the wave fields of multiple excitation patterns. Phys. Rev. E **83**, 056213 (2011)
18. N.P. Das, S. Dutta, Interaction of scroll waves in an excitable medium: reconnection and repulsion. Phys. Rev. E **91**, 030901(R) (2015)
19. S. Dutt, D.S. Ray, Electric field induced instabilities: waves and stationary patterns. Phys. Rev. E **73**, 026210 (2006)
20. S. Dutta, D.S. Ray, Thermodiffusion induced instabilities in reactive systems. Phys. Rev. E **75**, 066206 (2007)
21. O. Steinbock, J. Schütze, S.C. Müller, Electric-field-induced drift and deformation of spiral waves in an excitable medium. Phys. Rev. Lett. **68**, 248–251 (1992)
22. M. Vinson, S. Mironov, S. Mulvey, A. Pertsov, Control of spatial orientation and lifetime of scroll rings in excitable media. Nature **386**, 477–480 (1997)
23. C. Luengviriya, S.C. Müller, M.J.B. Hauser, Reorientation of scroll rings in an advective field. Phys. Rev. E **77**, 015201 (2008)
24. N.P. Das, S. Dutta, *Controlling Three Dimensional Vortices Using Multiple and Moving External Fields*. Phys. Rev. E **96**, 022206 (2017)
25. S. Dutta, O. Steinbock, Topologically mismatched pinning of scroll waves. J. Phys. Chem. Lett. **2**, 945–949 (2011)
26. Z.A. Jiménez, O. Steinbock, Stationary vortex loops induced by filament interaction and local pinning in a chemical reaction-diffusion system. Phys. Rev. Lett. **109**, 098301 (2012)
27. N.P. Das, D. Mahanta, S. Dutta, Unpinning of scroll waves under the influence of a thermal gradient. Phys. Rev. E **90**, 022916 (2014)
28. Z.A. Jiménez, Z. Zhang, O. Steinbock, Electric-field-controlled unpinning of scroll waves. Phys. Rev. E **88**, 052918 (2013)

One-Way Diffusion and Active Motion of Ionic Liquids in a Dissolution Process in Water

Noriko Oikawa and Rei Kurita

Abstract The dissolution dynamics of ionic liquid (IL) in ILs/water systems and at an IL/water/air interface is investigated. When a hydrophobic IL dissolves in water, it forms a droplet with a clear interface even though it is soluble to water. The transition between the two liquids remains sharp throughout the dissolution process, and it seems that the diffusion occurs unidirectionally, only outward from the droplet. The dissolution dynamics can be described as an activation process in which IL molecules escape from the droplet with a probability proportional to the surface area of the droplet. This distinctive feature of the dissolution dynamics may relate to alignment of the IL molecules at the IL/water interface. On the other hand, an active motion of water appears around the IL droplet, when hydrophobic ILs dissolve in a thin layer of water with contacting the air/water boundary.

1 Introduction

When two fluids, which are in different phases, are mixed, the transition between the two fluids remains sharp throughout the mixing process. Examples are found in crystals that dissolve in a liquid and droplets of liquid that evaporate into the air. In these systems diffusion occurs unidirectionally, only outward from the solutes, due to differences of the partial pressures. This mixing process could be regarded as a "one-way diffusion". For a mixing process between two miscible liquids, on the other hand, it is known that mass transfer takes place in both direction across the boundary between the two liquids and the concentration dynamics follow a diffusion equation [1]. However, it is known that some liquids do not follow this conventional

N. Oikawa (✉) · R. Kurita
Department of Physics, Tokyo Metropolitan University, Tokyo 192–0397, Japan
e-mail: oikawa@tmu.ac.jp

R. Kurita
e-mail: kurita@tmu.ac.jp

S.C. Müller et al. (eds.), *Complexity and Synergetics*,
https://doi.org/10.1007/978-3-319-64334-2_14

diffusion process during mixing, even when the system is at a thermodynamically homogeneous phase. When the surfactant dispersions are injected to other solvents, surfactant molecules typically generate mesoscale self-assembled structures such as micelles and lamellar [2, 3]. In case of *n*-butanol and isobutyric acid, the solute forms a droplet with a sharp interface in a solvent [4, 5] even without a surfactant. In these systems, the two miscible liquids remain in a non-equilibrium state until the solute dissolves into the solvent [6]. These are in contrast to the mixing behavior of traditional molecular solvents in which the dynamics follows a diffusion equation. Thus the mixing dynamics of the above systems might be related to the one-way diffusion realized in two miscible liquids. Although the one-way diffusion in miscible liquids has attracted many interests, there are only few reports on the detailed process of the mixing.

Ionic liquids (ILs) are a series of low temperature melting ionic compounds. Although ILs are composed merely of ions, hydrophobic alkyl chains attached mainly to the cation molecules lead to the low melting temperature by preventing IL molecules from packing closely as a crystal. ILs have found wide application in various fields such as chemical synthesis and electrochemistry [7–10]. Recently it has been reported that, unlike water and common organic solvents, ILs have a nanoscale structure composed of aggregates of polar and non-polar components [11–18]. It is recognized that ILs resemble ionic surfactants [19–22].

$[C_n\text{mim}][PF_6]$, 3-methylimidazolium hexafluorophosphate with different alkyl chain length $n = 2, 4, 6$ and 8, is a hydrophobic IL commonly used in the ILs research (see Fig. 1a). It is found that $[C_n\text{mim}][PF_6]$ form droplets in water while mixing. The droplet retains a sharp interface with a surface tension, gradually decreasing in droplet size until completely mixed. The shape of the droplet is stable against disturbance of flow. This behavior is in sharp contrast to the usual diffusive mixing process between two miscible liquids and can be assumed as a one-way diffusion. On the other hand, when $[C_n\text{mim}][PF_6]$ dissolve in thin layers of water with contacting the air/water boundary, dissolution dynamics induces an active motion of water around the $[C_n\text{mim}][PF_6]$ droplet. The amphiphilic feature of the cation molecules and the solubility of $[C_n\text{mim}][PF_6]$ make the surface tension of water spatially inhomogeneous, and the system becomes unstable.

In this letter, the one-way diffusion and the active motion observed in the $[C_n\text{mim}][PF_6]/H_2O$ system are reported. Due to the stability of the $[C_n\text{mim}][PF_6]$ droplet against flow, the dissolution dynamics can be obtained in detail. The mechanism of the dissolution dynamics will be discussed in relation with the mesoscale structure created at the $[C_n\text{mim}][PF_6]$/water interface.

2 Experimental

$[PF_6]^-$-based ILs, $[C_n\text{mim}][PF_6]$, of different alkyl chain lengths $n = 2, 4, 6$ and 8, are used in the experiments. The magnetic ionic liquid (MIL), 1-butyl-3-methylimidazolium tetrachloroferrate (bmim$[FeCl_4]$) [23, 24] is also used mainly to obtain

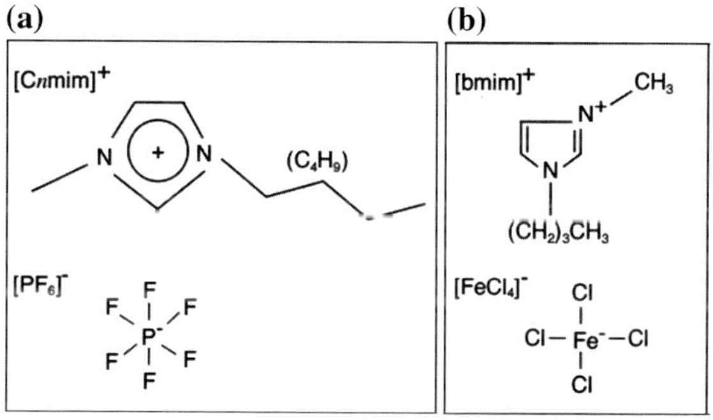

Fig. 1 Structure of the molecules of **a** the ILs ($[C_n mim][PF_6]$) and **b** the MIL ($[bmim][FeCl_4]^-$)

concentration profile of the droplet in the mixing process utilizing its brown color. The cation of the MIL has alkyl chains of $n = 4$. The images of the ILs molecules and the MIL molecules are shown in Fig. 1. The hydrophobic alkyl chains are attached in the imidazolium cations. All the samples were purchased from Kanto Chemical Co., Inc. The dissolution processes of the IL droplets were recorded with Panasonic HC-V520M High Definition Video Camera. For the experiments of dissolution dynamics, we used a petri dish of 7.8 cm diameter, which is sufficiently large compared to the size of the droplets. 30 g of deionized water was used as a solvent of the mixing. Thus the layer thickness of water in the petri dish is roughly 0.6 cm. The experiments have been performed at 18 °C, which is controlled by an air conditioning, except as described in the text. In experiments with changing temperature, the error of the temperature value was ±0.1 °C throughout each experiment.

3 Results and Discussion

50 μL of a MIL is gently injected into deionized water in a petri dish. We observe that the MIL forms a spherical droplet in water. Since the formation of a droplet is a characteristic feature of phase separation, it seems as if the MIL and water are immiscible. However, the saturation concentration of the MIL is $\phi_c \sim 25\%$ in water, and the experiment is performed on $\phi \sim 0.001\%$ ($\phi \ll \phi_c$), where ϕ is volume fraction of the MIL in the system. It seems unusual that the sharp interface remains even though the system is in a homogeneous phase at equilibrium. In order to quantitatively investigate the dissolution process of the droplet, we analyzed the temporal change of the image intensity around the droplet, which corresponds to the concentration of the MIL, due to the brown color of the MIL. The thickness ξ of the boundary region can be determined from the concentration profile measured across the droplet, as the width of the transition region where the intensity abruptly changes. It is confirmed that ξ is thin compared to the size of the droplet and remains constant with time as shown in Fig. 2. This indicates that a clear interface between MIL and water remains throughout the dissolution process. The neighboring water turns to a

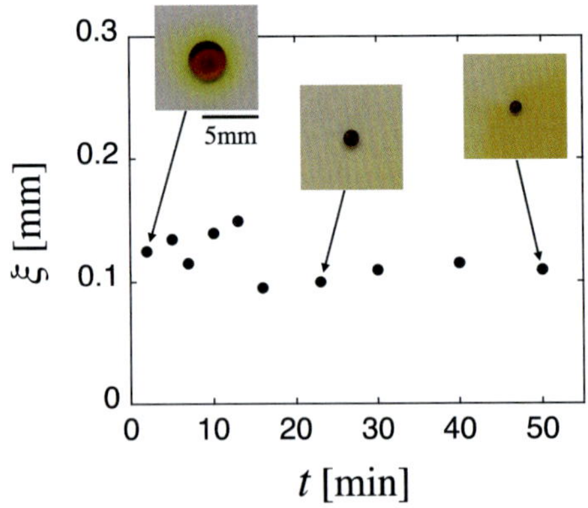

Fig. 2 Time dependence of the boundary thickness ξ of the MIL droplet and temporal change of the MIL droplet dissolving in water

light yellow-color as the MIL dissolves in water. Subsequently, the size of the droplet decreases with time as seen in the pictures in Fig. 2, and finally the MIL completely dissolves into water.

We go on to investigate ILs of $[C_n\text{mim}][PF_6]$. $[C_n\text{mim}][PF_6]$ is known to have a hydrophobic nanoscale structure in a $[C_n\text{mim}][PF_6]$/water mixture at equilibrium [13–15]. We inject $[C_4\text{mim}][PF_6]$ into water in the same manner as in the MIL experiment. Similarly to the MIL case, $[C_4\text{mim}][PF_6]$ forms a droplet in water and dissolves slowly while retaining the interface and the spherical shape of the droplet. When the $[C_4\text{mim}][PF_6]$/water system is cooled, the reversed process happens by phase separation of $[C_4\text{mim}][PF_6]$ from water. Thus the mixing of $[C_4\text{mim}][PF_6]$ and water is a thermodynamical phenomenon and not a chemical reaction process such as the decomposition of $[C_4\text{mim}][PF_6]$.

In order to examine whether the mixing process is the one-way diffusion or not, the water content of the $[C_4\text{mim}][PF_6]$ droplet, which had been undergoing the dissolution process in water by 20% decrease in volume, was measured with transmitted IR light intensity by using a spectrophotometer. As a result, the water content of the $[C_4\text{mim}][PF_6]$ droplet was estimated as less than 3%. If the $[C_n\text{mim}][PF_6]$ molecules and the water molecules exchange their position simultaneously in the dissolution process, the water content in the droplet should increase as much as the volume decrease of the droplet due to the dissolution. In the present system, however, the water content of the droplet was much smaller than the volume decrease of the droplet. Thus it seems that the one-way diffusion effectively occurs, in which mass transfer takes place unidirectionally only from the $[C_n\text{mim}][PF_6]$ droplets.

The one-way diffusion exhibited in $[C_4\text{mim}][PF_6]$/water system is further investigated quantitatively. The time evolution of the diameter d of the $[C_n\text{mim}][PF_6]$ droplet is examined at 80 °C for different alkyl chain lengths $n = 2, 6, 8$. d linearly decreases with time for all the cases as plotted in Fig. 3a. This leads us to the equation, $dV/dt = -aS$, where V, S and t are droplet volume, surface area and time, respectively. The coefficient a corresponds to the IL flux that dissolves from the surface of the droplet per unit time and unit area. The obtained results indicate that the dissolution of the IL droplet is an activation process. In Fig. 3b the temperature dependence of a is plotted with respect to $1/T$ on the basis of the Arrhenius plot. The dissolution speed increases as the temperature increases. $\ln(a)$ linearly depends on $1/T$, supporting the assumption that the dissolution occurs through an activation process. Since the nanostructures formed in ILs are induced by the solvophobic interaction of alkyl chains [20–22], it is thought that the alkyl chain length affects the dissolution dynamics of $[C_n mim][PF_6]$. The activation energy ΔE can be obtained by the equation $a = a_0 \exp(-\Delta E/k_B T)$. The slope in the $\ln(a)$–$1/T$ plot (Fig. 3b) corresponds to $(-\Delta E/k_B T)$. The activation energies per 1 mol for $[C_2mim][PF_6]$, $[C_4mim][PF_6]$, $[C_6mim][PF_6]$ and $[C_8mim][PF_6]$ are 36.9 kJ/mol, 41.5 kJ/mol, 43.9 kJ/mol and 51.3 kJ/mol, respectively. One finds that ΔE increases as n increases. Since longer alkyl chains lead to enhanced hydrophobicity, it is considered that the hydrophobicity controls the dissolution dynamics via a rearrangements of the alkyl chains at the interface.

When an energy barrier exists at the interface, it is advantageous for the system to minimize the interfacial surface area. We investigate the existence of an interfacial tension by making a collision of two $[C_4mim][PF_6]$ droplets. We extract edge lines of the droplets by digital image analysis for each time frame and calculate a total circumference L for the droplets. In Fig. 4, the temporal change of L is plotted. The edge line images of the droplets are also inserted in the figure at times when the droplets change their forms. One can see that the collided droplets merge with each other within a short time interval (less than 100 ms), and the form of the $[C_4mim][PF_6]$ droplet recovers the spherical shape immediately after the collision. This observation provides strong evidence that there exists an interfacial tension at

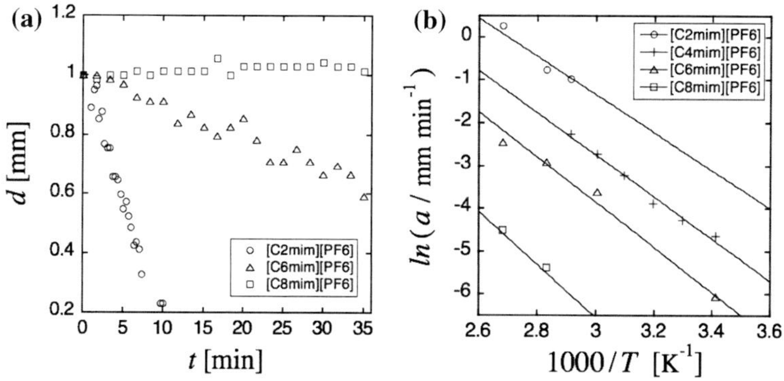

Fig. 3 **a** Time evolution of the diameter d of $[C_2mim][PF_6]$, $[C_6mim][PF_6]$ and $[C_8mim][PF_6]$ droplets in water obtained at 80 °C. d decreases proportionally with time. **b** Dependence of the slopes a in (a) on $1/T$. The *solid lines* are obtained by linear fitting

the [C_4mim][PF_6]/water interface. The interfacial tension of [C_4mim][PF_6] in water was measured by using the pendant drop method and determined as 19.04 [mN/m].

These results lead to an interpretation about the contrasting behavior in the dissolution dynamics of [$C_n$mim][PF_6]. The [$C_n$mim][PF_6] are primarily soluble to water when the system is at the homogeneous phase point. On the other hand, the hydrophobic interaction of alkyl chains likely induces a nanoscale structure at the interface with water. Since the characteristic time scale of the formation of the nanoscale structure is much shorter than that of the dissolution dynamics, the structure is created at the interface before the [$C_n$mim][PF_6] and water are mixed. It is thought that the structure formed at the interface plays a role of the energy barrier in the dissolution process of the droplet and also creates the interfacial tension of the droplet.

When [$C_n$mim][PF_6] dissolves in a thin layer of water with contacting the air/water boundary, another type of dissolution process is observed. 10 μL of [C_4mim][PF_6] was dropped onto 30 g of distilled water, which was poured into a petri dish of 15.5 cm diameter. Since the thickness of the water is small, the top part of the [C_4mim][PF_6] droplet is not covered completely by the water and is in contact with the air. In such a situation, [C_4mim][PF_6] expands to the surface of water first because of the amphiphilic nature of the cation molecules. Therefore, the surface tension of the water is strongly influenced at the same time when the [C_4mim][PF_6] dissolves. The system undergoes the following processes, which are represented as schematic images in Fig. 6. Cations evading from the droplet expand to the surface of water and decrease the surface tension around the droplet as mentioned above (Fig. 6a). At this stage water is pulled away from the droplet, and the droplet becomes isolated against water (Fig. 6b). After the cations, which are floating on the surface, dissolve into water, the imbalance of the surface tension is diminished (Fig. 6c), and water returns back toward the droplet (Fig. 6d). These inverted processes occur repeatedly and irregularly, and induce an active motion of the water surface around the droplet

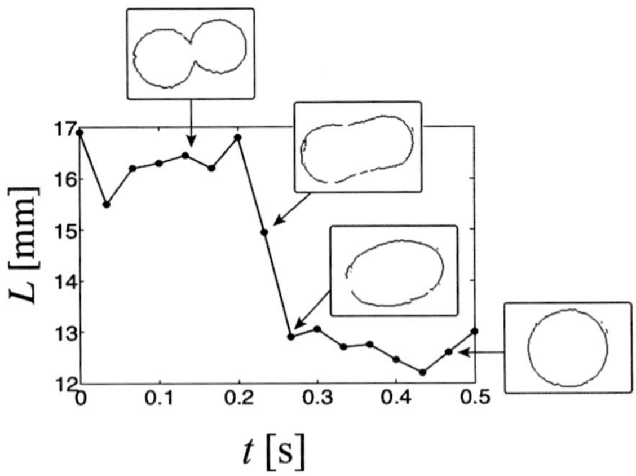

Fig. 4 Temporal change of the circumference L of [C_4mim][PF_6] droplets during the merging process of two droplets in water. The two droplets merge together when they collide, and recover the spherical shape immediately after the collision

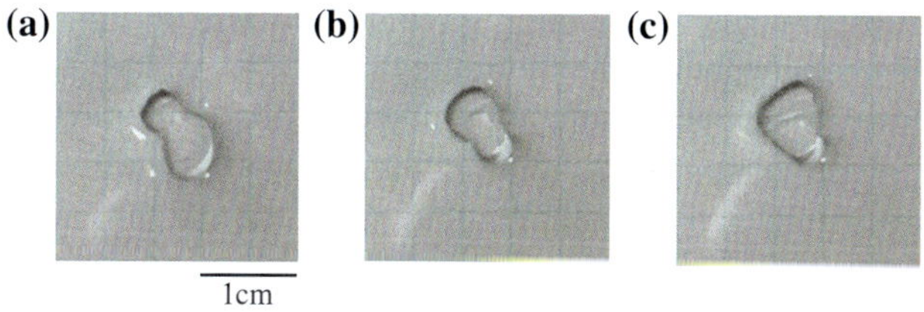

Fig. 5 Successive images of the temporal change of $[C_4mim][PF_6]$ droplet dissolving in a thin layer of water acquired at **a** 0 sec, **b** 0.6 sec, and **c** 1.2 sec. The droplet exists at the center of the dish

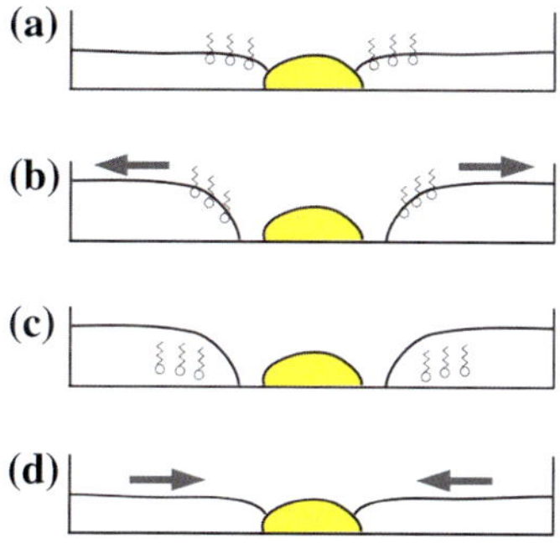

Fig. 6 The mechanism of the active motion observed in $[C_nmim][PF_6]$/water/air interface. **a** The cation molecules expand to the surface of the water and decrease the surface tension around the droplet. **b** Water is pulled apart from the droplet because of the imbalance of the surface tension. **c** The cation molecules dissolve into water and the imbalance of the surface tension is diminished. **d** Water returns back toward the droplet

as shown in Fig. 5. The active motion observed in the present system can be regarded as a camphor scraping pinned at a certain point in space.

4 Summary

The present study has investigated the dynamics of a mixing process between hydrophobic ILs and water. It is found that the ILs form a droplet in water in the dissolution process. The droplet retains a sharp interface with a surface tension, and the size of the droplet gradually decreases until it completely disappears. This is in contrast to the usual diffusive mixing process between two miscible liquids. The observed dissolution dynamics of $[C_nmim][PF_6]$ and MIL droplets indicate that the dissolution of the ILs obeys an activation process, and the one-way diffusion takes place, in which ions diffuse in one direction only from the IL droplets into water. The activation energy at the $[C_nmim][PF_6]$/water interface increases with increasing the length of the hydrophobic alkyl chains of the $[C_nmim][PF_6]$. Based on the dependence of the activation energy on the morphology of the cations, it is considered that a reorientation of the alkyl chains at the $[C_n mim][PF_6]$/water interface induces formation of a structure on the surface of the droplet. Indeed, recent studies have revealed that when hydrophobic ILs get in contact with water the charged parts of the IL molecules are oriented towards the water [13–15, 25, 26]. Thus, the nanoscale structure of ILs presumably influences macroscopic non-equilibrium dynamics in the mixing process of ILs and water. It is expected that the new aspects of the dissolution dynamics will lead to further understanding of mixing dynamics of miscible liquids.

References

1. A. Onuki, *Phase Transition Dynamics* (Cambridge University Press, Cambridge, 2002)
2. W. Helfrich, Lyotropic lamellar phases. J. Phys. Condens. Mat. **6**, A79–A92 (1994)
3. R. Strey, R. Schomäcker, D. Roux, F. Nallet, U. Olsson, Dilute lamellar and L_3 phases in the binary water-$C_{12}E_5$ system. J. Chem. Soc. Faraday Trans. **86**, 2253–2261 (1990)
4. J.A. Pojman, C. Whitmore, M.L.T. Liveri, R. Lombardo, J. Marszalek, R. Parker, B. Zoltowski, Evidence for the existence of an effective interfacial tension between miscible fluids: isobutyric acid-water and 1-butanol-water in a spinning-drop tensiometer. Langmuir **22**, 2569–2577 (2006)
5. B. Zoltowski, Y. Chekanov, J. Masere, J.A. Pojman, V. Volpert, Evidence for the existence of an effective interfacial tension between miscible fluids. 2. dodecyl acrylate-poly(dodecyl acrylate) in a spinning drop tensiometer. Langmuir **23**, 5522–5531 (2007)
6. T. Ban, T. Yamada, A. Aoyama, Y. Takagi, Y. Okano, Composition-dependent shape changes of self-propelled droplets in a phase-separating system. Soft Matter **8**, 3908–3916 (2012)
7. T. Welton, Room-temperature ionic liquids: solvents for synthesis and catalysis. Chem. Rev. **99**, 2071–2084 (1999)
8. T. Welton, Ionic liquids in catalysis. Coord. Chem. Rev. **248**, 2459–2477 (2004)
9. C.A.M. Afonso, L.C. Branco, N.R. Candeias, P.M.P. Gois, N.M.T. Lourenço, N.M.M. Mateus, J.N. Rosa, Efficient catalyst reuse by simple dissolution in non-conventional media. Chem. Commun. 2669–2679 (2007)
10. Q. Liu, M.H.A. Janssen, F. van Rantwijk, R.A. Sheldon, Room-temperature ionic liquids that dissolve carbohydrates in high concentrations. Green Chem. **7**, 39–42 (2005)
11. S. Hayashi, R. Ozawa, H. Hamaguchi, Raman spectra, crystal polymorphism and structure of a prototype ionic-liquid [bmim]Cl. Chem. Lett. **32**, 498–499 (2003)
12. H. Katayanagi, S. Hayashi, H. Hamaguchi, K. Nishikawa, Structure of an ionic liquid, 1-n-butyl-3-methylimidazolium iodide, studied by wide-angle x-ray scattering and Raman spectroscopy. Chem. Phys. Lett. **392**, 460–464 (2004)
13. A. Triolo, O. Russina, B. Fazio, R. Triolo, E.D. Cola, Morphology of 1-alkyl-3-methylimidazolium hexafluorophosphate room temperature ionic liquids. Chem. Phys. Lett. **457**, 362–365 (2008)
14. C. Hardacre, J.D. Holbrey, C.L. Mullan, T.G.A. Youngs, D.T. Bowron, Small angle neutron scattering from 1-alkyl-3-methylimidazolium hexafluorophosphate ionic liquids ($[C_n mim][PF_6]$, $n = 4$, 6, and 8). J. Chem. Phys. **133**, 074510 (2010)
15. M. Macchiagodena, L. Gontani, F. Ramondo, A. Triolo, R. Caminiti, Liquid structure of 1-alkyl-3-methylimidazolium-hexafluorophosphates by wide angle x- ray and neutron scattering and molecular dynamics. J. Chem. Phys. **134**, 114521 (2011)
16. C.E.S. Bernardes, M.E.M. da Piedade, J.N.C. Lopes, The structure of aqueous solutions of a hydrophilic ionic liquid: the full concentration range of 1-ethyl-3-methylimidazolium ethylsulfate and water. J. Phys. Chem. B **115**, 2067–2074 (2011)

17. R. Hayes, S. Imberti, G.G. Warr, R. Atkin, How water dissolves in protic ionic liquids. Angew. Chem. Int. Ed. **51**, 7468–7471 (2012)
18. S. Palchowdhury, B.L. Bhargava, Effect of cation asymmetry on the aggregation in aqueous 1-alkyl, 3-decylimidazolium bromide solutions: molecular dynamics studies. J. Phys. Chem. B **118**, 6241–6249 (2014)
19. T.L. Greaves, D.F. Kennedy, A. Weerawardena, N.M.K. Tse, N. Kirby, C.J. Drummond, Nanostructured protic ionic liquids retain nanoscale features in aqueous solution while precursor brønsted acids and bases exhibit different behavior. J. Phys. Chem. B **115**, 2055–2066 (2011)
20. J.N.A.C. Lopes, A.A.H. Pádua, Nanostructural organization in ionic liquids. J. Phys. Chem. B **110**, 3330–3335 (2006)
21. D. Xiao, J.R. Rajian, S. Li, R.A. Bartsch, E.L. Quitevis, Additivity in the optical kerr effect spectra of binary ionic liquid mixtures: implications for nanostructural organization. J. Phys. Chem. B **110**, 16174–16178 (2006)
22. L.A. Aslanov, Ionic liquids: liquid structure. J. Mol. Liq. **162**, 101–104 (2011)
23. S. Hayashi, H. Hamaguchi, Discovery of a magnetic ionic liquid [bmim]$FeCl_4$. Chem. Lett. **33**, 1590–1591 (2004)
24. H. Hamaguchi, R. Ozawa, Structure of ionic liquids and ionic liquid compounds: are ionic liquids genuine liquids in the conventional sense? Adv. Chem. Phys. **131**, 85–104 (2005)
25. I. Goodchild, L. Collier, S.L. Millar, I. Prokeš, J.C.D. Lord, C.P. Butts, J. Bowers, J.R.P. Webster, R.K. Heenan, Structural studies of the phase, aggregation and surface behaviour of 1-alkyl-3-methylimidazolium halide + water mixtures. J. Colloid Interf. Sci. **307**, 455–468 (2007)
26. A.L. Sturlaugson, A.Y. Arima, H.E. Bailey, M.D. Fayer, Orientational dynamics in a lyotropic room temperature ionic liquid. J. Phys. Chem. B **117**, 14775–14784 (2013)

Part IV
Biology

Pattern Formation in Marine Systems

Ulrike Feudel

Abstract Pattern formation processes in marine science are very versatile, the spatial scale of these patterns ranges from cm to about a 100 km. We demonstrate three mechanisms of pattern formation in the water column as well as in the sediment of the ocean. Plankton patterns result from an intricate interplay between biological growth and transport by ocean currents. Mesoscale hydrodynamic structures like vortices can act as incubators for plankton blooms, while transport barriers in the flow can lead to a segregation of species in certain spatial regions of the ocean characterized by the dominance of a particular plankton species. Spatial patterns of chemicals and microorganisms in the sediment can emanate from a Turing instability leading to inhomogeneous distributions of nutrients and bacteria. Marine aggregates agglomerations of plankton, bacteria and inorganic substances form preferential concentrations, i.e. inhomogeneous distributions in space, which in turn influence strongly aggregation and fragmentation processes and, hence, the export of carbon from the atmosphere to the bottom of the ocean.

1 Introduction

Several decades of research in nonlinear dynamics have resulted in comprehensive theories of pattern formation in systems characterized by nonlinear biological, chemical or physical interactions of system components and described by partial differential equations [1]. Already in 1952 Turing [2] derived in his seminal paper conditions leading to stationary inhomogeneous distributions of chemical substances caused by a diffusion instability. Today we know a large variety of examples of pattern formation in nature that range from fluid dynamics, e.g. Rayleigh-Bénard convection

U. Feudel (✉)
Institute of Chemistry and Biology of the Marine Environment,
Carl-von-Ossietzky University Oldenburg, Oldenburg, Germany
e-mail: ulrike.feudel@uni-oldenburg.de

S.C. Müller et al. (eds.), *Complexity and Synergetics*,
https://doi.org/10.1007/978-3-319-64334-2_15

[3, 4] and Taylor-Couette flows [5], via spatial and spatio-temporal patterns in chemical systems, e.g. the Belousov-Zhabotinsky reaction [6] to patterns in ecology [7, 8], e.g. vegetation patterns in semiarid areas [9] to name only a few. The emergence of stationary or time-dependent patterns in space are a result of an intricate interplay between nonlinear local processes and physical transport processes such as diffusion or advection processes. An appropriate description is often given in terms of partial differential equations by either reaction-diffusion systems (RDS) and reaction-advection-diffusion systems (RADS).

Interesting classes of patterns have been reported in marine science studying different processes in the open ocean, in tidal areas close to the coast as well as at the sediment-water interface. These patterns emanate from the interaction of growing marine organisms with their physical environment, either the transport of nutrients by hydrodynamic flows in the water column or by diffusion in the sediment [10]. Since the seminal paper by Abraham [11], the coupling of biological and physical processes in the open ocean has been addressed in various studies [12–14]. Several different aspects of plankton growth have been discussed, such as the emergence of sustainable plankton blooms [15], localized plankton blooms in vortices [16], the coexistence [17] and dominance [18, 19] of different plankton species. The patterns arising at the sediment-water interface include the spatial distribution of plants such as salt marsh vegetation and seagrass [20, 21], the formation of mussel beds [22] and the inhomogeneous distribution of bacteria within the sediment [23, 24].

In this contribution we discuss three different examples of pattern formation in marine science to illustrate the large variety of possible patterns. Thereby we focus on very different spatial scales in which the different patterns arise. As a first example we demonstrate in Sect. 2 how mesoscale hydrodynamic vortices of about 50–100 km in diameter influence the growth and competition of plankton species. We show that such vortices can act as incubators for plankton growth as well as possible causes for the segregation of different plankton species. In Sect. 3 we illustrate the emergence of Turing patterns on a spatial scale of about 1–10 cm in the sediment as a possible mechanism to explain the appearance of depth-dependent inhomogeneous distributions of bacteria. Finally, in Sect. 4 we discuss the role of preferential distributions of finite size inertial particles like marine aggregates for aggregation and fragmentation processes at scales below 1 cm. The latter processes are an essential part of the carbon cycle in the earth system responsible for part of the export of carbon to the deep ocean.

2 Plankton Patterns in the Ocean

One of the major requirements for the growth of phytoplankton—the plants in the world's oceans—is the availability of nutrients which is strongly dependent on the hydrodynamic flow patterns. Hence, one of the essential factors controlling the primary production are horizontal and vertical transport of nutrients. Horizontal transport is influenced by the mesoscale hydrodynamic flow structures such as vortices and jet currents, while vertical transport is often related to coastal upwelling. The latter occurs usually when wind-driven currents, in combination with the Coriolis force, produce Ekman transport, by which surface waters are driven away from the coast and are replaced by nutrient-rich deep waters. Due to this upwelling the growth of phytoplankton in these areas is strongly enhanced, subsequently leading to an increase of zooplankton and fish populations. On the one hand, horizontal stirring by mesoscale structures like vortices and jets is responsible for the redistribution of plankton and nutrients and may therefore enhance primary production [13, 15]. On the other hand, it influences competition and coexistence of different plankton species [17–19].

The basic equations which govern the dynamics of chemical substances and plankton species are reaction-advection-diffusion equations. Accordingly, three different processes are considered: (a) reaction stands for the biological growth of species and their consumption of inorganic nutrients, (b) advection accounts for the transport of plankton and chemical substances by hydrodynamic flows and (c) diffusion takes the small scale turbulent mixing into account which is usually not contained in the large scale velocity fields used to describe the flow patterns in the ocean. Food webs which represent the complex interactions between various species in an ecosystem are modeled by highly nonlinear ordinary differential equations which incorporate mathematical terms specifying growth, predator-prey interactions, competition, mutualism and mortality. Simpler versions of ecosystem dynamics restrict themselves to food chains where on each trophic level only one group of species is taken into account. For such a food chain consisting of nutrients N, phytoplankton P and zooplankton Z the equations would read:

$$\begin{aligned} \frac{\partial N}{\partial t} + v \cdot \nabla N &= F_N + D\Delta N \\ \frac{\partial P}{\partial t} + v \cdot \nabla P &= F_P + D\Delta P \\ \frac{\partial Z}{\partial t} + v \cdot \nabla P &= F_Z + D\Delta Z. \end{aligned} \tag{1}$$

A large variety of different models is used to study the biological interactions for the growth of the species F_N, F_P, F_Z. These include models exhibiting steady state and oscillatory behavior [25] as well as models particularly designed to model plankton blooms in terms of excitable systems [26]. The velocity fields v are either given by simple two-dimensional kinematic flows [15] or by ocean circulation models in more realistic setups [18, 27]. Since the diffusion term stands for eddy diffusion, all diffusion constants used for the different species obey the same value D. The Eq. (1) are solved numerically by a semi-Lagrangian algorithm [28]. This procedure allows for resolving filamental structures, which are important features of hydrodynamic flows.

In the following we demonstrate the versatile impacts of mesoscale hydrodynamic flow patterns on various aspects of plankton growth: (i) Mesoscale hydrodynamic structures can act as incubators for plankton growth and enhance productivity in the ocean. (ii) transport barriers to nutrients can facilitate the emergence of dominance

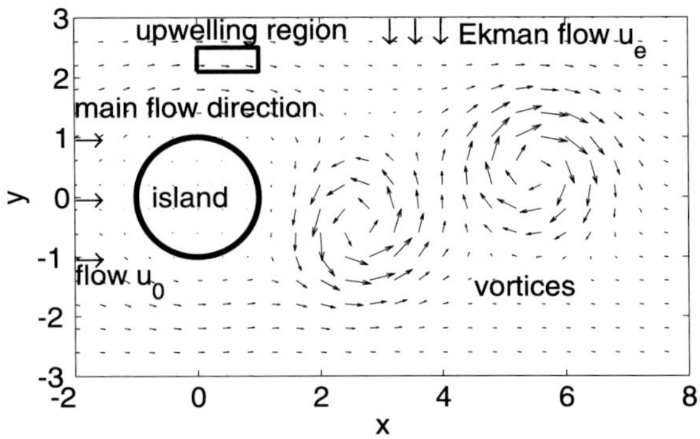

Fig. 1 Sketch of the flow in the wake of an island. The upper border of the picture denotes the coastline

patterns, i.e. a spatial separation of different regions in the ocean in which different species dominate the ecosystem. (iii) Flow patterns are possible structures to support biodiversity in the ocean leading to a coexistence of species which without flow would outcompete each other yielding a loss of diversity.

The hydrodynamic flow field acts like a forcing and is given externally. While marine oceanographers use large scale ocean models as the velocity field $v(\mathbf{x}, t)$ to investigate the growth and distribution of plankton, conceptual models which highlight the mechanisms of the interactions of flow fields and biological growth focus on kinematic models for two-dimensional flows using stream functions. The justification of employing only 2d velocity fields lies in the fact that vertical velocities in the ocean are often at least one order of magnitude smaller than the horizontal ones. Here we consider only the flow in the wake of an island which mimics the dynamics of a von Kármán vortex street (cf. Fig. 1). Such a situation can be found in the Canary current with the Canary islands located in the flow and an upwelling region at the African coast [29] or in a similar setup in the California Bight [30]. The details of the employed flow field can be found in [31]; its parametrization for the Canary island area is described in [29].

To illustrate the effects on plankton patterns we use here a simple food chain model [16] which is based on a three component model developed by Steele and Henderson [32]. An extension which focusses on the competition of different species can be found in [19]. The marine food chain model contains two different trophic levels consisting of nutrients N, phytoplankton P consuming nutrients, and zooplankton Z feeding on phytoplankton. The processes taken into account can be formulated as:

$$\frac{dN}{dt} = \text{upwelling} - \text{uptake} + \text{recycling}$$
$$\frac{dP}{dt} = \text{uptake} - \text{grazing} - \text{mortality}$$
$$\frac{dZ}{dt} = \text{grazing} - \text{mortality}.$$

Expressed in mathematical terms this reads:

$$\begin{aligned}\frac{dN}{dt} &= F_N = \Phi_N - \beta\frac{N}{k_N+N}P + \mu_N\Big((1-\gamma)\frac{\alpha\eta P^2}{\alpha+\eta P^2}Z + \mu_P P + \mu Z^2\Big)\\ \frac{dP}{dt} &= F_P = \beta\frac{N}{k_N+N}P - \frac{\alpha\eta P^2}{\alpha+\eta P^2}Z - \mu_P P \qquad (2)\\ \frac{dZ}{dt} &= F_Z = \gamma\frac{\alpha\eta P^2}{\alpha+\eta P^2}Z - \mu Z^2,\end{aligned}$$

where the upwelling term Φ_N accounts for an additional supply of nutrients due to vertical transport from deeper nutrient-rich layers in the ocean. For details of the biological modelling we refer to [16]. The parameters used in this model are taken from [27] and mimic the situation in the open ocean.

Let us now discuss the impact of mesoscale hydrodynamic structures in the flow (in our case vortices emerging in the wake of the island) on the growth of plankton and their dominance with respect to competitors. To this end we couple the biological and the hydrodynamical model and simulate the growth of plankton depending on the inflow of nutrients and plankton from the left into the observation area shown in Fig. 1. The role of the hydrodynamic flow consists to a large extent in a redistribution of nutrients leading to plankton patterns observable in satellite pictures. To enhance the effect of nutrient distribution on the emergence of these patterns, we assume the existence of an upwelling region (cf. rectangular between the island and the coastline, i.e., the upper boundary of the observation area in Fig. 1) in which a hundred times more nutrients are upwelled into the upper layer of the ocean compared to the remaining area. Due to the background flow those nutrients are transported through the observation area. As nutrients are the basis of the growth of plankton, we would expect that more nutrients would lead to more growth of plankton. This is indeed the case for a high inflow of nutrients and plankton into the observation area as reflected by the high phytoplankton abundances concentrated in the upwelling plume (Fig. 2, left panel). Under these inflow conditions no bloom within the vortex can be observed. On the contrary we obtain a strong plankton bloom within the vortex, when we assume a low input of plankton and nutrients into the observation area (Fig. 2, right panel). This bloom would be even observed when we would shut off the additional upwelling from the upwelling region. This latter dynamics is counterintuitive and shows that nutrients and plankton are entrained into the vortex in which the exchange of water with its surroundings is rather low. Due to this confinement in the vortex, plankton spends a rather long time within the vortex leading to an enhanced biological growth. This way the vortex acts like an incubator for a localized plankton bloom. Satellite pictures from various regions around the globe exhibit such localized bloom patterns (cf. http://www.nasa.org/oceancolor).

To unravel the mechanism how nutrients and plankton are transported into the vortex it is useful to employ methods from nonlinear dynamics which allow to gain insights into the details of the flow structures. Jung et al. [31] have analyzed the flow field and shown that there exists a chaotic saddle located in the space between the island and the vortex street. A chaotic saddle is an unstable invariant chaotic set possessing stable and unstable manifolds (foliations) in a similar way as fixed

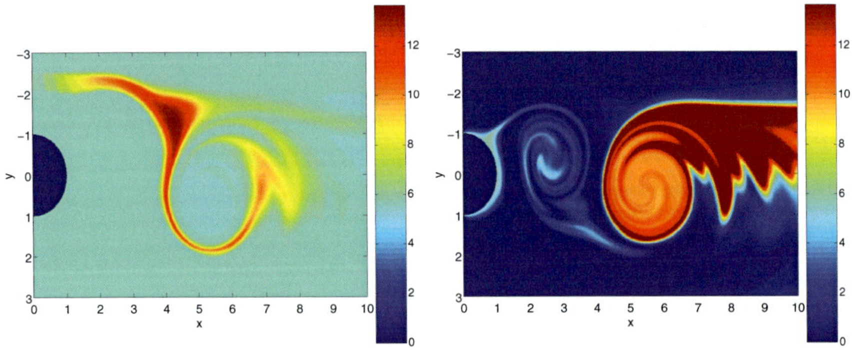

Fig. 2 Snapshot of the phytoplankton abundance with high inflow (*left panel*) and low inflow (*right panel*) from the surrounding ocean. Concentrations and abundances from low to high are denoted by colors from *blue* via *green* to *red*

points or periodic orbits of saddle type. Tracer particles in the neighborhood of the stable manifolds move towards the chaotic saddle while tracers close to the chaotic saddle leave its neighborhood along its unstable manifolds. Hence this chaotic saddle embedded in the flow can be considered as a "bridge" between the island and the vortex. Nutrients and plankton entering the observation area will slow down in the neighborhood of the island and spend some time there, where the velocity field is rather slow (the velocity goes to zero along the coast of the island). During that time plankton starts growing and are transported along the stable manifolds from the vicinity of the island towards the chaotic saddle and from there into the vortex along the unstable manifolds of the chaotic saddle. Therefore, the chaotic saddle builds the backbone of transport between the immediate neighborhood of the island and the interior of the vortex. This transport mechanism is visible in the filamental structures stretching from the island to the vortex in Fig. 2 (right panel).

To elucidate this transport mechanism further, we utilize the concept of finite time or finite size Lyapunov exponents (FTLE or FSLE). This method has been successfully applied to identify Lagrangian coherent structures like vortices and barriers to transport in general time-dependent hydrodynamic flows. The computation of finite size Lyapunov exponents (FSLE) is based on the calculation of the time needed for two tracers to reach a final prescribed distance δ_f when starting from an initial distance δ_0 [33, 34]. From the elapsed time, $\tau_\pm$, the FSLE is calculated as

$$\lambda_\pm(x, t, \delta_0, \delta_f) = \frac{1}{\tau_\pm} \log \frac{\delta_f}{\delta_0} \tag{3}$$

The positive subscripts indicate that the tracers are advected forward in time, while for the negative subscript they are advected backward in time. $\lambda_\pm$ is a scalar measure for the stretching/contracting rate in the flow given by the inverse of the separation time $\tau_\pm$. Maxima in the spatial distribution of λ_+ (λ_-), the positive (negative) or expanding (contracting) FSLE, approximate the underlying stable (unstable) manifold embedded in the flow [34]. We observe that the stable and unstable manifolds are intertwined around the vortex cores and at the island (Fig. 3) in such a way, that the "bridge" between the island and the vortex core becomes visible.

Note the white tracers in Fig. 3 symbolizing nutrients released in the upwelling region. They are entrained into one of the vortices along the stable manifolds of the chaotic saddle, but they are completely blocked to enter the interior of the second vortex. This blocking is realized by the unstable manifolds which act as transport barriers in the flow.

The organization of the flow by the time-dependent manifold structures has some other important consequences for the formation of plankton blooms when several species compete for the same nutrients. The manifolds which would redistribute the nutrients from upwelling regions, particularly the unstable manifolds which act as transport barriers, can "organize" preferred regions for certain species, i.e. the unstable manifolds segregate different species according to their nutrient needs yielding a spatial structure of dominance of different species in different regions. The emergence of such dominance structures has been shown explicitly in [19] using the same kinematic flow. Furthermore, this possible segregation of species is capable of creating niches for different species in the ocean and can be considered as a possible mechanism for maintaining a high level of biodiversity in the ocean providing local niches for the coexistence of many species [17].

Upwelling events are usually rather intermittent since they are related to the interplay of wind parallel to the coast, Ekman drift and the Coriolis force. Therefore, the wind patterns determine the strength of upwelling and subsequently the amount of nutrients provided for the growth of plankton. Upwelling events can last a very short time, say 1–2 days, but also for a week and longer. Taking into account that the growth of plankton is much slower than upwelling, one can imagine that short upwelling pulses would hardly have any effect and only longer and stronger upwelling would be capable of enhancing plankton growth. Furthermore, the time-dependent location of the manifolds would not always transport the upwelled nutrients to the regions where the plankton resides. If the upwelling happens in a inappropriate moment, then the upwelled nutrients could be blocked by transport barriers in form of unstable manifolds and do not get to the plankton community to boost a

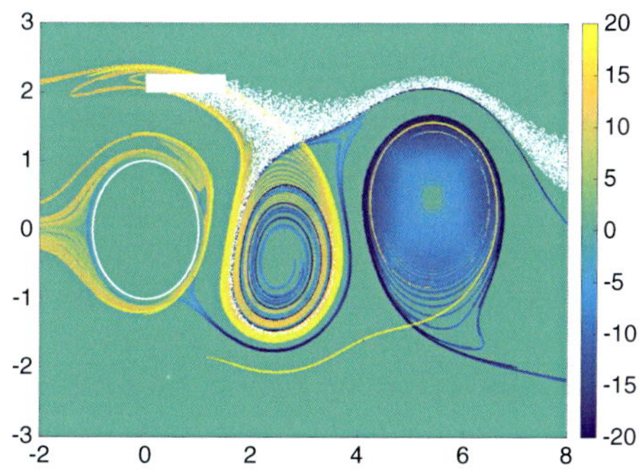

Fig. 3 Snapshot of the stable (*yellow*) and unstable (*blue*) manifolds in the wake of the island

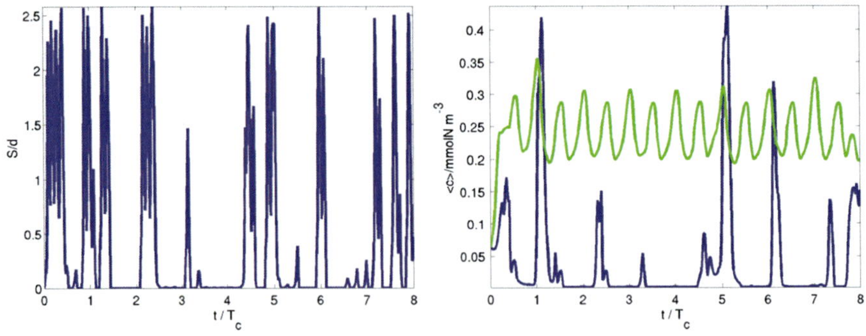

Fig. 4 Simulation of the competition of two phytoplankton species in the wake of an island with intermittent upwelling of nutrients. Though higher nutrient input happens more often, a dominance change is only occasionally visible. When there is no dominance change, the upwelling event occurred at a time instant when the flow would not transport the nutrients into the region where the phytoplankton resides

bloom or to change the dominance of plankton species. Using the simple kinematic flow outlined above one can demonstrate this behavior by replacing the constant upwelling Φ_N by an intermittent upwelling. A simple way of constructing such an intermittent upwelling is to take a chaotic time series for the upwelling strength Φ_N in Eq. (1) (cf. Fig. 4 left panel). Applying such an upwelling scenario results in dominance changes which occur only occasionally, as depicted in Fig. 4 (right panel).

In summary, these investigations show, that spatio-temporal plankton patterns in the ocean are a result of the complex interplay of biological growth processes and physical transport by hydrodynamic flows. Mesoscale structures like vortices and transport barriers play a fundamental role in the emergence of such patterns and can lead to localized plankton blooms, dominance patterns for competing species and even to a support of the coexistence of more species.

3 Bacteria Patterns in Sediments

Stationary spatial patterns in a marine environment can be observed as vegetation patterns at the sediment-water interface. Prominent examples are the patchy vegetation of seagrass and salt marsh vegetation [20, 21], the development of plankton films on sediment surfaces [10] or communities of benthic organisms within the sediment [35]. While all the patterns mentioned above are horizontally extended, other patterns related to bacterial activity can be found in the vertical, exhibiting inhomogeneous distribution patterns of different bacterial species in deeper sediment layers below the sediment-water interface [36, 37]. Recently, the formation of mussel beds has been described as a result of a pattern formation process based on equations which are similar to the Swift-Hohenberg equations [22].

Biogeochemical processes in sediments at the bottom of the ocean are characterized by a complex network of chemical reactions which are mediated by bacteria [38]. Therefore, the investigation of the various interactions between bacteria and their nutrients is an essential part of marine research. Observations of bacterial activity in different depths of the sediment reveal distinct heterogeneous distributions of different bacteria species and chemical substances [36]. Various mechanisms for the emergence of such patterns of microorganisms have been proposed, most of them are based on heterogeneous geophysical conditions as an external forcing for the growth of bacteria. However, one can show that such patterns can also emanate from a diffusion instability discovered by Turing [2] implying that no external influences are necessary to explain those patterns. They can arise solely due to nonlinear processes combined with physical transport. To demonstrate this, we consider a conceptual model describing the growth of a bacteria species feeding upon a nutrient in the sediment. We focus here on a mud flat in which advection processes can be neglected due to the very small grain size. Hence, there are only two different transport processes: diffusion as a passive and bioirrigation as an active transport process for nutrients. The latter is realized by higher organisms which move in the sediment and actively transport nutrients from the sediment-water interface into deeper layers of the sediment. A sketch of the model is represented in Fig. 5.

The mathematical formulation of this model reads as follows when we denote the bacteria by B and the nutrients by N:

$$
\begin{aligned}
\dot{B} &= \alpha\left(\gamma + (1-\gamma)\frac{B}{K+B}\right)\frac{BN}{L+N} - mB + \varepsilon + D_B \Delta B \\
\dot{N} &= -\beta\left(\gamma + (1-\gamma)\frac{B}{K+B}\right)\frac{BN}{L+N} + \sigma(N_0 - N) + D_N \Delta N
\end{aligned}
\tag{4}
$$

where α denotes the maximum growth rate and m the mortality of bacteria. The parameter γ stands for the fraction of active bacteria. The parameter ε introduces a small but constant input of bacteria from outside, i.e. neighboring regions. This term guarantees that bacteria would never die out corresponding to a realistic situation. The growth term of the bacteria depends on the nutrients N and repeats in the equation for the nutrients with an uptake rate β but with a negative sign because the gain term for bacteria is at the same time a loss term for nutrients. The active transport process by higher organisms is proportional to the difference between nutrient concentration in the water above the sediment N_0 and the concentration at the local site. Its strength, which is used as a bifurcation parameter is denoted by σ.

Turing patterns as stationary spatially inhomogeneous distributions of different chemical species arising from a diffusion instability are one of the most studied pattern formation processes in nature [1]. Though the theoretical basis has been outlined by Turing [2] already in 1952, it took almost 40 years to show their existence in experiment [39]. Turing derived mathematical conditions for the emergence of an instability of a homogeneous distribution of two interacting species in a reaction diffusion system (RDS) to find a possible mechanism for morphogenesis. He identified two conditions which have to be fulfilled for a diffusion instability to occur: (i) there should be an autocatalytic reaction and (ii) the two species have to have different mobilities represented by their diffusion constants. While the first condition can be realized for different chemical reactions, the second condition is difficult to fulfill, if the chemical experiment is taking place in a solution. Let us check those conditions in the sediment: (i) Autocatalytic reactions are equivalent to reproduction processes of bacteria, which involve a positive feedback of the individuals of the populations on their growth. Bacteria often occur in two different states, a lively state in which they actively decompose organic material, i.e. their nutrients, and a dormant state, in which they do not take part in the degradation process [40]. It has been shown that bacteria can communicate with each other based on signal molecules (e.g. cAMP) [41] and our model assumes such a communication to increase the number of active bacteria from the pool of dormant bacteria. This results in a self-reproduction process needed for a Turing instability to occur. (ii) The condition of different mobility of nutrients and bacteria is given by the fact that bacteria usually stick to the grains in the sediment matrix and move only very slowly if at all, while the nutrients diffuse freely. This way the conditions for a Turing instability can be fulfilled and we can expect to obtain Turing patterns for certain parameter ranges. Indeed, we find three different kinds of patterns (Fig. 6) when taking the strength of bioirrigation as a bifurcation parameter: (1) hot spots of bacterial activity with large abundances of bacteria inside the spots and corresponding low abundances of nutrients, (2) labyrinth patterns and (3) cold spots with low abundances of bacteria within the spots large abundances around them. Surprisingly the same kind of patterns have been observed in experiments with real bacterial communities from sediments which have been taken from tidal flats [24]. The experimenters called the cold spots honey comb patterns and the hot spots inverse honey combs. In nature, bioirrigation is not constant in space but decreases with increasing depth. If such a depth dependence is introduced one can find all the different patterns shown above in different depths resulting in a zonation of the patterns as observed in the sediment [23, 37]. The observed patterns are on a scale of a few cm.

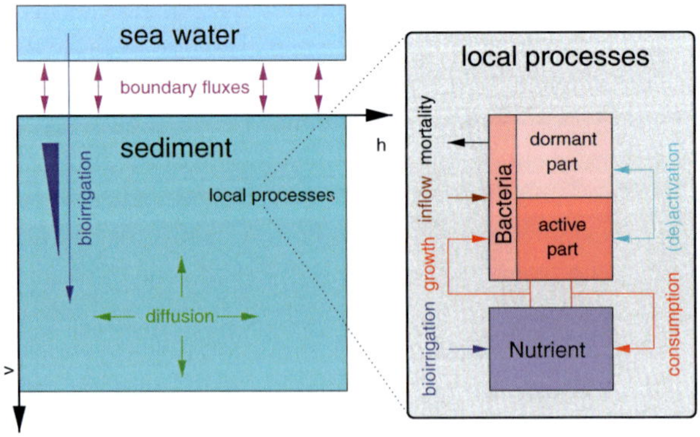

Fig. 5 Sketch of the sediment model

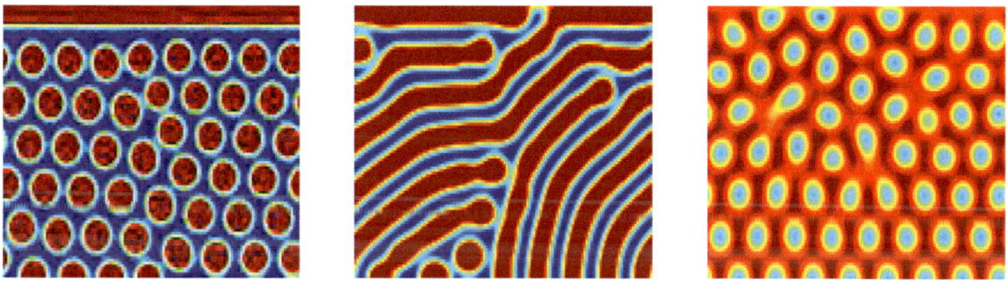

Fig. 6 Turing patterns arising in the sediment model: hot spots (*left panel*), labyrinth pattern (*middle panel*), cold spots (*right panel*)

Instead of a nutrient-consumer model as the one explained above, one can also analyze a more general setup of a predator-prey model from population dynamics. Let us consider the following model for a prey X and a predator Y developing in space and time and migrating diffusively with mobilities D_X, D_Y:

$$\dot{X} = \alpha B(1 - \frac{X}{K}) - \beta \frac{X}{\gamma + X} Y + D_X \Delta X$$
$$\dot{Y} = \delta \beta \frac{X}{\gamma + X} Y - \mu Y - \nu Y^2 + D_Y \Delta Y \qquad (5)$$

This model is a modified version of the famous Rosenzweig-MacArthur model [42] which describes the logistic growth of a prey with maximum growth rate α and carrying capacity K. The prey X is consumed by the predator Y with maximum rate β and half-saturation constant γ. The grazing term shows up again in the equation for the predator with the opposite sign since it is now a growth term, which is additionally multiplied by a conversion factor δ since only this fraction of prey is converted into biomass of the predator. Finally, the predator dies with natural mortality rate μ and due to grazing by predators on a higher trophic level with rate ν. This model exhibits a very versatile behavior; without the spatial interaction we find steady states, bistability and oscillations. Adding the migration to the model Turing patterns become possible as in the previously discussed model.

However, since the parameter space is rather large, one can ask what kind of behaviors can be obtained. To shed some light on achievable patterns we focus here only on the mathematical possibilities regardless of their biological relevance. In parameter space various bifurcations arise: several saddle-node, a Hopf and a Turing bifurcation. Furthermore, a cusp and a Turing-Hopf as higher co-dimension bifurcations occur. Finally there is a homoclinic bifurcation as a global bifurcation. Due to this complexity we expect a quite complicated behavior in space and time. To this end we perform a Monte-Carlo selection of parameters in the huge parameter space after identifying for each parameter a search interval which is biologically reasonable. As a result, we obtain lots of different space-time patterns which are interesting and have not been observed before. Figure 7 presents snapshots of some of those spatio-temporal patterns which are presumably all related to space-time chaos.

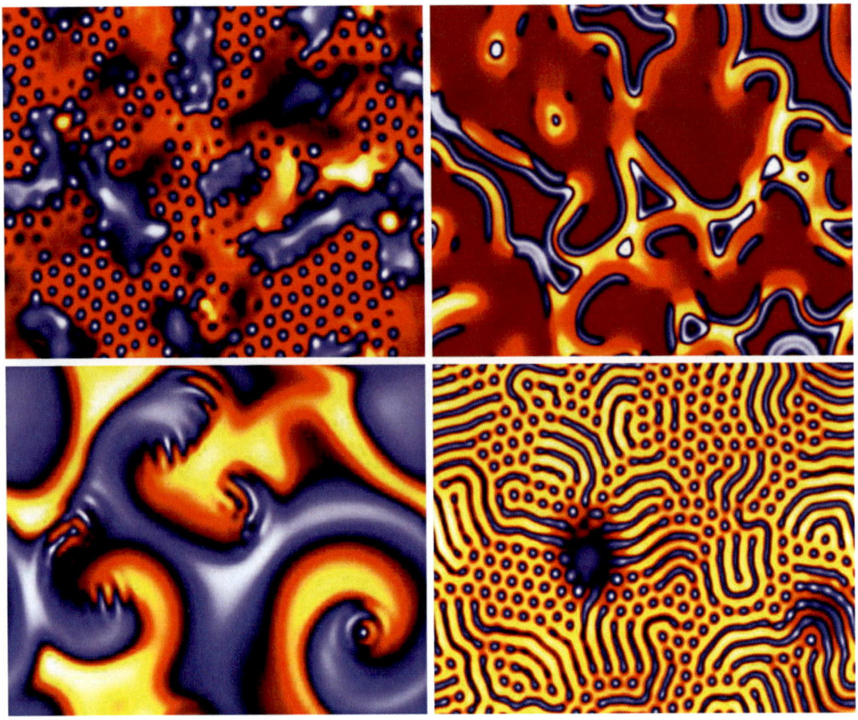

Fig. 7 Snapshots of spatio-temporal patterns for the predator prey system Eq. (5) in various regions in parameter space: mixture of hot and cold spot patterns embedded in an oscillatory background (*upper left panel*); spiral with a hot or a cold spot as the tip of the spiral (*lower left panel*); irregularly spreading domains of different concentrations of the species, where the domain walls represent wave-like structures (*upper right panel*); almost stationary spot and labyrinth patterns interspersed with homogeneous regions which are oscillating (*lower right panel*)

4 Inertial Particles in Flows

As a last example we turn to the smallest spatial scale in the ocean. Pattern formation processes on a scale below 1 cm concern the spatial distribution of marine aggregates, which we model as finite size particles. Marine aggregates are particles consisting of an inorganic kernel like sand or silt, to which plankton cells are attached. These particles also called marine snow are an integral part of the carbon cycle in the ocean [43]. Chemical substances produced by the attached plankton cells in their metabolism make the aggregates sticky, so that they can grow in size when colliding with other aggregates [44]. The larger aggregates sink down due to gravity and this process contributes to the export of carbon to the deep ocean. Additionally, those chemicals, called transparent exopolymers, attract bacteria which can again decompose the aggregates. Besides this degradation by bacteria, marine aggregates are fragmented by shear forces in the flow exceeding the binding forces of the aggregate. Aggregation by collision due to shear or differential sinking and fragmentation due to shear forces lead to dynamical sizes of the aggregates and finally to steady-state size distributions when aggregation and fragmentation events balance each other. Because of these changes in size, marine aggregates are not only passively transported by the flow but are active particles. Their sizes can be rather large when consisting mainly of plankton cells as in the open ocean or their density can be rather large compared to the fluid density when consisting mainly of inorganic material as in coastal areas. In both cases, they can not be treated as tracer particles but their finite volume needs to be taken into account.

While tracer particles are considered as point particles which follow precisely the flow and move along trajectories of fluid parcels, particles with a finite volume follow their own trajectories which are different from the flow. The simplest description of the movement of finite size particles in a hydrodynamic flow is given by the Stokes equation which considers only two forces acting on the particles: the Stokes drag and in case of the ocean and the atmosphere also gravity. Therefore the equations of motion of a particles with position **X** and velocity **V** are:

$$\begin{aligned}\frac{d\mathbf{X}}{dt} &= \mathbf{V}\\ \frac{d\mathbf{V}}{dt} &= \frac{1}{St}(\mathbf{u}-\mathbf{V}) - W\mathbf{e}\end{aligned} \tag{6}$$

where **u** denotes the velocity field of the hydrodynamic flow, St the Stokes number, W the settling velocity and **e** is a vector pointing in z-direction. The two parameters St and W depend on the properties of the particle particularly their size and density. The Stokes equations Eq. (6) are a simplification of the Maxey-Riley equations [45, 46] and are valid for the case of heavy particles whose density is much larger than the density of the fluid. Another assumption is that the particle Reynolds number is small, i.e. the interaction between particles can be neglected corresponding to a high dilution. To demonstrate the formation of spatial pattern for marine aggregates we concentrate here on this limit only, a more comprehensive description can be found in [47].

Let us now discuss aggregation and fragmentation of marine aggregates from a dynamical systems point of view. Since the aggregates are active particles as outlined above, their dynamics is more complicated than pure advection of finite-size particles. The difficulty lies in the different sizes of the aggregates being advected by the fluid. Aggregates of different sizes correspond to different parameters, namely different Stokes numbers St and settling velocities W in Eq. (6). Instead of one dynamical system one has to deal with a set of dynamical systems, each of them corresponding to a certain aggregate size. Moreover, the number of aggregates in each dynamical system is changing continuously due to aggregation and fragmentation events.

We now fix the density of the aggregates focussing on heavy particles and discuss only the impact of the size. To illustrate the patterns of marine aggregates we employ a very simple velocity field **u** describing four counter rotating vortices with a sinusoidal change in their strength (for details cf. [48]). Though this

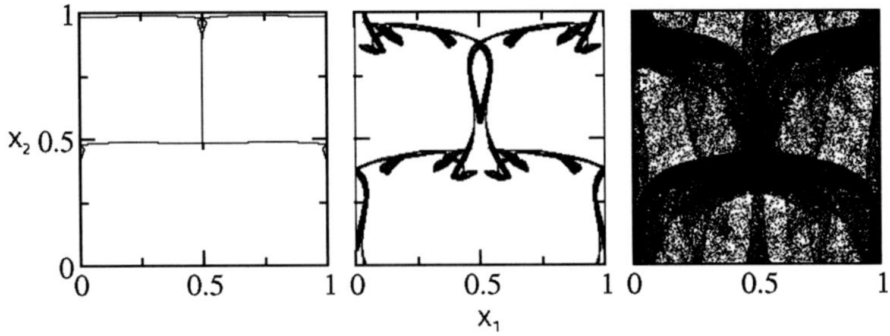

Fig. 8 Attractors for 3 different size classes in a convection flow **a** $St^{-1} = 7.0$, **b** $St^{-1} = 2.778$, **c** $St^{-1} = 2.253$

two-dimensional velocity field is time periodic and the flow is laminar, it has been shown that the motion of finite size particles in this flow can be chaotic and hence is called chaotic advection [49]. The most striking consequence of chaotic advection is, that the motion of the particles converges to attractors in configuration space. The shape of these attractors depends strongly on the size and the density of the aggregates (cf. Fig. 8) and ranges from fixed point attractors via limit cycles and quasiperiodic motion on tori to chaotic attractors [49, 50]. Some of them are rather localized in space while others are distributed over most of the configuration space. Moreover, the whole dynamics for all sizes consists of an overlay of all different attractors for the various sizes present in the system. Additionally, since aggregation and fragmentation events happen usually on much smaller time scales than convergence to the attractor, the overall dynamics is in general transient and only a blurred structure of the attractors will be observable.

This behavior has some important consequences for the dynamics: Aggregates of different size are located in different regions of the configuration space in the long-term limit. For this reason their collision rate is larger in highly populated regions than in regions with only a small number of particles. This fact highlights the importance of the distribution of the aggregates in space which is usually neglected in oceanographic approaches based on Smoluchowski equations [51]. The same applies to fragmentation processes, where the spatial distribution of shear determines in which areas of the configuration space the fragmentation rate is largest. In the classical oceanographic approaches this spatial dependence of fragmentation is again not taken into consideration. This simple example already demonstrates that the spatial distribution of aggregates plays an important role for aggregation and fragmentation processes in hydrodynamic flows and needs to be taken into account when computing size distributions in the open ocean or in tidal areas close to the coast.

Hydrodynamic flows in nature are of course not properly described by chaotic advection but by real turbulent flows. All results outlined above are still valid for such flows, but of course, due to the non-steady behavior of the flow, the particles do not converge to attractors. However, for a considerable range of Stokes numbers they are still concentrated in certain regions of space leading to patterns of preferential concentration [52, 53]. These patterns move with the flow and have the same impact on aggregation and fragmentation rates as discussed above [54]. For real ocean flows, the spatial scales which are usually resolved in hydrodynamic models are much larger than the ones considered above. For this reason either size-class based [51] or distribution-based [55, 56] models are mostly used to calculate size distributions of marine aggregates, which do not take the spatially inhomogeneous distribution of the aggregates into account. Nonetheless, the framework of considering individual particles outlined here can help to reformulate collision kernels and fragmentation rates of marine aggregates depending on their spatial distribution to be finally incorporated into Smoluchowski-like approaches to analyze size distributions of those aggregates. Despite the fact that such approaches are still not fully developed, this simple example of chaotic advection emphasizes the essential role of spatial patterns formed by marine aggregates in the biogeochemical cycles of the ocean.

5 Future Challenges to Analyze Marine Systems

We have demonstrated the importance of pattern formation mechanisms to explain a variety of observations in marine science. While in terrestrial ecology as well as in general population dynamics, the emergence of spatial patterns has been investigated for a long time [8, 57], such studies are rather rare in marine science. One reason could be that in the past the focus was on Turing patterns or, more general, on reaction-diffusion systems. The nutrient-bacteria patterns in the sediment discussed above, follows this way of analysis too. But in marine systems hydrodynamic flow patterns are, at least in the water column, more important than diffusion processes. However, they are difficult to handle. The easiest way to start these studies is to employ simple velocity fields which are periodic but still possess certain features of real ocean flows. Such kinematic flows have been used here to show, how plankton blooms are influenced by structures embedded in the hydrodynamic flow. We have shown, that mesoscale vortices can act as incubators for plankton blooms; transport barriers can lead to dominance patterns of different species or can even create niches for the coexistence of different species. Additionally, we have discussed the role of spatial patterns for aggregation and fragmentation of marine aggregates. For the latter again a simple kinematic flow has been used to illustrate the emerging patterns which have a large impact on the collision rates of finite size aggregates in the ocean. However, real ocean flows are not periodic but exhibit a complicated time dependence leading to a more complicated setup. Only rather recently new techniques in nonlinear dynamics have been established to tackle arbitrary time dependencies. These developments have paved the road to investigate patterns arising in reaction-advection-diffusion systems driven by real ocean flows. Concepts as finite time or finite size Lyapunov exponents [33, 34], the identification of Lagrangian coherent structures [58] open new ways to find patterns in systems with arbitrary time-dependent flows. Additionally some progress has been made in extending conventional bifurcation theory to parameters following certain trends [59]. So far those new techniques can be employed to study the impact of climate change on marine systems, their extension to bifurcations leading to pattern formation is a future challenge.

Acknowledgements I would like to thank all PhD students and postdocs in my group, in particular David Bastine, Martin Baurmann, Ksenia Guseva, Joeran Maerz, Matthias Sandulescu and Jens Zahnow, who have contributed to various aspects of this work and who provided figures to this review. Furthermore, I wish to thank Emilio Hernández-García, Cristobal López and Tamás Tél for their fruitful long-term collaboration in those topics.

References

1. M. Cross, P.C. Hohenberg, Pattern formation outside of equilibrium. Rev. Mod. Phys. **65**, 851–1112 (1993)
2. A.M. Turing, The chemical basis of morphogenesis. Philos. T. R. Soc. Lon. B **237 B.641**, 37–72 (1952)
3. W. Pesch, Complex spatiotemporal convection patterns. Chaos **6**, 348–357 (1996)
4. E. Bodenschatz, W. Pesch, G. Ahlers, Recent developments in Rayleigh-Benard convection. Ann. Rev. Fluid Mech. **32**, 709–778 (2000)
5. C.D. Andereck, S.S. Liu, H.L. Swinney, Flow regimes in a circular Couette system with independently rotating cylinders. J. Fluid Mech. **164**, 155–183 (1986)
6. I.R. Epstein, K. Showalter, Nonlinear chemical dynamics: oscillations, patterns, and chaos. J. Phys. Chem. **100**, 13132–13147 (1996)
7. M. Rietkerk, J. van de Koppel, Regular pattern formation in real ecosystems. Trends Ecol. Evol. **23**, 169–175 (2008)
8. E. Meron, *Nonlinear Physics of Ecosystems* (CRC Press, Boca Raton, 2015)
9. M. Rietkerk, M.C. Boerlijst, F. van Langevelde, R. HilleRisLambers, J. van de Koppel, L. Kumar, H.H.T. Prins, A.M. de Roos, Self-organization of vegetation in arid ecosystems. Am. Nat. **160**, 524–530 (2002)
10. J. Weerman, J. van Belzen, M. Rietkerk, S. Temmerman, S. Kefi, P.M.J. Herman, J. van de Koppel, Changes in diatom patch-size distribution and degradation in a spatially self-ortganized intertidal mudflat ecosystem. Ecology **93**, 608–618 (2012)
11. E. Abraham, The generation of plankton patchiness by turbulant stirring. Nature **391**, 577–580 (1998)
12. C. López, Z. Neufeld, E. Hernández-García, P. Haynes, Chaotic advection of reacting substances: Plankton dynamics on a meandering jet. Phys. Chem. Earth, B **26**, 313–317 (2001)
13. A. Martin, K. Richards, A. Bracco, A. Provenzale, Patchy productivity in the open ocean. Global Biogeochem. Cy. **16**, 1025 (2002)
14. T. Tél, A. de Moura, C. Grebogi, G. Károlyi, Chemical and biological activity in open flows: a dynamical systems approach. Phys. Rep. **413**, 91–196 (2005)
15. E. Hernández-García, C. López, Sustained plankton blooms under open chaotic flows. Ecol. Complex. **1**, 253–259 (2004)
16. M. Sandulescu, E. Hernández-García, C. López, U. Feudel, Kinematic studies of transport across an island wake, with application to the Canary islands. Nonlin. Processes Geophys. **14**, 443–454 (2007)
17. I. Scheuring, G. Károlyi, Z. Toroczkai, T. Tél, Á. Péntek, Competing populations in flows with chaotic mixing. Theor. Pop. Biol. **63**, 77–90 (2003)
18. A. Bracco, A. Provenzale, I. Scheuring, Mesoscale vortices and the paradox of the plankton. Proc. Roy. Soc. Lond. B **267**, 1795–1800 (2000)

19. D. Bastine, U. Feudel, Inhomogeneous dominance patterns of competing phytoplankton groups in the wake of an island. Nonlin. Processes Geophys. **17**, 715–731 (2010)
20. J. van de Koppel, D. van der Wal, J.P. Bakker, P.M.J. Herman, Self-organization and vegetation collapse in salt marsh ecosystems. Am. Nat. **165**, E1–E12 (2005)
21. T. van der Heide, T.J. Bouma, E.H. van Nes, J. van de Koppel, M. Scheffer, J.G.M. Roelofs, M.M. van Katwijk, A.J.P. Smolders, Spatial self-organized patterning in seagrasses along a depth gradient of an intertidal ecosystem. Ecology **91**, 362–369 (2010)
22. J. van de Koppel, M. Rietkerk, N. Dankers, P.M.J. Herman, Scale-dependent feedback and regular spatial patterns in young mussel beds. Am. Nat. **165**, E66–E77 (2005)
23. M. Baurmann, W. Ebenhöh, U. Feudel, Turing instabilities and pattern formation in a benthic nutrient-microorganism system. Math. Bio. Sci. Eng. **1**, 111–130 (2004)
24. R. Thar, M. Kühl, Complex pattern formation of a marine gradient bacteria explained by a simple computer model. FEMS Microbiol. Lett. **246**, 75–79 (2005)
25. A.M. Edwards, J. Brindley, Oscillatory behavior in a three component plankton population model. Dyn. Stab. Sys. **11**, 347–370 (1996)
26. J.E. Truscott, J. Brindley, Ocean plankton populations as excitable media. Bull. Math. Biol. **56**, 981–998 (1994)
27. C. Pasquero, A. Bracco, A. Provenzale, Coherent vortices, Lagrangian particles and the marine ecosystem, in *Shallow Flows*, ed. by W. Uijttewaal, G. Jirka (Balkema Publishers, Leiden, 2004), pp. 399–412
28. M. Sandulescu, C. López, E. Hernández-García, U. Feudel, Biological activity in the wake of an island close to a coastal upwelling. Ecol. Complex. **5**, 228–237 (2008)
29. M. Sandulescu, E. Hernández-García, C. López, U. Feudel, Kinematic studies of transport across an island wake, with application to the Canary islands. Tellus A **58**, 605–615 (2006)
30. C. Dong, E.Y. Idica, J.C. McWilliams, Circulation and multiple-scale variability in the Southern California Bight. Prog. Oceanogr. **82**, 168–190 (2009)
31. C. Jung, T. Tél, E. Ziemniak, Application of scattering chaos to particle transport in a hydrodynamical flow. Chaos **3**, 555–568 (1993)
32. J. Steele, E. Henderson, The role of predation in plankton models. J. Plankton Res. **14**, 157–172 (1992)
33. V. Artale, G. Boffetta, M. Celani, M. Cencini, A. Vulpiani, Dispersion of passive tracers in closed basins: beyond the diffusion coefficient. Phys. Fluids **9**, 3162–3171 (1997)
34. F. d'Ovidio, V. Fernández, E. Hernández-Garcia, C. López, Mixing structures in the mediterranean sea from finitesize Lyapunov exponents. Geophys. Res. Lett. **31**, L17203 (2004)
35. P.M.J. Herman, J.J. Middelburg, C.H.R. Heip, Benthic community structure and sediment processes on an intertidal flat: results from the ECOFLAT project. Cont. Shelf Res. **21**, 2055–2071 (2001)
36. S. Madani, F.J.R. Meysman, J.J. Middelburg, Biogeochemical modeling of sediments from the Santa Barbara Basin (California), in *BioGeoChemistry of Tidal Flats*, vol. 12, ed. by J. Rullkötter (Forschungszentrum Terramare, Wilhelmshaven, 2003), pp. 91–93
37. K. Bosselmann, M.E. Böttcher, M. Billerbeck, E. Walpersdorf, A. Theune, D. de Beer, M. Hüttel, H.-J. Brumsack, B. Bb Jørgensen, Iron-sulfur-mangaanese dynamics in intertidal surface sediments in the North Sea, in *BioGeoChemistry of Tidal Flats*, ed. by J. Rullkötter (Forschungszentrum Terramare, Wilhelmshaven, 2003), pp. 32–35
38. P. van Cappellen, J.-F. Gaillard, Biogeochemical dynamics in aquatic sediments, in *Reactive Transport in Porous Media*, ed. by P.C. Lichter, C.I. Steefel, E.H. Oelkers, Series in Reviews in Mineralogy, vol. 34, (Mineralogical Society of, America, 1996), pp. 335–376
39. V. Castets, E. Dulos, J. Boissonade, P. De Kepper, Experimental evidence of a sustained standing Turing-type nonequilibrium chemical pattern. Phys. Rev. Lett. **64**, 2953–2956 (1990)
40. L.H. Stevenson, A case for bacterial dormancy in aquatic systems. Microbial Ecol. **4**, 127–133 (1978)
41. A. Bruns, H. Cypionka, J. Overmann, Cyclic AMP and acyl homoserine lactones increase the cultivatioon efficiency of heterotrophic bacteria from the Central Baltic Sea. Appl. Environ. Microb. **68**, 3978–3987 (2002)

42. J.D. Murray, *Mathematical Biology I: An Introduction* (Springer, Berlin, 2002)
43. A. Alldredge, M.W. Silver, Characteristics, dynamics and significance of marine snow. Prog. Oceanogr. **20**, 41–82 (1988)
44. A. Alldredge, U. Passow, B.E. Logan, The abundance and significance of a class of large transparent organic particles in the ocean. Deep Sea Res. **I**(40), 1131–1140 (1993)
45. M.R. Maxey, J.J. Riley, Equation of motion for a small rigid sphere in a nonuniform flow. Phys. Fluids **26**, 883–889 (1983)
46. T.R. Auton, J.C. Hunt, M. Prud'homme, The force exerted on a body in inviscid unsteady non-uniform rotational flow. J. Fluids Mech. **197**, 241 (1988)
47. K. Guseva, U. Feudel, T. Tél, Influence of the history force on inertial particle advection: gravitational effects and horizontal diffusion. Phys. Rev. E **88**, 042909 (2013)
48. J.C. Zahnow, R.D. Vilela, U. Feudel, T. Tél, Coagulation and fragmentation dynamics of inertial particles. Phys. Rev. E **80**, 026311 (2009)
49. T. Nishikawa, Z. Toroczkai, C. Grebogi, T. Tél, Finite-size effects on active chaotic advection. Phys. Rev. E **65**, 026216 (2002)
50. J.C. Zahnow, R.D. Vilela, U. Feudel, T. Tél, Aggregation and fragmentation dynamics of inertial particles in chaotic flows. Phys. Rev. E **77**, 055301(R) (2008)
51. G.A. Jackson, A model of the formation of marine algal flocs by physical coagulation processes. Deep Sea Res. **37**, 1197–1211 (1990)
52. J. Bec, A. Celani, M. Cencini, S. Musacchio, Clustering and collisions of heavy particles in random smooth flows. Phys. Fluids **17**, 073301 (2005)
53. J. Ruiz, D. Macías, F. Peters, Turbulence increases the average settling velocity of phytoplankton cells. Proc. Nat. Acad. Sci. **101**, 17720–17724 (2004)
54. J.C. Zahnow, U. Feudel, What determines size distributions of heavy drops in a synthetic turbulent flow? Nonlin. Processes Geophys. **16**, 677–690 (2009)
55. I. Kriest, G.T. Evans, Representing phytoplankton aggregates in biogeochemical models. Deep Sea Res. **I**(46), 1841–1859 (1999)
56. J. Maerz, K. Wirtz, Resolving physically and biologically driven suspended particulate matter dynamics in a tidal basin with a distribution-based model. Estuar. Coast. Shelf S. **84**, 128–138 (2009)
57. J.D. Murray, *Mathematical Biology II: Spatial Modes and Biomedical Applications* (Springer, Berlin, 2003)
58. G. Haller, Lagrangian coherent structures. Annu. Rev. Fluid Mech. **47**, 137–162 (2015)
59. P. Ashwin, S. Wieczorek, R. Vitolo, P. Cox, Tipping points in open systems: bifurcation, noise-induced and rate-dependent examples in the climate system. Phil. Trans. R. Soc. A **370**, 1166–1184 (2012)

Nonlinear Behavior of a Self-Propelled Droplet Coupled with a Chemical Oscillatory Reaction

Nobuhiko J. Suematsu, Yoshihito Mori, Takashi Amemiya and Satoshi Nakata

Abstract Several self-propelled objects have been investigated and used to add functionalities mimicking biological systems. One promising approach is the introduction of nonlinear chemical reactions such as the Belousov-Zhabotinsky (BZ) reaction. In this work we placed an aqueous droplet of the BZ solution into an oil phase composed of monoolein and squalane. The BZ droplet moved spontaneously, and its speed oscillated periodically in synchronization with the redox state of the aqueous solution. This finding and measurements of the interfacial tension between water and squalane reveal that the oscillatory motion of the BZ droplet originated from the oscillation of the Br_2-concentration. This system has the potential to reflect the characteristics of nonlinear chemical reactions inside the aqueous droplet: not only periodical oscillation but also bifurcations, hysteresis, and responsiveness to the environment.

N.J. Suematsu (✉)
Graduate School of Advanced Mathematical Sciences, Meiji University,
4-21-1 Nakano, Tokyo 164-8525, Japan
e-mail: suematsu@meiji.ac.jp

N.J. Suematsu
Meiji Institute of Advanced Study of Mathematical Sciences,
Meiji University, 4-21-1 Nakano, Tokyo 164-8525, Japan

Y. Mori
Graduate School of Humanities and Sciences, Ochanomizu University,
2-1-1 Ohtsuka, Bunkyo-Ku, Tokyo 112-8610, Japan

T. Amemiya
Graduate School of Environment and Information Sciences, Yokohama National University,
79-7 Tokiwadai, Hodogaya-Ku, Yokohama 240-8501, Japan

S. Nakata
Graduate School of Science, Hiroshima University, 1-3-1 Kagamiyama,
Higashi-Hiroshima 739-8526, Japan

S.C. Müller et al. (eds.), *Complexity and Synergetics*,
https://doi.org/10.1007/978-3-319-64334-2_16

1 Introduction

Movement or locomotion is an important characteristic of living systems. Cells migrate [1], bacteria and microorganisms swim [2], one species of insects moves on water [3], and almost all animals walk. Recently, many small swimmers have been developed in non-living systems as biomimetic motors or locomotors [4]; examples include a solid disk or a droplet spontaneously moving on water [5–7], an oil droplet swimming within an aqueous phase containing a surfactant [8], and an oil droplet moving on a glass plate in an aqueous phase containing a surfactant [9]. Those moving objects are called "self-propelled objects" . Although their detailed mechanisms of movement are different from those of living matter, self-propelled objects are a promising system for understanding biological motion and collective behavior [10].

Self-propelled objects with many functionalities, e.g., responsiveness to their physical or chemical environment [11] and rhythmic motion [12], have been developed. However, these objects are much simpler than living matter, which shows a variety of functional movements [2]. One of the next steps in the development of self-propelled objects is to recreate the properties of nonlinear chemistry [13]. Nonlinear chemical systems with various characteristic behaviors have been realized, e.g., oscillation, bifurcation, hysteresis, pattern formation, and nonlinear response to the environment [14–16]. Thus, an object whose self-propelled motion reflects the characteristic properties of nonlinear chemistry would have the potential to exhibit a variety of functional movements.

Several constructions of a self-propelled object coupled to the Belousov-Zhabotinsky (BZ) reaction, a well-known nonlinear chemical reaction, have been reported. The first idea was suggested by Steinbock and Müller [17] and was demonstrated by Kitahata et al. [18]. After that, several types of self-propelled object coupled to BZ reaction have been constructed [19–21]. Kitahata et al. placed a BZ droplet on/in an oleic acid solution without any surfactant and observed reciprocal motion [19]. In this case, the motion is driven by the spatial heterogeneity of the redox state in the BZ droplet, i.e., a chemical wave. Other examples include self-propelled motion induced by a difference in the contact angle between reduced and oxidized states. Szymanski et al. reported swelling and constriction cycles synchronized with BZ oscillation [20]. Furthermore, Nakata et al. showed the sliding of a BZ gel on the BZ solution without a catalyst which is triggered by a chemical wave generated inside the gel [21]. Although these self-propelled objects are well coupled to the BZ reaction, reciprocation [19, 21] and stretching vibration [20] are hard to carry the object over long distance.

Here, we propose a strategy for creating a novel self-propelled object coupled with a nonlinear chemical reaction. We focus on a droplet system driven by an interfacial chemical reaction [22]. In this system, an interfacial tension gradient is generated by the fluctuation of the interfacial chemical reaction, which induces an inhomogeneous distribution of the surfactant on the interface (Fig. 1). If Marangoni flow enhances the inhomogeneity, i.e., if there is positive feedback process, the convective flow is stably generated inside the droplet, providing the droplet with a driving force [22]. In this system, self-propelled motion strongly depends on the rate of the interfacial chemical reaction. Therefore, if a nonlinear chemical reaction controls the reactant concentration, the droplet motion might reflect the properties of the nonlinear chemical reaction. To explore this idea, we focus on the moving droplet within an oil phase reported by Thutupalli et al. in 2011 [23]. A droplet of an aqueous bromine (Br_2) solution is driven within a squalane oil phase including monoolein (MO) as a surfactant. Thutupalli et al. suggested that this self-propelled motion is driven by the bromination of MO on the interface. Because Br_2 is an intermediate of the BZ reaction, this system is suitable for realizing our strategy. In a recent work by Herminghaus et al., they briefly demonstrate an oscillatory motion of a BZ droplet synchronized with the redox state [24]. However, these researchers do not discuss the coupling mechanism in detail. In a recent work, we controlled the chemical conditions and identified the appropriate conditions for achieving oscillatory motion [25, 26]. In that report, we discussed the coupling mechanism and concluded that the BZ reaction controlled the concentration of the reactant (Br_2), as mentioned above.

In this report, we first investigate the effect of the MO concentration on the droplet motion and discuss the mechanism of self-propelled motion based on oil/water interfacial tension measurements. We then report the characteristic behavior of a self-propelled BZ droplet, which reflects the properties of nonlinear chemical reactions.

2 Experimental Setup

Monoolein (MO) and squalane were purchased from Wako Pure Chemical (Kyoto, Japan) and used without further purification. Water was distilled using a Milli-Q filtering system. To prepare brominated MO (Br_2MO) solution, MO solution was exposed to Br_2 gas for over 12 h. The solution became dark brown in color.

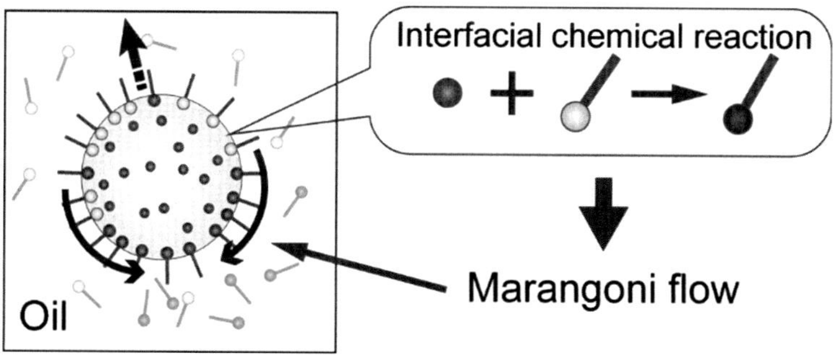

Fig. 1 Schematic illustration of the mechanism for droplet motion induced by an interfacial chemical reaction of the surfactant

An aqueous droplet (0.2 μl) was placed in squalane solutions of various MO concentrations (0.1—100 mM) using a micropipette. The motion of the droplet was observed every 1 s for 10 min using a microscope (SMZ 745-T or TS100, Nikon Inc., Japan), and at least three replicate experiments were conducted for each MO concentration. The videos were analyzed using image analysis software (ImageJ, National Institute for Health, USA) on a personal computer.

The interfacial tension between squalane and water was measured using the standard Wilhelmy method (Dyne Master DY-300; Kyowa Interface Science Co. Ltd., Japan) at room temperature (24 ± 1 °C). The concentration of the MO in the squalene was between 0.001 and 50 mM. The surface pressure of MO and Br_2MO (π-A isotherm) was measured using a surface pressure meter (Kyowa Interface Science Co. Ltd., Japan) at 25 ± 1 °C. To prepare the monolayer of MO or Br_2MO, a chloroform solution of the surfactant was dropped on a water phase, where the total amount was 1.62×10^{-8} mol. After the chloroform evaporated, the surface area was decreased from 210 to 40 cm^2 at a rate of 0.315 $cm^2\ s^{-1}$, and the sampling rate was 2.0 Hz.

3 Dependence of the Speed of Self-propelled Motion on Surfactant Concentration

The aqueous droplet in the squalane solution of MO moved spontaneously. The droplet settled downwards due to its weight and moved in two dimensions on the glass surface. In a previous work, the droplet motion was attributed to the chemical reaction of MO [23]. Because the MO concentration is therefore a key determinant of the droplet motion, we investigated the influence of the MO concentration on the speed of self-propelled motion. For simplicity, we used the following aqueous solution, not the BZ solution: 0.40 M BrO_3^-, 1.80 M H_2SO_4, and 4.0 mM Fe $(phen)_3^{2+}$. The average speed of the self-propelled motion increased as the MO concentration increased up to 10 mM and decreased thereafter (Fig. 2a).

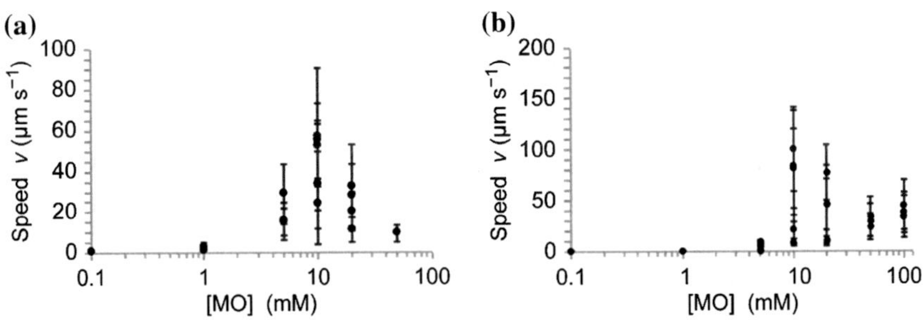

Fig. 2 Average speed of the self-propelled motion of an aqueous droplet (0.2 μl) as a function of MO concentration. The aqueous droplets contained **a** 0.40 M BrO_3^-, 1.80 M H_2SO_4, and 4.0 mM $Fe(phen)_3^{2+}$ or **b** 0.40 M BrO_3^-, 1.20 M H_2SO_4, and 4.0 mM $Fe(phen)_3^{2+}$. The error bars indicate the standard deviation for 10 min data

These results are quite different from those reported by Thutupalli et al., in which the speed increased monotonically and then plateaued as the MO concentration increased [23]. This discrepancy might be due to differences in the droplet size and composition. In Thutupalli et al.'s study, the droplets were approximately 100 μm in size and composed of 28 mM BrO_3^-, 50 mM H_2SO_4, 400 mM $CH_2(COOH)_2$, and 2.7 mM $Fe(phen)_3^{2+}$, which was the BZ solution. In contrast, the droplets in our experiment were approximately 1 mm in diameter, and the BZ solution was not used. The composition affects the droplet motion. Indeed, when the H_2SO_4 concentration was varied, the relationship between the average speed and MO concentration changed (Fig. 2b).

4 Driving Force of Droplet Motion

The driving force of the self-propelled motion is believed to originate from the difference in the oil/water interfacial tensions of the MO and Br_2MO surfactants [23]. However, no detailed measurements of these interfacial tensions have been reported. Herein, the interfacial tension between squalane and pure water was measured with varying concentrations of MO or Br_2MO. For both surfactants, the interfacial tension decreased monotonically with increasing surfactant concentration (Fig. 3a). Over the concentration range studied, the interfacial tension of the MO solution was lower than that of the Br_2MO solution.

Based on the interfacial tension measurements, the surface concentration was calculated using the Gibbs adsorption isotherm as follows [27, 28]:

$$\Gamma = -\frac{C}{RT}\left(\frac{\partial \gamma}{\partial c}\right), \tag{1}$$

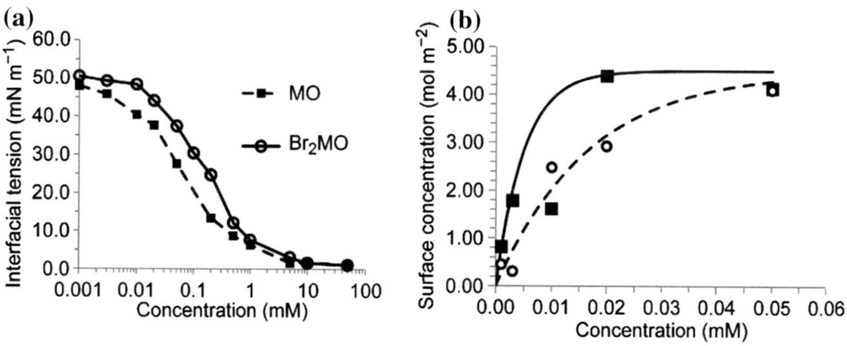

Fig. 3 **a** Squalane/water interfacial tension as a function of surfactant concentration (*filled square*: MO, *open circle*: Br_2MO). **b** Surface concentration of MO and Br_2MO as a function of its bulk concentration. The solid and *broken lines* are the *fitting curves* for MO and Br_2MO, respectively

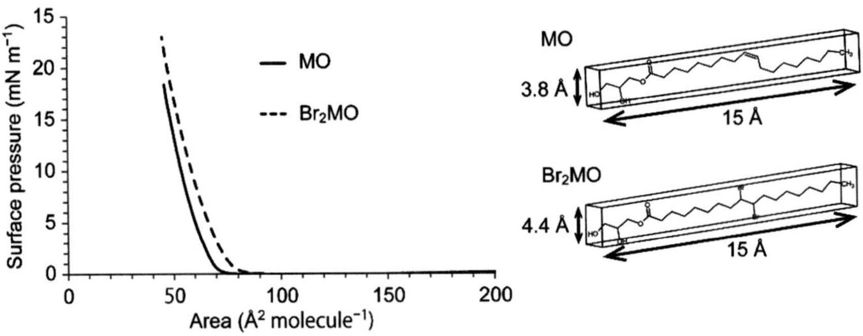

Fig. 4 *π-A curve* for MO and Br_2MO. The total mass of the surfactant on the water chamber was 1.62×10^{-8} mol. The barrier moved at a rate of 0.315 $cm^2\ s^{-1}$

where c (mol m^{-3}) is the bulk concentration of the surfactant, γ (N m^{-1}) is the interfacial tension, Γ (mol m^{-2}) is the surface concentration of the surfactant, R (J mol^{-1} K^{-1}) is the gas constant, and T (K^{-1}) is the absolute temperature. The surface concentrations of MO and Br_2MO increased with increasing bulk concentration and then plateaued (Fig. 3b). The interfacial tension of Br_2MO was higher than that of MO because the surface concentration of MO higher than that of Br_2MO, which was caused by the adsorption of less Br_2MO to the interface.

The adsorption can be roughly estimated based on the size of the molecules. To estimate the molecular size, we measured the surface pressure of MO and Br_2MO. The surface pressure began to increase at 70 $Å^2$ molecule^{-1} for MO and 80 $Å^2$ molecule^{-1} for Br_2MO (Fig. 4). Therefore, the volume of the Br_2MO was estimated to be larger than that of MO, with a ratio of 0.87. To check the validity of these parameters, we utilized a molecular model and calculated its size as a 15-Å-wide rectangle that is 3.8 Å and 4.4 Å in height for MO and Br_2MO, respectively. The areas were estimated to be 57 $Å^2$ molecule^{-1} for MO and 66 $Å^2$ molecule^{-1} for Br_2MO, making the ratio of areas of Br_2MO and MO is 0.87. Thus, although the experimental results were overestimations for both molecules, the ratio between MO and Br_2MO was close to the calculated values.

The difference in the oil/water interfacial tensions of MO and Br_2MO is the origin of the Marangoni flow on the interface of the aqueous droplet. Although dynamic interfacial tension is necessary to estimate the driving force of the droplet motion [23], the equilibrium interfacial tension (shown in Fig. 3a) is also an important indicator. The speed of the droplet motion is proportional to the Marangoni coefficient $M = \partial\gamma/\partial\phi$, where ϕ is the fraction of Br_2MO on the interface [23]. Assuming that the interfacial tension is proportional to ϕ, M becomes constant for the whole parameter region and is determined by the difference in the interfacial tensions of MO and Br_2MO ($\Delta\gamma$). Figure 3a shows that M increases with increasing bulk concentration up to 0.1 mM and then decreases. The value of M is still a coefficient used to determine grad(γ), the driving force for Marangoni flow, which must be considered when estimating the driving force for the droplet motion.

The value of grad(ϕ) is determined by the interfacial chemical reaction of MO, adsorption-desorption of the surfactants, and diffusion.

Our experimental results (Fig. 2) reveal that the optimum concentration of MO is 10 mM. Figure 3a indicates that the value of M is the highest at a bulk concentration of 0.1 mM, in conflict with the result in Fig. 2. This mismatch might be due to the different values of grad(ϕ). Applying a simplification, ϕ becomes almost 1 for the whole interface of the droplet, and the value of grad(ϕ) is low in the case of a low bulk concentration because of the slow adsorption process. This is why the droplet motion was slow in the case of a low bulk concentration of MO. On the other hand, the value of M becomes low and ϕ approaches 0 due to the fast adsorption-desorption process at high bulk concentrations, making the motion slow. Thus, the local maximum observed for the speed may result from the competition between M and grad(ϕ).

5 Oscillatory Motion of a BZ Droplet

In the aqueous droplet of the BZ reaction, the redox state spontaneously oscillated over time with a period that depends on the droplet size. Smaller droplets had a shorter period, and no oscillation was observed for droplets smaller than almost 100 μm in diameter for the aqueous droplet composed of 0.40 M BrO_3^-, 0.60 M H_2SO_4, 0.20 M $CH_2(COOH)_2$, 0.03 M Br^-, and 2.0 mM $Fe(phen)_3^{2+}$. In the aqueous droplet with a diameter of approximately 1 mm (0.2 μl), the redox state oscillated spontaneously, and periodic color changes were observed (Fig. 5a). In that case, the droplet speed also oscillated and synchronized in phase with the color change [25, 26]. Specifically, when the droplet was blue, the droplet motion was fast. The speed and blue color reached a local maximum at the same time. Thus, the speed of the droplet motion was affected by the chemical environment in the droplet.

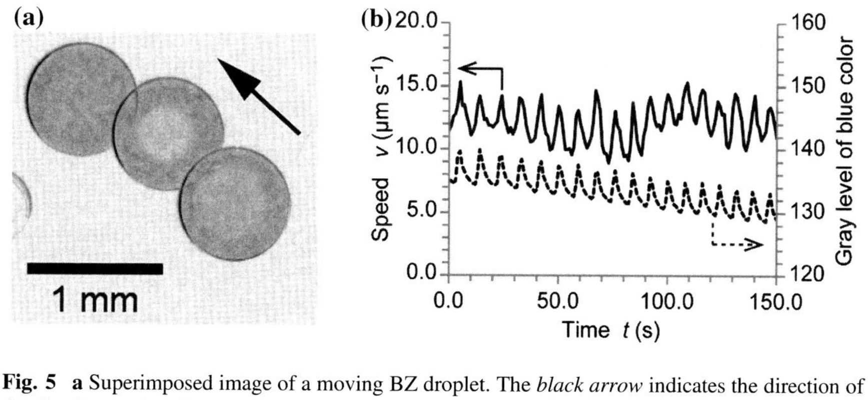

Fig. 5 **a** Superimposed image of a moving BZ droplet. The *black arrow* indicates the direction of the droplet motion. Images were acquired every 30 s. **b** Time series of droplet speed and color ([MO] = 10 mM). The droplet color was estimated by the intensity of the *blue component*. Thus, high blue values indicate the oxidized state and low the reduced state

The speed of self-propelled motion is determined by the droplet size and surfactant and reactant concentrations, as theoretically investigated [22, 23]. In the BZ reaction, Br_2 is an intermediate, and its concentration also oscillates in time [29]. Numerical calculation of the BZ reaction using a modified Oregonator model indicated that the Br_2 concentration drastically increases as the redox state transitions from the reduced to the oxidized state [30]. This numerical result reveals that the Br_2 concentration is high and the droplet motion is fast in the oxidized state of the droplet. Our experimental observations, shown in Fig. 5, agree well with the results of these theoretical and numerical studies. Therefore, the oscillatory motion of the BZ droplet was realized by the oscillation of the concentration of Br_2 in the aqueous phase. This result indicates that the self-propelled motion generated by the interfacial chemical reaction of the surfactant is suitable for coupling to nonlinear chemical reaction and that the droplet behavior strongly reflects the characteristics of the chemical environment inside the aqueous phase.

6 Effect of MO on the BZ Reaction

In the previous section, it was shown that the concentration of the intermediate of the BZ reaction (Br_2) affected the droplet motion. This finding reveals that consumption of the intermediate influences the BZ reaction. Indeed, the period of the

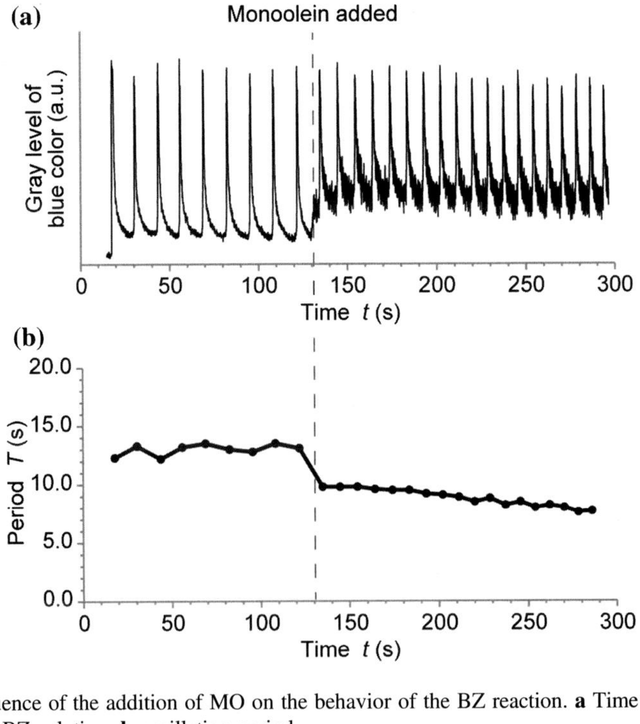

Fig. 6 Influence of the addition of MO on the behavior of the BZ reaction. **a** Time series of the color of the BZ solution; **b** oscillation period

BZ oscillation depended on the droplet size. As mentioned above, smaller droplets exhibited shorter oscillation periods. Assuming that the MO surfactant does not dissolve well into the aqueous phase, the MO only reacts with Br_2 on the oil/water interface; thus, the rate of the chemical reaction is proportional to the specific surface area of the droplet. The interfacial chemical reaction therefore has a stronger influence in the case of smaller droplets.

To confirm the effect of MO, the oscillation of the BZ reaction in the bulk phase was observed before and after the addition of MO. Upon the addition of MO, the oscillation period decreased (Fig. 6). This result agrees with the observations for the BZ reaction in the droplet, where the period was shorter than that in the bulk phase.

Numerical calculations support these experimental observations. The Oregonator is a well-known mathematical model for the BZ reaction and can numerically reproduce several of its characteristics. The original version of the Oregonator is constructed with five chemical reactions, as shown below.

$$\mathrm{A} + \mathrm{Y} \rightarrow \mathrm{X} + \mathrm{P} \tag{O1}$$

$$\mathrm{X} + \mathrm{Y} \rightarrow 2\mathrm{P} \tag{O2}$$

$$\mathrm{A} + \mathrm{X} \rightarrow 2\mathrm{X} + 2\mathrm{Z} \tag{O3}$$

$$2\mathrm{X} \rightarrow \mathrm{A} + \mathrm{P} \tag{O4}$$

$$\mathrm{Z} \rightarrow f\mathrm{Y} \tag{O5}$$

where A is BrO_3^-, P is HOBr, X is $HBrO_2$, Y is Br^-, and Z is the oxidized catalyst. Although the original version of the Oregonator neglects the effect of Br_2, a modified version including this effect has been reported [30]. In the modified model, following three reactions are added:

$$\mathrm{P} \rightarrow \mathrm{Y} \tag{O6}$$

$$\mathrm{Y} + \mathrm{P} \leftrightarrow \mathrm{U} \tag{O7}$$

$$\mathrm{U} \rightarrow \text{atmosphere} \tag{O8}$$

where U is Br_2, and Eq. (O8) describes the effect of the evaporation of Br_2. In the case of the droplet system, evaporation can be neglected because the aqueous phase is completely covered by the oil phase. Instead of the evaporation process, we must consider the effect of the dissolution to the oil phase and the interfacial chemical reaction of Br_2 with MO. Assuming that the interfacial chemical reaction is linear and much faster than the dissolution, we can rewrite Eq. (O8) as follows:

$$\mathrm{U} + \mathrm{S} \rightarrow \mathrm{Q} \tag{O9}$$

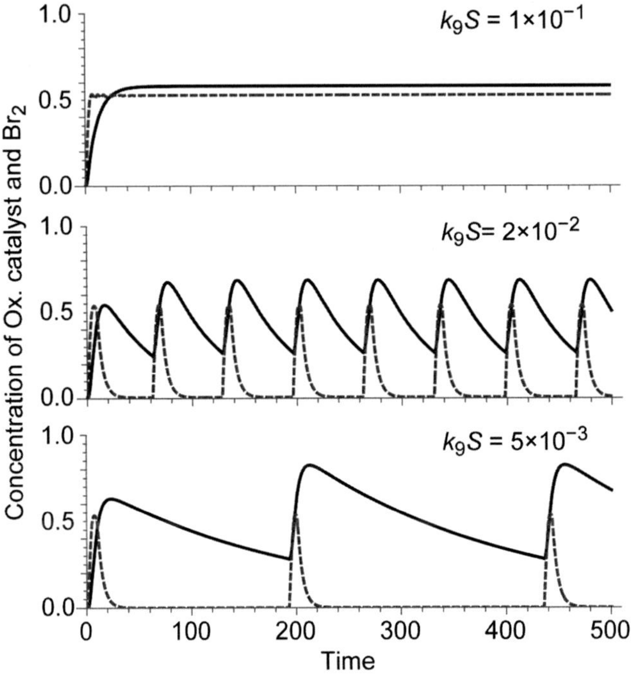

Fig. 7 Numerical calculation using the modified Oregonator model. As the consumption rate of Br_2 (k_9S) increases, the period of the oscillation decreases and bifurcates to the steady-state. The solid and broken lines indicate the concentration of Br_2 and oxidized catalyst, respectively. The parameters are $A = 0.40$, $B = 0.20$, $H = 1.20$, $f = 1.5$, $k_1 = 2H^2$, $k_2 = 3 \times 10^6 H$, $k_3 = 48H$, $k_4 = 3 \times 10^3$, $k_5 = B$, $k_6 = 0.5B$, $k_7 = 8 \times 10^9$, and $k_{-7} = 110$

where S is the surfactant (MO) and Q is the product of the interfacial chemical reaction (Br_2MO). Using Eqs. (O1–O7) and Eq. (O9), we numerically calculated the modified Oregonator model. Here, the reaction rate of Eq. (O9) (k_9) is an unknown coefficient. The results indicated that the period of the oscillation decreased with increasing k_9S (Fig. 7), where S, concentration of MO, was regarded as constant due to the large excess of MO. This finding, namely, that the addition of MO shortens the period, is in agreement with our experimental observations.

7 Conclusion

We suggested a strategy for constructing a functional self-propelled object that exhibits the characteristics of a nonlinear chemical reaction. The crux of this strategy is that a nonlinear chemical reaction controls an interfacial reaction of the surfactant, which generates the driving force for the self-propelled motion. To demonstrate this strategy, an aqueous droplet of Belousov-Zhabotinsky (BZ) solution was placed in a squalane monoolein (MO) solution, producing oscillatory motion of the aqueous droplet due to the oscillation of the Br_2 concentration induced by the BZ reaction. The droplet motion and BZ reaction are interdependent through the Br_2 concentration. Our strategy is widely applicable for constructing self-propelled objects coupled to characteristic chemical reactions. We believe that the self-propelled BZ droplet provides a foundation for the future development of functional self-propelled droplets.

Acknowledgements The authors thank Ms. K. Ito and Ms. A. Deguchi for their dedication to conducting many experiments. Ms. Ito observed and analyzed the self-propelled motion of the BZ droplets, and Ms. Deguchi measured the surface pressure of MO and Br_2MO. This work was supported in part by the Research Project Grant (B) of the Institute of Science and Technology of Meiji University (N.J.S.) and by the JGC-S Grant for Young Researchers (N.J.S.). This report is constructed based on the oral presentation on the Symposium "Complexity and Synergetics" in Hannover (2015), which was supported by the VW-foundation.

References

1. L. Bosgraaf, P.J.M. van Haastert, The ordered extension of pseudopodia by amoeboid cells in the absence of external cues. PLoS ONE **4**, e5253 (2009)
2. L. Alvarez, B.M. Friedrich, G. Gompper, U.B. Kaupp, The computational sperm cell. Trends Cell Biol. **24**, 198–207 (2014)
3. J.W.M. Bush, D.L. Hu, Walking on water: biolocomotion at the interface. Ann. Rev. Fluid. Mech. **38**, 339–369 (2006)
4. Y. Hong, D. Velegol, N. Chaturvedic, A. Sen, Biomimetic behavior of synthetic particles: from microscopic randomness to macroscopic control. Phys. Chem. Chem. Phys. **12**, 1423–1435 (2010)
5. S. Nakata, Y. Iguchi, S. Ose, M. Kuboyama, T. Ishii, K. Yoshikawa, Self-rotation of a camphor scraping on water: new insight into the old problem. Langmuir **13**, 4454–4458 (1997)
6. K. Nagai, Y. Sumino, H. Kitahata, K. Yoshikawa, Mode selection in the spontaneous motion of an alcohol droplet. Phys. Rev. E **71**, 065301 (2005)
7. S. Tanaka, Y. Sogabe, S. Nakata, Spontaneous change in trajectory patterns of a self-propelled oil droplet at the air-surfactant solution interface. Phys. Rev. E **91**, 032406 (2015)
8. T. Toyota, N. Maru, M.M. Hanczyc, T. Ikegami, T. Sugawara, Self-propelled oil droplets consuming "fuel" surfactant. J. Am. Chem. Soc. **131**, 5012–5013 (2009)
9. Y. Sumino, N. Magome, T. Hamada, K. Yoshikawa, Self-running droplet: emergence of regular motion from nonequilibrium noise. Phys. Rev. Lett. **94**, 068301 (2005)
10. T. Vicsek, A. Zafeiris, Collective motion. Phys. Rep. **517**, 71–140 (2012)
11. N.J. Suematsu, S. Nakata, Self-propelled object response to environment. Curr. Phys. Chem. **5**, 21–28 (2015)
12. N.J. Suematsu, Y. Miyahara, Y. Matsuda, S. Nakata, Self-motion of a benzoquinone disk coupled with a redox reaction. J. Phys. Chem. C **114**, 13340–13343 (2010)
13. S. Nakata, M. Nagayama, H. Kitahata, N.J. Suematsu, T. Hasegawa, Physicochemical design and analysis of self-propelled objects that are characteristically sensitive to environments. Phys. Chem. Chem. Phys. **17**, 10326–10338 (2015)
14. I.R. Epstein, K. Showalter, Nonlinear chemical dynamics: oscillations, patterns, and chaos. J. Phys. Chem. **100**, 13132–13147 (1996)

15. K. Asakura, R. Konishi, T. Nakatani, T. Nakano, M. Kamata, Turing pattern formation by the CIMA reaction in a chemical system consisting of quaternary alkyl ammonium cationic groups. J. Phys. Chem. B **115**, 3959–3963 (2011)
16. J.J. Tyson, Analytic representation of oscillations, excitability, and traveling waves in a realistic model of Belousov-Zhabotinskii reaction. J. Chem. Phys. **66**, 905–915 (1977)
17. O. Steinbock, S.C. Müller, Radius-dependent inhibition and activation of chemical oscillations in small droplets. J. Phys. Chem. A **102**, 6485–6490 (1998)
18. H. Kitahata, R. Aihara, N. Magome, K. Yoshikawa, Convective and periodic motion driven by a chemical wave. J. Chem. Phys. **116**, 5666–5672 (2002)
19. H. Kitahata, N. Yoshinaga, K.H. Nagai, Y. Sumino, Spontaneous motion of a droplet coupled with a chemical wave. Phys. Rev. E **84**, 015101 (2011)
20. J. Szymanski, J. Gorecki, M.J.B. Hauser, Chemo-mechanical coupling in reactive droplets. J. Phys. Chem. C **117**, 13080 (2013)
21. S. Nakata, M. Yoshii, S. Suzuki, R. Yoshida, Periodic reciprocating motion of a polymer gel on an aqueous phase synchronized with the Belousov-Zhabotinsky reaction. Langmuir **30**, 517–521 (2014)
22. N. Yoshinaga, K.H. Nagai, Y. Sumino, H. Kitahata, Drift instability in the motion of a fluid droplet with a chemically reactive surface driven by Marangoni flow. Phys. Rev. E **86**, 016108 (2012)
23. S. Thutupalli, R. Seemann, S. Herminghaus, Swarming behavior of simple model squirmers. New J. Phys. **13**, 073021 (2011)
24. S. Herminghaus, C.C. Maass, C. Krüger, S. Thutupalli, L. Goehring, C. Bahr, Interfacial mechanisms in active emulsions. Soft Matter **10**, 7008–7022 (2014)
25. N.J. Suematsu, Y. Mori, T. Amemiya, S. Nakata, Oscillation of speed of a self-propelled Belousov-Zhabotinsky droplet. J. Phys. Chem. Lett. **7**, 3424–3428 (2016)
26. N.J. Suematsu, Self-propelled motion coupled with chemical reaction. Presented at Symposium "Complexity and Synergetics", Hannover, 8–11 July 2015
27. C.-H. Chang, E.I. Franses, Adsorption dynamics of surfactants at the air/water interface: a critical review of mathematical models, data, and mechanisms. Colloid Surfaces A **100**, 1–45 (1995)
28. A.W. Adamson, P. Alice, *Physical Chemistry of Surfaces*, 6th edn. (Wiley, New York, 1997)
29. R.J. Field, P.M. Boyd, Bromine-hydrolysis control in the cerium ion-bromate ion-oxalic acid-acetone Belousov-Zhabotinskii oscillator. J. Phys. Chem. **89**, 3707–3714 (1985)
30. R.J. Field, H.-D. Försterling, On the oxybromine chemistry rate constants with cerium ions in the Field-Körös-Noyes mechanism of the Belousov-Zhabotinskii reaction: the equilibrium HBrO2 + BrO3- + H + <=> 2BrO2· + H2O. J. Phys. Chem. **90**, 5400–5407 (1986)

All About Cells

Lisa Kolb and Susanne Krömker

Abstract Investigating the longterm behaviour of cells in the lymph nodes releasing an activating substance into an intercellular space is a demanding task for mathematical modelling and numerical simulations of reaction-diffusion systems coupled with ordinary differential equations on unstructred 3D grids. To explore the vast amount of computed spatio-temporal data the visualization has to be fast and concise. Persistent homology in discrete Morse theory is a mighty tool for meaningful data reduction. Isosurfaces at characteristic thresholds can be extracted from the data without missing the crucial changes in the topology of the manifold. Visualization results are presented for single and multiple cells with details on the topological concept and computational efforts of the preprocessing steps.

1 Mathematical Model of Cytokine Signaling

Cells are the building blocks of life, the characteristics of living things. Especially the cells of the immune system are of interest for biologists and physicians, and here, the lymph nodes play an important role. Therein three types of T lymphocytes can be differentiated: T helper cells that have already been activated, responding T helper cells that have not yet been activated and regulatory T cells (see [1] and [2]). Stimulation of T helper cells by an antigen make them release Interleukin-2 (IL-2) on a synapse. The increasing concentration of IL-2 in the intercellular space can activate the responding T helper cells by an increase of the IL-2 concentration around them. Once activated they proliferate and differentiate into effector T cells.

L. Kolb
Carl Zeiss Microscopy GmbH, Kistlerhofstraße 75, 81379 Munich, Germany

S. Krömker (✉)
Interdisziplinäres Zentrum für Wissenschaftliches Rechnen – IWR,
Universität Heidelberg, Im Neuenheimer Feld 205, 69120 Heidelberg, Germany
e-mail: kroemker@iwr.uni-heidelberg.de

S.C. Müller et al. (eds.), *Complexity and Synergetics*,
https://doi.org/10.1007/978-3-319-64334-2_17

The regulatory T cells function as inhibitors by absorbing IL-2. A mathematical model for the interaction of cellular signaling with diffusion processes (see [3]) can be analysed in terms of macroscopic effects of equations on a microscopic scale as was done earlier in [4]. The cells and their interactions can be described by four functions. There is the concentration of IL-2 itself, which is a function u of time t and position $x \in \mathbb{R}^3$. This function is only defined on the intercellular space Ω and its boundaries. Furthermore, for each T cell three numbers R, B and E are of interest, which are all connected to the number of receptors. The sum $R + B$ is the total number of receptors of a cell. Above a certain number of receptors the cell is surely activated, below another critical number it is not.

Let Ω be the domain, the intercellular area, Γ_i then is the boundary of cell i and Γ_{out} is the outer boundary of Ω. The function $u(t,x) : [0,T] \times \Omega \to \mathbb{R}$ models the concentration of IL-2 at time t and position x, and the functions $R_i : [0,T] \to \mathbb{R}$ model the number of receptors, $B_i : [0,T] \to \mathbb{R}$ model the number of built receptor-complexes, and $E_i : [0,T] \to \mathbb{R}$ model the number of internalized complexes for cell T_i. These functions are not explicitly known but mathematically modelled by a reaction-diffusion equation for the IL-2 concentration in the intercellular space with no-flux conditions at the outer boundary, D being a matrix with diffusion coefficients.

$$\begin{aligned} \partial_t u(t,x) &= D\Delta u(t,x) - k_d u(t,x) && \forall (t,x) \in (0,T] \times \Omega \\ D\partial_n u(t,x) &= q_i(t,x) - k_{on} R_i(t) u(t,x) + k_{off} B_i(t) && \forall (t,x) \in (0,T] \times \Gamma_i \\ \partial_n u(t,x) &= 0 && \forall (t,x) \in (0,T] \times \Gamma_{out} \end{aligned}$$

A mathematical cell in this context can be a small and repeated spatial domain for mappings of any kind, the term is used in homogenization as well as in topology. In the cell problem for homogenization it makes sense to change the no-flux to periodic boundary conditions, which means that everything that leaves on one side of Γ_{out} re-enters on the opposite side. This differential equation (PDE) is coupled with ordinary differential equations (ODEs) for the various biological cell types

$$\begin{aligned} \partial_t R_i(t) &= w_i^0 + w_i^1 \frac{B_i(t)^3}{K^3 + B_i(t)^3} - k_{on} R_i(t) \tilde{u}_i(t) - k_{iR} R_i(t) \\ &\quad + k_{off} B_i(t) + k_{rec} E_i(t) \\ \partial_t B_i(t) &= k_{on} R_i(t) \tilde{u}_i(t) - (k_{off} + k_{iB}) B_i(t) \\ \partial_t E_i(t) &= k_{iB} B_i(t) - (k_{rec} + d_{deg}) E_i(t) \\ \tilde{u}_i(t) &= \frac{\int_{\Gamma_i} u(t,s) \mathrm{d}s}{|\Gamma_i|} \end{aligned}$$

and completed with initial conditions $u(0,x) = 0 \;\; \forall\, x \in \Omega$ and $R_i(0) = R_i^0$, $B_i(0) = B_i^0$, and $E_i(0) = E_i^0$ for all cells i.

As an approach to realistic multicellular settings a model of $6 \times 6 \times 6 = 216$ biological cells in 3D space with periodic boundary conditions is implemented and the

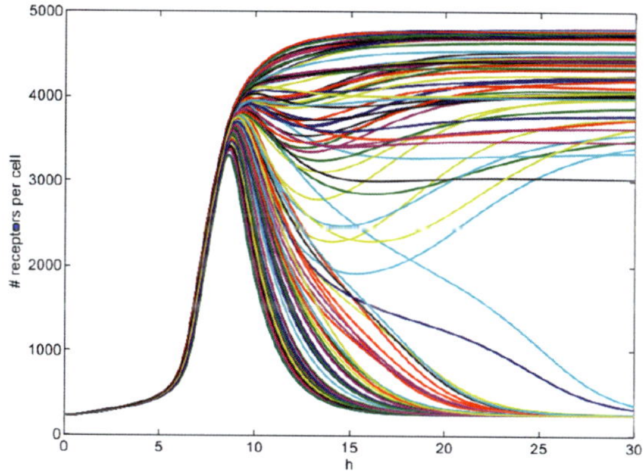

Fig. 1 Number of receptors per T cell over time

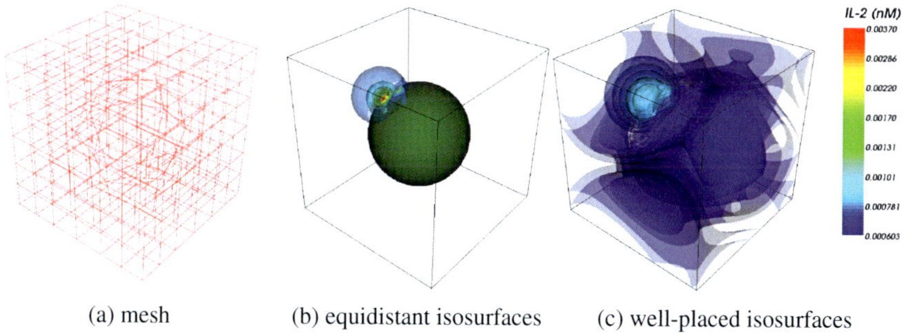

Fig. 2 **a** Mesh of a single biological cell with IL-2 concentration using **b** 20 isosurfaces distributed uniformly; **c** shows the result after applying persistent homology

production and diffusion of concentrations in intercellular space is computed (see [5]). The system is time-dependent. Initially, the concentration of IL-2 is assumed 0 everywhere. Figure 1 shows that the number of receptors for all cells increases, but after about 10 hours they start to differentiate. Finally a steady state is reached and the number of receptors in the activated cells stays high.

There have been earlier simulations of T cells in the lymph nodes in 2D as described in [2]. In 3D the distribution of IL-2 is not restricted to the plane with 2D obstacles and can emerge into this third dimension. For simplicity in the analysis of the homogenization and in numerical simulation it is assumed that the cells have a fixed position. The system is discretized by defining a grid around each biological cell as in Fig. 2a. The used grid is mainly rectangular, but slightly deformed around the enclosed biological cell. Since only the intercellular space is modelled no lattice points are placed within the biological cell.

The simulation output data is stored in the file format of the Visualization Toolkit (VTK) containing information about the geometric location of the data, in this case stored in an unstructured grid for the domain around the spherical biological cells.

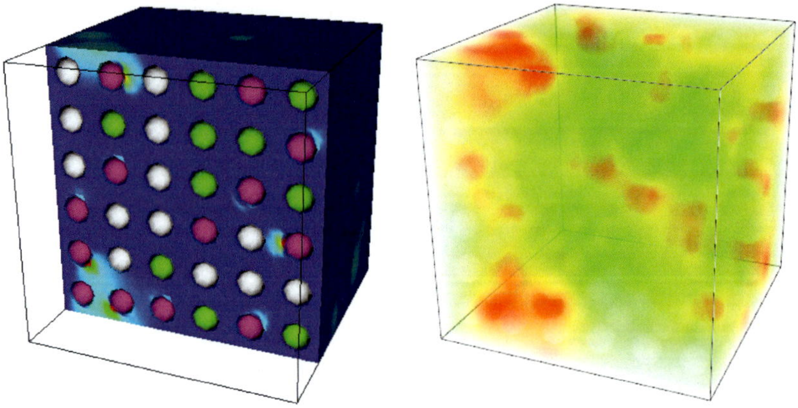

(a) cutting plane (b) volume rendering

Fig. 3 Intercellular space of 216 cells with regulatory T cells

The position of the vertices and connectivity form 3D mathematical cells. Here only hexahedrons are used, which are, as the grid is mostly rectangular, mainly cubes. Finally a scalar value for the IL-2 concentration is stored for each vertex.

2 Topological Methods Used for Visualization

In mathematically terms, a set of mathematical, not biological, cells build up a topological space by attaching them to each other according to predefined rules to end up with a complex of a certain dimension. The easiest such complex is a simplicial complex formed by vertices (0-cells), straight lines (1-cells), triangular faces (2-cells), terahedrons (3-cells) and possibly higher dimensional cells. This resembles a discretization of any spatial domain. In the case of the simulated biological data, a scalar field is given on an unstructured 3D grid that is mainly rectangular. This is why a so-called cubical complex can be used for a discretized domain Ω. The vertices are again 0-cells, the edges between the vertices (usually of unit lenght) are 1-cells, the squares are 2-cells and the cubes are 3-cells.

Definition 1 (*Faces, cofaces and complex*) Let $\mathcal{K} = \mathcal{K}(\Omega)$ be such a set of p-cells in Ω. A p-cell $\alpha^{(p)}$ is face of a q-cell $\beta^{(q)}$ if $p < q$ and the vertices of α are a subset of the vertices of β. Write: $\alpha^{(p)} < \beta^{(q)}$. Then β is called a coface of α. A set of cells S is a complex if for all $\alpha \in S$ all faces of α are also in S.

One way of visualizing the high resolution data from the simulated system of differential equations is by using cutting planes passing through the domain Ω as can be seen in Fig. 3a. The secretory T cells are displayed in pink, regulatory T cells are green, and responding T cells are gray. The problem with this type of visualization is that the data can only be seen on a 2D plane. Visualizing the whole data using this method requires either multiple pictures or a film in which the cutting plane traverses the whole domain. Another method is to use volume rendering as can be seen in Fig. 3b. However, it takes a lot of time to render a single projection in acceptable quality. Since exploration of the spatial data needs transformations in real time, this method turned out to be not suitable.

The amount of data produced during the simulations is that huge that gaining insight is possible only with fast and good visualization tools revealing the important features. The idea is to use isosurfaces for visualizing the IL-2 concentration. These isosurfaces can be rendered with VTK fast enough to allow for any kind of transformations in real time. Imagine the gas bubbles in sparkling water. The interface between gas and water can be located at a jump in density, and an isosurface at a threshold value within this jump may be extracted from given data. This isosurface marks the boundary of all bubbles. They may not all be spherical, have different sizes and are nonuniformly distributed. When stirred they may also separate or join over time forming smaller or bigger cavities. In topology the exact form and location is neglected but changes, e.g., in the number of cavities will be registered. Unfortunately, it is difficult to determine suitable isovalues, and nested opaque isosurfaces hide each other, whereas displaying transparent isosurfaces raises other difficulties. The goal is to get rid of these disadvantages by acquiring information about the topology of the data. Isovalues then can be located at characteristic and persistent changes in the data.

In topology, distances between points are no longer important. The only interest is in the relation to a neighborhood for every point in space. Imagine a rather complicated surface which can be deformed into something much simpler by disregarding the exact location of each and every vertex. Such a deformation should be a continuous mapping that works back and forth and is called a homeomorphism. A topological invariant like the Euler characteristic $\chi(\Omega)$ just counts the number of vertices V, subtracts the number of edges E and adds the number of faces F, and do not change under such transformations. As can be easily seen, a bounded simply connected, compact subset Ω of the real Euclidean plane $\mathbb{R}^2$, that has no holes, is homeomorphic to a square. In case of starting with Ω being a square, it has $\chi(\Omega) = V - E + F = 4 - 4 + 1 = 1$. Subdividing the square by a diagonal edge adds another edge but also another face, and still $\chi(\Omega) = V - E + F = 4 - 5 + 2 = 1$. A triangle Δ counts for $\chi(\Delta) = V - E + F = 3 - 3 + 1 = 1$. Any continuous bijective transformation, e.g., scrunching a sheet of paper and describing its shape with connected triangles, will not change this topological invariant.

Homotopy gives a weak definition of two geometric objects to be equal and resembles a continuous morphing. A map $f : X \to Y$ is said to be a homotopy equivalence if there is a continous map $g : Y \to X$ such that $g \circ f \simeq id_X$ and $f \circ g \simeq id_Y$ and the spaces are said to be homotopy equivalent, $X \simeq Y$. The point and a disk are homotopy equivalent. The point and the circular line, however, are not. The idea is that spaces may be deformed and stretched, but it is not allowed to cut out holes. In the deformation only so-called free pairs may be removed.

Definition 2 (*Free pair*) The pair (α, β) is called a free pair, if a cell $\beta^{(p)}$ of a cubical complex has a free face $\alpha^{(p-1)}$, that is if α is a face of β and α has no other cofaces.

By removing such a free pair from a complex $\mathscr{K}$, a subcomplex $\mathscr{K}'$ is created. Two complexes $\mathscr{K}$ and $\mathscr{K}'$ are equivalent if a sequence of collapses (or the contrary: expansions) can transform one into the other, which is nothing else but a continuous morphing, previously called a homotopy.

Now imagine a 2D grid with scalar data that can be visualized as a landscape with mountains and valleys with their specific height. The landscape is filled with water. As long as the water level is below the lowest valley, the landscape is dry. Once the water starts filling the lowest valley, a lake is formed and the topology of the water surface changes. Once the water level exceeds a saddle between two hills and thereby combines two valleys such that two lakes join, the topology changes again. As long as no new valleys are filled and the water level does not exceed mountains, the topology does not change. This also works in the case of a 3D grid with the scalar values of neighboring vertices. Consider the lower level cuts, which contain all cells with values lower than a threshold. The idea is to analyse how the topology of these lower level cuts changes as the threshold value varies, and then use these critical values as isovalues.

Definition 3 (*Lower star*) The star of a cell $\alpha^{(p)}$ in a cubical complex is the set of all cofaces $\beta^{(p+1)}$ of α, the lower star of a cell $\alpha^{(p)}$ in a cubical complex is the set of all cofaces of α with scalar value less or equal to the scalar value of α.

Definition 4 (*Lower level cut of a cubical complex*) Let Ω be a discrete (rectangular) grid and let $g(x) : \Omega \rightarrow \mathbb{R}$ be a scalar function defined on Ω. A lower level cut of Ω is the set of all data points with scalar value less than a threshold t, that is $\Omega_t = \{x \in \Omega | g(x) \leq t\}$. A cell α is in the lower level cut if all its vertices $x \in \alpha$ have scalar values less than the threshold:

$$\mathscr{K}_t = \{\alpha \in \mathscr{K} | g(x) \leq t, \forall x \in \alpha\} .$$

Figure 4a shows an example of a lower level cut for the threshold value $t = 6$ with only 0-cells and 1-cells, whereas in (b) is shown the lower level cut for $t = 8$ which is more connected and also contains a 2-cell. All cells of the lower level cut for a lower threshold are contained in the one of a higher threshold.

An isosurface I is defined as the set of all points in the domain Ω with the same scalar value t, that is $I(t) = \{x \in \Omega | g(x) = t\}$ and thereby is the boundary of a lower level cut. It changes when the threshold value is varied. For t less than the smallest scalar data value, $\mathscr{K}_t = \emptyset$. By increasing t, cells are added to $\mathscr{K}_t$ so that $\mathscr{K}_{t_1} \subset \mathscr{K}_{t_2} \ \forall \ t_1 < t_2$. For t bigger than the maximum scalar value, $\mathscr{K}_t$ is the whole data set. Such a sequence of nested complexes is called a filtration. The simulated data is a scalar field on the vertices of a grid forming a cubical complex. Lower level cuts or their boundaries shall now be compared to each other.

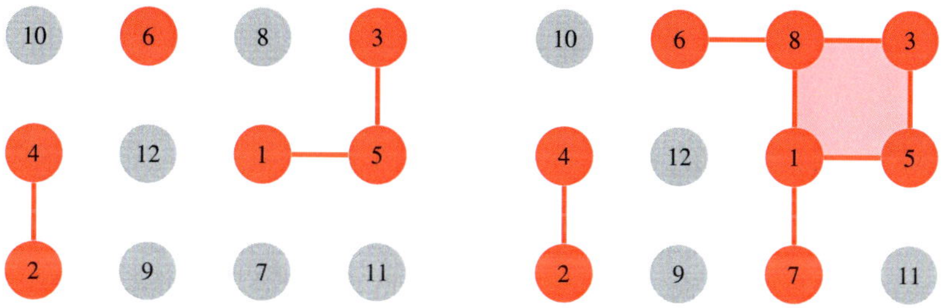

(a) lower level cut for $t = 6$ (b) lower level cut for $t = 8$

Fig. 4 The *red cells* show examples for lower level cuts

3 Discrete Morse Theory

Morse theory was originally defined only for smooth manifolds that are mathematical spaces which locally look like the $\mathbb{R}^n$ and are differentiable. For further readings see [6] and [7]. A Morse complex on the data has the same topology as the cubical complex, but contains only the critical cells (see [8]). As our data is given on a discrete grid of points, this Morse theory is not applicable. Instead, a discrete Morse theory which is defined for cell complexes as introduced by Forman ([9, 10]) is used here.

Definition 5 (*Discrete Morse function*) A function $f : \mathscr{K} \to \mathbb{R}$ is called discrete Morse function if for all $\alpha^{(p)} \in \mathscr{K} : f$ is less or equal $f(\alpha)$ on at most one coface of α and f is bigger or equal $f(\alpha)$ on at most one face of α ([10], p. 11):

$$\#\{\beta^{(p+1)} > \alpha | f(\beta) \leq f(\alpha)\} \leq 1 ,$$
$$\#\{\gamma^{(p-1)} < \alpha | f(\gamma) \geq f(\alpha)\} \leq 1 .$$

Figure 5 illustrates the definition, where the functions assign values to 0-cells (vertices) and 1-cells (arcs). Then (a) is an example for a discrete Morse function, whereas (b) is no such function, as the 1-cell with value 0 has two faces both of which have a value higher than zero.

Fig. 5 Example of a discrete Morse function

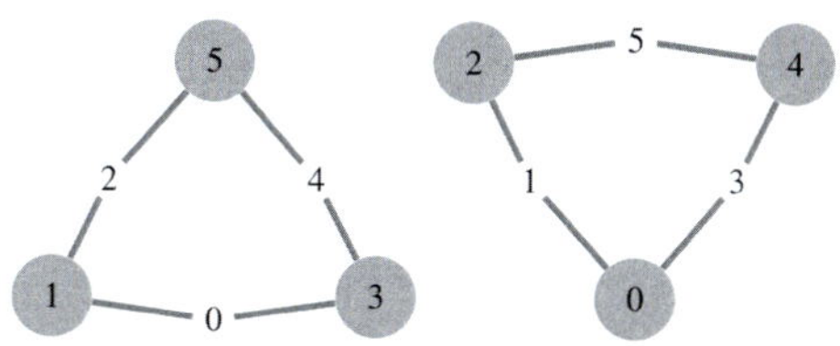

(a) discrete Morse function (b) no such function

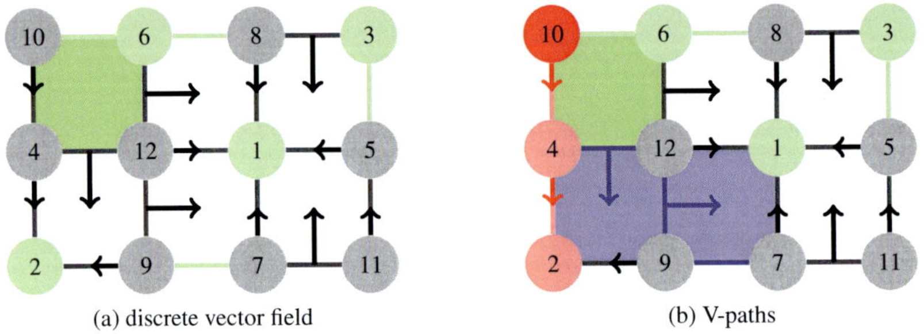

Fig. 6 Pairs (*arrows*) and critical cells (*green*) in a discrete vector field; **b** V-paths (*red*) from 0-cell (10) to 0-cell (2) and (*blue*) from 1-cell (12.4) to 1-cell (9.7)

Definition 6 (*Critical cells*) A cell $\alpha^{(p)}$ is critical if all its cofaces have strictly higher values and all faces have strictly lower values.

The function in Fig. 5a has two critical cells: one with value 0, as both cofaces have values bigger than 0 and the cell itself has no faces. The other one with value 5 is also critical, as both faces have values lower than 5 and the cell has no cofaces.

Definition 7 (*Discrete vector field*) A discrete vector field V is a set of pairs ($\alpha^{(p)}$, $\beta^{(p+1)}$) of cells in $\mathscr{K}$ with $\alpha^{(p)} < \beta^{(p+1)}$, such that every cell appears in at most one pair of V (see [10], Definition 3.3).

When graphically showing discrete vector fields, arrows are often used to mark the pairing as shown in Fig. 6a.

Definition 8 (*V-path*) A V-path is a sequence of cells:

$$\alpha_0^{(p)}, \beta_0^{(p+1)}, \alpha_1^{(p)}, \beta_1^{(p+1)}, \alpha_2^{(p)}, \ldots, \beta_{r-1}^{(p+1)}, \alpha_r^{(p)}$$

with $(\alpha_i, \beta_i) \in V,\ \beta_i > \alpha_{i+1}$, and $\alpha_i \neq \alpha_{i+1}\ \forall\ i = 0, \ldots, r-1$. A V-path is trivial if $r = 1$. It is closed if $\alpha_r = \alpha_0$ (see [10], p. 19).

Figure 6b shows two V-paths in a vector field. Graphically speaking, you get the V-paths by following the arrows.

Definition 9 (*Gradient vector field*) A discrete Morse function defines a discrete gradient vector field by pairing $\alpha^{(p)} < \beta^{(p+1)}$ if $f(\beta) \leq f(\alpha)$. The critical cells are exactly those which do not appear in a pair.

Along a V-path in the gradient vector field, the Morse function decreases continuously (see [10], Theorem 3.4). A discrete vector field is a gradient vector field of a Morse function if and only if there are no non-trivial closed V-paths (see [10], Theorem 3.5).

Theorem 1 *Let $\mathcal{K}$ be a cubical complex with a discrete Morse function. Then $\mathcal{K}$ is homotopy equivalent to a closure-finite weak topology (CW)-complex with exactly one p-cell for every critical p-cell in $\mathcal{K}$ ([9], Corollary 3.5).*

The neighborhood of cells in a Morse complex is defined by the boundary map:

Definition 10 (*Boundary map*) Let $\mathcal{M}_p \subseteq C_p(\mathcal{K}, \mathbb{Z}_2)$ be the subgroup of a chain C_p of p-cells generated by the critical p-cells. Let $\Gamma(\beta, \alpha)$ be the set of all V-paths from faces of β to α and

$$c_{\alpha,\beta} = \begin{cases} 1, \text{ if } \#\Gamma(\beta,\alpha) \text{ odd} \\ 0, \text{ if } \#\Gamma(\beta,\alpha) \text{ even.} \end{cases}$$

The boundary map is then defined as follows:

$$\partial_p : \mathcal{M}_p \rightarrow \mathcal{M}_{p-1}$$

$$\partial_p \beta = \sum_{\alpha \in \mathcal{M}_{p-1}} c_{\alpha,\beta}\alpha$$

Definition 11 (*Morse chain complex*) A (cubical) Morse chain complex in 3D space is a sequence of spaces and maps of the following form:

$$0 \rightarrow \mathcal{M}_3(X, \mathbb{Z}_2) \stackrel{\partial_3}{\rightarrow} \mathcal{M}_2(X, \mathbb{Z}_2) \stackrel{\partial_2}{\rightarrow} \mathcal{M}_1(X, \mathbb{Z}_2) \stackrel{\partial_1}{\rightarrow} \mathcal{M}_0(X, \mathbb{Z}_2) \rightarrow 0.$$

Homology groups are the quotients

$$H_p(\mathcal{M}, \partial) = \frac{Ker\, \partial_p}{Im\, \partial_{p+1}}.$$

Then for all p it holds that $H_p(\mathcal{M}, \partial) \cong H_p(\mathcal{K}, \mathbb{Z}_2)$. In other words the field $\mathbb{Z}_2 = \{0, 1\}$ is just needed to answer whether a cell belongs to a chain or not, and in short the homotopy of the homology groups reduces the more difficult idea of being the boundary of an object in a higher dimension to the simple idea of being a perfect surface or having cut out holes.

As the 3D scalar data is given in the form of a VTK file, its cubical complex is implicitly given. At first a discrete vector field on that cubical complex has to be constructed. Then its critical points are used to build a Morse chain complex. The implementation details follow [11] on 2D or 3D grayscale images from CT scans and had to be adapted to work for 3D curvilinear grids and with different range of scalar data. The idea of the algorithm is to iteratively construct the vector field by using simple homotopy expansion. The algorithm thus starts with the 0-cell with the lowest scalar value and computes the lower level cut for $t = t_{min}$. Then cells are added to the complex by gradually increasing t. For 0-cell x_i the lower star $L(x_i)$ is examined (see definition 3). The algorithm assumes that the scalar values on all data points are unique so that $\mathscr{K}_{i+1} = \mathscr{K}_i \cup L(x_i)$. With a slight change of the data this is achieved easily. Thus, every data point (0-cell) can be examined independently, as every cell is contained in the lower star of exactly one 0-cell.

Afterwards the Morse chain complex can be computed and its cells coincide with the critical cells. The algorithm follows the V-paths from the faces of the critical cells to other critical cells using a breadth-first search. The number of paths between two critical cells then defines the boundary map. If critical cells can be reached by more than one path, they appear once for each distinct path stored in a list of faces. The algorithm traverses all critical cells. For each critical p-cell, a p-cell is added to the Morse chain complex. For $p = 0$ nothing else needs to be done, as 0-cells do not have a boundary and thus there are no V-paths to follow. For all other cells $\gamma^{(p)}$ their faces $\alpha^{(p-1)}$ are determined.

With that, the Morse chain complex is fully determined and the homology groups and corresponding Betti numbers can be computed as follows. The pth homology group corresponds to the set of p-cells that have no boundary ($Ker\ \partial_p$) and that are not the boundary of a $(p+1)$-cell ($Im\ \partial_{p+1}$). The rank of the pth homology group $H_p(X, \partial)$ is called pth Betti number and corresponds to the number of closed p-dimensional loops that are p-cycles which are not filled. When considering 3D objects, the 0th Betti number is the number of connected components, the 1st Betti number is the number of handles and the 2nd Betti number the number of enclosed voids. For implementation details see the thesis of [12].

The algorithm for persistent Betti numbers is from [13], who introduced the concept of persistent homology. The idea is to iteratively add cells to the complex. For every cell added, exactly one Betti number changes by exactly one. If for example a 0-cell is added, the number of connected components increases and, therefore, also the 0th Betti number. When adding a 1-cell to the complex, two different things can happen. The first possibility is that two different formerly unconnected components are connected by the 1-cell. The result of this is that the 0th Betti number decreases by one. Otherwise, the 1-cell connects two 0-cells that are already connected. In this case, the 1st Betti number increases, as this 1-cell then closes a 1-cycle. This works analogously for 2- and 3-cells. Cells that create a cycle are called creators and cells that fill in a cycle are called destroyers. Creator and destroyer cells build pairs. The persistence is then some sort of difference between the creator and destroyer cells. The longer a cycle exists in the filtration, the higher is its persistence value. By removing cells that have low persistence values, small fluctuations in the data can be ignored.

With the introduction of persistent homology we return to the initial idea to use isosurfaces for the visualization of IL-2 concentration. Persistent Betti numbers support the selection of isovalues which are placed such that every important topological feature is represented in the set of rendered isosurfaces. See Sects. 4.2 and 4.3 for the application of this idea.

Table 1 Sizes of cubical and Morse complex

	Single cell	Cell synapse	216 cells
Size of VTK file (kB)	543	4489	13498
Number of 0-cells (cubical)	2170	15282	50437
Number of 1-cells (cubical)	5976	43824	146556
Number of 2-cells (cubical)	5472	41856	141264
Number of 3-cells (cubical)	1664	13312	44928
Total number (cubical)	15282	114274	383185
Number of 0-cells (Morse)	12	68	587
Number of 1-cells (Morse)	46	101	720
Number of 2-cells (Morse)	42	35	350
Number of 3-cells (Morse)	6	0	0
Total number (Morse)	106	204	1657
Morse cells to cubical cells (%)	0.7	0.18	0.4

4 Visualization Results

The results focus on three simulations: (i) a single biological cell that emits IL-2 on its whole surface, (ii) a single biological cell emitting IL-2 on a synapse, that is on a tiny point on its surface, and (iii) a simulation of 216 biological cells. This last example is much more complex than the previous two. From a biological perspective it is of greater interest, as the interaction between the cells can be observed.

Table 1 shows the different sizes of the example data. The total number of cells in the cubical complex is, depending on the number of data points, extremely high. The total number of cells in the corresponding Morse complex is much lower, less than 1%, but still contains the same topological information.

Table 2 shows the execution times on a laptop computer with Intel Core i3 M370 @ 2.40 GHz processor. The table shows that the times for transforming the data to a cubical complex and for processing the lower stars are highest. These times strongly depend on the number of data points, as a loop over all data points is needed in both cases. The cubical complex has to be built only once for every grid. Afterwards, execution times are much lower because there are fewer cells in the iteration. The rendering times are low enough to allow user interaction. All the examples shown in this section can be interactively rotated and zoomed in real time. The times for I/O are not listed.

Table 2 Execution times for different data examples

	Single cell (s)	Cell synapse (s)	216 cells (s)
Cubical complex	0.04	2.39	8.56
Vector field	0.04	2.52	8.99
Morse complex	0	0.03	0.3
Betti numbers	0	0.01	0.06
Rendering	0.13	0.16	0.65
Total	0.22	5.42	19.48

4.1 Single Cell

The first simulation considered is of a single biological cell that emits IL-2 on its whole surface. The data is given in two different resolutions. The concentration of IL-2 decreases very fast as the distance from the biological cell grows. Therefore, the concentration is lowest in the corners of the grid cube. Isosurfaces in this data would all be spheres around the biological cell if the domain were not restricted. As the domain is a cube, some of the spheres are cut off at the boundary of the domain.

The algorithm for lower star computations determines the critical cells as shown in Fig. 7a. These critical points reflect the topological changes of the lower level cuts as the threshold value increases. For this example, 8 critical 0-cells in the corners, 12 critical 1-cells along the edges and 6 critical 2-cells on the sides can be expected. The critical cells highly depend on the grid of the data. This can be seen in Fig. 7b. It shows the critical points for the same data, but for a higher resolution. One can see that there are more critical cells. Here, the used grid produces local minima around the cell. These can be removed by using information about the persistence. As the corresponding p-cycles do not persist for a big range of threshold values, they have a low persistence value. Figure 7c shows only those critical cells that have a persistence value higher than 0.1. The resulting critical cells are then the same as for the lower resolution and correspond with what is expected.

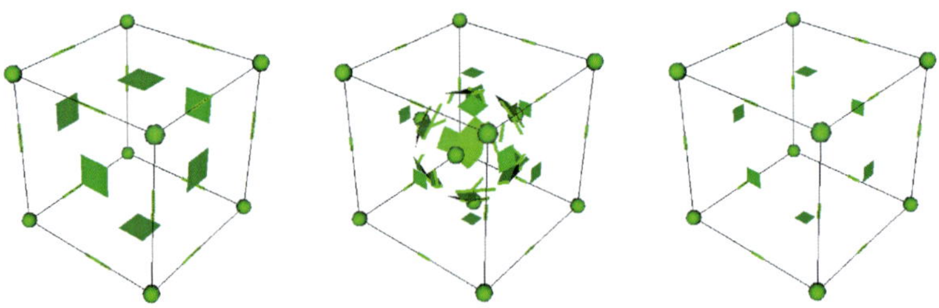

(a) low resolution (b) high resolution (c) high res. persistent cells

Fig. 7 Single cell computed with different resolutions. Critical 0-cells are shown as *spheres*, 1-cells as *lines* and 2-cells as *squares*

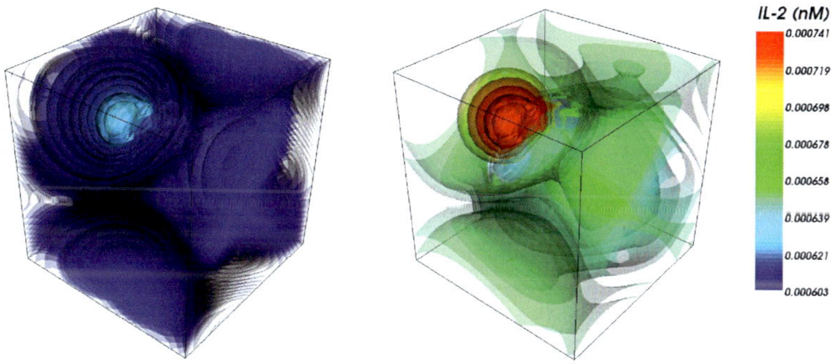

(a) 57 isosurfaces at critical values (b) 18 persistent isosurfaces

Fig. 8 Single cell, isosurfaces at **a** critical or **b** persistent values and color scale with 20% quantile compared to Fig. 2c

4.2 *Single Cell Emitting on a Synapse*

A model closer to a real biological cell let it secrete only on a tiny point on its surface at a synapse. Isosurfaces are now truncated spheres around the synapse as can be seen in Fig. 2b. Periodic boundary conditions are used, as all isosurfaces that end on one side continue on the opposite side. The isovalues are chosen uniformly over the range of the data from around 0.0006 to 0.0037. This example demonstrates quite well that the data is not uniformly distributed, as all isosurfaces are very close to the synapse. Using information about the critical cells of the data to determine isovalues results in more uniformly distributed isosufaces. Figure 8a shows 57 isosurfaces of 204 cells with critical isovalues (see in Table 1). It can be seen that the isosurfaces are distributed over the whole domain of the data, but it is still difficult to distinguish between that many isosurfaces.

Selecting only the most persistent critical cells reduces the number of isosurfaces. Figure 8b shows only the isosurfaces that correspond to cells with persistence values higher or equal to 0.1. In that case 18 isosurfaces are drawn. The example with transparent isosurfaces demonstrates better the high concentration values around the synapse, but it is somewhat difficult to distinguish the isosurfaces. For opaque isosurfaces it is easier to see the exact position of the isosurfaces, but the very high concentrations close to the synapse cannot be seen as the corresponding isosurfaces are hidden.

4.3 *Multiple Cells*

An even more realistic example shows 216 cells arranged regularly in a $6 \times 6 \times 6$ grid. Almost a third of them are T helper cells that emit IL-2 on a synapse, like in the second example. Next there are as many regulatory T helper cells, and the rest are

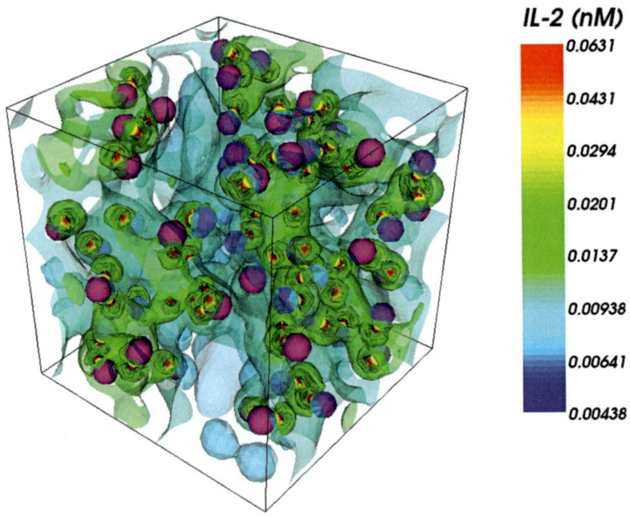

Fig. 9 The intercellular space is filled with 18 isosurfaces for values chosen according to a logarithmic scale; 71 *spheres* are secretory T helper cells (*pink*)

responding T helper cells that can be activated. Now the coupling through the diffusing IL-2 concentration in the intercellular space can be simulated. Figure 9 shows 18 isosurfaces using a logarithmic scale for the coloring with changing transparencies for the various values. The secreting T helper cells are shown in pink. As they emit IL-2, concentration values are highest around their synapses.

The next step is to choose the isosurfaces according to the critical values of the data. As the data is very complex, the total number of critical cells in the Morse chain complex is 1657 (see Table 1). It is important to choose those critical values that actually represent important changes in the topology. In this case it is not possible to make suggestions on the geometric shape of the isosurfaces. Close to the secreting T helper cells they will also be truncated spheres around the synapse, but for lower

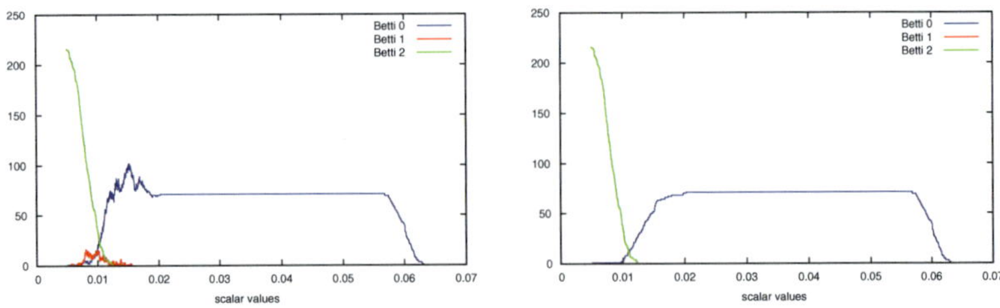

(a) all critical cells considered (b) only persistent critical cells considered

Fig. 10 Plot of the Betti numbers for the upper level cuts for different threshold values, **b** shows only Betti numbers produced by persistent critical cells

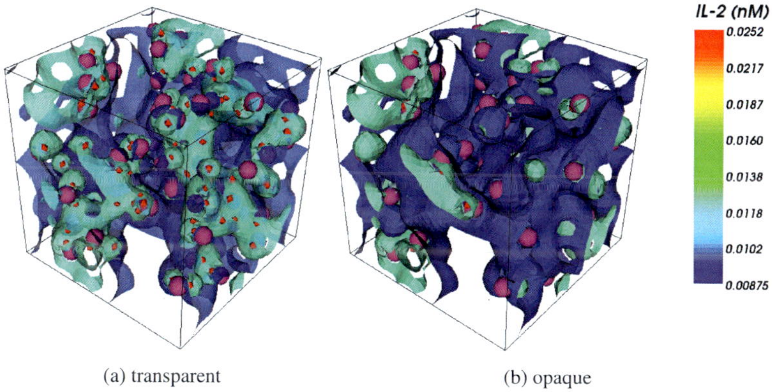

Fig. 11 Multiple cells, three selected isosurfaces; color scale with quantile

isovalues they will start to merge. Upper level cuts can be used instead of the lower level cuts. Figure 10a shows the plot that corresponds to the upper level cuts. This plot is to be read from right to left. For scalar values that are higher than the maximum value, all Betti numbers are zero. As the value decreases, the upper level cuts are points or balls around the synapses of the emitting T cells. Therefore, the 0th Betti number, which is the number of connected components, increases to a value of 71, as there are 71 secreting T helper cells. For a threshold value of about 0.02 some of the balls around the T cells start to reach the boundary of the domain, which leads to an increase of the 0th Betti number as new components appear. For the same value, the first balls around different synapses merge. At values slightly above 0.01, the first biological cells are surrounded by the upper level cuts. For a value close to zero, the whole domain is in the upper level cut. As there are 216 biological cells, the 2nd Betti number, which is the number of enclosed voids, increases to this value.

In Fig. 10b only those critical cells with persistence values higher or equal to 0.05 are considered. The persistence values in this case are computed using a simple difference between the critical values of creator-destroyer-pairs. It can be seen that for high values the plot looks exactly the same. For lower values, the plot appears smoother. The 1st Betti number stays zero for all threshold values. The 0th Betti number initially increases until it reaches a plateau. It does not become higher than 71, as seen in Fig. 10a. This plot shows that these new components do not persist for a long range of scalar values, but are connected to other components for only slightly smaller threshold values.

The Fig. 11a and b show isosurfaces that are picked according to decisive values of the plot in Fig. 10. They both show three isosurfaces. One isovalue is chosen at a value with persistence slightly higher than 0.02, where the number of connected components is still highest. The second isovalue is chosen as the value where the 1st Betti number increases to 1. For that value, the 2nd Betti number is 26. The third isosurface is chosen at the value 0.01 where the upper level cut is connected and 51 biological cells are already surrounded.

5 Conclusion and Outlook

The implemented algorithms are fast and efficient to extract topological information about the data. However, computing persistence values that perfectly mirror the most important topological changes is difficult since the values range over several orders of magnitude. Another peculiarity of the data are the periodic boundary conditions. Critical cells are added if the level cut crosses the boundary of the domain, resembling the same level cut. These additional connected components could be removed by identifying points on opposite sides of the domain.

Up to now only the steady state of the data is visualized by applying topological methods for choosing relevant isovalues and rendering steady state isosurfaces. A challenging task would be to visualize the data over time. This requires to be able to follow single isosurfaces over several time steps, but the data shows big jumps. The temporal resolution of the output, although computed fine enough for numerical stability reasons, was too sparse to see a continuous development. Moreover, following a single isosurface over time is almost impossible, as most isosurfaces that appear in one time step do no longer exist in the next one. At times close to the steady state, the values between different time steps stay closer together and it might thus be possible to follow a single isosurface.

The question of topological changes over time might be answered by computing the Morse complex with corresponding (persistent) Betti numbers for all time steps and see if the Betti number distribution stays the same in each time step.

Acknowledgements For always introducing nice ideas to be visualized, I want to especially thank Elfriede Friedmann. She investigated the cytokine signalling together with the German Cancer Research Center, dkfz, Heidelberg, and worked together with Daniel Gerecht who did the numerical investigations within his PhD thesis and provided us with the simulation data. The visualizations with standard VTK-methods like cutting planes and volume visualizations were carried out by Marcus Schaber.

References

1. K. Thurley, D. Gerecht, E. Friedmann, T. Höfer, Three-dimensional gradients of cytokine signaling between t cells. PLoS Comput. Biol. **11**(4), e1004206 (2015)
2. D. Busse, M. de la Rosa, K. Hobiger, K. Thurley, M. Flossdorf, A. Scheffold, T. Höfer, Competing feedback loops shape IL-2 signaling between helper and regulatory T lymphocytes in cellular microenvironments. Proc. Natl. Acad. Sci. USA **107**(7), 3058–3063 (2010)
3. E. Friedmann, PDE/ODE modeling and simulation to determine the role of diffusion in long-term and- range cellular signaling. BioMed. Cent. Biophys. **8**(1), 1–21 (2015)
4. A. Marciniak-Czochra, M. Ptashnyk, Derivation of a macroscopic receptor-based model using homogenization techniques. SIAM J Math. Anal. **40**(1), 215–237 (2008)
5. T. Carraro, E. Friedmann, D. Gerecht, Coupling vs decoupling approaches for PDE/ODE systems modeling intercellular signaling. J. Comput. Phys. **314**, 522–537 (2016)
6. J.W. Milnor, *Morse Theory*, vol. 51. (Princeton University Press, 1963)
7. J. Smoller, *Shock Waves and Reaction-Diffusion Equations*, vol. 258 (Springer, New York, 1994)

8. H. Edelsbrunner, J. Harer, A. Zomorodian, Hierarchical Morse complexes for piecewise linear 2-manifolds, in *Proceedings of the Seventeenth Annual Symposium on Computational Geometry*, Association for Computing Machinery (2001), pp. 70–79
9. R. Forman, Morse theory for cell complexes. Adv. Math. **134**(1), 90–145 (1998)
10. R. Forman, A user's guide to discrete Morse theory. Sém Lothar Combin. **48**, 1–35 (2002)
11. V. Robins, P.J. Wood, A.P. Sheppard, Theory and algorithms for constructing discrete Morse complexes from grayscale digital images. IEEE Trans. Pattern Anal. Mach. Intel **33**(8), 1646–1658 (2011)
12. L. Kolb, *Visualizing high-resolution numerical data with isosurfaces using topological methods, in Bachelor's thesis* (Heidelberg University, Germany, 2013)
13. A. Zomorodian, Computational topology, in *Algorithms and Theory of Computation Handbook*, ed. by M.J. Atallah, M. Blanton (Chapman & Hall/CRC, New York, 2010), pp. 1–31

Propagation and Aggregation of Motile Cells of *Escherichia coli* Pattern

Tatsunari Sakurai, Tohru Tsujikawa and Daisuke Umeno

Abstract A concentric pulse by motile cells of *Escherichia coli* (*E. coli*) propagates and the cells aggregate to form self-organized patterns. We summarize experimental and numerical results on the self-organized pattern formation of *E. coli* to elucidate some aspects of its mechanism. Our presentation includes experiments on *E. coli* patterns, as well as numerical simulations on the basis of a reaction-diffusion-chemotaxis model. We find good agreement for one-dimensional propagating fronts in observation and simulation. However, corresponding results for two-dimensional circular bacterial clusters have still not been obtained.

1 Introduction

For living micro-organisms, cells are working as smart artificial swimming devices at the microscale. For example, an individual cell of *Escherichia coli* (*E. coli*) has the ability to move [1] and, in a large community, shows a very beautiful self-organized pattern formation [2, 3] in semi-solid agar. Increased attention has been focused on the elucidation of the configurational structure in microscopic behavior, such as motilities of individual cells and macroscopic self-organized order in collective pattern formations.

T. Sakurai (✉)
Department of Creative Arts & Media, Yamaguchi College of Arts, 1-7-1, Ogori Mirai-machi, Yamaguchi-shi 754-0032, Yamaguchi, Japan
e-mail: sakurai@yamaguchi-jca.ac.jp

T. Tsujikawa
Faculty of Engineering, University of Miyazaki, 1-1, Gakuen Kibanadai-nishi, Miyazaki 889-2192, Japan

D. Umeno
Department of Applied Chemistry and Biotechnology, Chiba University, 1-33, Yayoi-cho, Inage-ku, Chiba 263-8522, Japan

S.C. Müller et al. (eds.), *Complexity and Synergetics*,
https://doi.org/10.1007/978-3-319-64334-2_18

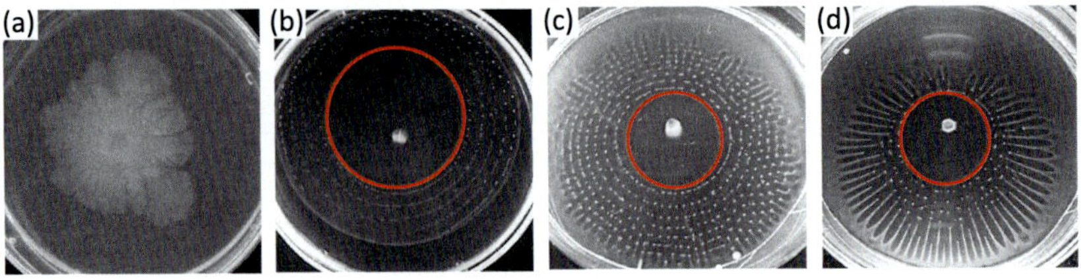

Fig. 1 Cooperative patterns formed by wild type *E. coli* (RP437). The *gray scale* corresponds to the cell density. **a** Eden pattern on 1.0% hard agar. **b–d** Aggregated patterns outside the *red circle* in 0.2% semi-solid agar; **b** initial concentration of carbon source (nutrient) is 2 mM succinate; **c** 3 mM succinate; **d** 4 mM succinate

Figure 1 represents macroscopic *E. coli* patterns. When these are inoculated at the center on harder agar (more than 0.5% agar), they grow on the agar and then assemble to form Eden-like patterns [4] as shown in Fig. 1a. Such patterns are similar to cooperative growth patterns in bacterial colonies of *B. subtilis* [5, 6] and can be explained on the basis of reaction-diffusion systems [7–9]. On the other hand, when on semi-solid agar (less than 0.5% agar), cells of *E. coli* can move into the agar, then they swim by using flagellar filaments: Thus, an individual cell of *E. coli* behaves as a self-propelled object and the motile cells of *E. coli* form complex patterns (see Fig. 1b–d). Details of the pattern formation will be shown in Sect. 3.

A motile cell of *E. coli* swims mostly straight for about 0.8 s (running time), then tumbles and changes the direction randomly, and subsequently swims again for about

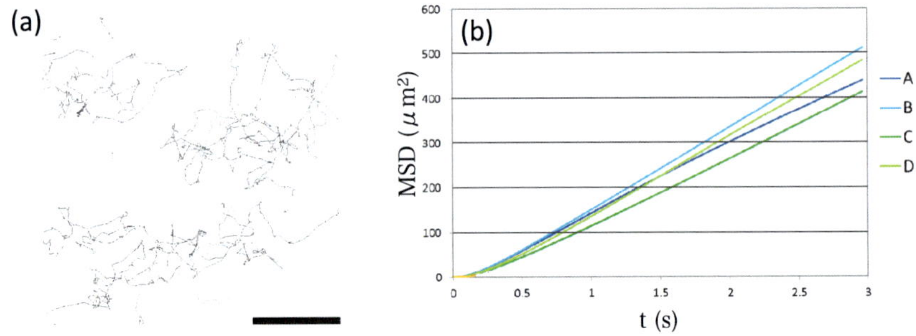

Fig. 2 Typical example of motile cells of *E. coli*. **a** Trajectory of several cells on a 2D plane in 0.2% semi-solid agar. Scale bar: 50 μm. **b** Mean square displacement (MSD) of different strains under different conditions. Lines A and B: wild type strain (RP437). Lines C and D: ΔTar strain (RP5869), which is knock-outed a chemotaxis receptor for aspartate. Lines A and C: 2 mM succinate as an initial concentration of carbon source. Lines B and D: 10 mM succinate. All lines are very similar to one another. This means the effective diffusion coefficients of chemotactic and non-chemotactic cells under different initial concentrations of the carbon source are almost same in our case

0.8 s [10, 11]. Figure 2 demonstrates a random migration (a) and its mean square displacements (MSD) (b) of different strains (chemotactic and non-chemotactic cells) and at different initial concentrations of the carbon source, which is succinate. We can estimate the effective diffusion coefficient of cells as a macroscopic parameter from the MSD plot.

Another feature is chemotaxis. A motile cell of *E. coli* senses chemo-attractants by chemotactic receptors. When a cell swims up the gradient of an attractant, the run time becomes longer. On the other hand, when a cell swims down the gradient, it becomes shorter or does not change. Cells, on the average, move toward the region of higher concentration of the chemo-attractants [10, 12–14]. Figure 3 shows the velocity of the propagating pulse of wild type cells (RP437) with a chemotactic receptor for aspartate and of the front of disk-like pattern [15] of non-chemotactic cells (RP5869) without the receptor for aspartate. Even if each effective diffusion coefficient is not dependent on both strains and the initial concentration of succinate as shown in Fig. 2b, the velocity of chemotactic cells is faster than that of non-chemotactic cells. And it decreases, as the initial concentration of succinate increses. These characteristics are non-trivial in reaction-diffusion systems. Furthermore under certain conditions, cells of *E. coli* make the concentration field of chemo-attractant by excreting chemicals (here aspartate) themselves. The relation between the self-making chemo-attractant concentration field and the chemotactic feature that feels the field, may play an important role to establish the self-organizing pattern formation of *E. coli* (Fig. 1b–d).

Some models based on the reaction-diffusion-chemotaxis systems with the minimal mechanism have been used to understand the above non-trivial pattern formation [16–21]. In these models, the patterns were very similar to the final ones resulting in experiments. But they did neither represent any propagating pulse (mathematical model reproposed by Aotani et al. [20] represented a propagating pulse at higher initial concentration of nutrient in 2D). They, however, did not represent a propagating pulse in 1D and at lower initial concentration of nutrient in 2D, and also did not clarify the two boundaries shown in Fig. 7, nor the relationship between the velocity of the propagating pulse of chemotactic cells and the initial concentration of nutrient (see Fig. 3). How can we revise the models to solve these problems?

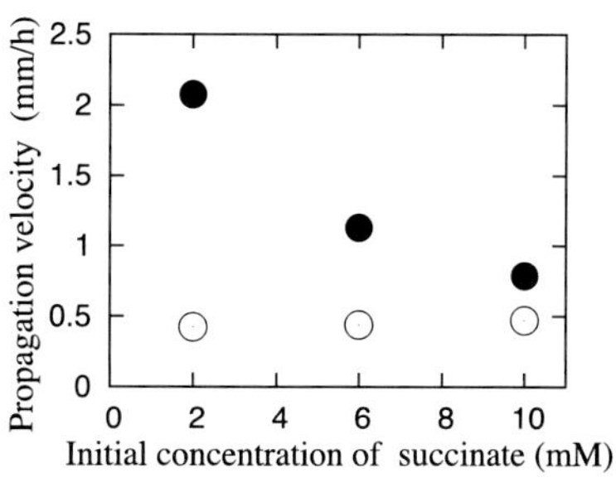

Fig. 3 Velocity of propagating pulse of wild type cells (RP437), *closed circles*, and front of disk-like pattern [15] of non-chemotactic cells (RP5869), *open circles*

In this paper, we summarize the self-organized patterns of *E. coli* experimentally and propose a model by using a reaction-diffusion-chemotaxis mechanism to understand this kind of pattern formation.

2 Methods

In the pre-culture stage, wild cells of *E. coli* (RP437) were grown in 5 ml solution of 0.1 ml/ml M9 salts, 1 mM magnesium sulfate and 0.1 mM calcium chloride for making a M9 minimal medium, 20 μg/ml each of L-threonine, L-leucine, L-histidine and L-methionine for growth, and 5 mM α-ketoglutalate as a carbon source for more than 24 h by using a shaking incubator (37 °C, 200 rpm). Then the condensed *E. coli* suspension was separated from the pre-cultured solution by a centrifugal separator with 7200 rpm for 2 min. The 5 μl suspension was inoculated at the center of a ϕ 90 mm asept petri dish on 10 ml of a similar medium containing 0.2 or 0.24% agar with a different concentration of succinate instead off α-ketoglutalate as a carbon source and 50 μg/ml tetrazolium red as a indicator to enhance image contrast of *E. coli* patterns. The dish was covered for keeping the experimental conditions. All experiments were carried out in a box (W600 mm × H400 mm × D450 mm) controlled at 25 °C. The patterns were recorded from below by cameras (Mintron: MTV-63V1N, MTV-63W1N, and Canon EOS-Kiss).

3 Pattern Formation of *E. coli*

We now present some of the well-known patterns by Budrene and Berg [2, 3], as well as our experimental results. Different patterns appear in a certain range of carbon source (succinate) concentration. At lower or higher initial concentration of succinate, a circular pulse corresponding to higher density of cells, named swarm ring, propagates outwards from the inoculation point in 0.2% semi-solid agar, to be noticed in Fig. 4, where 6 mM succinate is used as the initial concentration. The density rapidly increases at the front of the pulse and slowly decreases at its back (see Fig. 4d–f). The velocity of the pulse is almost constant in the later stage after 40 h, as can be seen in the space-time plot of Fig. 4g.

At certain concentrations of succinate, the circular propagating pulse also appears in 0.2% semi-solid agar with 2 mM succinate as the initial concentration. The velocity is also constant in the later stage (after 30 h) as shown in Fig. 5. Figure 6 shows the enlarged pictures at the right middle region of Fig. 5c. Under this condition, the cells aggregate in the propagating concentric ring and create spots behind the pulse. The spots of high cell density can move soon after aggregating, because the aggregated spots contain vigorously motile cells [2]. After a while, however, the cells lose the ability to swim, when the concentration of succinate decreases at the corresponding location. Thus, the cells in the spots change to non-motile ones. The spots remain

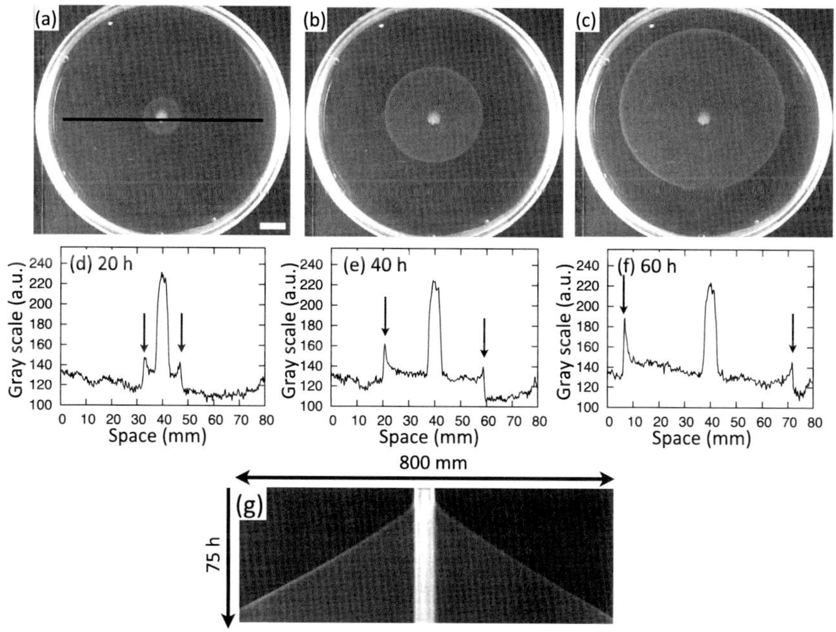

Fig. 4 Propagating pulse, named swarm ring, in 0.2% agar with 6 mM succinate as initial concentration. **a, b**, **b, e** and **d, f** are at 20 h, 40 h and 60 h after inoculation. **d–f** are intensity profiles corresponding to cell density along the *black line* in (**a**). The *arrows* show the front of the propagating pulse. **g** Space-time plot of the propagating pulse along the *black line* in (**a**). *White scale bar* in (**a**): 10 mm. These results correspond to patterns in [2, 3]

then at the same location. Finally, they make beautiful patterns representing the density of non-motile cells (see Fig. 1b–d).

To understand the characteristics of the pattern, we carried out experiments on the velocity of the propagating pulse and the position of the first aggregated spots, as they depend on the initial concentration of succinate in 0.24% semi-solid agar (Fig. 7). There are two boundaries between a propagating pulse without spots (Fig. 4) and with spots (Fig. 5) around 2 and 5 mM initial concentration of succinate. At lower initial concentration, the velocity is getting larger with increase of the initial concentration, but the relationship becomes negative at higher initial concentration. At certain initial concentrations only, between 2 and 5 mM succinate, the aggregating patterns appear and the radius, where the first aggregated spots evolve, decreases with increase of the initial concentration.

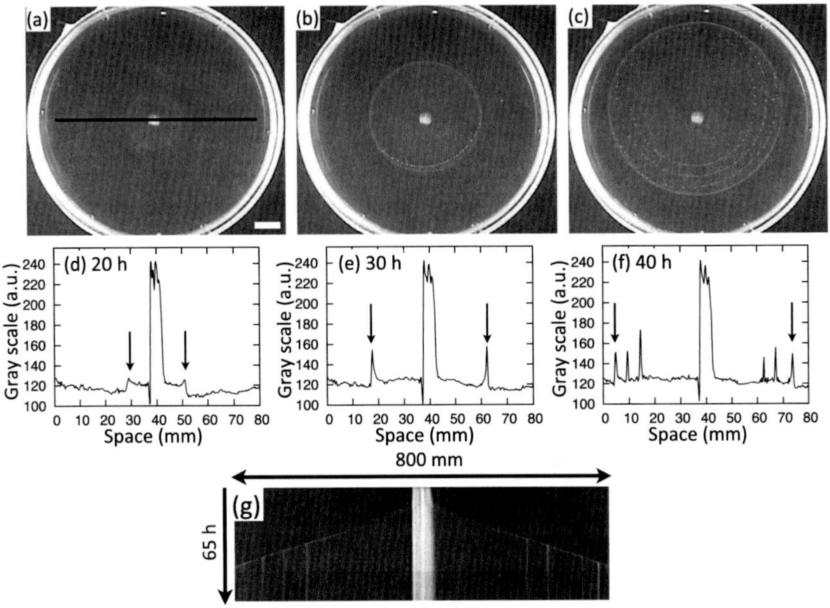

Fig. 5 Propagating pulse and aggregated spots, in 0.2% agar with 2 mM succinate as initial concentration. **a**,**d**, **b**, **e** and **d**, **f** are at 20, 30 and 40 h after inoculation. **d–f** are intenisity profiles along the *black line*in (**a**). The *arrows* show the front of the propagating pulse. **g** Space-time plot of the propagating pulse along the *black line* in (**a**). *White scale bar* in (**a**): 10 mm. These results correspond to patterns in [2, 3]

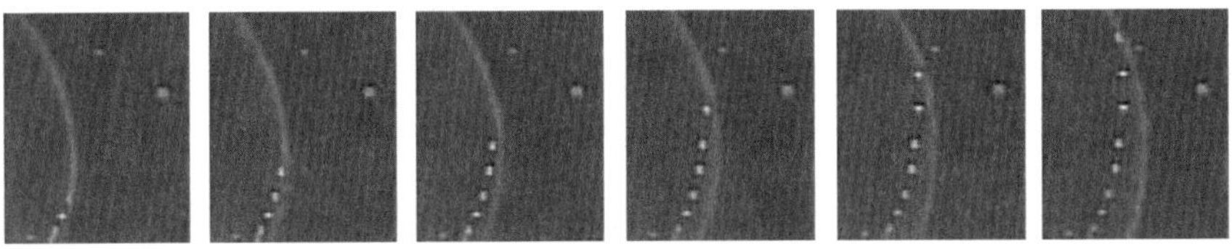

Fig. 6 Enlarged pictures at the *right middle* region of Fig. 5c. Time interval between pictures is 0.5 h

4 Reaction-Diffusion-Chemotaxis Model with a Minimal Mechanism; Numerical Results

Figure 8 shows a schematic diagram of a minimal mechanism for understanding the *E. coli* pattern formation.

The mechanism leads naturally to the following equations:

$$\frac{\partial u}{\partial t} = g(u,n) - i(u,n) + \nabla \cdot \left(D_u(u,n)\nabla u\right) + \alpha(\rho)\nabla \cdot (\chi(u,n)\nabla c(\rho,n)),$$

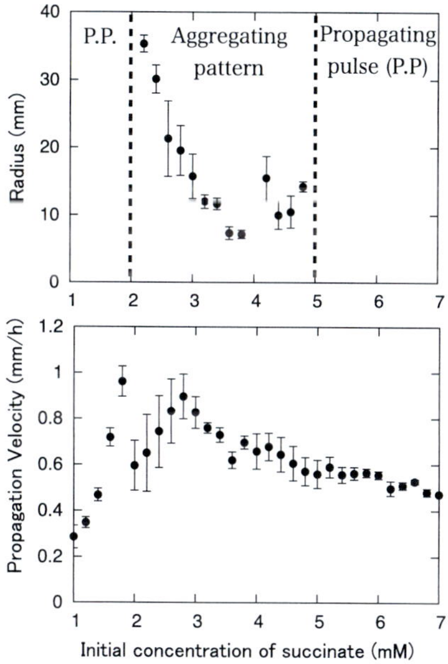

Fig. 7 Velocity of the propagating pulse (*lower graph*) and radius of first aggregated spots in 0.24% agar. The radius corresponds to one of the *red circles* in Fig. 1b–d

$$\frac{\partial v}{\partial t} = \qquad -i(u,n),$$
$$\frac{\partial \rho}{\partial t} = p(u,n) - b\rho + D_\rho \nabla^2 \rho,$$
$$\frac{\partial n}{\partial t} = -\beta g(u,n) - \gamma p(u,n) + D_n \nabla^2 n,$$

where u and v are the populations of active and inactive cells of *E. coli*; ρ and n are the concentration of chemo-attractant (aspartate) and nutrient (succinate). The population of active cells increases by consuming nutrient: growth term $g(u,n)$. The random migration and attraction by gradients of chemo-attractant of active cells leads to a diffusion term: the third term and chemotactic term: the fourth term on the right hand side of the first equation, respectively. Inactive cells do none of these things. In the case of nutrient shortage, the active cells of *E. coli* convert to the inactive state: conversion term $i(u,n)$. Active cells of *E. coli* secrete the chemo-attractant: $p(u,n)$

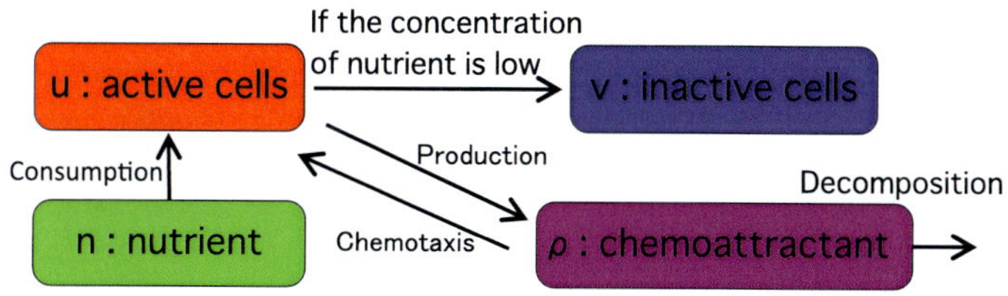

Fig. 8 Schematic diagram of a minimal mechanism for *E. coli* patterns

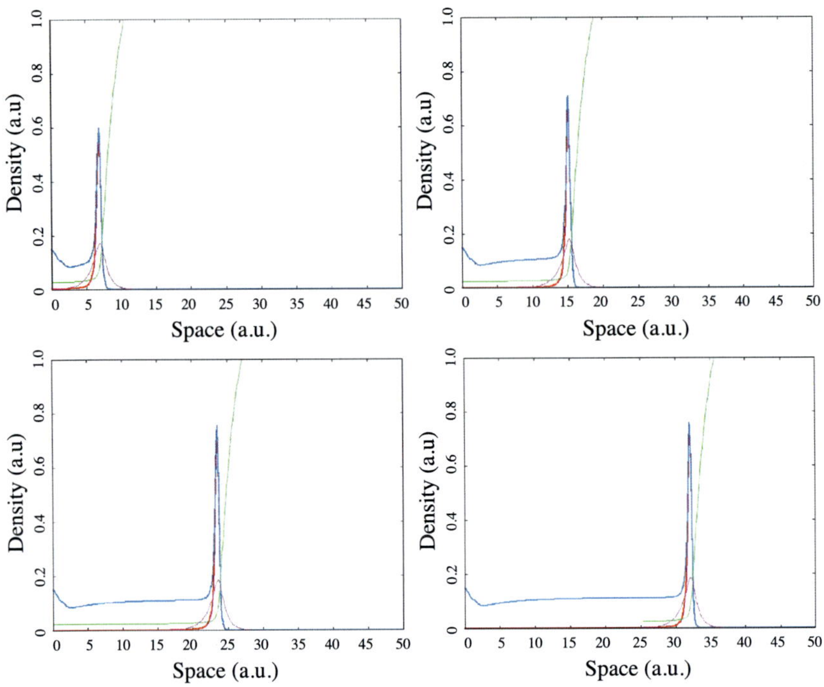

Fig. 9 Profiles of u (density of active cells, red line), $u + v$ (density of all cells, *blue line*), n (concentration of nutrient, *green line*), and ρ (concentration of chemo-attractant, *purple line*). Time interval between figures is 20 (a. u.). Initial values: $u(0, x) = \frac{1}{2}(1 - \tanh(2(x - 1)))$, $v(0, x) = 0$, $\rho(0, x) = 0$, $n(0, x) = 1.3$

which attracts active cells themselves and the chemo-attractant decomposes itself: $b\rho$. Then the nutrient is consumed for increasing the numver of active cells and by producing the chemo-attractant.

Here we use the following functions:

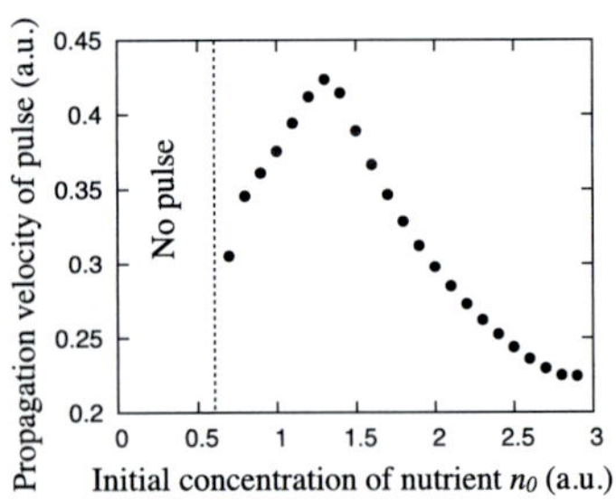

Fig. 10 Velocity of the propagating pulse in 1D

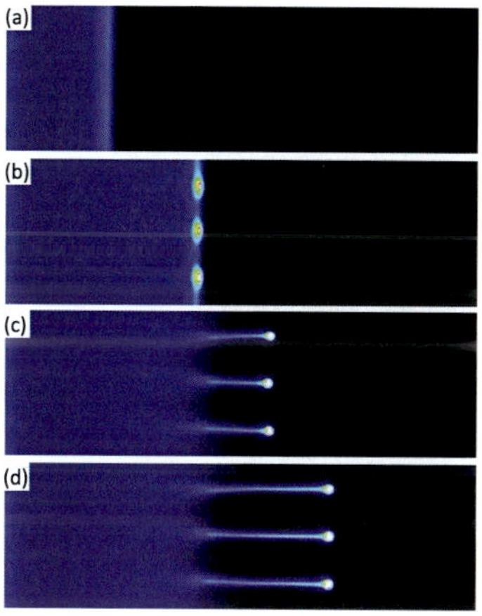

Fig. 11 Typical numerical results in 2D. The color corresponds to the density of cells (u + v). Time interval between figures is 20 (a. u.). Initial values: $u(0, x, y) = \frac{1}{2}(1 - \tanh(2(x-1)))$, $v(0, x, y) = 0$, $\rho(0, x, y) = 0$, $n(0, x, y) = 1.3$

$$g(u,n) = \frac{1}{2}(1 + \tanh(100(u - 0.02)))\, nu,$$
$$i(u,n) = \frac{0.2}{1 + \frac{n}{0,5}}u,\ p(u,n) = anu,$$
$$\alpha(\rho)\nabla \cdot (\chi(u,n)\nabla c(\rho,n)) = \alpha(\rho)\nabla \cdot (u\nabla\rho),$$
$$\alpha(\rho) = 0\ (\rho \geq 0.2) \text{ or } 10\ (\rho < 0.2),\ \beta = 1,\ \gamma = 1,$$
$$a = 10,\ b = 2,\ D_u(u,n) = 0.5,\ D_\rho = 1,\ D_n = 1.$$

We can reproduce the propagating pulse of concentration of cells (u + v) in 1D by using the proposed model depicted in Fig. 9. The profile of the propagating pulse is similar to those of the experimental results in Figs. 4d–f and 5d–f. It is the first numerical representation of a propagating pulse in *E. coli* patterns. Next we present the velocity of this pulse. The positive and negative slope as a function of nutrient (Fig. 10) is in good agreement with the experimental results as shown in Fig. 7.

Figure 11 shows a typical numerical result in 2D. A pulse of condensed cells propagates in the first stage (Fig. 11a), and then cells aggregate along pulse front (Fig. 11b). The pattern (Fig. 11c, d) is similar to one of the experimental results (Fig. 1d). The propagating pulse, however, disappears and the aggregated spots are just moving. This behavior does not correspond to any experimental results.

5 Concluding remarks

We summarized experimental results of self-organized pattern formation of *E. coli* cells and numerical analyses of our proposed model to understand the mechanism. We reported that the velocity of a propagating pulse increases and decreases at lower and higher initial concentration of nutrient, respectively (Fig. 7). Our proposed model using reaction-diffusion-chemotaxis equations represented the propagating pulse and this way the relationship between the velocity and the initial concentration in 1D numerical calculations (shown in Figs. 9 and 10). All models including ours, however, have not yet revealed the secrets of the propagating pulse in 2D. Budrene and Berg have also asked "*How is the identity of the swarm ring maintained?*" to understand the complexity of pattern evolution [2, 3]. The mechanism of the propagating pulse remains still an open question. We hope that our recent experimental and numerical studies will contribute to understanding the self-organized pattern formation of motile cells.

Acknowledgements This research was partially supported by JSPS KAKENHI Grant Numbers, 25103008, 16K21725 and 26400173, and by the JSPS Core-to-Core Program "Non- equilibrium dynamics of soft matter and information". TS thanks his students, Mr. Tomohiro Nakamura, Mr. Hiroshi Iechika and Mr. Hironori Kawakami for useful help and discussion on various aspects of this work.

References

1. H.C. Berg (ed.), *E. coli in Motion*, (Springer, Heidelberg, 2004)
2. E.O. Budrene, H.C. Berg, Complex patterns formed by motile cells of Escherichia coli. Nature **349**, 630–633 (1991)
3. E.O. Budrene, H.C. Berg, Dynamics of formation of symmetrical patterns by chemotactic bacteria. Nature **376**, 49–53 (1995)
4. R. Tokita, T. Katoh, Y. Maeda, J. Wakita, M. Sano, T. Matsuyama, M. Matsushita, Pattern formation of bacterial colonies by *Escherichia coli*, J. Phys. Soc. Jpn. **78** 074005-074005-6 (2009)
5. M. Matsushita, J. Wakita, H. Itoh, I. Rafols, T. Matsuyama, H. Sakaguchi, M. Mimura, Interface growth and pattern formation in bacterial colonies. Physica A **249**, 517–524 (1998)
6. M. Mimura, H. Sakaguchi, M. Matsushita, Reaction-diffusion modelling of bacterial colony patterns. Physica A **282**, 283–303 (2000)
7. E. Ben-Jacob, O. Schochet, A. Tenenbaum, I. Cohen, A. Czirok, T. Vicsek, Generic modelling of cooperative growth patterns in bacterial colonies. Nature **368**, 46–49 (1994)
8. I. Golding, Y. Kozlovsky, I. Cohen, E. Ben-Jacob, Studies of bacterial branching growth using reaction-diffusion models for colonial development. Physica A **260**, 510–554 (1998)
9. Y. Kozlovsky, I. Cohen, I. Golding, E. Ben-Jacob, Lubricating bacteria model for branching growth of bacterial colonies. Phys. Rev. E **59**, 7025–7035 (1999)
10. H.C. Berg, D.A. Brown, Chemotaxis in Escherichia coli analysed by three-dimensional tracking. Nature **239**, 500–504 (1972)
11. H.C. Berg (ed.), *Random Walks in Biology* (Princeton Univ Press, Princeton, 1993)
12. J. Adler, Chemotaxis in bacteria. Science **153**, 708–16 (1966)
13. H.C. Berg, L. Turner, Chemotaxis of bacteria in glass capillary arrays. Biophys. J. **58**, 919–930 (1990)
14. M. Eisenbach (ed.), *Chemotaxis*, (Imperial College Press, London, 2004)
15. T. Sakurai, T. Tsujikawa, in preparation
16. L. Tsimring, H. Levine, I. Aranson, E. Ben-Jacob, I. Cohen, O. Shochet, W.N. Reynolds, Aggregation patterns in stressed bacteria. Phys. Rev. Lett. **79**, 1859–1862 (1995)
17. E. Ben-Jacob, I. Cohen, O. Shochet, I. Aranson, H. Levine, L. Tsimring, Complex bacterial patterns. Nature **373**, 566–567 (1995)

18. R. Tyson, S.R. Lubkin, J.D. Murray, A minimal mechanism for bacterial pattern formation. Proc. R. Soc. Lond. B **266**, 299–304 (1999)
19. A.A. Polezhaev, R.A. Pashkov, A.I. Lobanov, I.B. Petrov, Spatial patterns formed by chemotactic bacteria Escherichia coli. Int. J. Dev. Biol. **50**, 309–314 (2006)
20. A. Aotani, M. Mimura, T. Mollee, A model aided understanding of spot pattern formation in chemotactic E. coli colonies. Jpn. J. Indust. Appl. Math. **27**, 5–22 (2010)
21. J.H. Miller (ed.), *Experiments in Molecular Genetics* (Cold Spring Harbor Laboratory Press, New York, 1972)

From Synchronised to Desynchronised Glycolytic Oscillations in Individual Yeast Cells

André Weber, Yury Prokazov, Werner Zuschratter and Marcus J.B. Hauser

Abstract The cell density dependence of the glycolytic oscillations of intact yeast cells of the strain *Saccharomyces carlsbergensis* was investigated at both the levels of cell populations and of individual cells. On the collective, macroscopic population level the oscillation amplitudes gradually decreased with decreasing cell densities and below the critical density the collective dynamics became quiescent. However, single cells kept their oscillatory behaviour for low and very low cell densities even when cells were far away and isolated from each other. The lower the cell density of a population, the more the glycolytic oscillations of the individual cells became independent from that of their neighbours. This leads to a broadening of the distribution of oscillation frequencies in low-density populations, and also to a loss of entrainment among the individual cells.

A. Weber (✉) · M.J.B. Hauser
Biophysics Group, Institute of Experimental Physics, Otto-von-Guericke University Magdeburg, Universitätsplatz 2, 39106 Magdeburg, Germany
e-mail: andre.weber@lin-magdeburg.de

M.J.B. Hauser
e-mail: marcus.hauser@ovgu.de

A. Weber · Y. Prokazov · W. Zuschratter
Special Laboratory for Electron and Laserscanning Microscopy, Leibniz Institute for Neurobiology, Brenneckestraße 6, 39120 Magdeburg, Germany
e-mail: yury.prokazov@lin-magdeburg.de

W. Zuschratter
e-mail: zuschratter@lin-magdeburg.de

M.J.B. Hauser
Department of Biometry and Medical Informatics, Otto-von-Guericke University Magdeburg, Leipziger Straße 44, 39120 Magdeburg, Germany
e-mail: marcus.hauser@ovgu.de

S.C. Müller et al. (eds.), *Complexity and Synergetics*,
https://doi.org/10.1007/978-3-319-64334-2_19

1 Introduction

Synchronisation is a frequent manifestation of self-organisation, where individuals coordinate their behaviour in time, leading to the emergence of collective, macroscopic dynamics [1–3]. For instance, in synchronised populations of cells, all individuals adhere to a collective, common rhythm. Typical examples of synchronisation in biological systems are the blinking rhythms of male fireflies [4, 5], the synchronisation of circadian rhythms in mammals [6–8], the synchronisation of cells in the segmentation clock of vertebrates [9], the immune response to malaria parasites [10, 11], and the macroscopic, collective glycolytic oscillations of yeast cells [12]. Although the physiological function of glycolytic oscillations in yeast cells is not yet known [13], it is conjectured that for unicellular organisms, such as yeast, life within a community is beneficial for their long-term survival [14]. Efficient cell-to-cell communication is a prerequisite for the organisation of communities, and such communication among individuals may be provided by metabolic oscillations.

Glycolysis is a fundamental pathway in the energy metabolism of eukaryotic cells and it may show oscillatory dynamics. In fact, glycolytic oscillations of NADH and other metabolites in populations of yeasts have been intensively studied [15–19]. In dense populations, the individual cells synchronise their metabolism to a joint oscillatory mode [16, 20]. This has been demonstrated in experiments, where two sub-populations that oscillated with the same frequency, but at opposite phases, were blended and a new, collective oscillatory time-trace was recovered after a few oscillatory cycles [18, 21].

The coupling of individual yeast cells is based on the relay and subsequent diffusion of the messenger molecule acetaldehyde [18] through the extracellular medium, where it is absorbed by the other cells. In addition to studies in stirred cell suspensions, the propagation of waves of glycolytic activity has been observed in spatially extended media, such as settled cell suspensions [22] or gel-entrapped cells [23].

Already early on, it has been observed that the dynamics of a cell population depends on cell density. At high cell densities yeast cells show synchronous, coherent oscillations. Once the cell density falls below a critical threshold, the yeast population no longer displays any collective oscillations, but rather remains quiescent [24]. The transition between collective oscillatory and stationary dynamics at the population level can follow two different pathways. The first is called ‘amplitude death’ [25] in the field of Dynamical Systems and ‘dynamical quorum sensing’ in biological literature [26, 27] and consists in the simultaneous cessation of oscillatory dynamics in every cell of the population when the cell density drops below the critical threshold. Thus, the stationary dynamics at the collective level is identical to that of all cells of the population, namely quiescent. Alternatively, the transition between oscillatory and quiescent dynamics at the collective, macroscopic level may follow a so-called ‘Kuramoto transition’ [28, 29]. The collective oscillatory signal is generated at high cell densities, where the oscillations of all individual cells are coherent and synchronised both in phase and oscillation frequency. At low densities, the individual cells remain oscillatory, but they lose their coherence so that each of them oscillates with its own phase. As the collective behaviour is the sum of the signals of the individual cells, phase-incoherent oscillations will lead to stationary, quiescent collective dynamics [30, 31].

Theoretical studies on oscillators coupled through an external medium show that the mechanism of transition between synchronous collective oscillations and quiescence on the macroscopic level depends on both, cell density and the strength of the coupling between individual oscillators (or cells). In fact, the Kuramtoto type of transition is found to take place at low coupling strengths. On the other hand, amplitude death (dynamical quorum sensing) occurs when the cells are strongly coupled to each other [25].

Both types of transitions, i.e., amplitude death and Kuramoto transition, have been reported to occur in yeast cell colonies. As the occurrence of amplitude death (dynamical quorum sensing) requires strong coupling among the cells, such dynamics is usually observed in stirred yeast cell suspensions [26] or in suspended and stirred beads of an oscillatory reaction [27]. Here, the coupling among cells is strong, since stirring provides for an homogeneous concentration field of the messenger acetaldehyde, such that each cell is effectively coupled to all others through the extracellular medium. On the other hand, early reports that individual cells remain oscillatory in unstirred cell suspensions at low cell densities [32] have recently been confirmed in experiments using either *S. carlsbergensis* cells immobilised on coverslips [33] or *S. cerevisiae* fixed in a microfluidic device [34].

The present study aims at investigating details of the transition from synchronous to asynchronous dynamics in a colony of immobilised cells. Under these conditions, at the population level the transition between synchronous and quiescent dynamics occurs via desynchronisation of the oscillations of the individual cells. Here, we wish to shed some more light into the mechanisms leading to a Kuramoto transition as observed in colonies of immobilised *Saccharomyces carlsbergensis* cells.

2 Materials and Methods

2.1 Experimental Procedures

Cells of the yeast *Saccharomyces carlsbergensis* were cultivated aerobically until the glucose in the medium was exhausted at the transition from the logarithmic to the stationary growth phase [20]. The cells were harvested, washed and stored as described by Weber et al. [33]. The wet cells were suspended in phosphate buffer as a 20% (weight/volume, w/v) suspension and stirred at 23 °C until they showed NADH oscillations. For the experiments, cell suspensions containing different cell densities were prepared and transferred to the batch chamber (for details, see Ref. [33]). The batch chamber consists of two parts between which the coverslip is clamped. 100 μl well-aerated yeast suspension of a chosen cell density were placed into the chamber. After about 30 min, all cells sedimented and adhered to the poly-D-lysine-coated coverslip when their plasma membranes got immobilised by contact with the polylysine coating.

After sedimentation and immobilisation of the cells the experiment was started. 3 mM KCN were added to induce anaerobiosis by blocking the activity of the mitochondria. About 10 min later, 52 mM glucose were added to trigger the glycolysis and induce the metabolic oscillations in the cells.

The intracellular dynamics was monitored through the autofluorescence of NADH which serves as an indicator for glycolytic activity [20], since NADH is an intracellular metabolite directly involved in glycolysis and energy metabolism. The autofluorescence of NADH is characterised by absorption and emission maxima at $\lambda_{ex} = 340$ nm and $\lambda_{em} = 460$ nm, respectively [35].

The NADH autofluorescence from single yeast cells was measured with an inverted Nikon Ti Eclipse microscope, equipped with a 100×/0.5–1.3 plan fluor lens and a position sensitive single photon counting photomultiplier tube [36–38]. For excitation of intracellular NADH a 8 MHz pulsed frequency-tripled Nd:vanadate laser tuned at 355 nm was used. A dichroic mirror discriminated between excitation and emission. The emission light of yeast cells was filtered by a long-pass (LP 442 nm) and a bandpass filter (FF01–440/40) and detected by the photomultiplier. For all studied samples, the laser intensity was adjusted such that 3,000–65,000 fluorescence photons/s reached the detector, thus making sure that it operated under optimal conditions. The positions where the photons were detected on the photomultiplier were binned into frames of 512×512 pixels, resulting in a resolution of 0.33 μm/pixel in the object space. The field of view had a diameter of 169 μm. The photon flux was integrated over 2 s time intervals, which was a sufficient sampling rate to analyse glycolytic oscillations with periods T ranging from approximately 24 to 70 s.

2.2 Image Analysis

The immobilised yeast cells were randomly distributed on the coverslip. In our experiments, we analyse the NADH fluorescence signal stemming from domains of different sizes within the field of view of the microscope. We analysed the collective population signal as well as the fluorescence time-series for each of the individual cells. The macroscopic, collective population signal was obtained (i) by the fluorescence acquired over the entire field of view and (ii) as the average of the baseline-subtracted fluorescence of all individual cells found in the regions of interest (ROIs). In these two cases, all incident photons stemming from the entire field of view were summed up every second.

Furthermore, we also monitored the fluorescence signal from each single cell. To this purpose the exact position of each cell in the population was determined as region of interest. At any time, the single cell fluorescence F_{cell} is the mean value of the intensity of the fluorescence signal detected in the area occupied by an individual cell

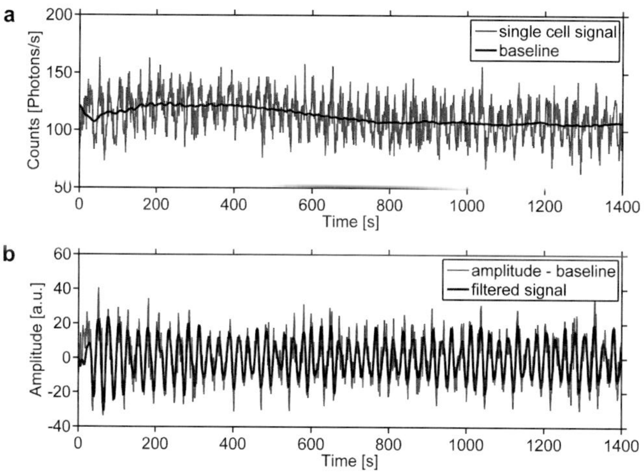

Fig. 1 Baseline correction to eliminate spurious drifts and trends as well as bandpass filtering to the frequency range of glycolytic oscillations. **a** For each individual cell signal the baseline was computed as the walking average of the fluorescence data using a 100 s wide window. **b** The baseline was subtracted from the fluorescence signal of the yeast cell. The oscillation frequency of each cell was determined by a fast Fourier transform. Noise was reduced by applying a Fourier bandpass filter that cuts off frequencies higher than 0.050 Hz and lower than 0.014 Hz. These filtered time-series were used for the calculation of the phases ϕ_i

$$F_{cell} = \frac{\sum c}{a}, \tag{1}$$

where c is the number of incident photons originating from the area a occupied by the individual cell. The temporal sequence $F_{cell}(t)$ yields the time evolution of the fluorescence of an individual cell.

The time-series of the fluorescence of each cell was subjected to a baseline subtraction such that any spurious drifts or longterm trends were eliminated. The baseline was computed as the walking average of the fluorescence data using a time window of 100 s (Fig. 1a). After baseline subtraction, we obtained oscillations around zero (Fig. 1b) and determined their frequencies by a fast Fourier transform. These baseline subtracted single cell fluorescence signals were averaged to obtain the collective signal of all ROIs.

Noise reduction was achieved by bandpass filtering. The applied Fourier bandpass filter cut off frequencies higher than 0.05 Hz and lower than 0.014 Hz. Thus, the frequencies of the glycolytic oscillations were preserved in the filtered time series $x_i(t)$.

The phase $\phi_i(t)$ of each oscillating cell i

$$\phi_i(t) = arctan\left(\frac{\widetilde{x}_i(t)}{x_i(t)}\right) \tag{2}$$

was computed through the Hilbert transform

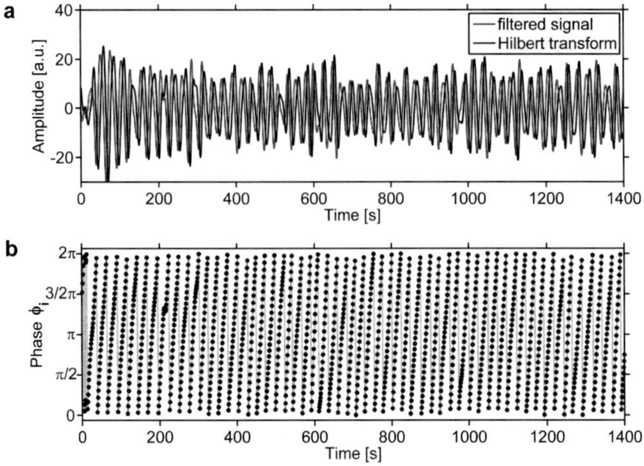

Fig. 2 Hilbert transform and phase of the bandpass filtered fluorescence signal of a single cell. For each cell of a population, **a** the Hilbert transform and **b** the phase ϕ_i were calculated from the bandpass filtered oscillations

$$\widetilde{x}_i(t) = \frac{1}{\pi} \int_{-\infty}^{\infty} \frac{x_i(t')}{t - t'} dt' \tag{3}$$

of the filtered single cell signal $x_i(t)$ (Fig. 2).

The order parameter K, introduced by Shinomoto and Kuramoto [31],

$$K(t) = \left| N^{-1} \sum_i^N exp(i\phi_i(t)) - \left\langle N^{-1} \sum_i^N exp(i\phi_i(t)) \right\rangle \right|, \tag{4}$$

was chosen for measuring the phase synchronisation at each time point. An order parameter $K = 1$ indicates a complete synchronisation (complete coherence) among the cells (oscillators) in a population, whereas $K = 0$ means that the cells oscillate at random phases.

3 Results

Intact yeast cells of the strain *S. carlsbergensis* reveal that the dynamics of glycolytic oscillations depends on the cell density of the population. Cell populations with densities varying from $\rho = 0.001$ to 0.8% were investigated.

The collective population signal has been analysed in two ways, as described in Sect. 2.2. First, the fluorescence signal was collected over the entire field of view, yielding the raw macroscopic signal. Second, we determined the area and position of each cell (as a ROI) in the field of view, and collected only photons stemming from

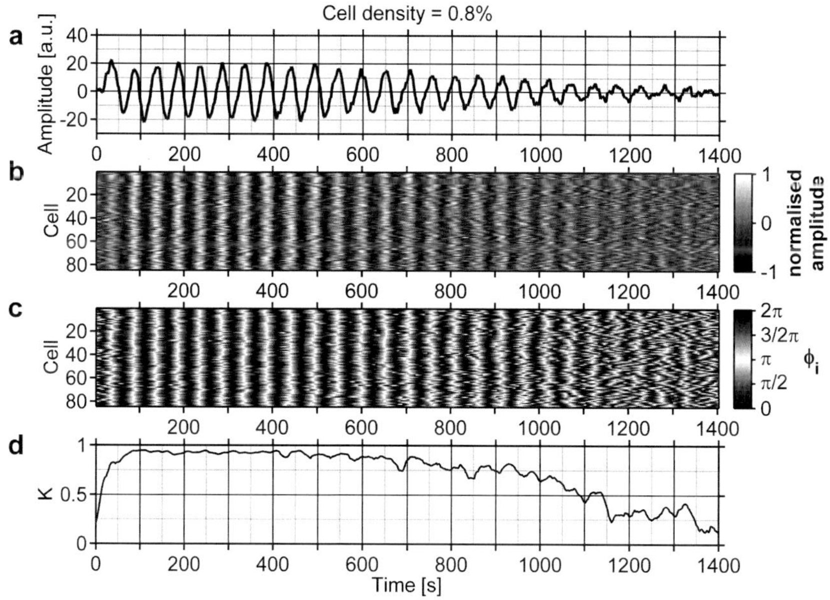

Fig. 3 Collective and individual dynamics of a population of *S. carlsbergensis* cells at a high cell density of 0.8%. The oscillations show a common rhythm. Oscillations start at $t = 0$ s about 80 s after an aliquot of 52 mM glucose was added to the starved and immobilised cells. During the first ~50 s, the cells start to oscillate and to synchronise with each other. **a** The baseline corrected time-series of the averaged single cell signals (ROIs) represents the collective fluorescence signal, which shows well-developed glycolytic oscillations. **b** The oscillations of the individual cells are highly coherent and synchronised in their amplitudes. At $t > 1000$ s the oscillation amplitudes decay and the cells lose synchrony. For visualisation the individual oscillations were filtered and their highest amplitudes were normalised to the range $[-1, 1]$. Note that cells are numbered randomly. **c** The phases of the oscillations of the individual cells are synchronised with each other. Again, for $t > 1000$ s the cells desynchronise. **d** For $50\text{ s} \leq t \leq 600$ s the time-dependent Kuramoto order parameter K is higher than 0.9, indicating a very high coherence of the oscillations of the individual cells

the cells while neglecting any photons that stem from the cell-free background. The collected photons were averaged and the time-series baseline-corrected. These time-series are referred to as baseline-corrected averaged single cell traces. In addition to these collective signals, the microscopic dynamics of every individual yeast cell in the population were tracked, too.

An aliquot of 52 mM glucose was added to the cell medium containing starved yeast cells to induce glycolytic oscillations. By convention, time $t = 0$ s denotes the onset of the glycolytic oscillations approximately 40–200 s after glucose addition. The oscillatory response of the yeast cells was long-lasting, however, always transient, since the experiments were performed under batch conditions.

Dense yeast cell populations are known to show collective glycolytic oscillations, which are clearly detectable on a macroscopic scale, as depicted in Fig. 3. The oscillations of the individual cells were highly synchronised in their amplitudes (Fig. 3b)

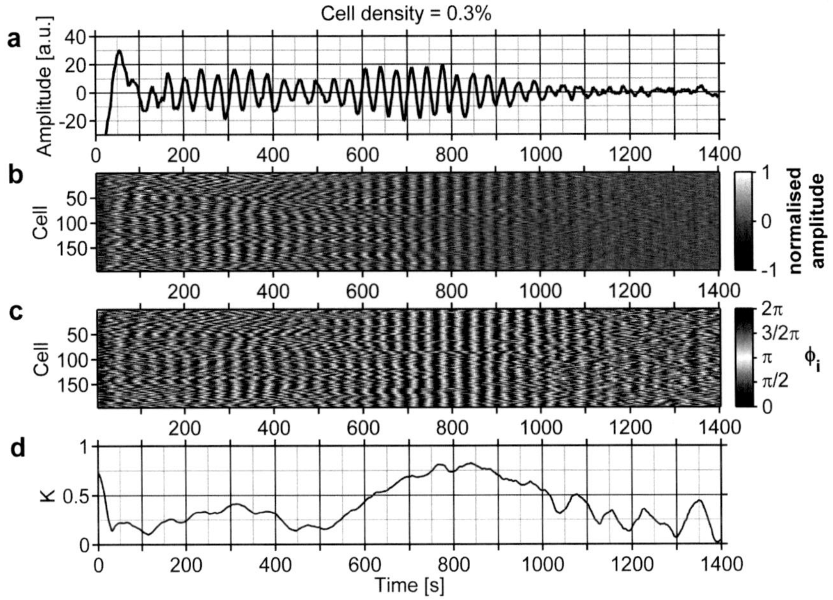

Fig. 4 Collective and individual dynamics of *S. carlsbergensis* at an intermediate cell density of 0.3% shows a transition from synchronous to desynchronous dynamics. Oscillations start at $t = 0$ s about 200 s after an aliquot of 52 mM glucose was added to the cells. **a** The time-series of the collective fluorescence signal oscillates with changing amplitudes. Two episodes of oscillations of larger amplitude occur, the first between $200\text{ s} \leq t \leq 400$ s and the second between $600\text{ s} \leq t \leq 800$ s. **b** The individual cells remain oscillatory during the experiment. At $t = 600$ s a transient episode of synchronised, collective oscillations sets in and the population signal attains its highest amplitudes. For $t \geq 1000$ s the single cell oscillation amplitudes dampen, which causes a decaying population signal. Note that the cells are numbered in an random order. **c** At the beginning and the end of the experiment most of the cells oscillate at their own periods and phases (ϕ_i), which leads to lower amplitudes in the collective signal (*a*). The synchronisation episode at $600\text{ s} \leq t \leq 1000$ s is caused by a temporary entrainment of the oscillations of the individual cells. **d** The time-dependent Kuramoto order parameter K reflects the degree of synchronisation among the cells

and phases (Fig. 3c). The phase synchronisation is reflected by the time dependent Kuramoto order parameter (Fig. 3d), which, after a short induction time, attained values above 0.9, indicating a high degree of synchronisation and coherence of the cellular oscillations.

Macroscopic population signals were monitored in experiments using different cell densities. For intermediate cell densities in the range 0.3–0.01% the macroscopic and collective oscillations were detectable and the cells gradually lost their synchrony in periods and phases as the cell density decreased.

When the cell densities fall below $\rho \leq 0.3\%$, desynchronisation begins to set in (Fig. 4, where $\rho = 0.3\%$). In the baseline-corrected time-series, the desynchronisation manifests itself in macroscopic oscillations whose amplitudes may change in time (Fig. 4a). On the other hand, on the level of individual cells, one observes that each cell shows glycolytic oscillations. However, for a long time (i.e., up to $t \sim 550$ s)

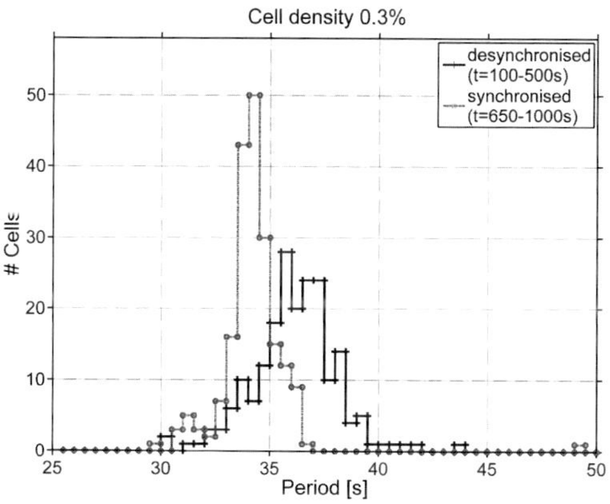

Fig. 5 Histograms of oscillation periods of individual cells in a population of intermediate cell density ($\rho = 0.3\%$) for episodes of desynchronised (*black line*) and synchronised (*gray line*) oscillations. Single cell oscillations during the episode of desynchronisation ($100\,\text{s} \leq t \leq 500\,\text{s}$, Fig. 4) show a fairly broad distribution of oscillation periods. During phase synchronisation ($650\,\text{s} \leq t \leq 1000\,\text{s}$, Fig. 4) the oscillation periods decrease and the distribution of periods narrows since most cells entrain their oscillations to a common period

these oscillations are desynchronised in both amplitude (Fig. 4b) and phase (Fig. 4c). After this induction time, the amplitude of collective oscillations increased (Fig. 4a), indicating the formation of a transient episode of large-amplitude oscillations. These are due to a temporal synchronisation of the oscillations of the individual cells (e.g., from $600\text{ s} \leq t \leq 1000\text{ s}$ in Fig. 4b, c). The time-dependent order parameter K reflects this behaviour: it lies between 0.2 and 0.4 for times $t \leq 550\,\text{s}$ indicating that the oscillation phases of the cells are desynchronised (Fig. 4d). For times $t \geq 600\,\text{s}$ cells synchronised their oscillations for about 400 s and K reached values higher than 0.75.

The distribution of the oscillation periods of the individual cells was determined for the two episodes of distinct macroscsopic oscillations shown in the time-series of the population of intermediate cell density ($\rho = 0.3\%$, Fig. 4). The distribution of the oscillation periods during the episode of desynchronised cell oscillations (that lasts from $\approx 150\text{ s} \leq t \leq 450\text{ s}$, Fig. 4) is relatively broad (Fig. 5), whereas the distribution narrows considerably during the episode of coherent, synchronous oscillations (from $\approx 600\text{ s} \leq t \leq 1000\text{ s}$, Fig. 4). Furthermore, we observe that the synchronised oscillations present a shorter mean period ($T = 33$ s) than that of the desynchronised oscillations, whose mean period is about $T = 37$ s (Fig. 5).

As the cell densities are decreased even further, it becomes more and more difficult to detect the collective, macroscopic glycolytic oscillations because the amplitude of these oscillations shrink. This is the case once the cell density gets as low as 0.01% (Fig. 6). At such low densities, any possible macroscopic oscillations succumb to the inherent noise such that it becomes impossible to decide whether coherent oscillations still occur. In fact, the raw macroscopic fluorescence signal suggests the absence of collective oscillations (Fig. 6a).

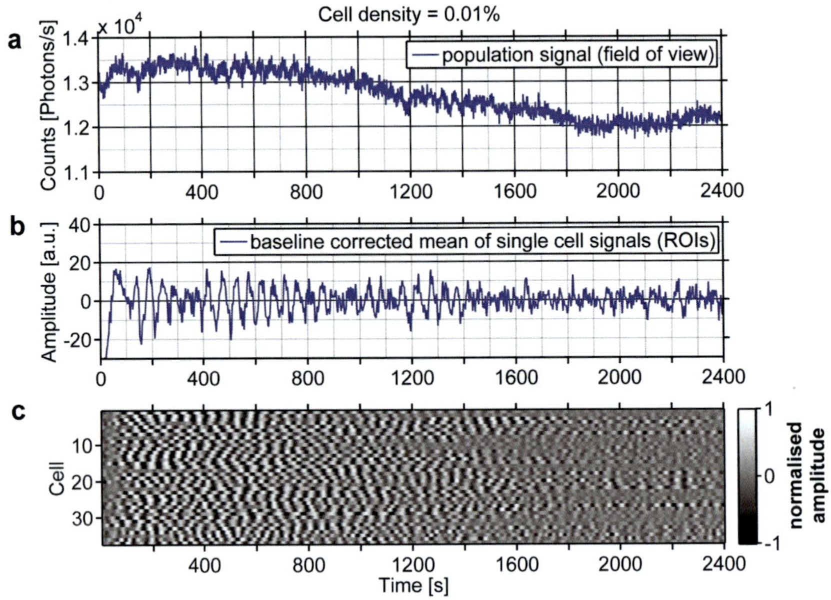

Fig. 6 Macroscopic population signal of the whole field of view compared to the averaged and baseline corrected dynamics of the cell population at a low cell density of 0.01%. **a** The time-series of the macroscopic, collective fluorescence signal of the whole field of view is noisy and consists of quiescent and possibly oscillatory phases. Transient reminiscences of oscillations can be spotted at $400\,\text{s} \leq t \leq 800\,\text{s}$. **b** The oscillations of individual cells were averaged after detrending. The oscillation amplitudes change in time and eventually the collective oscillations cease because individual cells oscillate out of phase. **c** Even at low cell densities the individual cells remain oscillatory when the macroscopic collective signals of (**a**) and (**b**) are already quiescent, i.e., at $t > 1400\,\text{s}$

However, the inspection of the dynamics at the level of individual cells reveals that the cells continue to oscillate (Fig. 6c) even when the macroscopic oscillations ceased, i.e., at $t < 400\,\text{s}$, or $800\,\text{s} < t < 1200\,\text{s}$ (Fig. 6a). The collective signal computed after baseline-subtraction (Fig. 6b) is considerably less noisy than the raw macroscopic signal. Here, short episodes of oscillatory, low-amplitude dynamics and epochs of quiescence are observed. These episodes are reflected by events of some coherence (at $400\,\text{s} \leq t \leq 800\,\text{s}$ and $t \approx 1200\,\text{s}$ in Fig. 6b) or decoherence ($800\,\text{s} < t < 1200\,\text{s}$ in Fig. 6b), respectively, of the phases and frequencies of the oscillations of the individual cells (Fig. 6c). For times $t > 1600\,\text{s}$ no collective oscillations were detectable because only a small number of cells continued to oscillate while others oscillated with decaying amplitudes due to the batch conditions.

Yeast populations of very low cell density ($\rho \leq 0.01\%$) are characterised by a loss of synchronisation of the glycolytic oscillations of the individual cells. The dynamics of individual cells reveals that oscillations in each of them may begin at different points in time (Fig. 7). Despite of this cell-to-cell variability, the cells remained oscillatory, but they oscillate with increasingly more independent rhythms (Fig. 7b) the lower the cell density is. In sparse populations ($\rho \leq 0.01\%$) the oscilla-

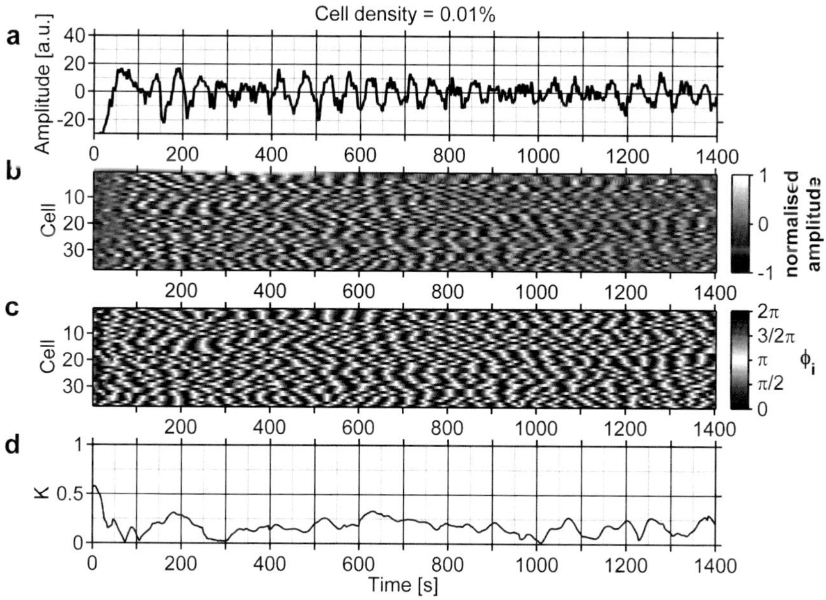

Fig. 7 Collective and individual dynamics of *S. carlsbergensis* at a low cell density of 0.01% with loss of entrainment. **a** The collective fluorescence signal shows oscillations with small amplitudes and epochs of transient quiescence at $t \sim 300$ s and $t \sim 900$ s. **b** The individual cells remain oscillatory. The oscillation amplitudes of the individual cells are decoherent. Note that the time spans where individual cells show pronounced oscillatory amplitudes vary considerably from cell to cell. **c** Each cell oscillates with its own phase and oscillation period. Slight and transient entrainment in the oscillation phases of some cells occasionally give rise to the reminiscences of oscillations, e.g. at $400\,\text{s} \leq t \leq 800$ s. **d** The time-dependent Kuramoto order parameter K remains low at all times, indicating desynchronisation of the cells

tion phases of the cells are incoherent (Fig. 7c), and consequently the order parameter is low ($0 \leq K \leq 0.3$, Fig. 7d). The baseline-subtracted collective signal (Fig. 7a) reflects the dephasing of individual oscillations in episodes of quiescence (e.g., at $250\,\text{s} < t < 400\,\text{s}$) and shows episodes of increasing oscillation amplitudes, when individual oscillations become transiently more synchronised.

The frequency spectra and corresponding histograms of oscillation periods obtained from experiments performed at cell densities of $\rho = 0.8\%$, 0.3% and 0.01%, respectively, are shown in Fig. 8. We analysed the oscillations of cells during the time interval $200\,\text{s} \leq t \leq 600$ s. As the cell density of the population diminishes the distribution of the oscillation frequencies widens. At very low cell densities this leads to the situation where cells become decoupled and oscillate independently with their own, individual frequencies. Figure 8 provides evidence for these findings: At a high cell density of $\rho = 0.8\%$ basically all cells shared a common frequency of $f \approx 0.0194$ Hz and oscillated in concert. The glycolytic oscillations of the intermediate cell density of $\rho = 0.3\%$ are characterised by a broader frequency distribution where most cells in the population share a common frequency band around $f \approx 0.028$ Hz, but individual

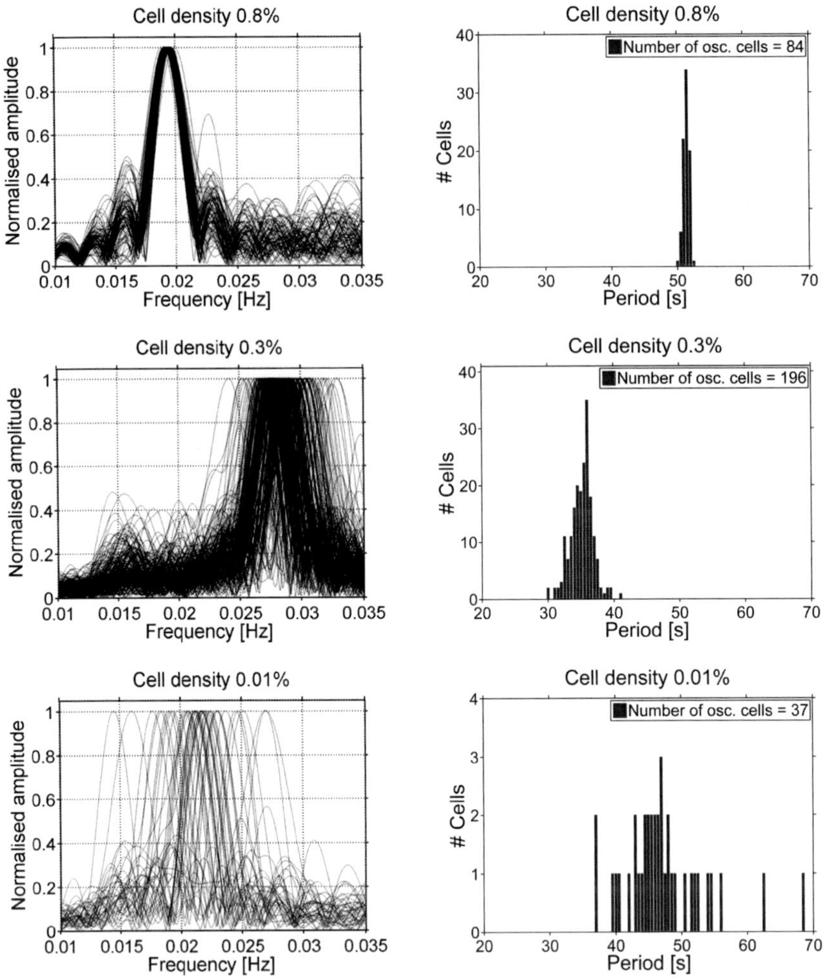

Fig. 8 Frequency spectra and histograms of oscillation periods of individual cells in populations of cell densities of 0.8, 0.3, and 0.01%. Fourier spectra of the oscillations of individual cells were calculated for the time intervals $200\,\text{s} \le t \le 600\,\text{s}$. At high cell density ($\rho = 0.8\%$) all cells oscillate with the same frequency which leads to a narrow distribution of the periods (*top row*). The lower the cell density of the population was chosen, the more cells lose coherence in oscillation periods. For $\rho = 0.3\%$ (*middle row*) the distribution of the oscillations periods of the cell widens, and for even lower densities ($\rho = 0.01\%$, *bottom row*) nearly all cells oscillate with their own frequencies. Note that the number of cells for $\rho = 0.8\%$ is relatively small because only a few representative cells were selected for the analysis

cells differed. In case of low cell densities ($\rho \le 0.01\%$) most of the cells oscillated with their individual frequencies, which were broadly distributed (Fig. 8).

Note that the frequency of oscillations can vary from experiment to experiment (population to population), even when all experimental parameters are kept constant;

this is most probably due to tiny differences in the individual preparations of cells. In other words, while the frequency distribution always remains narrow, a certain variability of the position of the center of the distributions is found for populations under the same nominal conditions.

4 Discussion

We have studied the dynamics of glycolytic oscillations in *Saccharomyces carlsbergensis* cell populations in dependence of the cell density. The collective, macroscroscopic dynamics and the dynamics of each cell were monitored at the cellular level. We observe that at a critically low cell density the collective oscillations cease and give way to quiescent dynamics, whereas on the single cell level the individual cells continue to oscillate. The oscillatory dynamics of individual cells persists even for very low cell densities (down to $\rho = 0.001\%$), where the cells are almost isolated.

The mechanism leading to the transition of the collective dynamics from oscillatory to quiescent (and vice-versa) is due to a loss (gain) of coherence among the glycolytic oscillations taking place in the individual cells. Below the critical density $\rho_{crit} \sim 0.3\%$, the cells continue to display glycolytic oscillations, but these oscillations begin to lose their synchrony in phase and frequency. This desynchronisation scenario is typical for a Kuramoto transition to incoherence (or desynchronisation) [28, 30, 31, 33].

The coupling between cells is provided by the messenger molecule acetaldehyde. During glycolytic oscillations the cells release acetaldehyde periodically to the extracellular medium, where the concentration of this messenger is detected by other cells. When and where the extracellular concentration of acetaldehyde is sufficiently high, yeast cells start to oscillate in synchrony with the oscillating concentration of acetaldehyde in the extracellular medium. The magnitude of acetaldehyde oscillations hence provides for the strength of the coupling between cells of a population. At populations of high cell densities a sufficiently high magnitude of extracellular acetaldehyde is reached very briefly after addition of glucose, leading to synchronised oscillations of the cells and to a clear time-series of collective oscillations. By contrast, at very low cell densities the extracellular concentration of acetaldehyde always remains too low to induce synchrony in the glycolytic oscillations of the individual cells.

An interesting behaviour is observed in yeast populations of intermediate density, e.g., in the experiment run at a cell density of $\rho = 0.3\%$ (Fig. 4), where desynchronisation sets in. Here, oscillations at cellular level set in shortly after addition of glucose (at $40\text{ s} \leq t \leq 200\text{ s}$) leading to collective oscillations of low amplitude since only a part of the cells oscillates in concert, as evidenced by the relatively low order parameter of $K \sim 0.3$. It takes a while (until about $t = 550$ s, Fig. 4) until the extracellular concentration of acetaldehyde becomes coherent in time and space. Once this is achieved, we observe an episode of relatively high synchronisation among the cells of the population (in the interval $550\text{ s} \leq t \leq 1000\text{ s}$, Fig. 4). It is noteworthy that amplitudes of the oscillations of the single cells have already begun to decay as the macroscopic collective signal attains its highest amplitudes.

The time-series at the onset of desynchronisation (e.g., at $\rho = 0.3\%$, Fig. 4), allow to study some characteristics of the oscillations of individual cells when these are either synchronised or desynchronised. We found that in the event of desynchronised oscillations the variability in oscillation period was larger than at times where the cellular oscillations are synchronised (Fig. 5). Furthermore, the synchronisation of cells led to a slightly shorter mean period of the intracellular glycolytic oscillations as compared to the periods of oscillations when the cells were desynchronised (Fig. 5). Hence, events of synchrony are caused by the entrainment of cells to a period that is slightly shorter than their free-running oscillatory periods.

In summary, we report on the mechanism of the transition between collective glycolytic oscillations and quiescence in a population of immobilised yeast cells. We found that while the collective dynamics of the populations changed from oscillatory to stationary, the individual cells remained oscillatory. However, the transition in the macroscopic, collective dynamics is caused by the loss of phase and frequency coherence of the individual cells. Hence, the transition between oscillatory and quiescent dynamics in immobilised yeast cells occurs via a Kuramoto transition. This dynamic transition differs significantly from the behaviour reported from stirred yeast cell suspensions, where the cessation of collective oscillations at the population level is due to a dynamical quorum sensing (or amplitude death) phenomenon [26], where all cells stop oscillating immediately, once the cell density is lowered below the critical threshold. Obviously, yeast cells can follow both routes, depending on the strength of cell-to-cell coupling: at high coupling strength between the cells, the dynamical quorum sensing (or amplitude death) mechanism is favoured, whereas at low coupling strength, the transition occurs via a Kuramoto-type transition of synchronous to desynchronous oscillations of the individual cells.

Acknowledgements We wish to thank Mr. Kristian Loewe for help with preparing the figures. Financial support by the Federal Ministry of Education and Research (TCAM4Life grant number 13N12675) and the Deutsche Forschungsgemeinschaft (DFG SFB 854 TPZ) is acknowledged.

References

1. A.T. Winfree, *The Geometry of Biological Time*, 2nd edn. (Springer, New York, 2001)
2. J.-R. Kim, D. Shin, S.H. Jung, P. Heslop-Harrison, K.-H. Cho, A design principle underlying the synchronization of oscillations in cellular systems. J. Cell Sci. **123**, 537–543 (2010)
3. J. Stark, C. Chan, A.J.T. George, Oscillations in the immune system. Immun. Rev. **216**, 213–231 (2007)
4. J. Buck, E. Buck, Mechanism of rhythmic synchronous flashing of fireflies. Science **159**, 1319–1327 (1968)
5. G.M. Ramírez-Avila, J.-L. Deneubourg, J.-L. Guisset, N. Wessel, J. Kurths, Firefly courtship as the basis of the synchronization-response principle. Europhys. Lett. **94**, 60007 (2011)
6. D. Gonze, S. Bernard, C. Walterman, A. Kramer, H. Herzel, Spontaneous synchronization of coupled circadian oscillators. Biophys. J. **89**, 120–129 (2005)

7. J.S. O'Neill, M.H. Hastings, Increased coherence of circadian rhythms in mature fibroplast cultures. J. Biol. Rhythms **23**, 483–488 (2008)
8. A.B. Webb, N. Angelo, J.E. Huettner, E.D. Herzog, Intrinsic, nondeterministic circadian rhythm generation in identified mammalian neurons. Proc. Natl. Acad. Sci. U.S.A. **106**, 16493–16498 (2009)
9. O. Pourquié, Vertebrate segmentation: From cyclic gene networks to scoliosis. Cell **145**, 650–663 (2011)
10. I.M. Rouzine, F.E. McKenzie, Link between immune response and parasite synchronization in malaria. Proc. Natl. Acad. Sci. U.S.A. **100**, 3473–3478 (2003)
11. Y. Su, S. Ruan, J. Wei, Periodicity and synchronization in blood-stage malaria infection. J. Math. Biol. **63**, 557–574 (2010)
12. P. Richard, The rhythm of yeast. FEMS Microbiol. Rev. **27**, 547–557 (2003)
13. F.A. Chandra, G. Buzi, J.C. Doyle, Glycolytic oscillations and limits on robust efficiency. Science **333**, 187–192 (2011)
14. Z. Palková, L. Váchová, Life within a community: benefit to yeast long-term survival. FEMS Microbiol. Rev. **30**, 806–824 (2006)
15. L.N. Duysens, J. Amesz, Fluorescence spectrophotometry of reduced phosphopyridine nucleotide in intact cells in the near-ultraviolet and visible region. Biochim. Biophys. Acta **24**, 19–26 (1957)
16. A. Betz, B. Chance, Phase relationship of glycolytic intermediates in yeast cells with oscillatory metabolic control. Arch. Biochem. Biophys. **109**, 585–594 (1965)
17. A. Ghosh, B. Chance, Oscillations of glycolytic intermediates in yeast cells. Biochem. Biophys. Res. Commun. **16**, 174–181 (1964)
18. P. Richard, B.M. Bakker, B. Teusink, K. van Dam, H.V. Westerhoff, Acetaldehyde mediates the synchronization of sustained glycolytic oscillations in populations of yeast cells. Eur. J. Biochem. **235**, 238–241 (1996)
19. S. Danø, P.G. Sørensen, F. Hynne, Sustained oscillations in living cells. Nature **402**, 320–322 (1999)
20. B. Hess, A. Boiteux, Mechanism of glycolytic oscillation in yeast. I. Aerobic and anaerobic growth conditions for obtaining glycolytic oscillation, Hoppe Seyler. Z. Physiol. Chem. **349**, 1567–1574 (1968)
21. A.K. Ghosh, B. Chance, E.K. Pye, Metabolic coupling and synchronization of NADH oscillations in yeast cell populations. Arch. Biochem. Biophys. **145**, 319–331 (1971)
22. J. Schütze, T. Mair, M.J.B. Hauser, M. Falcke, J. Wolf, Metabolic synchronisation by travelling waves in yeast cell layers. Biophys. J. **100**, 809–813 (2011)
23. J. Bolyó, T. Mair, G. Kuncová, M.J.B. Hauser, Spatio-temporal dynamics of glycolytic waves provide new insights into the interactions between immobilized yeast cells and gels. Biophys. Chem. **153**, 54–60 (2010)
24. J. Aldridge, E.K. Pye, Cell density dependence of oscillatory metabolism. Nature **259**, 670–671 (1976)
25. D.J. Schwab, A. Baetica, P. Mehta, Dynamical quorum-sensing in oscillators coupled through an external medium. Physica D **241**, 1782–1788 (2012)
26. S. De Monte, F. d'Ovidio, S. Danø, P.G. Sørensen, Dynamical quorum sensing: Population density encoded in cellular dynamics. Proc. Natl. Acad. Sci. U.S.A. **104**, 18377–18381 (2007)
27. A.F. Taylor, M.R. Tinsley, F. Wang, Z. Huang, K. Showalter, Dynamical quorum sensing and synchronization in large populations of chemical oscillators. Science **323**, 614–617 (2009)
28. A. Pikovsky, M. Rosenblum, J. Kurths, *Synchronization: A Universal Concept in Nonlinear Sciences* (Cambridge Univ. Press, Cambridge, UK, 2004)
29. S.H. Strogatz, From Kuramoto to Crawford: exploring the onset of synchronization in populations of coupled oscillators. Physica D **143**, 1–20 (2000)
30. Y. Kuramoto, Self-entrainment of a population of coupled non-linear oscillators. Lect. Notes Phys. **39**, 420–422 (1975)
31. S. Shinomoto, Y. Kuramoto, Phase transition in active rotator systems. Prog. Theor. Phys. **75**, 1105–1110 (1986)

32. M.A. Aon, S. Cortassa, H.V. Westerhoff, K. van Dam, Synchrony and mutual stimulation of yeast cells during fast glycolytic oscillations. J. Gen. Microbiol. **138**, 2219–2227 (1992)
33. A. Weber, Y. Prokazov, W. Zuschratter, M.J.B. Hauser, Desynchronisation of glycolytic oscillations in yeast cell populations. PLoS One **7**, e43276 (2012)
34. A.-K. Gustavsson, D.D. van Niekerk, C.B. Adiels, F.B. du Preez, M. Goksör, J.L. Snoep, Sustained glycolytic oscillations in individual isolated yeast cells. FEBS J. **279**, 2837–2847 (2012)
35. J.R. Lakowicz, *Principles of Fluorescence Spectroscopy* (Springer, New York, 2006)
36. Y. Prokazov, E. Turbin, A. Weber, R. Hartig, W. Zuschratter, Position sensitive detector for fluorescence lifetime imaging. J. Instr. **9**, C12015 (2014)
37. R. Hartig, Y. Prokazov, E. Turbin, W. Zuschratter, Wide-field fluorescence lifetime imaging with multi-anode detectors. Meth. Mol. Biol. **1076**, 457–480 (2014)
38. M. Vitali, F. Picazo, Y. Prokazov, A. Duci, E. Turbin, C. Götze, J. Llopis, R. Hartig, A.J.W.G. Visser, W. Zuschratter, Wide-field multi-parameter FLIM: long-term minimal invasive observation of proteins in living cells. PLoS One **6**, e15820 (2011)

Part V
Economy

Financial Market Models

Lisa Borland

Abstract The dynamics of financial markets are discussed. After a brief introduction of the price formation process, we review the statistical features (also known as "stylized facts") of stock return time series, which exhibit fat tails and intermittent periods of higher or lower volatility. Several models aimed at understanding the mechanisms that lead to these seemingly ubiquitous features of financial markets are then reviewed. Those models have largely been developed within the Econophysics community but we emphasize here that they all contain elements consistent with a Synergetic approach.

1 Introduction

Financial markets are nonlinear non stationary complex systems, with a constantly evolving unique history of price and transaction data. It is impossible to run experiments akin to those in a physics lab, but on the other hand we have an increasing amount of data, since every market interaction that takes place on the electronic exchanges are recorded, resulting in terra bytes of new information each day. These empirical observations are fodder for all kinds of analyses. On a high level of statistical observation, they show a slew of so-called stylized facts which appear to be stable across instruments, geographies and time, hence quite universal. Over the past decades there has been effort from the Econophysics community to analyze the data and to come up with models of various aspects of financial markets and economies, such as the dynamics of wealth formation in an economy, multi agent models of price formation, and the dynamics of volatility and correlations to name a few.

This chapter will be a review of some of these interesting models that stem from statistical physics. We emphasize and highlight that a main common feature among

L. Borland (✉)
Cerebellum Capital, San Francisco, CA, USA
e-mail: lborland@stanford.edu

S.C. Müller et al. (eds.), *Complexity and Synergetics*,
https://doi.org/10.1007/978-3-319-64334-2_20

them is that they all take into account cooperative effects one way or another, very much in the spirit of Synergetics [1].

2 Microscopic Dynamics of Price Formation

There is no such thing as "the price" of a stock. Instead the price formation process is the result of the individual actions of market participants, each with their own view and utility. They act on a spectrum of time scales; they base their buy or sell decisions on their individual information set for their own individual intent. These buy and sell orders are submitted to the market place and it is ultimately the collection and interaction of this supply and demand that drive the microscopic dynamics of price formation, in conjunction of course with the rules of individual exchanges and regulations (which have evolved over the years and continue to do so).

Specifically for example in the US, in the mid nineties transactions could only occur on the NYSE or Nasdaq, whereas today we have a fragmented market consisting of "dark pools" (where traders cannot see the orders of others) and 12 "lit" exchanges where traders can place their orders to buy or sell a given quantity of a stock at a given price. The orders on lit exchanges can be seen by market participants (hence the name). They can be organized into a so-called limit order book according to the price-time priority. An order has a time, price, size and action (for example buy, sell, cancel among others) associated to it. If two orders to buy come in with the same price, the one that came in first gets placed ahead in the order book queue. However if an order comes it to buy at a higher price, it goes ahead of the other orders regardless of when it entered the book. Orders to sell are handled in the same fashion. Size doesn't affect priority and the different amounts available to be bought or sold at different prices constitute the full order book. At any given moment at the top of the book there is a best price to buy a certain amount and a best price to sell a certain amount. The difference between them is called the spread. If an order comes in that crosses the spread a transaction will occur. All long as there is enough volume at the top of the book, then that is the price that the transaction will occur at, otherwise some of the order will be filled at worse prices as the liquidity at deeper levels of the book gets consumed.

This brings us to the point here: there is no well-defined price. Is it the price to sell or the price to buy? Is it the mid of those? Is it some kind of volume weighted mid, depending on how many orders to buy or sell at a given price? Often, the last recorded transaction price is what is used here, but it is easy to see how this could be problematic if a stock doesn't trade very frequently. All said, the price formation process is clearly a complex one, driven by events involving local and global interactions, resulting in an evolving order book dynamic. In my opinion the most insightful and detailed studies attempting to understand the dynamics of price formation on this level, as well as the market impact of trading, have been done within the Physics community for example by Bouchaud et al. [2, 3] and Lillo et al. [4]. They reveal that the processing of supply and demand in markets has long range memory, and is also related to the origin of market fluctuations among many other interesting findings. Another succinct model of the order book was developed by Cont et al. [5] who formulate a stochastic equation for the mid price based on the order book dynamics. More recently, some authors model the order flow dynamics of bids and asks via self-exciting Hawkes processes [6, 7], leading to a nice framework where questions such as optimal trade execution for example can be studied.

In spirit and in analogy to physics, all these models eluded to above can be seen as microscopic models, based on underlying empirical observations of the actual order placement and execution process (namely the order book). Ultimately though, the price once formed evolves as a stochastic process, and is typically modeled by a Langevin equation. This level of description is in a sense analogous to a mesoscopic view in physics, and can be seen itself as the result of a vastly complex system with a multitude of interacting participants. This view lends itself to other classes of models attempting to describe the collective dynamics of those participants. In this chapter we shall review several classes of both microscopic and mesoscopic models which can all be seen to fit very well in the spirit of Synergetics.

3 Stylized Facts of Markets

Returns (or relative price changes) of stocks can be calculated over different time scales τ, and if the distributions of these are plotted out it is clear that they are far from Gaussian, but rather are well fit with power law tails in such a way that the power law behavior persists from time-scales ranging from intraday to the order of a few weeks. On daily time-scales the power law tail is about 3, often referred to as the cubic law of finance [8, 9]. The kurtosis of these distributions decays in a regular

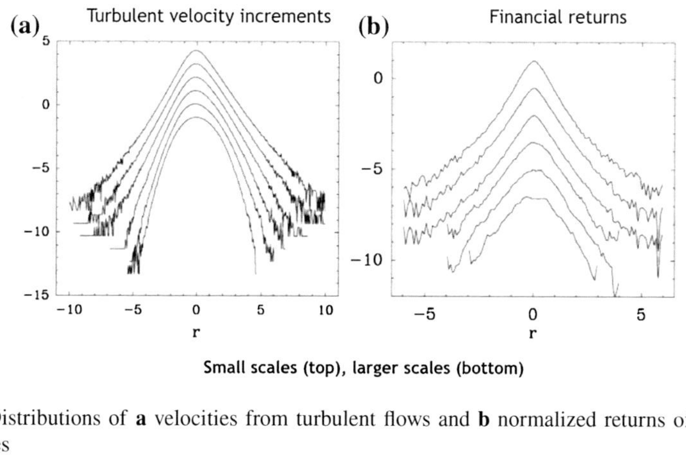

Fig. 1 Distributions of **a** velocities from turbulent flows and **b** normalized returns of financial time-series

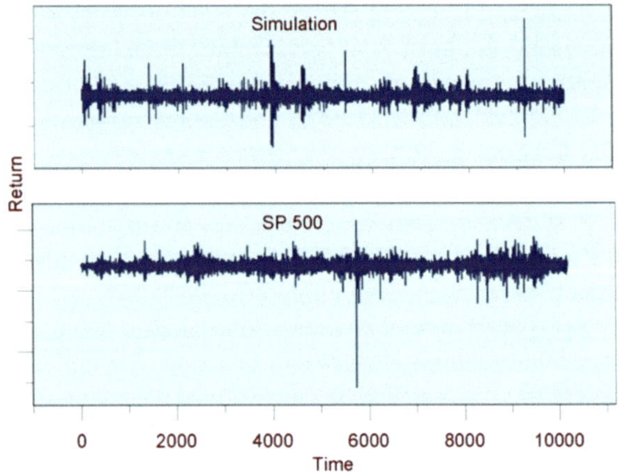

Fig. 2 Real market returns (*below*) and results of a market model simulation (the multi-timescale model of Eq. (16))

fashion, roughly as $\tau^{-0.2}$ [10]. These distributions bear many similarities to those of turbulent systems (see Fig. 1).

Volatility (defined either as the standard deviation of returns over some past window, or simply by a proxy such as the absolute value of returns) exhibits clustering behavior such that there appear to be regimes of higher or lower volatility (see Fig. 2). Furthermore, there is memory in volatility in the sense that the auto-correlation of absolute returns is very strong, decaying slowly as a power law. More subtle statistical features are inherent, such as a behavior analogous to the Omori law for earthquakes in that a large volatility shock will be followed by aftershocks at a certain rate. Furthermore, large negative returns are indicative of higher volatility, an effect known as the leverage effect [11]. There is also time-reversal asymmetry in the sense that future volatility conditioned on past observations is not symmetric [12]. Finally, financial time series exhibit multi-fractal scaling of moments (see for example [13]). Interestingly, all of these so-called stylized facts are not only observed for stock returns, but also for other financial instruments such as commodities and currencies, and they are observed across geographies. Realistic market models should ideally capture the basics of these features.

4 Wealth Models as Predator-Prey

The Lotka-Volterra equation is one that arises in the field of nonlinear dynamics and Synergetics quite frequently. It is used to describe the positive and negative feedback loops between interacting species, where one preys on the other. It is therefore very exciting that one of the first successful models for describing the fact that wealth among members of the society follows a Pareto distribution (and in addition connecting this to the fluctuations in financial markets) is based on that equation [14–16].

That model also recovers realistic features of financial markets such as bubbles and crashes as well as volatility clustering.

The basic ingredients of the model are to introduce feedback between individual and collective wealth fluctuations of a collective set of traders. The central feedback loop consists in computing the market price of the stock as the sum of the individual wealths w_i invested in the stock by the traders, and then determining fluctuations of a given trader's wealth as their previous wealth multiplied by the stock return. Explicitly, consider the total population of wealth

$$\bar{w}(t) = \sum_{i=1}^{N} w_i(t) \tag{1}$$

where w_i is the wealth of investor i (in other words, value of the stock owned by investor i at time t). At time $t + 1$, the wealth becomes:

$$w_i(t+1) = \lambda(t)w_i(t) + a(t)\bar{w}(t) - c(t)\bar{w}(t)w_i(t) \tag{2}$$

The first term states that wealth at time $t + 1$ is proportional to wealth at time t multiplied by the random factor $\lambda(t)$ which corresponds to relative gains or losses over the last period. The second term represents a coupling of the individual's wealth to the global wealth of the society. Hence $a(t)\bar{w}(t)$ represents things like social services, for example. The final term corresponds to the competition between each individual i and the other members of society, and plays the part of limiting growth of $\bar{w}(t)$ to values that are sustainable for the current conditions and resources.

The Solomon–Levy model leads to a power law for the distribution of individual wealth namely

$$P(w) \propto w^{-1-\alpha} \tag{3}$$

where α is typically between -1 and -2. In [16] the very interesting conclusion is drawn that any quantity which is a sum of random increments proportional to the wealths w_i will have fluctuations described by a Levy distribution of index β equal to the exponent α of the wealth power distribution. Since the individual investments are stochastically proportional to the investors wealth, the stock market fluctuations will be described by a truncated-Levy distribution of index equal to the measured exponent $\alpha = 1.4$ (which results in a tail index close to 3 as eluded to above). This is an amazing and non-trivial result: Based on simple notions of competition, local and global feedback of the wealth of members of society, a mechanism for describing the distribution of stock market fluctuations is designed.

Note that Eq. (2) is a mesoscopic description of the stock market; It was in fact written down by the authors as just such as description of the outcome of many simulation runs of their microscopic model [14]. The microscopic model looks at individual investors with various ways of deciding how much stock to buy or sell at a given time. Simulations of that collective group of investors then gave rise to the dynamics of wealth fluctuations which are described by Eq. (2). We'll discuss other such models that have many interacting agents in the next section.

5 Many Interacting Agents, Spin Models and Minority Games

Models with a large number of interacting agents have had success in reproducing the emergent properties observed in real markets, such as bubbles and crashes, volatility clustering and fat tailed distributions. A very simple class of agent-based models is the El Farol bar problem. The bar has a limited number of seats, less than the total number of patrons who would like to come. The goal is to learn which days to go to the bar in order to get a seat. Agents have a set of predictors based on their past experience and via reinforcement learning they use their best predictors more and replace their bad predictors. A simple version of the El Farol bar problem is known as the minority game [17]. Agents are divided into two groups and are only rewarded if they are in the minority. This leads to a global frustration as not every agent can be in the minority. So, in simple terms, the minority game describes how a system evolves according to trying to maximize the payoff of each individual agent under the given boundary conditions. When applied to modeling financial markets, the analogy is that similar dynamics may be going on because it is often better to be in the minority when buying or selling. In spite of the extreme simplicity of the model in this context, the minority game model is nevertheless an interesting framework.

Another approach of interacting agents which is much more realistic is the one proposed by Lux and Marchesi [18]. The individual trading agents are simulated, including an explicit price formation process. Agents are modeled as different types of traders interacting in a speculative market: "noise traders" and "fundamentalists". Fundamentalists base their action on fundamental valuation of the stock. The noise traders base their trading decisions on price data and flows, which leads to herding behavior. They react to the recent past of the market and can have either positive or negative expectations of the future based on that past. The dynamic of the model is that traders compare profits gained by the noise traders and fundamentalists, and then switch their own strategy to that which was more profitable in the recent past. Depending on whether traders want to buy or sell, supply or demand will be infused into the market and the price will be adjusted according to the excess demand. In addition, the dynamics of the fundamental value of the stock follows a standard log-normal Brownian motion with uncorrelated Gaussian noise.

Lux and Marchesi formulate the state dependent transition probabilities that describe, for each group of traders, the probability of switching to the other group. For noise traders there is also the internal switching between a pessimistic and optimistic view of the market. They obtain

$$P(+|-) = v_1 \exp(U_1) \quad (4)$$

$$P(-|+) = v_1 \exp(-U_1) \quad (5)$$

with:

$$U_1 = \alpha_1 x + (\alpha_2/v_1)p'(t)/p \quad (6)$$

where the α's and v's are constants and x corresponds to the majority opinion of other noise traders (flows) and $p'(t)$ is the recent price trend. In addition, for the traders switching from a noise trading (N) chartist strategy to a fundamental (F) strategy they postulate

$$P(N|F) = v_2 \exp(U_2) \quad (7)$$

$$P(F|N) = v_2 \exp(-U_2) \quad (8)$$

with: $U_2 = \alpha_3 \Delta w$ where Δw corresponds to the profit differential. Finally, the price gets adjusted up or down based on supply and demand according to excess demand being either on the buy or sell side.

Based on these rather simple yet realistic dynamics, a theoretical analysis as well as simulations show that the most important features of real financial markets emerge as a consequence. In particular they find that the market is on average efficient in the sense that the price on average reflects a fundamental equilibrium. The amount of pessimistic and optimistic traders is roughly even, and in equilibrium both the noise traders and the fundamentalists do equally well. Though the system always tends towards a stable equilibrium, it exhibits autocorrelated fluctuations around that fundamental equilibrium and simulations of the model show on-off intermittency of fluctuations. This is very similar to the properties of real market fluctuations, where volatility shows memory and clustering.

Another class of models are spin based models in which analogies are made between the interactions and dynamics of spin systems and financial markets. Some interesting spin models have been proposed for example [19, 20]. These models can reproduce (under certain parameter settings) features such as the large fluctuations of financial markets, but need to be tuned specifically. Another spin based model [21] that doesn't need tuning stands out as particularly simple and intuitive, the approach of which we summarize here. In that paper a very simple model is formulated, which nevertheless includes terms corresponding to strategic interactions between real market players. To be more precise, the spins correspond to 'traders' who can have two states which represent positive or negative expectations of the market. The model aims to capture two conflicting forces of financial markets: (1) Do what others do, such as in the herding behavior of the Lux-Marchesi model, and (2) Do what the minority does, for example if traders have knowledge about fundamental values and as modeled in the minority game.

Explicitly, Bornholdt proposes a model with $i = 1, \ldots, N$ spins such that $S_i(t + 1) = 1$ with probability

$$p = (1 + \exp(-2\beta h_i(t)))^{-1} \quad (9)$$

and $S_i(t+1) = -1$ with probability $1-p$. The local field h is specified by

$$h_i(t) = J_{ij}S_j\alpha C_i(t)\frac{1}{N}\sum_{j=1}^{N} S_j(t) \tag{10}$$

The first term is a local Ising coupling with nearest neighbor interactions $J_{ij} = J$ and $J_{ij} = 0$ for other pairs. The second term is a global coupling to the magnetization of the system with a coupling $\alpha > 0$, where $C_i(t)$, represents the strategy of agent i with respect to the magnetization. In particular, this second coupling allows for the case of spins frustrated across scales, seeking ferromagnetic order locally, but anti-ferromagnetic order globally. Depending on the value and dynamics of C, the model captures scenarios where chartists (herders) or fundamentalists (minorities) dominate, and more interestingly where they switch group memberships. In that case it is shown that a whole slew of realistic features of financial markets is recovered, such as intermittent volatility clustering, power-law scaling of returns and auto-correlation of absolute returns. Most interestingly, these results do not depend on tuning, and the author concludes (much in the spirit of Synergetics) that there is self-organization of the spin system to the financial market-like regime dominated by metastable phases with intermittent disorder.

6 Statistical Feedback and Correlation Dynamics

The above models are useful as frameworks to think about the dynamics of market participants and the emergence of the stylized facts one observes. For certain applications (such as for example risk management or the pricing of options) a slightly higher level view of price formation can be useful, namely in terms of modeling the price itself as a stochastic process.

Most of traditional mathematical finance is based on a Brownian equation of motion for stock prices:

$$dS(t) = \mu S dt + S\sigma d\omega \tag{11}$$

where ω is drawn from a Gaussian distribution with zero mean and variance one. This type of model is very useful because it allows for the analytic calculation of many important derivative instruments such as options, for which Black and Scholes [22] received the Nobel prize in Economics in 1997. However, that model is too simple to capture all of the anomalous statistics observed in real financial time series. In an attempt to rectify that, several modifications to the standard Black–Scholes model of price returns have been proposed in the literature, and they all have in common that they somehow extend either the assumption of a constant volatility term σ in Eq. (11), or the source of the noise term ω. For example there is the stochastic volatility model of Heston [23] where σ itself is modeled as a mean reverting stochastic process, and the Levy models where the noise ω is assumed to be drawn form a fat tailed Levy distribution. Those models are a little more realistic than the standard model but both have the shortcoming that they convolve too quickly to a Gaussian distribution, meaning that they do not capture the persistence of fat tails of return distributions over the time-scales observed in reality, where returns over time-scales ranging up to about 2 weeks or longer all still exhibit tails. One model which does really well in this sense (and also earned a Nobel prize for Engle [24]) is the GARCH model, which incorporates memory into σ. In fact, the memory of volatility is a key feature that reproduces many known stylized facts of financial price series. Motivated among other things by this, we proposed a model [25, 26] within the framework of nonextensive statistical physics [27] in which the volatility term follows a statistical feedback process in the sense that it depends on the probability of past observations, explicitly:

$$dS = \mu S dt + \sigma S d\Omega \tag{12}$$

where

$$d\Omega = P(\Omega)^{\frac{1-q}{2}} d\omega. \tag{13}$$

In this equation, P corresponds to the probability distribution of Ω, which simultaneously evolves according to the corresponding nonlinear Fokker–Planck equation [28, 29]

$$\frac{\partial P}{\partial t} = \frac{\partial P^{2-q}}{\partial \Omega^2}. \tag{14}$$

It can be solved exactly yielding

$$P = \frac{1}{Z(t)}(1 - (1 - q)\beta(t)\Omega(t))^{\frac{1}{1-q}} \tag{15}$$

The precise form of the coefficients Z and β are given in [25]. Equation (15) recovers a Gaussian in the limit $q \to 1$ while exhibiting power law tails for $q > 1$. In that case, our model is exactly equivalent to the Black–Scholes model.

The statistical feedback term P can be seen as capturing the market sentiment. Intuitively, this means that if the market players observe unusually high deviations of $\Omega(t)$ (which is essentially equal to the de-trended and normalized log stock price) from the reference value $\Omega(0)$, then the effective volatility will be high because in such cases $P(\Omega)$ is small, and the exponent $1 - q$ a negative number. Conversely, traders will react more moderately if Ω is close to its more typical or less extreme values. As a result, the model exhibits intermittent behavior consistent with that observed in the effective volatility of markets. In practice, q can be obtained empirically from a fit to the data. Remarkably, $q = 1.4 - 1.5$ fits very well to return distributions of very many financial instruments, corresponding to a tail index of about 3.

This non-Gaussian statistical feedback model allowed us to derive closed form option pricing formulae [25, 26] that fit very well to real market prices over many time horizons, and we used the model in real life trading situations. For further reading about this topic, summarized successes, applications and shortcomings of the model we refer to [30]. In spite of the success at pricing options and other derivatives such as credit default swaps, as a model of real returns the statistical feedback formulation has the drawback that returns relative to a particular initial time constitute the memory in the volatility; instead we took inspiration from that model and proposed that the volatility depends on returns over multiple time-scales [13]. The intuition is that different traders pay attention to different time-scales. For example, some only care about returns on an intraday or daily level, others are more focused on monthly or whatever the rebalancing frequency is of their trades. Explicitly, this multi-timescale model can be written as

$$\Delta y = \sigma(t)\Delta\omega \tag{16}$$

$$\sigma(t) = \sigma_0 \sum_{\tau=0}^{T} \sqrt{\left(\frac{g}{\tau^\alpha \sigma_0^2}(y_t - y_{t-\tau})^2\right)} \tag{17}$$

where g and α can be calibrated to fit empirical data ($g = 0.85$ and $\alpha = 1.15$) [13] and σ_0 corresponds to the baseline volatility. Without the summation and setting $t - \tau = 0$, this equation can be shown to be of the same form as our statistical feedback model. That model has the advantage of analytic tractability for options pricing, whereas Eq. (16) does not allow for that. However it can be simulated and shown to fit real data remarkably well, reproducing a slew of known stylized facts [13].

So far we have largely concentrated on modeling the dynamics of financial market returns over time. We have not really said much about the cross sectional dynamics of stock returns, i.e. the dynamics of the correlation structure of markets. After the financial crisis of 2008 it was interesting to study the cross-sectional dynamics of the distribution of stock returns, in particular to understand what happens during times of panic. Understanding how correlations behave during such times can be extremely important for managing financial risks, for example. We noticed [31] that, during panic times, the dispersion (standard deviation) of stock returns cross-sectionally increases as does the time-series volatility. However, the kurtosis (corresponding to the tails of the cross sectional distribution) decreases (see Fig. 3). In addition, correlations tend to 1. In order to model these effects we proposed a simple synergetic framework for the correlation dynamics, and coupled that to the multi time-scale stochastic model for stock returns of Eq. (16), as described in the following.

In the spirit of Synergetics, we define as a proxy for the collective behavior of all stocks in the market the variable s:

$$s = \frac{s_{up} - s_{down}}{s_{up} + s_{down}} \tag{18}$$

where s_{up} is the number of stocks that have positive returns over a given interval, and s_{down} is the number of stocks that have negative moves on that same interval (for example a day). If $s = 0$ then roughly the same number of stocks moved up as down, and the assumption is that the stocks had little co-movement and so were uncorrelated. If all stocks move together either up or down, though, the value of s will be $+1$ or -1 and the stocks will have high correlation. So, the following picture emerges: If $s = 0$ there is no correlation, and we are in a disordered state. However if $s \neq 0$ then there is correlation and we are in an ordered state; in other words, s is the order parameter. Empirically we have observed rather drastic changes in the cross-sectional distribution of stocks in the times of panic versus more normal market conditions [31]. In particular, histograms of s in both periods show that in normal times, s is unimodal, and in panic times we obtain a bimodal distribution (see Fig. 3). These observations are consistent with the frame-work of a phase transition leading to self-organization in panic times, following a schematic set of dynamics similar to those of a ferromagnetic system: The variable s of Eq. (18) can namely be seen as analogous to the magnetic moment in a ferromagnetic spin system. In such a system, the total magnetic moment depends on the orientation of the individual magnetic spins comprising the system. It is proportional to

$$m = \frac{m_{up} - m_{down}}{m_{up} + m_{down}} \tag{19}$$

where m_{up} and m_{down} denote the number of spins lined up and down respectively. The distribution of possible outcomes of this macroscopic quantity is given by

$$P(m) = N \exp(F(m, T)) \tag{20}$$

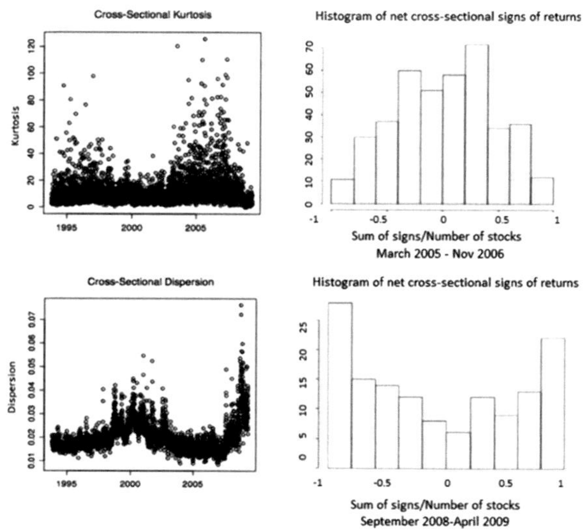

Fig. 3 Features of the cross-sectional distribution of stocks. Kurtosis is high in normal times, low in panic, while dispersion shows the opposite behavior. The co-movement of stocks is bi-modal in times of panic, uni-modal in normal times

where T is the temperature, N is a normalization factor and F is the free energy of the system. Depending on the value of T, the magnetic system will either be in an ordered or disordered state. Assuming one can perform a Taylor expansion of F and invoking symmetry arguments, the corresponding Langevin equation takes the form

$$\frac{dm}{dt} = -m - m^3 + W_t \tag{21}$$

where W_t is thermal noise. The coefficient a can be written as

$$a = \alpha(T - T_c) \tag{22}$$

where T_c is the so-called critical temperature. One can envision these dynamics as motion in a potential well V given by $V(m) = -F(m)$. If $T > T_c$, the only minimum is the trivial one at $m = 0$. However, for $T < T_c$ there are two real roots appearing, yielding non-zero values of m. Clearly, m can be positive or negative, depending on which minima is reached by the system. This is referred to as symmetry breaking. Due to the noise, the dynamics can also drive m from one minimum to the other. Because the value of T determines whether the system is in the disordered state ($m = 0$) or the ordered state ($m \neq 0$), it is called the control parameter. The probability distribution of the system in the disordered state will be a unimodal one, while the probability distribution of m in the ordered state will be bimodal. As T passes from above to below T_c, or vice-versa, there is clearly a phase transition: the state of the system is drastically altered. In this type of symmetric system, the phase transition is referred to as a second order one.

Along exactly the same lines, we model the dynamics of s by

$$\frac{ds}{dt} = -as - bs^3 + W_t. \tag{23}$$

with

$$a = \alpha(\sigma_c - \sigma_0) \tag{24}$$

where W_t is a Gaussian noise term and σ_0 corresponds to the baseline volatility level of stocks as in Eq. (16). This volatility is assumed constant across all instruments, and essentially measures the general uncertainty in the environment, so in this sense acts much as the temperature in the magnetic system. Note that it is the feedback effects in the system which induce stock-specific variations in volatility over time, and can largely explain most of the excess volatility observed in stock time-series, whereas the parameter σ_0 is not driving the stock-specific dynamics, but instead describes a "global" level of risk. The quantity σ_c would correspond to a critical level of uncertainty, below which the market is in a normal phase, and above which we have the onset of panic. Much as in the case of ferromagnetism, where the control parameter T can be tuned externally above or below the critical temperature, in our model the uncertainty level σ_0 captures the external environment. In a sense it represents the general perception of risk in the public mind. This volatility increases to values larger than σ_c due to factors such as exogenous jumps (news) affecting all stocks so that σ_0 becomes $\sigma_0 + \sigma_{shock}$. (However, an extension of the model that includes coupling between different stock returns in the volatility feedback term of Eq. (16) [13] allows for a framework where σ_0 can also increase due to endogenous, idiosyncratic jumps which may for example be induced by a large negative return [32].)

Our hypothesis is that financial markets appear to exhibit a phase transition from the disordered to ordered state, after crossing a critical level of risk perception. Putting the dynamics together, we have the multi time-scale feedback process for each stock k

$$dy_i^k = \sigma_i^k d\omega_i^k \tag{25}$$

with $k = 1 \cdots N$, and the volatility of each stock k given by Eq. (16) The random variables ω_i^k are drawn from a Gaussian distribution, uncorrelated in time such that $< \omega_i^k \omega_{i+\tau}^k >= \delta(i, i+\tau)\tau$, yet amongst themselves at a given time point i across stocks k, they are correlated with correlation $|s|$. The macroscopic order parameter s is therefore just a signature of the cross-stock correlations, whose dynamic behavior manifests itself in the order parameter equation (23). The coefficients must always be such that $|s| \leq 1$ which can be imposed by running the dynamics of s on a real valued variable $\hat{s}$ such that $s = \tanh(\hat{s})$.

Conceptually, this model captures the main properties observed in the data. Across time, we already showed that the main stylized facts are reproduced. Across stocks, if $\sigma_0 < \sigma_c$, correlations fluctuate around $s = 0$ and we expect to see a unimodal distribution of s. The cross-sectional kurtosis should be rather high since there is no mechanism to cause either stocks or stock volatilities to have any co-movement at all, so at each time point it is as if the cross-sectional returns are drawn from a Gaussian process with stochastic volatility, yielding a fat-tailed distribution as the superposition. Then as the market crashes with $\sigma_0 > \sigma_c$, the system enters a phase transition. The order parameter s becomes $s \neq 0$ and the system enters the ordered phase with high co-movement. Because the random variables ω_i^k are now correlated across stocks, cross-sectional returns will be more similar and the distribution will have lower kurtosis. Additionally, due to the fact that the phase transition is triggered by an external shock in volatility, all stocks will tend to have higher volatilities and higher cross-sectional dispersion.

Simulations of this model for the joint stochastic process of stocks were performed (see Fig. 4) and we refer the reader to [31] for details on the implementation. To summarize those results, the baseline volatility was assumed to be $\sigma = 0.2$, and at a certain time a volatility shock $\sigma_{shock} = 0.6$ (consistent with levels observed in the VIX volatility index in late 2008) was applied to the system. This induced the phase transition from the disordered state where correlation among stocks are relatively low, centered around zero, to a highly ordered state where the correlations are different from zero. We found that the main features of financial markets are captured within this framework. The order parameter s goes from 0 (the disordered state) to $s \approx 0.8$

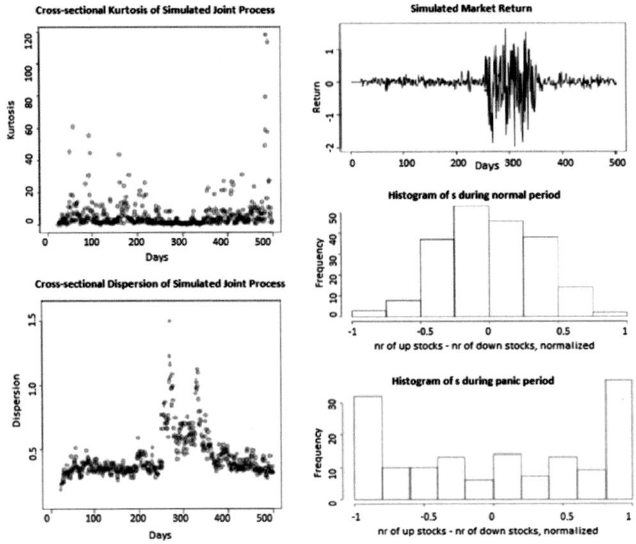

Fig. 4 Simulations of Eq. (25) with correlations evolving according to Eq. (23). The same features as for real markets are found: Kurtosis is high in normal times, low in panic, while dispersion shows the opposite behavior. The co-movement of stocks is bi-modal in times of panic, uni-modal in normal times

(the ordered state) at the time of the volatility shock. When the shock subsides, it returns to the disordered state again. As expected, the market volatility rises when s is in the ordered state, which corresponds to the panic phase. In addition, the cross-sectional dispersion rises during the market panic, while the cross-sectional kurtosis drops close to zero. The correlation between the two quantities is in this example -17%, consistent with empirical observations that also showed a strong negative correlation. Histograms corresponding to the distribution of the order parameter s in the normal market phase as well as in the panic phase are in excellent agreement with the empirical observations of the real market data, namely unimodal in the normal phase, and clearly bimodal during the panic time. Encouraged by these findings we extended this joint stochastic model to include skew and explored its self-similar properties on different time-scales, as presented in [32]. Additional studies of interest in a similar vein aiming to understand and explore the joint dynamics of stocks as well as correlations in time of panic were done by others [33–39].

This work was initially done following the financial crisis of 2008. After the Flash Crash of May 2010 we extended the work to look at intraday properties and signatures. On May 6, 2010, there was a stock market crash in the US equity markets, which started at 2:32 pm. Eastern Time and lasted for a little over half an hour. We analyzed market behavior in a time frame surrounding this event, and found quite similar results [32] to that of the larger time-scale crisis of 2008. Figure 5 shows a marked jump in 5 min volatility right after the intra-day crash, and it is also seen that the distribution of s is clearly bi-modal as a consequence. More recently in July

Fig. 5 *Top* Volatility (over 5 min time scales) on the day of the flash crash. *Bottom* Co-movement of stocks in bi-modal following the flash crash

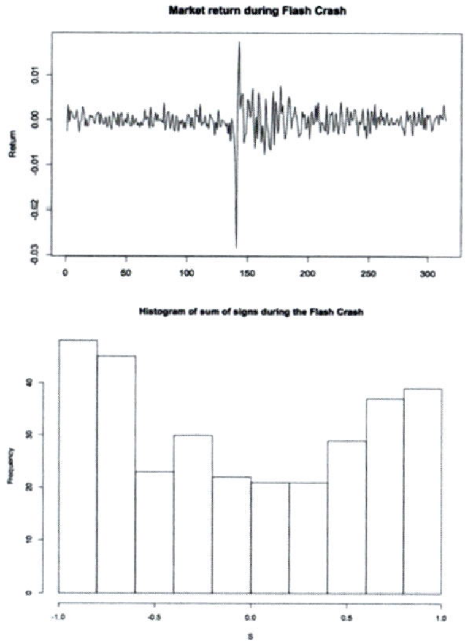

2015 we performed an updated analysis as follows: The Euro Vix levels jumped up indicating external fear in those geographies, probably due to worries about Greece, whereas the Vix levels in the United States markets stayed moderate. We looked at

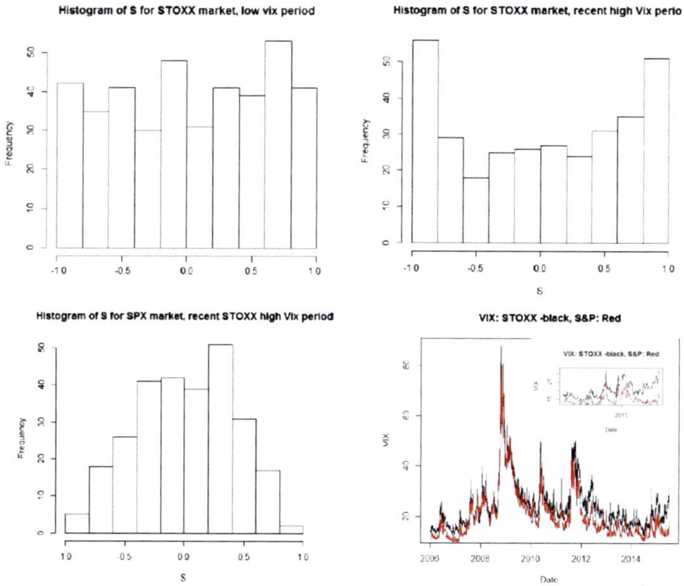

Fig. 6 Panic in Europe in 2015, calm in the US. There are less stocks in the Euro Stoxx universe so the effect is less pronounced but still there

the distribution of s Eq. (18) in the S&P 500 and the Euro Stoxx for January–July 2015 and the results (Fig. 6) confirmed the notions of this model: Bi-modal for Stoxx, where panic prevailed, uni-modal for the S&P 500.

7 Conclusion

We have reviewed a set of models, all capture some aspects of real financial markets. Key common components are feedback and a coupling between the external system and the local dynamics. Though typically discussed in an Econophysics setting it might be useful, as we have done here, to point out that concepts and notions from Synergetics are inherent to these models. It makes intuitive sense that this is the case, as the dynamics of financial markets are due to the complex interactions of humans (and algorithms), dispersed across time scales, time zones, and geographies. These local interactions of market participants quite amazingly give rise ultimately to a collective consensus of notions such as a fair price or a market sentiment—which again feed back to the decisions made by those same humans and algorithms, very much in the essence of the Synergetics framework.

References

1. H. Haken, *Synergetics: An Introduction* (Springer, Heidelberg, 1977)
2. J.-P. Bouchaud, Y. Gefen, M. Potters, M. Wyart, Fluctuations and response in financial markets: The subtle nature of random price changes. Quant. Finan. **4**, 176–190 (2004)
3. J.-P. Bouchaud, J.D. Farmer, F. Lillo, How markets slowly digest changes in supply and demand, in *Handbook of Financial Markets: Dynamics and Evolution* (Elsvier, Amsterdam, 2009), pp. 57–160
4. F. Lillo, J.D. Farmer, R.N. Mantegna, Econophysics: master curve for price-impact function. Nature **421**, 129–130 (2003)
5. R. Cont, S. Stoikov, R. Talreja, A stochastic model for order book dynamics. Oper. Res. **58**, 549–563 (2010)
6. E. Bacry, J.-F. Muzy, Hawkes model for price and trades high-frequency dynamics. Quant. Finan. **14**, 1147–1166 (2014)
7. A. Alfonsi, P. Blanc, Dynamical optimal execution in a market-impact Hawkes price model, in *Finance and Stochastics* (Springer, Heidelberg, 2015), pp. 1–36
8. P. Gopikrishnan, V. Plerou, L.A. Nunes Amaral, M. Meyer, H.E. Stanley, Scaling of the distribution of fluctuations of financial market indices. Phys. Rev. E **60**, 5305 (1999)
9. X. Gabaix, P. Gopikrishnan, V. Plerou, H.E. Stanley, A theory of power-law distributions in financial market fluctuations. Nature **423**, 267 (2003)
10. J.-P. Bouchaud, M. Potters, *Theory of Financial Risks and Derivative Pricing* (Cambridge University Press, Cambridge, 2004)
11. J.-P. Bouchaud, A. Matacz, M. Potters, Leverage effect in financial markets: The retarded volatility model. Phys. Rev. Lett. **87**, 228701 (2001)
12. P.E. Lynch, G.O. Zumbach, Market heterogeneities and the causal structure of volatility. Quant. Finan. **3**, 320–331 (2003)
13. L. Borland, J.-P. Bouchaud, On a multi timescale statistical feedback model for volatility fluctuations. J. Invest. Strateg. **1**, 65–104 (2011)
14. H. Levy, M. Levy, S. Solomon, *Microscopic Simulation of Financial Markets: from Investor Behavior to Market Phenomena* (Academic Press, San Diego, 2000)

15. M. Levy, S. Solomon, New evidence for the power-law distribution of wealth. Physica A **242**, 90–94 (1997)
16. S. Solomon, Stochastic Lotka-Volterra systems of competing auto-catalytic agents lead generically to truncated Pareto power wealth distribution, truncated levy-stable intermittent market returns, clustered volatility, booms and crashes, in *Decision Technologies for Computational Finance: Proceedings of the Fifth International Conference Computational Finance*, ed. by A.-P.N. Refenes, A.N. Burgess, J.E. Moody (Springer, New York, 1998), pp. 73–86
17. D. Challet, M. Marsili, Y.-C. Zhang, *Minority Games: Interacting Agents in Financial Markets* (OUP Catalogue, 2013)
18. T. Lux, M. Marchesi, Volatility clustering in financial markets: a microsimulation of interacting agents. Int. J. Theor. Appl. Finan. **3**, 675–702 (2000)
19. R. Cont, J.-P. Bouchaud, Herd behavior and aggregate fluctuations in financial markets. Macroecon. Dyn. **4**, 170–196 (2000)
20. D. Chowdhury, D. Stauffer, A generalized spin model of financial markets. Eur. Phys. J. B **8**, 477–482 (1999)
21. S. Bornholdt, Expectation bubbles in a spin model of markets: intermittency from frustration across scales. Int. J. Mod. Phys. C **12**, 667–674 (2001)
22. F. Black, M. Scholes, The pricing of options and corporate liabilities. J. Polit. Econ. **81**, 637–659 (1973)
23. S.L. Heston, A closed-form solution for options with stochastic volatility with applications to bond and currency options. Rev. Finan. Stud. **6**, 327–343 (1993)
24. T. Bollerslev, R.F. Engle, D.B. Nelson, ARCH models, in *Handbook of Econometrics*, vol. 4, ed. by R.F. Engle, D. McFadden (Elsevier Science, Amsterdam, 1994)
25. L. Borland, Option pricing formulas based on a non-Gaussian stock price model. Phys. Rev. Lett. **89**, 098701 (2002); L. Borland, A theory of non-Gaussian option pricing. Quant. Finan. **2**, 415–431 (2002)
26. L. Borland, J.-P. Bouchaud, A non-Gaussian option pricing model with skew. Quant. Finan. **4**, 499–514 (2004)
27. C. Tsallis, J. Stat. Phys. **52**, 479 (1988); E.M.F. Curado, C. Tsallis, J. Phys. A **24**, L69 (1991); **24**, 3187 (1991); **25** 1019 (1992)
28. C. Tsallis, D.J. Bukman, Anomalous diffusion in the presence of external forces: exact time-dependent solutions and their thermostatistical basis. Phys. Rev. E **54**, R2197(R) (1996)
29. L. Borland, Microscopic dynamics of the nonlinear Fokker–Planck equation: a phenomenological model. Phys. Rev. E **57**, 6634 (1998)
30. L. Borland, Non-Gaussian option pricing: successes, limitations and perspectives, in *Anomalous Fluctuation Phenomena in Complex Systems, Plasmas, Fluids and Financial Markets*, ed. by C. Riccardi, H.E. Roman (Special Review Book for Research Signpost, Transworld Research Network, Kerala, India, 2008), pp. 311–333
31. L. Borland, Statistical signatures in times of panic: markets as a self-organizing system. Quant. Finance **12**, 1367–1379 (2012)
32. L. Borland, Y. Hassid, Market panic on different time-scales (2010), arXiv:1010.4917
33. T. Preis, D. Kenee, H.E. Stanley, D. Helbing, W.E. Ben-Jacob, Quantifying the behavior of stock correlations under market stress. Nat. Sci. Rep. **2**(752) (2012)
34. T. Kaizoji, Power laws and market crashes. Prog. Theor. Phys. Suppl. **162**, 165–172 (2006)
35. F. Lillo, R. Mantegna, Variety and volatility in financial markets. Phys. Rev. E **62**, 6126–6134 (2000)
36. M. Munnix, T. Shimada, R. Schafer, F. Leyvraz, T.H. Seligman, T. Guhr, H.E. Stanley, Identifying states of a financial market. Sci. Rep. **2**, 644 (2012)
37. P. Ferreira, A. Dionisio, S.M.S. Movahed, Stock market comovements: nonlinear approach for 48 countries (2015), arXiv.org[q-fin], arXiv:1502.05603
38. G. Raffaelli, M. Marsili, Dynamic instability in a phenomenological mode of correlated assets. J. Stat. Mech. **2006**, 8001 (2006)
39. D. Sornette, *Why Stock Markets Crash: Critical Events in Complex Financial Systems* (Princeton University Press, Princeton, 2002)

Structural Change in (Economic) Time Series

Christian Kleiber

Abstract Methods for detecting structural changes, or change points, in time series data are widely used in many fields of science and engineering. This chapter sketches some basic methods for the analysis of structural changes in time series data. The exposition is confined to retrospective methods for univariate time series. Several recent methods for dating structural changes are compared using a time series of oil prices spanning more than 60 years. The methods broadly agree for the first part of the series up to the mid-1980s, for which changes are associated with major historical events, but provide somewhat different solutions thereafter, reflecting a gradual increase in oil prices that is not well described by a step function. As a further illustration, 1990s data on the volatility of the Hang Seng stock market index are reanalyzed.

1 Introduction

In time series analysis, the point of reference is that of a stationary stochastic process; i.e., a process for which the sequence of first and second-order moments is constant ('weak stationarity'), or even the sequence of the entire marginal distributions ('strict stationarity'). In practice, many time series exhibit some form of nonstationarity: changing levels, changing variances, changing autocorrelations, or a combination of some or all of these aspects. These phenomena are then called structural changes or structural breaks and the associated statistical methodology is sometimes called change point analysis. Such phenomena may be seen as 'complex' in the sense of this Volume, in that classical models with constant coefficients are rejected by the data.

C. Kleiber (✉)
Faculty of Business and Economics, University Basel, Peter Merian-Weg 6, 4002 Basel, Switzerland
e-mail: christian.kleiber@unibas.ch

S.C. Müller et al. (eds.), *Complexity and Synergetics*,
https://doi.org/10.1007/978-3-319-64334-2_21

Structural change methodology is widely used in economics, finance, bioinformatics, engineering, public health, and climatology, to mention just a few fields of application. An interesting recent contribution [1] disputes the existence of a 'little ice age' for parts of Central and Northern Europe between the 14th and 19th century. Using structural change methodology (of the type used below) on temperature reconstructions spanning several centuries, Kelly and Ó Gráda find no evidence for sustained falls in mean temperatures prior to 1900, instead several relevant series are best seen as white noise series. One explanation for the contradiction to the established view is the climatological practice of smoothing data prior to analysis. When the raw data are in fact uncorrelated, such preprocessing can introduce spurious dependencies (the 'Slutsky effect').

More than 25 years ago, a bibliography on structural change methodology and applications published in an economics journal [2] already lists some 500 references, and the literature has grown rather rapidly since then. More recently, a bibliography available with the R package **strucchange** provides more than 800 references, ending in 2006 [3]. Recent surveys of the methodology include [4–6]. Much of this methodology relies quite heavily on functional central limit theorems (FCLTs), an excellent reference is [7].

Apart from practical relevance of the associated issues, one reason for the large number of publications is that the notion of 'structural change' can be formalized in many different ways. In terms of statistical hypothesis tests, the null hypothesis of 'no structural change' is reasonably clear (model parameters are constant), but the alternative can mean many things: a parameter (or several parameters) change(s) its (their) value(s) abruptly (once, twice, or more often), or it changes gradually according to a stochastic mechanism (e.g., via a random coefficient model), or it switches randomly among a small number of states (e.g., via a hidden Markov model), etc. There are many further possibilities.

The available methodology therefore incorporates ideas from a variety of fields: linear models, sequential analysis, wavelets, etc. In economics, there is comparatively greater interest in changes in regression models, whereas in many other fields of application interest is focused on changes in a univariate time series. A further dividing line is on-line (sequential) analysis of a growing sample versus off-line (retrospective) analysis of a fixed sample.

This chapter provides some basic ideas of change point methodology along with empirical examples. It is biased towards least-squares methods, methodology used in economics and finance as well as availability in statistical software. The following section outlines selected methods in the context of a simple signal-plus-noise model. For reasons of space, the exposition is confined to retrospective analysis of changes in univariate time series. In Sect. 3, several recent algorithms are explored for dating structural changes in a series of oil prices. Section 4 dates volatility changes in Hang Seng stock market index returns, thereby revisiting data formerly studied in [8]. The final section provides some references for sequential analysis of structural change and also for more complex data structures.

2 Some Basic Ideas in Change Point Analysis

To fix ideas, consider a signal-plus-noise model for an observable (univariate) quantity y_i,

$$y_i = \mu_i + e_i, \quad i = 1, \dots, T,$$

where μ_i is the (deterministic) signal and e_i is the noise, with $\mathrm{E}[e_i] = 0$ and $\mathrm{Var}[e_i] = \sigma^2$. As noted above, this chapter is confined to changes in a univariate time series. However, for many methods, there is a regression version with $\mu_i = x_i^\top \beta_i$, where x_i is a vector of covariates and β_i the corresponding set of regression coefficients. In the classical setting, the e_i form a sequence of independent and identically distributed (i.i.d.) random variables, but many more recent contributions, especially in economics and finance, consider dependent processes.

2.1 *Testing for Structural Change*

In terms of statistical testing, the null hypothesis of interest is $H_0 : \mu_i = \mu_0$ for all i; i.e., the signal exhibits no change. (For the regression version, the corresponding null hypothesis is $H_0 : \beta_i = \beta_0$ for all i.) Under the null hypothesis, natural estimates of μ_0 are the recursive estimates $\hat{\mu}_k = k^{-1} \sum_{i=1}^{k} y_i$, $k = 1, \dots, T$; i.e., the sequence of sample means computed from a growing sample. The corresponding recursive residuals are $\tilde{e}_i = y_i - \hat{\mu}_{i-1}$, $k = 2, \dots, T$.

A classical idea is to study the fluctuations of partial (or cumulative) sums (CUSUMs) of these recursive residuals and to reject the null hypothesis of parameter stability if their fluctuations are excessive. In order to assess significance, introduce an *empirical fluctuation process* indexed by $t \in [0, 1]$ as the process of partial sums of the recursive residuals via

$$\tilde{S}_T(t) = \frac{1}{\hat{\sigma}_T \sqrt{T}} \sum_{i=1}^{[Tt]} \tilde{e}_i, \quad 0 \leq t \leq 1, \tag{1}$$

where $[Tt]$ denotes the integer part of Tt and $\hat{\sigma}_T^2$ some consistent estimate of σ^2. This object is often called the *Rec-CUSUM process* as it is based on recursive residuals. It is well known that under the above assumptions this empirical fluctuation process can be approximated by a Brownian motion, $B(t)$, $0 \leq t \leq 1$, and hence the enormous literature on properties of this stochastic process can be used to assess the fluctuations in the recursive residuals. Excessive fluctuation is determined from the crossing probabilities of certain boundaries, this is the approach proposed in the seminal paper by Brown et al. [9].

However, from a regression point of view, the ordinary least-squares (OLS) residuals $\hat{e}_i = y_i - \hat{\mu}_T$ are perhaps a more natural starting point, leading to the test statistic

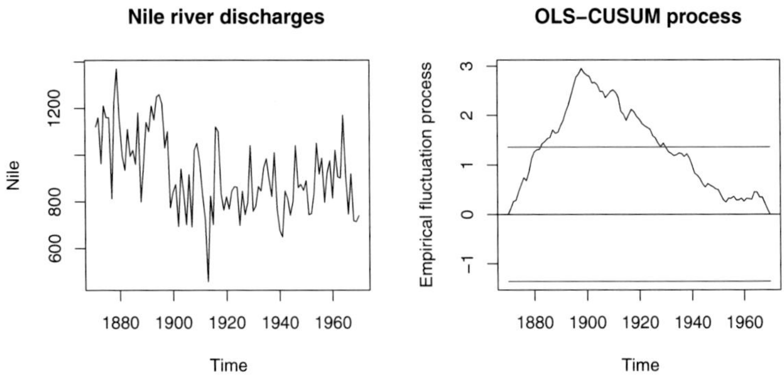

Fig. 1 Nile river discharges: Raw data (*left*) and OLS-CUSUM process (*right*)

$$\max_{k=1,\ldots,T} \left| \frac{1}{\hat{\sigma}_T \sqrt{T}} \sum_{i=1}^{k} \hat{e}_i \right|. \tag{2}$$

As OLS residuals are correlated and sum to zero by construction, the limiting process corresponding to the *OLS-CUSUM process* is no longer a Brownian motion. Instead, the limiting process is now a Brownian bridge, $B^0(t)$, with $B^0(t) = B(t) - t\,B(1), 0 \le t \le 1$, and the relevant limiting quantity for assessing significant deviation from the null hypothesis is

$$\sup_{0 \le t \le 1} |B^0(t)|, \tag{3}$$

the supremum of the absolute value of a Brownian bridge on the unit interval [10]. This object is well known in the statistical literature, and quantiles of its distribution provide critical values for a test based on (2).

To briefly illustrate the machinery, consider a time series of measurements of the annual flow of the river Nile at Aswan, for the period 1871–1970. This series is part of any binary distribution of R [11] under the name `Nile` and has been used repeatedly in the statistical literature on change point methods. The following illustrations make use of the R package **strucchange**, which among other things implements structural change detection using empirical fluctuation processes and related techniques. It should be noted that the original paper [3] describing the software documents the first release of the package, but many methods were added in subsequent years, including the methods for dating structural changes that are used in the next section. The package is still actively maintained, but the main developments happened some 10 years ago.

Figure 1 plots the time series (left panel) and the corresponding OLS-CUSUM process (right panel) along with a boundary indicating the 5% critical value for the test statistic (2). It is seen that the empirical fluctuation process crosses the boundary, and hence the hypothesis of a constant level is rejected at the 5% level.

The test statistic (2), often called the OLS-CUSUM test, measures the maximal absolute deviation from zero of the corresponding OLS-CUSUM process. There are many variations of this idea. For example, it is possible to use other functionals of the empirical fluctuation process, such as the range or some average of the fluctuations. It is also possible to study moving instead of cumulative sums, leading to moving sum (MOSUM) processes. Or, instead of the fluctuations in the residuals, one can directly assess the fluctuations in the estimates themselves; there are again recursive and moving versions [12]. In the univariate case considered here, the latter idea is equivalent to CUSUMs or MOSUMs of the residuals, but in the regression case it leads to new procedures. It is also possible to assess fluctuations in first-order conditions of fitting methods other than least squares, for example likelihood methods [13].

Also, applications in economics and finance often involve dependent data, so that the machinery described above requires adjustments. These involve the *long-run variance*,

$$\omega^2 = \lim_{T\to\infty} \mathrm{Var}(\tilde{S}_T(1)).$$

If a consistent estimator $\hat{\omega}^2$ of ω^2 is available, then $\hat{\omega}^{-1}\tilde{S}_T(t)$ or $\hat{\omega}^{-1}\hat{S}_T(t)$ can, in many settings of interest, again be approximated by a Brownian motion or a Brownian bridge.

2.2 Dating Structural Changes

Having found evidence for the presence of structural change it is of interest to estimate the change points themselves. In economics, basic references for dating structural changes are Bai and Perron [14, 15], who among other things provide a method for obtaining confidence intervals for the break dates. The associated point estimation issue—the segmentation of the sample into homogeneous parts—dates back at least to Bellman and Roth [16].

The model of interest is now a step function for the signal. With $m+1$ segments (corresponding to m breaks), this is

$$y_i = \mu_j + e_i, \qquad \tau_{j-1}+1 \le i \le \tau_j, \quad j = 1,\dots,m+1. \tag{4}$$

Here j is the segment index and $\{\tau_1,\dots,\tau_m\}$ denotes the set of the break points. By convention, $\tau_0 = 0$ and $\tau_{m+1} = T$.

Given the break points $\tau_1,\dots,\tau_m$, the least squares estimate of μ_j is the sample mean of the observations pertaining to segment j. The resulting aggregate residual sum of squares is given by

$$RSS(\tau_1,\dots,\tau_m) = \sum_{j=1}^{m+1} rss(\tau_{j-1}+1,\tau_j), \tag{5}$$

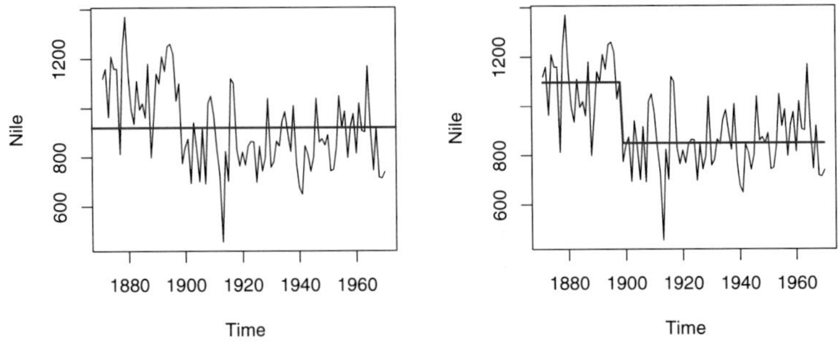

Fig. 2 Nile river discharges: Model with constant level (*left*) versus model with two regimes (*right*)

where $rss(\tau_{j-1}+1, \tau_j)$ is the residual sum of squares for segment j. The problem of dating structural changes is to find the break points $\hat{\tau}_1, \ldots, \hat{\tau}_m$ that minimize the objective function,

$$\{\hat{\tau}_1, \ldots, \hat{\tau}_m\} = \text{argmin}_{\{\tau_1,\ldots,\tau_m\}} RSS(\tau_1, \ldots, \tau_m), \tag{6}$$

over all partitions $\{\tau_1, \ldots, \tau_m\}$ with $\tau_j - \tau_{j-1} \geq T_{\min}$. Here $T_{\min}$ is a bandwidth parameter to be specified by the user, it defines the minimal segment length. For a given number m of break points, their optimal location can be found using a dynamic programming algorithm. The number of change points m itself can be determined via information criteria, the `breakpoints()` function in **strucchange** employs the Bayesian Information Criterion (BIC). Below, this algorithm is referred to as `breakpoints`. More details on the implementation may be found in [17].

Returning to the Nile river flows, Fig. 2 (left panel) provides the fit for a traditional autoregressive model of order one (AR(1)) with constant parameters. Clearly, the fit is quite poor in that for the first part of the series the data are almost always above the fitted mean level, whereas after approximately the year 1900 they are mostly below. In contrast, Fig. 2 (right panel) provides a model with a changing level, the BIC suggesting a model with a single break corresponding to the year 1898. There is a simple explanation for this break: the opening of the Aswan dam in 1898. It is worth noting that after modelling the break there is no need for further modelling of any dependence about the changing level: the dependence implied by the fitted AR(1) process (left panel), with an autoregressive parameter of 0.51, is spurious and stems from the neglected data feature of a changing level.

3 Dating Changes in a Commodity Price Series

This section revisits an empirical example presented in [17], namely dating structural breaks in a time series of oil prices. That paper considered a quarterly index of import prices of petroleum products obtained from the German Federal Statistical Office—

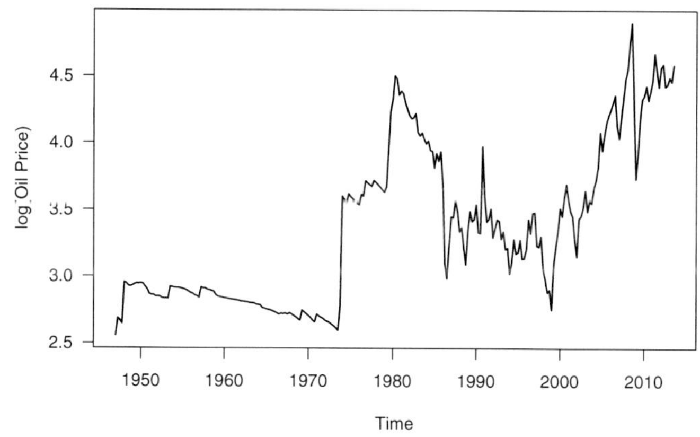

Fig. 3 Quarterly spot prices for West Texas Intermediate (WTI) *Source* Federal Reserve Bank of St. Louis (FRED database)

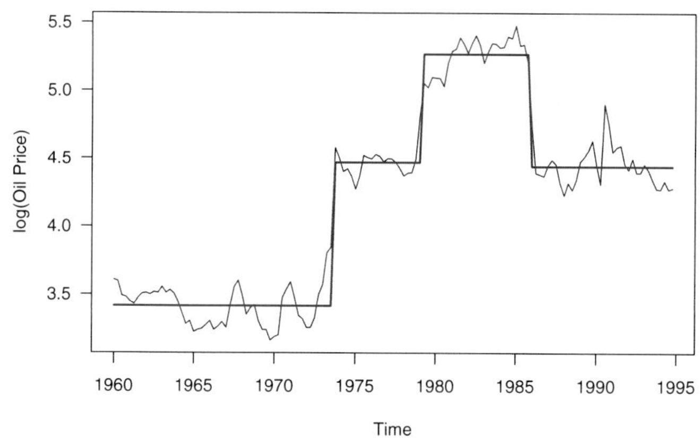

Fig. 4 German oil price data, 1960(1)–1994(4) (base year: 1991) *Data source* German Federal Statistical Office. Segmentation as in [17]

hereafter referred to as the German oil price data—for the period 1960(1)–1994(4) (base year: 1991). The present paper uses a much longer series, a quarterly time series of spot prices for West Texas Intermediate (WTI)—hereafter referred to as the WTI data—from 1947(1) to 2013(3). It is publicly available from the FRED database of the Federal Reserve Bank of St. Louis, more specifically from https://research.stlouisfed.org/fred2/series/OILPRICE/. The series is deflated using the GDP deflator (base year: 2009), which is available from https://research.stlouisfed.org/fred2/series/GDPDEF/. This deflated version is given in Fig. 3 (data are in logarithms). The task is to compare a change point model for the WTI data with the corresponding segmentation for the older German oil price data, and also to try out several more recent dating algorithms, on which more below.

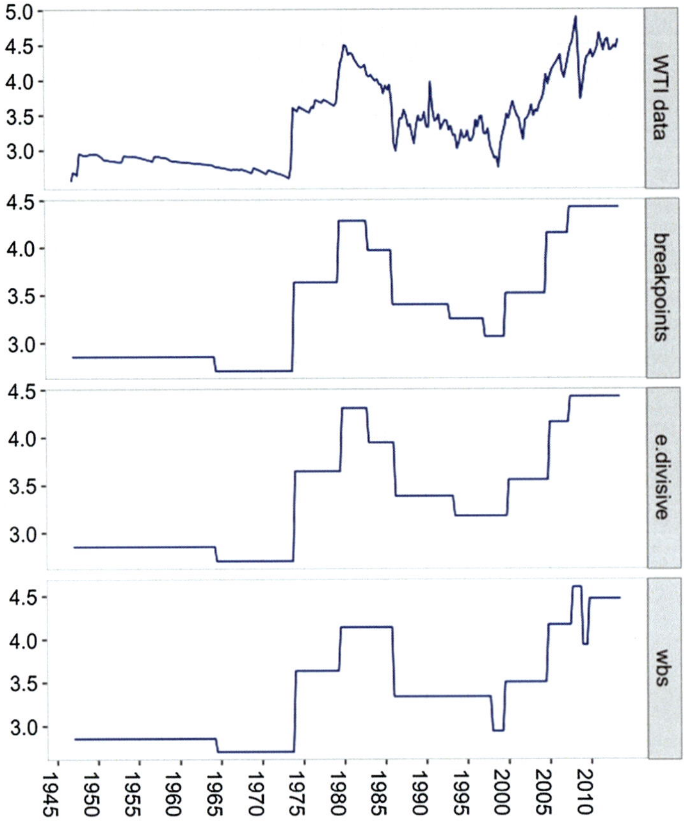

Fig. 5 WTI data (*top panel*) and segmentations obtained from three algorithms: `breakpoints`, `e.divisive` and `wbs`

For ease of reference, the older series along with a segmentation with three regimes is given in Fig. 4 (data are again in logarithms). The three breaks are for the quarters 1973(3), 1979(1) and 1985(1). The first two breaks correspond to two major historical events, the first oil crisis (the Arab oil embargo following the Yom Kippur war) and the beginning of the Iranian revolution. The break in 1985(1) may be seen as resulting from demand shifts, quarrels within OPEC, and the entry of several new suppliers (namely Great Britain, Mexico, and Norway) in international oil markets [17].

Repeating the exercise with the newer WTI data and a minimal segment size of 10 quarters, the BIC now favors a segmentation with 10 regimes. The resulting solution is provided in the second panel of Fig. 5. There is good agreement for the breaks corresponding to major historical events, here estimated at 1973(4) and 1979(2). Beginning in the second half of the 1990s, there is a gradual trend in the newer series that is not described well by a step function. Also, the two oil price series differ visibly in the first half of the 1980s, leading to two estimated change points for the newer series, at 1982(4) and 1985(4). These differences result from the fact that the older series is a price index while the newer is for a single product; also, there appear to be exchange rate effects in the older series.

We next compare the least-squares-based solution with two recent methods. The first method [18] uses ideas from cluster analysis combined with a nonparametric form of ANOVA based on so-called *energy statistics* [19]. The latter are functions of distances between statistical observations in Euclidean spaces (and beyond), the name derives from an analogy with Newton's gravitational potential energy. It should be noted that this method assesses differences in entire distributions, not just level shifts. However, there is a variant that assesses only changes in the mean; the relevant settings for this variant are used below. An implementation is available in the R package **ecp** [20]. The package offers several methods, here only the algorithm named `e.divisive` there is used, a form of hierarchical clustering. The second method is *wild binary segmentation* [21] (hereafter: WBS), a stochastic algorithm that uses ideas from the wavelets literature. The setup analyzed in the original paper is (4) with i.i.d. Gaussian noise. An implementation is available in the R package **wbs**.

All three procedures require specification of a trimming parameter, for the least-squares approach and `e.divisive` this is the minimal segment length (or 'cluster size', in the terminology of the **ecp** package), details differ from method to method. For `e.divisive`, the minimal segment size was again set to 10 quarters, here yielding 9 breaks. For WBS, which does not need a minimal segment size, the maximum number of breaks was fixed at 10 for comparability reasons. The solutions are provided in the third and fourth panel of Fig. 5. There is good agreement for the major historical events, while the algorithms differ somewhat for the second half of the series. This partly reflects the problems with this part of the series mentioned above. Overall, WBS tries to adapt to smaller details towards the end of the series. It is also worth noting that, using the settings recommended by the authors of the software, the WBS algorithm favors a solution with no fewer than 35 breaks. Clearly, not all of the estimated breaks will be of economic interest. This appears to be a problem in some financial applications, where alarms can be frequent with long series. As an example, not all changes identified in [21] for the S&P 500 index will likely be of practical relevance. Figure 5 provides an overall comparison of all solutions. The display highlights similarities and differences of the solutions obtained from the algorithms. Notably the `breakpoints` and `e.divisive` solutions are very similar. In contrast, `wbs` is more faithful to the more lively part towards the end of the series. `breakpoints` has one further change point in the second half of the 1990s (namely for the quarter 1997(1)), the other breaks differ, with one exception, by at most two quarters. For example, the break corresponding to the first oil crisis is in 1973(4) according to `breakpoints` and in 1974(1) according to `e.divisive`. For the Iranian revolution break, the algorithms give 1979(2) and 1979(4), respectively.

4 Dating Changes in the Volatility of a Stock Market Index

The previous section considered changes in the mean of a time series. It is also possible to study changes in other characteristics of the data, for example variances or autocorrelations. With financial time series, for example stock returns, assessing risks is a central issue; both squared and absolute returns may be seen as measures of risk. Assessing changes in such transformed returns can be viewed as an indirect check of structural change in GARCH-type models of volatility.

As a brief empirical illustration, we revisit an example from [8]. They consider four stock market indices (FTSE, Hang Seng, Nikkei, S&P 500) with an eye on changes associated with the Asian and Russian financial crises in the second half of the 1990s. Here we just consider one of these series, the Hang Seng index for the period 1989–01–04 to 2001–10–19, giving $T = 3338$ observations. For comparability reasons, the data are taken from Datastream. The data for the segmentation algorithm are the Hang Seng absolute returns (for which more breaks are found than for the more common squared returns). The original paper documents only the maximal number of breaks used but not the minimal segment size. Here we use a minimal segment size corresponding to 10% of the length of the series, which should permit recovery of the segmentation from [8]. This is only partly possible, however.

Figure 6 provides a plot of the absolute returns along with two models. The original paper suggests that a segmentation with three breaks, for the dates 1992–07–03, 1995–01–24 and 1997–08–15, is optimal. The 3-breaks solution found by `breakpoints` (the dashed line in the plot) differs, it finds 1995–06–14, 1997–08–

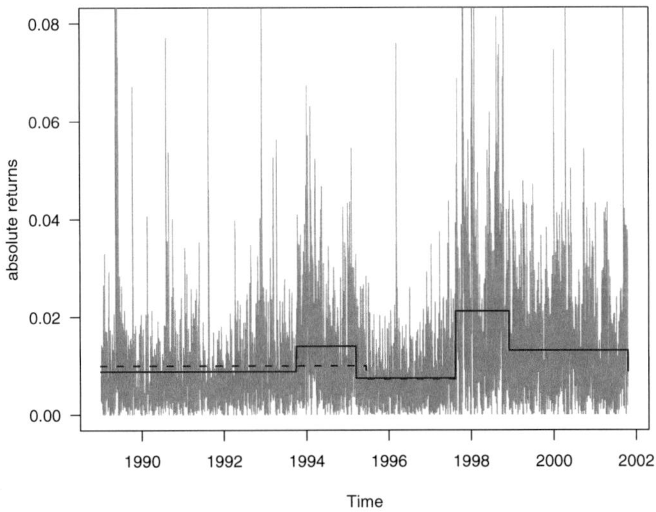

Fig. 6 Segmentation of Hang Seng absolute returns with `breakpoints`. *Dashed line* maximal number of breaks 3, minimal segment size 10%. *Solid line* maximal number of breaks 5, minimal segment size 10%

15 and 1998–11–30, so only the last break from the paper is recovered. Also, the `breakpoints` solution has no break prior to 1995 but a new break after 1997.

It is also worth noting that the method finds a further break upon increasing the maximum number of breaks. Setting the latter to five breaks, a segmentation with four breaks is found. Interestingly, the new break at 1993–10–01 is before 1995, although not overly close to the 1992–07–03 break of [8]. Further experiments with the minimal segment length (down to 5% of the length of the series) and the admissible number of breaks (up to 10) suggest that the results are quite sensitive to the settings of these parameters. The only break that is practically always found is for August 1997, it is often estimated at 1997–08–15 and is associated with the Asian financial crisis.

5 Discussion and Outlook

This chapter has illustrated some basic ideas in change point analysis. The exposition was confined to retrospective methods for univariate time series. (Most of) The methods described have extensions to regression models [4]. Many further topics had to be excluded, notably the timely topic of on-line monitoring and also structural change in multivariate or functional data. For on-line monitoring, in some fields referred to as surveillance or 'quickest detection' problems, see the recent survey [22] and references therein, for an exposition of associated optimality issues in a financial setting see [23]. Multivariate and functional data are briefly addressed in the recent survey [6], where further references may be found.

The literature will likely continue to grow rapidly, for several reasons: The classical methods are largely confined to linear models fitted via least squares methods. Nonlinear models for discrete-valued data are needed in some applications, but here the literature is still relatively small. Also, the growing number of large data sets demands improvements on the algorithmic side. Unfortunately, many recent methods are not readily available in statistical software, which to some extent hinders progress. A further big challenge is the unification of this widely scattered literature.

Computational Details

All results were obtained using R 3.2.4, with the packages **strucchange** 1.5-1, **ecp** 2.0.0, and **wbs** 1.3, on PCs running Mac OS X, version 10.10.5. Some plots were drawn using the package **ggplot2** 2.1.0 [24].

References

1. M. Kelly, C. O'Grada, Change points and temporal dependence in reconstructions of annual temperature: did Europe experience a little Ice Age? Ann. Appl. Stat. **8**, 1372–1394 (2014)
2. P. Hackl, A. Westlund, Statistical analysis of "structural change": an annotated bibliography. Empir. Econ. **14**, 167–192 (1989)
3. A. Zeileis, F. Leisch, K. Hornik, C. Kleiber, **strucchange**: An R package for testing for structural change in linear regression models. J. Stat. Softw. **7**, 1–38 (2002)
4. P. Perron, Dealing with structural breaks, in *Palgrave Handbook of Econometrics: Volume 1: Econometric Theory*, ed. by K. Patterson, T.C. Mills (Palgrave Macmillan, London, 2006), pp. 278–352
5. A. Aue, L. Horváth, Structural breaks in time series. J. Time Ser. Anal. **34**, 1–16 (2013)
6. L. Horváth, G. Rice, Extensions of some classical methods in change point analysis. TEST **23**, 219–255 (2014)
7. M. Csörgő, L. Horváth, *Limit Theorems in Change-Point Analysis* (Wiley, Hoboken, NJ, 1997)
8. E. Andreou, E. Ghysels, Detecting multiple breaks in financial market volatility dynamics. J. Appl. Econ. **17**, 579–600 (2002)
9. R.L. Brown, J. Durbin, J.M. Evans, Techniques for testing the constancy of regression relationships over time. J. R. Stat. Soc. Ser. B **37**, 149–163 (1975)
10. W. Ploberger, W. Krämer, The CUSUM test with OLS residuals. Econometrica **60**, 271–285 (1992)
11. R Core Team, *R: A Language and Environment for Statistical Computing* (R Foundation for Statistical Computing, Vienna, 2016)
12. C.-M. Kuan, K. Hornik, The generalized fluctuation test: a unifying view. Economet. Rev. **14**, 135–161 (1996)
13. A. Zeileis, A unified approach to structural change tests based on ML scores, F statistics, and OLS residuals. Economet. Rev. **24**, 445–466 (2005)
14. J. Bai, P. Perron, Estimating and testing linear models with multiple structural changes. Econometrica **66**, 47–78 (1998)
15. J. Bai, P. Perron, Computation and analysis of multiple structural change models. J. Appl. Econ. **18**, 1–22 (2003)
16. R. Bellman, R. Roth, Curve fitting by segmented straight lines. J. Am. Stat. Assoc. **64**, 1079–1084 (1969)
17. A. Zeileis, C. Kleiber, W. Krämer, K. Hornik, Testing and dating of structural changes in practice. Comput. Stat. Data An. **44**, 109–123 (2003)
18. D.S. Matteson, N.A. James, A nonparametric approach for multiple change point analysis of multivariate data. J. Am. Stat. Assoc. **109**, 334–345 (2014)
19. M.L. Rizzo, G.J. Székely, DISCO analysis: a nonparametric extension of analysis of variance. Ann. Appl. Stat. **4**, 1034–1055 (2010)
20. N.A. James, D.S. Matteson, **ecp**: An R package for nonparametric multiple change point analysis of multivariate data. J. Stat. Softw. **62**, 1–25 (2014)
21. P. Fryzlewicz, Wild binary segmentation for multiple change-point detection. Ann. Stat. **42**, 2243–2281 (2014)
22. M. Frisén, Optimal sequential surveillance for finance, public health, and other areas. Seq. Anal. **28**, 310–337 (2009)
23. A.N. Shiryaev, Quickest detection problems in the technical analysis of financial data, in *Mathematical Finance—Bachelier Congress 2000, Paris, June 29–July 1, 2000*, ed. by H. Geman, D. Madan, S. Pliska, T. Vorst (Springer, Heidelberg, 2002), pp. 487–521
24. H. Wickham, ***ggplot2****: Elegant Graphics for Data Analysis* (Springer, Heidelberg, 2009)

Part VI
Brain and Coordination Dynamics

Coarse-Graining to Investigate Cerebral Cortex Dynamics

The Case for Experimental Simplification of Biological Phenomena

Kentaroh Takagaki and Frank W. Ohl

Abstract Advances in multi-channel/multi-detector recordings and data analysis over the last decades have led to an explosion in the exploration of complex neural dynamics in mammalian cortex. Powerful methods have been applied to investigate such dynamics, including connectivity measures (correlation, causality, resting state synchrony, etc.), spatiotemporal pattern analyses, and finite-element modelling based on model neurons. These methods were initially applied to data from simple experimental models such as invertebrate neurons/ganglia/tecta, cell cultures, and organotypic slice preparations. Advances in the field have triggered the expanded use of such measures on more complex data, for example to mammalian ex vivo preparations, anesthetized preparations, and mammalian awake behaving preparations. With the increasing surgical, behavioral, and physiological complexity of the preparations themselves, less invasive measurement methods such as optical recordings, massively implanted arrays, or fMRI and other electromagnetic methods must be used to ensure robustness; however, these measures tend to feature lower signal-to-noise ratios, and are often prone to various biases. Furthermore, the high dimensionality of the data itself leads directly to potential errors in programming of analysis algorithms and overinterpretation of statistically significant but biologically insignificant findings. Given this situation, we advocate for the complementary use of the classical biological approach: the use of simplified preparations which may be limited in scope, but which highlight fundamental principles. We illustrate this approach with three experimental examples which use experimental and observational approaches to coarse-grain dynamic spatiotemporal activity patterns, to make coarse-graining observations of clinically relevant oscillations, and to coarse-grain complex behavior in mammalian discrimination learning.

K. Takagaki (✉) · F.W. Ohl
Leibniz Institute for Neurobiology, Magdeburg, and Institute for Biology, Otto-von-Guericke University, Magdeburg, Germany
e-mail: kentaroh.takagaki@lin-magdeburg.de

S.C. Müller et al. (eds.), *Complexity and Synergetics*,
https://doi.org/10.1007/978-3-319-64334-2_22

1 Putting the Mesoscopic Complexity of the Cerebral Cortex into Perspective: A Comparison with the Heart

In order to highlight the complexity problem we have described above and to illustrate the complexity of the cerebral cortex in particular, we start by comparing the physiological complexity of the cerebral cortex with that of the heart.

The heart is perhaps the best characterized mammalian organ in terms of its physiology and dynamics. Half a century of work has led to the characterization of the dynamics of, for example, basic pathological dynamics in atrial fibrillation [1, 2]. However, even in this extremely dramatic and common pathological dynamic, we are just now starting to unravel the causal influences of these atrial dynamics on the ventricle [3], to give an example. Even in the heart, we are far from being able to make quantitative predictions regarding the dynamic outcomes of even very basic pharmacological interventions, such as how an individual case of atrial fibrillation will respond dynamically to digitalis treatment.

Accepting that we have established that even the most basic of cardiac dynamics are just starting to be characterized, we can now illustrate the difficulty of handling cortical dynamics by comparing the cerebral cortex as a physiological system to the heart in terms of degrees of freedom.

We believe that it is justified to claim that the brain is astronomically more complex compared to the heart, in terms of its degrees of freedom (Table 1). One can start with the number of active elements, the basic unit of each dynamic system. It can be argued that the heart has only on the order of 10^0 active elements, since individual cardiomyocytes join into huge cardiac syncytia which are coupled both on the electrical and the cytoplasmic levels. Even if one counts individual cardiomyocyte nuclei as the basic unit, the number of nuclei in the heart is at least one order of magnitude less than the brain (Table 1). Next, one can think of the complexity of individual constituent units of each dynamic system. Whereas cardiac cells can be essentially classified into nodal cells and myocytes, even the most rudimentary classification of neurons will result in 10^1 transmitter classes and 10^2 anatomical morphological classes.

Spatially speaking, there is also a vast difference in degrees of freedom between the two dynamic systems. For example, the cardiac syncytium can essentially be handled as a warped two-dimensional excitable medium [2]. In contrast, the cortical mantle features not only a two-dimensional spatial distribution, but also a handful of topological degrees of freedom in the columnar (depth) dimension, as well as numerous proximal, short-range, and long-range projections, which can be considered as further dimensions in the cortical matrix, thus increasing the spatial dimensionality of the dynamic system. This difference in spatial dimensionality is clearly manifested by the types of clinical measurements utilized: for the heart, a 12-lead EKG will allow the description of most dynamic patterns, whereas even the coarsest measure of brain activity, the 10–20 EEG, features 19 leads, and this only allows unambiguous documentation of the simplest of dynamic patterns, such as epileptiform activity patterns and sleep slow waves.

Table 1 The brain is more complex than the heart. Various anatomical and functional measures are given to illuminate the magnitude of this difference

	Heart	Brain
Number of cells	10^0~ (nuclei: 10^{10})	10^{11}
Types of excitable cells	2	10^2–10^3
Spatial network dimensions	<2 spatial dimensions	>4 spatial dimensions
Connectivity	10^1/myocyte	10^3~/neuron
Temporal scales	1 temporal scale, fixed frequency for the most part	multiple irregular timescales spanning ≥6 orders of magnitude
Physiological oscillatory modes	Single	Desynchronized (δ, θ, α, β, γ, ...)
Physiological outcomes of dynamic activity	Blood is pumped	Various sensations, motor planning, emotional states, hypothesizing, calculating, remembering, planning, daydreaming, empathizing, etc.
Dynamic pathological disturbances	Blood is not pumped effectively (atrial fibrillation, ventricular fibrillation, conduction block, ventricular tachycardia, atrial tachycardia/bradycardia, etc.)	Seizures (innumerable different types), loss of consciousness, amnesia, psychiatric disturbances (innumerable types), movement disorders, sleep disturbances, coma, migraine, etc.

Related to spatial dimensionality on the level of unit connectivity, the difference in degrees of freedom is also obvious: whereas cardiomyocytes are basically all interconnected as syncytia, cortical neurons form an interwoven network of mind-boggling complexity, as with the oft-cited number of 10^3 connections per neuron.

We can also illustrate the complexity of the cortex as a system based on the complexity of its functional outputs.

In terms of temporal dynamics, whereas the heart operates physiologically at a regulated rate of approximately 60–120 beats per minute, the cortical mantle operates not only at varying frequencies, but on at least 6 temporal scales: the sub-millisecond scale of STDP, the millisecond scale of spikes and spike rates, the second scale of metabolic changes as seen in fMRI and SPECT, the minute-to-hour scale of classical plasticity paradigms such as LTP and LTD, the daily scale of circadian activity patterns, and the days-to-decades scale of learning and forgetting as well as of cortical development and degeneration.

As physiologists, we cannot fail to mention the vast differences in the functional phenomenological complexity of the two dynamic systems. While the essence of cardiac function is limited to a unidirectional pumping of blood, the physiological essence of the cortical mantle encompasses a wild array of behavioral and neurological functions, spanning sensation in various modalities, planning and patterning of motor output for a vast array of both skeletal and smooth muscles throughout the body, resting states such as sleep or posterior basic rhythm, as well as the myriad of cognitive and emotional states which make our lives so rich.

In terms of pathologies based on disruption patterns of dynamics, there is also a marked difference in the complexity of disruption patterns. Whereas the heart as an organ has only a handful of common failure modes, even a rudimentary list of common neurological and psychiatric disturbances would easily fill this page.

Taking all of the above comparisons together and summarizing them, it is a priori clear that the dynamics of the cerebral cortex are vastly more complex than that of the heart and have an astronomically higher degree of freedom, both on the constitutional substrate level and on the physiological phenomenological level. If one takes this together with the fact that even the vastly simpler cardiac dynamics have been explained to a limited extent, as argued above, it is no wonder that describing the neural dynamics of the cerebral cortex is a Sisyphean task.

Recent decades have seen the exponential growth of ever more precise measurement methods spanning from three-dimensional anatomical-physiological methods such as multiphoton imaging and MRI, to two-dimensional functional imaging methods such as single photon calcium and voltage-sensitive dye imaging, to metabolic methods such as intrinsic signal imaging, PET and fMRI, to electrical methods such as high-density EEG and massively parallel electrophysiology. We have seen a concurrent explosion in computational power, optimized algorithms (such as the vastly influential fast Fourier transform), and computational paradigms such as parallelization. Given such a large amount of information and information processing power, it is tempting to attempt bottom-up simulations of the brain from ground principles.

Valiant attempts have been made at bottom-up simulations of cortical dynamics from the unit level, most recently through the Blue Brain project [4]. However, the difficulty of the problem is illustrated by the result that even this most heroic of efforts by the most technically advanced computational neuroscientists has only succeeded in providing a partial reconstruction of a specific area of the cortex of a particular species, and only reproduces activity patterns resembling the most basic state transitions and cellular parameters. How does one then tackle the daunting task of understanding the dynamics of the brain? We argue that even in this age of massively parallel recordings and unprecedented information processing power, given the vast complexity of neural dynamics, that there is still persisting value in coarse-graining on a phenomenological level, i.e. to deal with biologically simplified systems.

Indeed, if one takes a step back, extrapolating from corner cases and simplified/special systems to understand general principles is a time-honored and vastly successful approach in biology: be it the study of genetics and reproduction in yeast or *E. coli* [5], the study of functional physiology in isolated organ preparations [6], or the study of evolutionary and adaptive principles in exotic ecosystems [7]. These approaches have provided insights which would not have been possible by directly studying human genetics, in vivo human physiology, or the full complexity of the Linnaean taxonomy, respectively.

For the field of neural dynamics, an equivalent biological approach is coarse-graining of observations. Indeed, this strategy has been fruitfully exploited, and most of the ground-breaking studies in neurodynamics have been on simplified preparations. For example, while anesthetics had long been used in animal neurophysiology for practical reasons of animal handling during electrophysiological recording experiments, it was the notion that some anesthetics effectively function as "openers" for feedback loops in the sense of engineering control theory that has paved the way for entirely new approaches in experimental and theoretical neurophysiology [8]: With the experimental availability of techniques for opening and closing feedback loops in brain circuitry, far-reaching new conceptualizations of neuronal mass activity in paleocortex and neocortex became possible by applying concepts of (linear) control theory to (linearized) neuronal dynamics (for review, see [9, 10]). A similarly fundamental phenomenon of brain dynamics, the coordination of multiple rhythms, has been principally elucidated by the simple experiment of coordinating periodic finger movement [11]. It was possible to demonstrate with this approach that "synergetic" concepts, developed for describing self-organization of systems far from thermodynamic equilibrium [12], can be used to identify coordinated movements is this type of scenarios with order parameters emerging in a synergetic system of competing oscillators (for review see [13, 14]).

In this chapter, we advocate a renewed focus on biological and experimental coarse-graining, that is, a renewed focus on experimentally simplified preparations. We elaborate on the argument that advances in measurement and computation can best be taken advantage of by simultaneously coarse-graining our preparations experimentally, such that they manifest core biological phenomena in a simplified system, and provide robust experimental observations on a level amenable to in silico science. We illustrate this approach through three experimental systems which we have utilized in our recent work.

2 Example 1: Coarse-Graining of Emergent Dynamic Spatial Activity Patterns: Spiral Waves in the Epileptic Neocortex

A classical method of coarse graining in mesoscopic neurophysiology is to use preparations with simplified state dynamics, which are evoked using pharmacological agents. This can be either with explicit pharmacological blocking of certain receptors, or implicit global state changes with anesthetic agents.

Pharmacological studies of cortical dynamics have traditionally focused on agents which create clear oscillatory or epileptiform states, such as interictal-like spiking elicited by the $GABA_A$ inhibitor bicuculline. Although the dynamics of such states tend to be vastly simplified, the clear spiking/oscillatory nature of the preparation is conducive to the analysis of spatial connectivity and propagation (e.g. [15]). One can also block certain pharmacological subcircuits of the cortex in vivo to try to elucidate the roles of different subcomponents (e.g. [16]). Of note, given the complex dynamics of the cortex, even the most delicate pharmacological

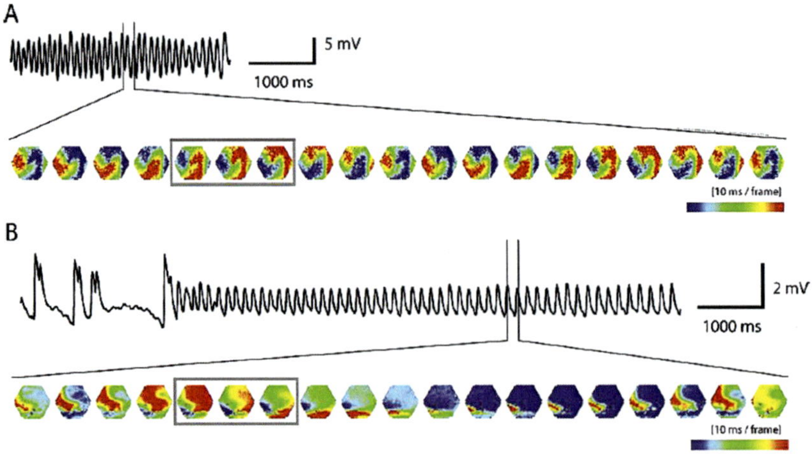

Fig. 1 Spiral activity in the neocortex as seen with voltage-sensitive dye imaging. Two examples of stable spiral dynamics are demonstrated, first in an organotypic slice preparation treated with carbachol and bicuculline (**a**), and next in an epileptiform intact cortex from an epileptic Mongolian gerbil, recorded through a craniotomy. From [19], with permission from Elsevier

blockades of the cortical network tend to lead to massively simplified and coarse-grained dynamic patterns, amenable to reductionistic analysis.

One recent line of work using a weakly coupled oscillatory state induced by bicuculline and the cholinergic agonist carbachol, for example, has shown that spiral dynamics can serve as an organizing principle for mesoscopic activity within the cortical mantle [17]. This is an example where experimental coarse graining of cortical dynamics has illustrated a new dynamic principle in cortical organization. Although one may question the extent to which such organizing principles are relevant for non-pharmacologically treated intact cortex, we are guardedly optimistic about the general significance of our findings given that the phenomenon, which was first discovered in organotypic slice preparations, has now been seen in in vivo pharmacologically treated preparations as well as in sleep-like states of light anesthesia [18]. We have also observed such activity in epileptiform cortical activity patterns of Mongolian gerbils with epilepsy (Fig. 1b; [19]). Since the spiral dynamics are much less stable in vivo than in drug-treated slice preparations, one can very well imagine that without initial coarse-grained study in non-physiological organotypic slice preparations, the phenomenon had very little chance of being observed at all.

Anesthetic states provide a somewhat more physiological means to curtail the dynamic degrees of freedom inherent in the cortex. Given the “burstiness” of anesthetized cortex in general (true anesthetic agents invariably lead to cortical depression and rebound bursting), these simplified preparations allow study of the transmission of large bursting impulses of population activity (e.g. [20]).

By modulating the depth of anesthesia, one can attain dynamics which are somewhat less simplified, yet still stable. For example, slow withdrawal of pentobarbital anesthesia leads to sleep-like cortical states with theta and delta wave domination; we have recently shown that such states can sustain short in vivo bouts of spiral dynamics [18], which we speculate may play a role to stabilize recurrent cortical dynamics during memory consolidation in sleep. Modulating the depth of urethane anesthesia leads to a biphasic state of slow-wave deep sleep-like activity and fast desynchronized REM sleep-like activity, and one can take advantage of such simplified dynamics to study the lateral propagation of mesoscopic activity [21]. Based on such findings, one can speculate as to the extrapolability to analogous physiological states of deep sleep and REM/wakefulness.

3 Example 2: Coarse-Graining of Human Mesoscopic Brain Dynamics: Posterior Basic Rhythm in the Development of Autism

Pharmacological simplifications such as in the previous section are rarely possible for studying the human brain, where pharmacological interventions are reserved for medical indications. Even if a medically necessary pharmacological treatment fortuitously simplifies cortical dynamics, such as say with pregabalin-induced increase in slow-wave sleep [22], various uncontrollable factors such as co-medication and disease state will necessarily complicate interpretation of results.

Although it is increasingly fashionable to measure complex metrics of cortical dynamics such as resting state connectivity, for example, these measures feature a low signal-to-noise ratio, and are much more susceptible to potential multiple testing and other statistical errors. Given the complexity of these measures, we believe that there is enduring value in focusing instead on the simplest of human brain dynamics and conducting coarse-grained observations at this level.

One classical method of coarse-graining human cortical dynamics is to drive the system with entraining stimuli. Stimuli driving strong dynamics are usually regular, such as photic stimulation eliciting epileptiform oscillatory bursting. This is used regularly in clinical practice to provoke and document non-overt epileptiform dynamics in patients with epileptic habitus [23]. Another classical experimental approach is to use continuous tone sequences which drive regular population activity in auditory areas of the brain; by occasionally changing these driving stimuli, for example by dropping stimuli or inserting "oddball" stimuli within the driving train stimulus, one can investigate excursions from a mesoscopically entrained and stable brain dynamic [24].

Another approach is to use highly synchronized brain states to the advantage of investigative questions. To this end, we have recently focused on one of the most classic and one of the most robust of human brain dynamics, the posterior basic EEG rhythm of Berger [25]. This very stable dynamic emerges in the posterior pole

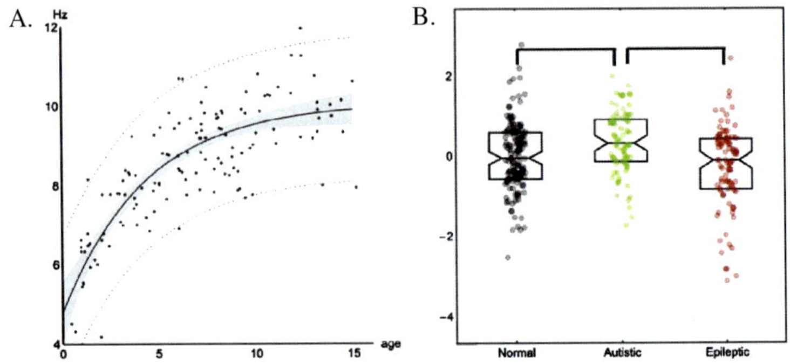

Fig. 2 The development of the posterior basic rhythm (PBR) is accelerated in autistic children. **a** Development of PBR in normal children (n = 134) who were referred to our center for evaluation of inattentiveness and other issues, which turned out to have a non-psychiatric diagnosis. **b** Corrected PBR frequency in developing children shows statistically significant increase in autistic children (*green*), and statistically significant decrease (developmental delay) in epileptic children (*red*). We believe that these statistically significant results may also be biologically and clinically significant. From [26], with permission from Elsevier

at around several months of age at a frequency of 3–4 Hz, and gradually develops over the course of 8–10 years of life to reach the adult "alpha" range of 8–12 Hz. Functionally, this stable dynamic is characterized by high amplitude in a quiet restful state with eyes closed, and is attenuated by eye opening. We focused on this most simple of cortical dynamics recorded by EEG, and retrospectively analyzed this state in children with autism and normal controls (We also studied epileptic children, given the high incidence of overt and covert epileptiform activity reported in autistic children.) By quantifying even the most simple characteristics of this state such as base oscillation frequency, we concluded that children with autism (on a retrospective cohort base) show faster development of the posterior basic rhythm than their non-autistic peers (Fig. 2).

4 Example 3: Coarse-Graining of Complex Behavior

Another dimension of complexity which must be dealt with in neuroscience is the diversity of behavior. For example, even the simplest tasks in experimental human psychology feature a large degree of hidden complexity, in that the subject is verbally appraised of the task context (e.g. sitting in a chair in a quiet room), the modality in question (e.g. looking at the monitor), as well as the general timescale of the experiment (e.g. "you will be asked to answer questions for about 30 min"). Such covert contextual cues cannot be given to animals, much less advanced linguistic descriptions of task contingencies.

A further dimension of complexity is the unitary nature of experience. Even for relatively simple tasks, it has been shown that operant responses are dependent

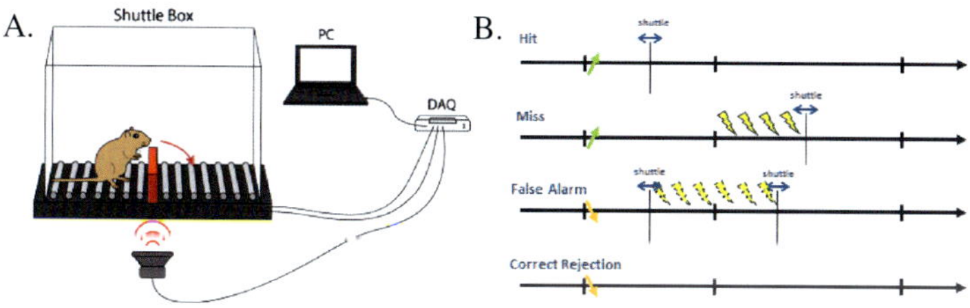

Fig. 3 The Mongolian gerbil auditory Go/NoGo shuttlebox task. **a** The animal performs the task in a two-chamber shuttlebox, with hurdle. We use a hurdle here (in contrast to the tunnels more common in shuttleboxes for mice), to take advantage of the species-specific tendency of gerbils to jump. The grid floor delivers a mildly aversive footshock to provide negative reinforcement to the animal. **b** Timeline of four possible behavioral outcomes. For Go trials, the animal will receive a certain tone, in this case a rising FM sweep (*light green*). If the animal shuttles to the other compartment within a fixed period of time, this is a correct "hit" trial (*first row*). If the animal fails to shuttle in response to a Go tone, this is considered a "miss" trial, and it will receive footshock until it escapes. For a NoGo tone (in this case a falling FM sweep; *orange*), the animal must remain in its current shuttlebox compartment (*fourth row*; "correct rejection"). If it shuttles, it will receive a mild footshock (*third row*; "false alarm")

upon the proximal history of reward [27]. Such history effects are further complicated by learning paradigms, where the animal's behavior changes progressively. Sometimes, sudden changes in behavior can mark the animals' mastery of the operant contingency, what we refer to the "aha" moment [28].

How can one simplify complex behaviors, and coarse grain them such that their dynamics are amenable to analysis? We have approached simplification of rodent auditory discrimination/categorization by using a Go/NoGo shuttlebox paradigm (Fig. 3). Using this task, we are able to simplify behavioral response contingencies into the four categories of signal detection theory, namely, hit, miss, false alarm, and correct rejection. The key point in terms of behavioral coarse-graining here is that the discrimination is manifested in a purely binary decision: whether to jump (Go) or to stay (NoGo). This binary nature reduces the complexity of the task when compared to more common rodent sensory two-choice discrimination tasks such as T maze or two nosepoke choice tasks, as these tasks are not strictly binary; they actually constitute a trinary choice task: go left, go right, or do not go, which complicates the behavioral readout in terms of signal detection theory.

By coarse-graining the behavioral paradigm in a manner tailored to our question (i.e. by simplifying our task to mirror our theoretical framework of signal detection theory), we were able to demonstrate spatiotemporal patterns in cortical ECoGs which predict, in a manner not time-locked to sensory stimulus, the animal´s behavioral contingency in single trials based on the framework of signal detection theory [28].

5 Summary

In this chapter, we have argued that experimental coarse-graining is still necessary to effectively tackle the vast complexity presented by the cerebral cortex, even with modern advances in computational capacity and algorithmic efficiency. We have supported our arguments by comparing cortical dynamics to the vastly simpler yet nonetheless highly complex dynamics of the heart. We illustrate our approach by describing three experimental systems which have provided us insights into cortical dynamics which would not have been possible through the more fashionable approach of attempting bottom-up induction from single-neuron dynamics or by using complex methods to analyze fine-grained data.

Acknowledgements We would like to thank Prof. Stefan Müller for inviting us to present in the Hannover symposium, and Profs. Walter J. Freeman, Gholam Motamedi, Henning Scheich, and Jian-Young Wu for inspiring our work. This work was supported by the Alexander von Humboldt Foundation and Deutsche Forschungsgemeinschaft (DFG) grant SPP 1665. We thank Jennifer Tegtmeier for the illustration in Fig. 3.

References

1. J. Jalife, Deja vu in the theories of atrial fibrillation dynamics. Cardiovasc. Res. **89**, 766–775 (2011). https://doi.org/10.1093/cvr/cvq364
2. S.V. Pandit, J. Jalife, Rotors and the dynamics of cardiac fibrillation. Circ. Res. **112**, 849–862 (2013). https://doi.org/10.1161/CIRCRESAHA.111.300158
3. M. Masè, M. Marini, M. Disertor, F. Ravelli, Dynamics of AV coupling during human atrial fibrillation: role of atrial rate. Am. J. Physiol.—Heart Circ. Physiol. **309**, H198–H205 (2015). https://doi.org/10.1152/ajpheart.00726.2014
4. H. Markram, E. Muller, S. Ramaswamy, M.W. Reimann, M. Abdellah, C.A. Sanchez et al., Reconstruction and simulation of neocortical microcircuitry. Cell **163**, 456–492 (2015). https://doi.org/10.1016/j.cell.2015.09.029
5. B. Alberts, A. Johnson, J. Lewis, D. Morgan, M. Raff, K. Roberts et al., *Molecular Biology of the Cell*, 6th edn. (Carland Science, New York, 2014)
6. J.E. Hall, *Guyton and Hall Textbook of Medical Physiology*, 13th edn. (Saunders, Philadelphia, 2015)
7. C. Darwin, J. Huxley, *The Origin of Species: 150th Anniversary Edition* (Signet, New York, 2003)
8. W.J. Freeman, Effects of surgical isolation and tetanization on prepyriform cortex in cats. J. Neurophysiol. **31**, 349–357 (1968)
9. W.J. Freeman, *Mass Action in the Nervous System* (Academic Press, New York, 1975)
10. W.J. Freeman, *Neurodynamics: An Exploration in Mesoscopic Brain Dynamics* (Springer, London, New York, 2000)
11. J.A.S. Kelso, J.P. Scholz, G. Schöner, Nonequilibrium phase transitions in coordinated biological motion: critical fluctuations. Phys. Lett. A **118**, 279–284 (1986). https://doi.org/10.1016/0375-9601(86)90359-2
12. H. Haken, *Synergetics—An Introduction. Nonequilibrium Phase Transitions and Self-Organization in Physics, Chemistry and Biology* (1983), http://www.springer.com/us/book/9783642964695. Accessed 26 July 2016

13. H. Haken, *Principles of Brain Functioning: A Synergetic Approach to Brain Activity, Behavior and Cognition*, Softcover reprint of the original 1st edn. (Springer, Berlin, New York, 1996)
14. J.A.S. Kelso, *Dynamic Patterns: The Self-organization of Brain and Behavior*, Revised edn. (MIT University Press, Cambridge, 1997)
15. L. Bai, X. Huang, Q. Yang, J.Y. Wu, Spatiotemporal patterns of an evoked network oscillation in neocortical slices: coupled local oscillators. J. Neurophysiol. **96**, 2528–2538 (2006)
16. M.F.K. Happel, M. Jeschke, F.W. Ohl, Spectral integration in primary auditory cortex attributable to temporally precise convergence of thalamocortical and intracortical input. J. Neurosci. Off. J. Soc. Neurosci. **30**, 11114–11127 (2010). https://doi.org/10.1523/JNEUROSCI.0689-10.2010
17. X. Huang, W.C. Troy, Q. Yang, H. Ma, C.R. Laing, S.J. Schiff et al., Spiral waves in disinhibited mammalian neocortex. J. Neurosci. **24**, 9897–9902 (2004)
18. X. Huang, W. Xu, J. Liang, K. Takagaki, X. Gao, J. Wu, Spiral wave dynamics in neocortex. Neuron **68**, 978–990 (2010). https://doi.org/10.1016/j.neuron.2010.11.007
19. K. Takagaki, C. Zhang, J.-Y. Wu, F.W. Ohl, Flow detection of propagating waves with temporospatial correlation of activity. J. Neurosci. Methods **200**, 207–218 (2011). https://doi.org/10.1016/j.jneumeth.2011.05.023
20. W. Xu, X. Huang, K. Takagaki, J.Y. Wu, Compression and reflection of visually evoked cortical waves. Neuron **55**, 119–129 (2007)
21. T. Wanger, K. Takagaki, M.T. Lippert, J. Goldschmidt, F.W. Ohl, Wave propagation of cortical population activity under urethane anesthesia is state dependent. BMC Neurosci. **14**, 78 (2013). https://doi.org/10.1186/1471-2202-14-78
22. T. Roehrs, T. Roth, Drug-related sleep stage changes: functional significance and clinical relevance. Sleep Med. Clin. **5**, 559–570 (2010). https://doi.org/10.1016/j.jsmc.2010.08.002
23. D.L. Schomer, F. Lopes da Silva (eds.), *Niedermeyer's Electroencephalography: Basic Principles, Clinical Applications, and Related Fields*, 6th edn. (LWW, Philadelphia, 2010)
24. M.I. Garrido, J.M. Kilner, K.E. Stephan, K.J. Friston, The mismatch negativity: a review of underlying mechanisms. Clin. Neurophysiol. **120**, 453–463 (2009). https://doi.org/10.1016/j.clinph.2008.11.029
25. H. Berger, Über das Elektrenkephalogramm des Menschen. Eur. Arch. Psychiatry Clin. Neurosci. **87**, 527–570 (1929). https://doi.org/10.1007/BF01797193
26. K. Takagaki, J. Russell, M.T. Lippert, G.K. Motamedi, Development of the posterior basic rhythm in children with autism. Clin. Neurophysiol. **126**, 297–303 (2014). https://doi.org/10.1016/j.clinph.2014.04.022
27. L.P. Sugrue, G.S. Corrado, W.T. Newsome, Matching behavior and the representation of value in the parietal cortex. Science **304**, 1782–1787 (2004). https://doi.org/10.1126/science.1094765
28. F.W. Ohl, H. Scheich, W.J. Freeman, Change in pattern of ongoing cortical activity with auditory category learning. Nature **412**, 733–736 (2001)

Coordination Dynamics and Synergetics: From Finger Movements to Brain Patterns and Ballet Dancing

Armin Fuchs and J.A. Scott Kelso

Abstract Starting with a brief review of the original Haken-Kelso-Bunz model and its generalizations from the 1980s, we discuss three examples from more than three decades of our research on coordination dynamics. From the 1990s, we show how movement coordination can be used to probe the brain of individual subjects and how coordination patterns in behavior are also manifested in brain signals. From the 2000s, we present an experiment on social coordination in brain and behavior and introduce an analysis technique for EEG signals recorded in such settings. Most recently, we recorded the performance of a professional ballet dancer, where we found the coordination patterns of in-phase and anti-phase as elementary building blocks in complex movements.

1 Introduction

Coordination dynamics, as a quantitative field established in the mid 1980s, opened a new level of applications to synergetics outside the hard core sciences. Hermann Haken's proposition that phase transitions far from thermal equilibrium play a much bigger role than widely thought was open for a new test. The concepts of synergetics had been applied with great success to the laser and hydrodynamic instabilities, systems where the mesoscopic or even microscopic dynamics of the subsystems are

A. Fuchs (✉) · J.A. Scott Kelso
Center for Complex Systems & Brain Sciences and Department of Physics,
Florida Atlantic University, 777 Glades Road, Boca Raton, FL 33431, USA
e-mail: fuchs@ccs.fau.edu

J.A. Scott Kelso
e-mail: kelso@ccs.fau.edu

J.A. Scott Kelso
Intelligent Systems Research Centre, Ulster University,
Derry Londonderry BT48 7JL, Northern Ireland

S.C. Müller et al. (eds.), *Complexity and Synergetics*,
https://doi.org/10.1007/978-3-319-64334-2_23

well known and from which the order parameters guiding the macroscopic dynamics can be derived—a so-called bottom-up approach. In coordination dynamics this situation is reversed: There are transition phenomena in movements that have been found experimentally and the goal is to establish a dynamics of order parameters at the macroscopic level and then find a description on a lower level from which the macroscopic dynamics can be derived. Such an approach is a top-down example of synergetics. The two approaches, top-down and bottom-up, are of course complementary [1].

Here we start with a brief review of the original Haken-Kelso-Bunz (HKB) model [2] and its generalizations and then focus on three applications of coordination dynamics, namely to probe the human brain of individual subjects, to relate brain and behavior in social coordination, and to establish the building blocks of complex movements in professional ballet dancing.

2 The Origin of Coordination Dynamics

In what ways can you move your two index fingers rhythmically? According to the lore of coordination dynamics, most people without special training are able to produce two stable movement patterns at low rates, i.e. flexion of one finger while the other extends (anti-phase) and simultaneous flexion and extension of both fingers (in-phase). Surprisingly, only the in-phase movement can be performed beyond a certain frequency of the movement that may vary across individuals. When a movement starts in anti-phase and the movement rate is increased, the coordination pattern switches spontaneously and involuntarily to in-phase; if the movement starts in in-phase no switch is observed. These observations by one of us [3, 4] led to a fruitful collaboration with Hermann Haken in Stuttgart and to publications (e.g., [2, 5]) that are the pillars of coordination dynamics as a quantitative science. The importance of these publications is at least twofold: First, they take the phenomenon described above from the real world of two fingers doing some seemingly trivial wagging movements to an abstract level of identifying order parameters, and deriving them using the theory of coupled nonlinear oscillators and stochastic dynamics. The movement rate becomes a control parameter (that leads to destabilizing the coordination between the limbs), and the relative phase ϕ is established as the order parameter, i.e. the relevant variable for the description of the two rhythmically moving fingers at a macroscopic level. The switching that occurs when the movement speeds up is seen as a ball moving in a potential landscape in an overdamped fashion that changes its shape as a function of rate expressed in the famous equations

$$\begin{aligned} \dot{\phi} &= -\tfrac{dV(\phi)}{d\phi} + \sqrt{Q}\xi_t = -a \sin\phi - 2b \sin 2\phi + \sqrt{Q}\xi_t \\ < \xi_t > &= 0; \quad < \xi_t \xi_{t'} > = \delta(t-t'); V(\phi) = -a\cos\phi - b\cos 2\phi \end{aligned} \tag{1}$$

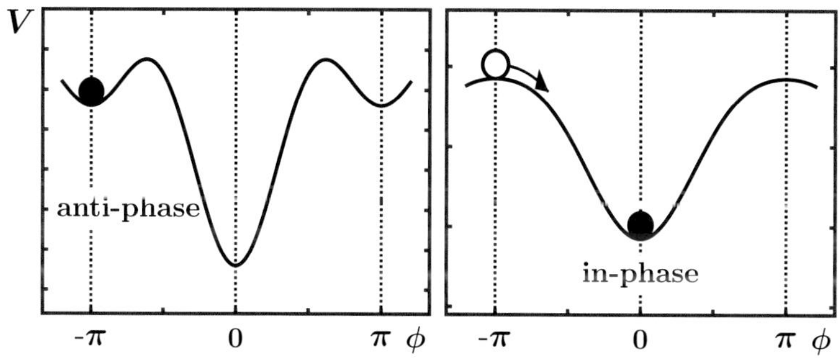

Fig. 1 Potential landscape from (1) for slow movement rates ($b/a = 0.7$, *left*) with minima for anti-phase and in-phase. For high rates ($b/a = 0.15$, *right*), where only one minimum exists, the only stable movement pattern is in-phase

where $\sqrt{Q}\xi_t$ is Gaussian white noise of strength Q and b/a represents the movement rate. A decrease in b/a corresponds to an increase in rate with a critical value of $b/a = 0.25$. The shape of the potential function below and above the critical rate is shown in Fig. 1. Second, and maybe even more important, the theory that was formulated in these initial papers more than 30 years ago modeled the switch in the movement patterns from the viewpoint of synergetics as a second order phase transition and made predictions that could be tested experimentally (see [6, 7], for reviews). On the deterministic side, an oscillator, known as the hybrid, was established that follows the amplitude-frequency relation found in human finger movements [8]; the model was extended to situations where the individual components are not symmetric (see below) and to gaits and gait changes in quadrupeds [9]. On the stochastic side it was shown that the switch exhibits the features of a phase transition in non-equilibrium systems: The system has to become unstable before the switch actually occurs. Such an instability leaves footprints like an increase in the variance of the movement (enhanced fluctuations) and a longer time to recover from a perturbation (critical slowing down) when the critical point is approached but before the actual transition takes place. The experimental discovery of these hallmarks showed that the spontaneous change from anti-phase to in-phase is not simply a replacement of one motor program by another but a self-organized pattern forming process that occurs in systems far from thermal equilibrium. Moreover, the dynamic principle seems to be universal: Whether it is two fingers, two hands, an arm and a leg [10, 11] or even the legs of two different people watching each other [12], it is always the anti-phase movement that becomes unstable at increasing rate and the coordination pattern switches to in-phase.

In the original work, where the interacting limbs were two fingers with the same eigenfrequencies, the system in terms of the oscillating limbs has a symmetry with respect to an exchange of the components. When we look at the coordination between an arm and a leg or the syncopation of one finger with an external stimulus, this symmetry is no longer present, i.e. we are now dealing with a system of oscillators with different eigenfrequencies. If the equation for the relative phase (1) is derived

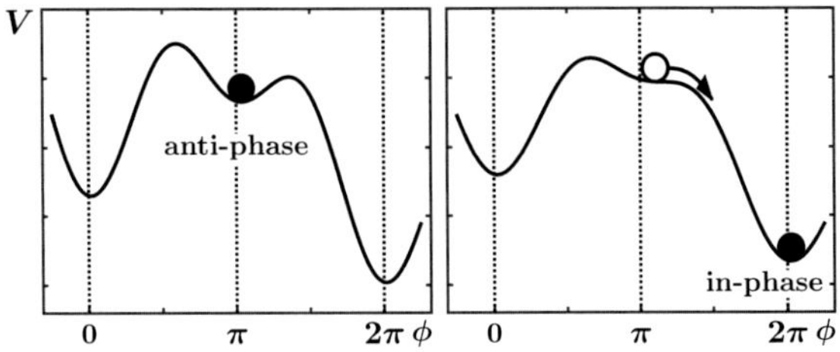

Fig. 2 When the interacting components are not symmetric, the potential landscape is tilted due to a linear term $\delta_\omega \phi$. The fixed points are shifted away from 0 and π, the system becomes unstable at slower rates (here $b/a = 0.4$), and the transition has a preferred direction

for such a system, an additional term, usually termed δ_ω, appears in the equation of motion and becomes a term $\delta_\omega \phi$ in the potential function [13]. The two main effects of this term for the dynamics of the system are: First, the coordination patterns are no longer strictly in-phase or anti-phase, i.e. at a relative phase of 0 or π, respectively, but the stable states are shifted with respect to these values. As shown in Fig. 2 the potential function is now tilted. As a consequence, the states $\phi = 0$ and $\phi = 2\pi$ are no longer the same. For a system that is initially in an anti-phase pattern at $\phi \approx \pi$ it is more likely to switch to $\phi \approx 2\pi$ than $\phi \approx 0$. Even though after the switch it is not possible to tell whether a system is close to 0 or 2π, by following the continuous relative phase it can be detected whether the direction was to the left or right. The coordination patterns with broken symmetry between arms and legs were studied intensively using the device shown in Fig. 3 with one of the authors[1] as the animal under investigation. This device, known as MAC (multi-articulation coordination), allows for perturbing the movements by applying a brake for a short moment and for manipulating the degree of symmetry breaking. The latter is achieved by using weights that are either attached to the wrist, which decreases the amount of symmetry breaking or on the ankle, which actually increases it. The predictions derived from the theoretical model have been found in good agreement with the experimental results [10, 11].

3 Coordination Patterns in Brain Signals

In the early 1990s coordination dynamics was used for probing the human brain, where the syncopation-synchronization paradigm became the vehicle of choice in many experiments. In contrast to the bimanual setup used in many of the behavioral studies, here the subjects are instructed to flex (or extend) their index finger in between two beats of a metronome, i.e. to syncopate with the stimulus. As the rate of the metronome increases, this behavior becomes unstable and the subjects switch to a pattern synchronized with the beats [13]. During the trials the magnetic fields orig-

[1] It should be pointed out that this is not a recent picture.

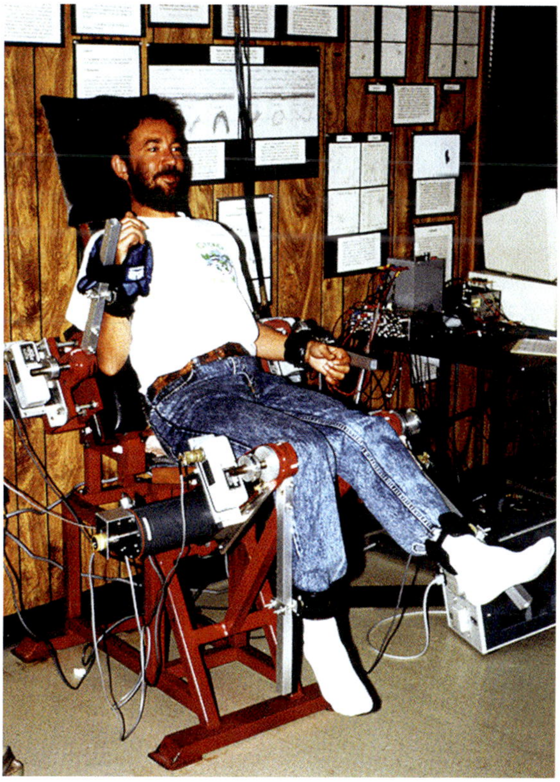

Fig. 3 A subject in the MAC (Multi-Articulation Coordination). This device allows for perturbations of the movements by applying a brake for a short moment and for varying the degree of symmetry breaking by attaching weights to either the wrist or ankle. As the legs naturally have a smaller eigenfrequency than the arms, the degree of symmetry breaking can either be further increased or reduced with loads at the ankles or wrists, respectively

inating from electrical activity in the brain were recorded using magnetoencephalography (MEG), a technology in its infancy at the time. In the first experiment a magnetometer with 37 SQuIDs (Superconducting Quantum Interference Devices) was positioned over left auditory and motor cortex covering roughly 2/3 of a hemisphere. The hallmarks of non-equilibrium phase transitions previously found in behavioral data were now detected in the brain signals [14–16]. The experiment was later repeated with a 143-channel device with full-head coverage that allowed for a better comparison of brain activity during the coordination task and signals from control condition of auditory and motor only [17]. One of the main findings from these experiments remains the switch in the phase of the first Fourier component of the brain signal, shown in Fig. 4, accompanying the switch from syncopation to synchronization in the behavior.

During the last decade, social coordination, i.e. the interaction between two or more people has become an expanding field where not only behavioral measures are recorded from interacting individuals but also brain signals using a dual-EEG for instance as shown in Fig. 5. In this experimental setting two subjects (we call them Red and Blue) are facing each other and are instructed to perform a certain coordination pattern say in-phase with their right index finger when possible. For the first 20

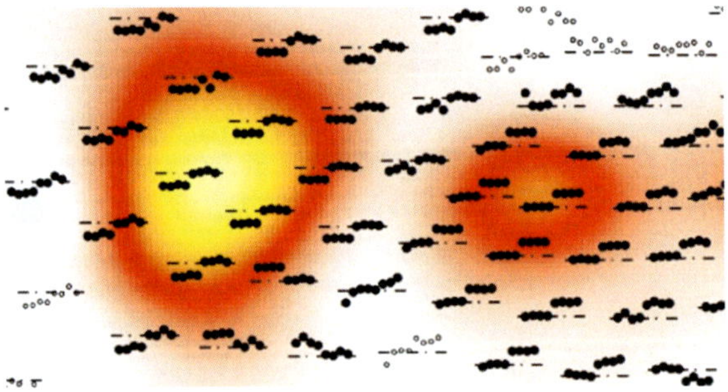

Fig. 4 Phase shift by π in the relative phase between the Fourier component of the MEG signal at the coordination frequency and the stimulus in sensors over the left hemisphere that coincides with the behavioral switch from syncopation to synchronization. The underlying color represents the signal power

seconds of each trial an opaque screen prevents them from seeing the other's finger and they move at a comfortable rate. Then the screen turns transparent allowing the subjects to synchronize their movements. During the trials EEG is recorded from both subjects using caps with 60 electrodes each. In earlier work [18–20] certain wave patterns were identified in the α-range around 10 Hz that are tied to coordinated action. Here we introduce a procedure for the segmentation of multi-channel EEG recordings that are band-pass filtered, in our case 7–13 Hz, and apply it to the data recorded in this experiment.

A short sequence of one second from a 60-channel EEG recording is shown as a butterfly plot in Fig. 6 together with the spatial patterns of the electric potential at

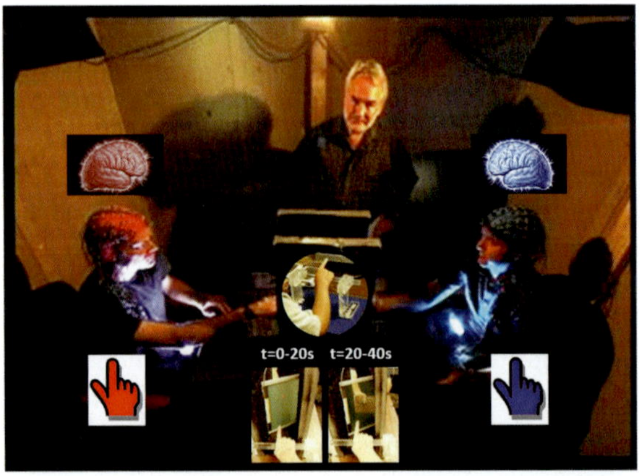

Fig. 5 Setup for a dual-EEG social coordination experiment: Two subjects are instructed to perform a certain coordination pattern with their index fingers. For the first 20 seconds they cannot see the other's movement and therefore cannot coordinate. Then the screen turns transparent and a coordination pattern is established. During the trials the movements and EEG signals from both subjects are recorded

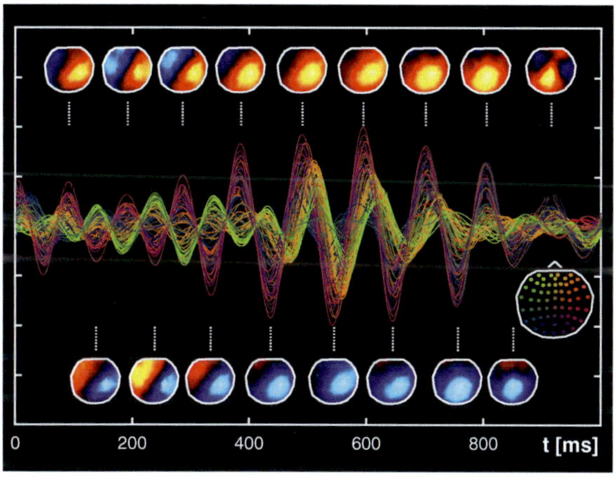

Fig. 6 Butterfly plot of 60 EEG channels for one second from the recording of one subject together with the spatial patterns of the electric potential on the scalp at the maxima and minima. There are segments where the signals only reverse polarity while the spatial shape remains unchanged and other time intervals where the pattern goes from dipolar to a single maximum or minimum and on to a quadrupole. *Insert* Color coding for the electrode locations

the maxima and minima and an insert on the right, which shows the color coding of the electrode locations on the scalp. Aside from the oscillation there is also a change in the shape of the pattern from a dipole to a pattern with a single maximum or minimum changing to a quadrupolar shape at the end of the time series. Such changes in the electric potential at the scalp originate from a change in the underlying electrical activity in the cortex, primarily in the macrocolumns of the gray matter. It is our goal to identify segments of the EEG signal where the shape of the spatial pattern is essentially constant while only reversing polarity and relate them to the coordination behavior.

In a first step we reduce the dimensionality of the system by applying a principal component analysis (PCA). In PCA a set of patterns is calculated that are best suited to represent the variance in the recordings whilst drastically reducing the number of time series that are needed. PCA is performed by calculating the eigenvalues and eigenvectors of the covariance matrix given by

$$C_{ij} = \frac{1}{T}\int_0^T dt\,\{e_i(t) - \bar{e}_i\}\{e_j(t) - \bar{e}_j\} \tag{2}$$

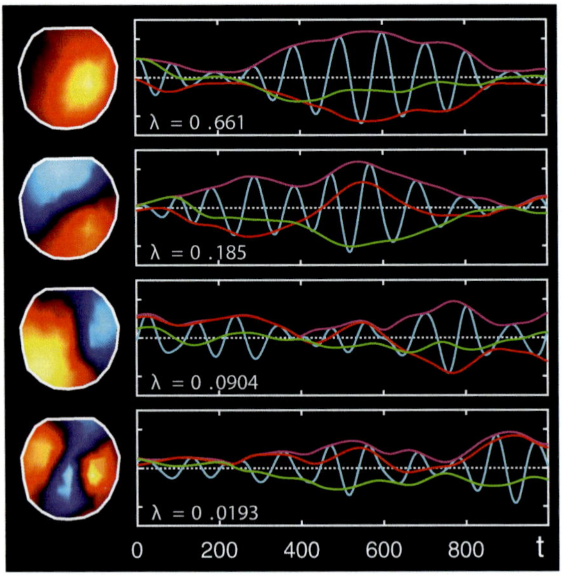

Fig. 7 Principal component analysis (PCA) applied to the EEG data shown in Fig. 6. The spatial patterns corresponding to the four largest eigenvalues (*left column*) carry almost 96% of the variance in the signal. Time series of the corresponding amplitudes (*cyan*) show oscillations around 10 Hz like the data from the single electrodes. This fast oscillation is removed by calculating complex slowly varying amplitudes shown in red and green for their real- and imaginary part, respectively. The magnitude of these functions represents the envelope plotted in magenta

where $e_k(t)$ represents the time series recorded from electrode k and $\bar{e}_k$ is its temporal mean. As the recordings were done with 60 electrodes for each subject, C is a 60×60 matrix. The eigenvectors corresponding to the largest eigenvalues of this matrix represent the spatial patterns that carry most of the variance in the EEG signal. The patterns calculated from the time series in Fig. 6 are shown on the left of Fig. 7. The corresponding eigenvalues λ (if properly normalized) quantify how much of the variance is carried by a given pattern. The four eigenvalues for the patterns in Fig. 7 sum up to 0.956, which means the four patterns represent almost 96% of the variance. Because C is a real symmetric matrix its eigenvectors are orthogonal and the amplitude for a pattern k at a time t, $p_k(t)$, can be computed by projecting the EEG signal vector $\boldsymbol{e}(t)$ onto the eigenvector $\boldsymbol{v}^{(k)}$

$$p_k(t) = \boldsymbol{e}(t) \cdot \boldsymbol{v}^{(k)} \tag{3}$$

The amplitudes for the first four modes are plotted in cyan in Fig. 7 next to the patterns. Like the time series from the electrodes they show oscillations around a frequency of 10 Hz. The second step of the procedure is intended to eliminate these fast oscillations and replace them by slowly varying time series like the envelope plotted in magenta in Fig. 7. To this end we perform a Fourier transform on the amplitudes $p_k(t)$ leading to complex valued functions $\tilde{p}_k(\omega)$. As the $p_k(t)$ are real, their

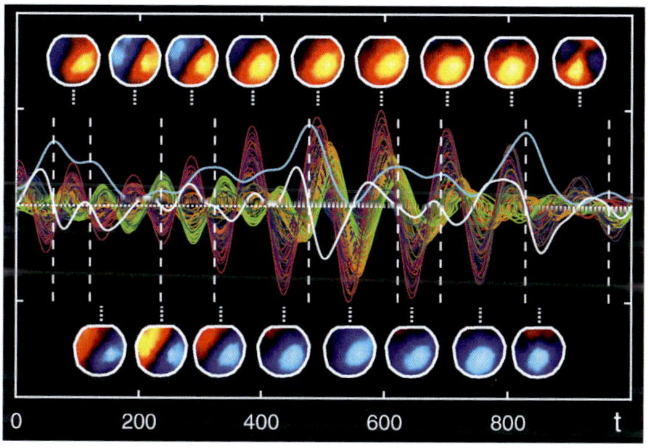

Fig. 8 Segmentation of the EEG data shown in a butterfly plot and patterns at the minima and maxima of the signal. The speed along the trajectory in the 8-dimensional space is plotted in *cyan* and its derivative in *white*. The time points where the derivative intersects the horizontal axis with a negative slope mark the boundaries of segments (*vertical white dashed lines*) within which the shape of the pattern is relatively constant

Fourier coefficients for ω and $-\omega$ are complex conjugates $\tilde{p}_k(-\omega) = \tilde{p}_k^*(\omega)$. Next we set the coefficients for all negative frequencies to zero, shift the remaining functions by 10 Hz to the left and apply an inverse Fourier transform. This process eliminates the 10 Hz component from the signal and leads to complex valued amplitudes $a_k(t)$ whose real- and imaginary part are plotted in red and green, respectively, in Fig. 7. The envelope of the amplitudes $p_k(t)$ is given by the magnitudes $|a_k(t)|$.

The dynamics recorded by EEG is now described by the four complex amplitudes $a_k(t)$, which represent the journey of a trajectory in an 8-dimensional space. If this trajectory stays within a small region of that space for a certain time the pattern is stationary. If the trajectory moves around the pattern changes. In analogy to lower dimensions we can define a velocity for the trajectory as the derivatives of the real- and imaginary parts of $a_k(t)$ and a scalar speed, $s(t)$, as the square root of the sum of these derivatives squared

$$s(t) = \sqrt{\sum_k \left\{ \left[\frac{d}{dt} \mathfrak{R}\{a_k(t)\} \right]^2 + \left[\frac{d}{dt} \mathfrak{I}\{a_k(t)\} \right]^2 \right\}} \tag{4}$$

This speed is plotted in cyan in Fig. 8 together with the original data in a butterfly plot and spatial pattern at the maxima and minima. Changes in the spatial shape occur where $s(t)$ has maxima (dashed vertical white lines), which are found as the time points where the temporal derivative of $s(t)$ (plotted in white) intersects the horizontal axis with a negative slope. The segments in-between the vertical lines are time intervals where the shape of the pattern does not change or changes only slightly.

Now we apply this segmentation to the dual-EEG recordings described above and relate it to the coordination behavior of the subjects. As mentioned, for the first 20

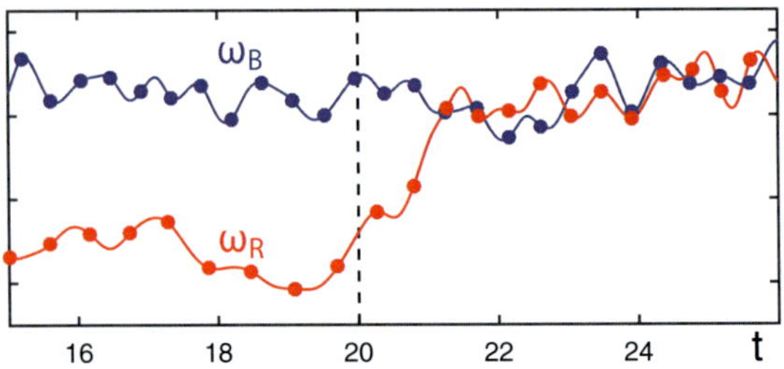

Fig. 9 Angular frequencies for the *red* and *blue* subject, ω_R and ω_B for a time span from $t = 15$ s to $t = 26$ s. The screen turns transparent at $t = 20$ s indicated by the *dashed black vertical line*. As in most of the trials *Red* starts at a lower rate and speeds up to *Blue*

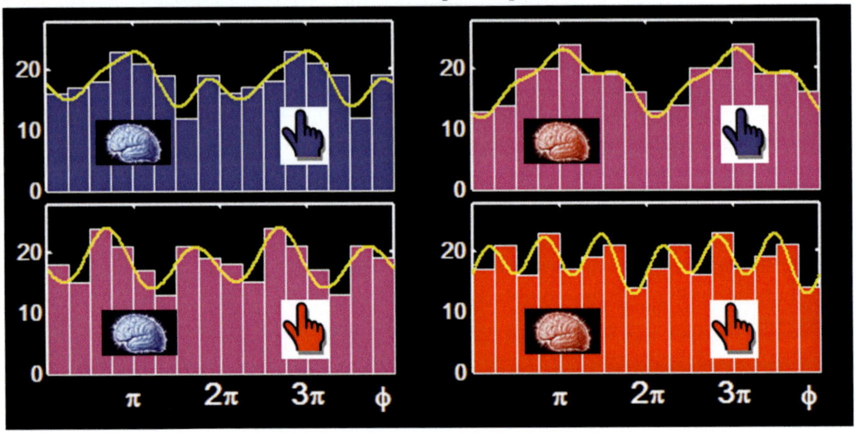

Fig. 10 Histograms of the locations measured by the phases where the segment boundaries fall within the movement cycles for all four brain-finger combinations. Only the boundaries from the *red* brain show a relation to the movement of the *blue* finger

seconds of each trial the view of the partner's finger was blocked by an opaque screen and each individual moved at her own rate. In order to synchronize the movement after the screen has turned transparent these rates have to become the same and there are various ways how this can happen: The faster one may slow down, the slower one speed up or they both change and meet in the middle or elsewhere. Here, we look at the data from 20 trials of a pair, where the red subject, in almost all trials, started at a slower movement rate and sped up to the blue subject, whose frequency stayed pretty much the same. The angular frequencies from a typical trial around the time when the screen turns transparent (from 15 to 26 s) are shown in Fig. 9.

We aim to determine whether there is any relation between the segment boundaries extracted from the two EEG data sets and the coordination behavior by determining where these boundaries fall into the movement cycle. This was done by calculating the phase of the movements at each boundary. There are four cases: The phase of the blue (red) finger at the boundaries from the blue (red) brain, and the phase of the blue (red) finger at the boundaries from the red (blue) brain. As expected, no relation was found for the first 20 seconds when the subjects moved individually without vision of the other's finger movements. But also after movement synchrony had been established for times greater than about 2 seconds after the screen had turned transparent, no relation could be seen. Only during the transition interval from $t = 20$ s to $t = 21.5$ s the histograms exhibited in Fig. 10 were found, showing that the segment boundaries from the red brain are more than twice as likely to fall around a phase of π of the blue finger than a phase of 0 or 2π (right upper diagram). This finding is an interesting reflection of the behavioral pattern where Red adjusts her movement rate to the blue partner, whereas Blue seems to simply move at her own rate largely independent of what the Red is doing, hinting at a master-slave or leader-follower relation in both brain and behavior.

4 Coordination Patterns in Complex Movements: Ballet Dancing

In the Fall of 2014 we invited the God daughter of one of us (JASK), the professional ballet dancer (Makaila Wallace from Ballet B.C.) and asked her to perform a choreography of her choosing lasting about 20 seconds in a number of different experimental conditions, e.g. with and without music, fast versus slow, expressing different emotions, etc. The performances were recorded using a Vicon motion capture system with eight infrared cameras and 32 infrared markers attached to the dancer's body that allowed for a reconstruction of the trajectories of the markers in 3-dimensional space. Snapshots of such trajectories together with stick figure representations of the dancer during two short time intervals of the performance[2] are shown in Fig. 11.

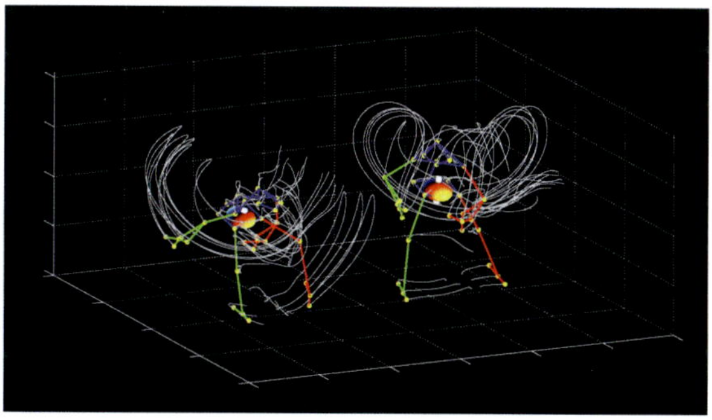

Fig. 11 Snapshots of the stick figure representing the dancer together with trajectories of the 32 infrared markers attached to different parts of her body. Each of the traces corresponds to a duration of one second. The colored spheres represent the six degrees of freedom for the locations and orientations of the dancer's body center in space

[2]An animated version of this figure as well as other movies from the analysis of dancing can be found at http://clifford.ccs.fau.edu/~coordinationofdancing.

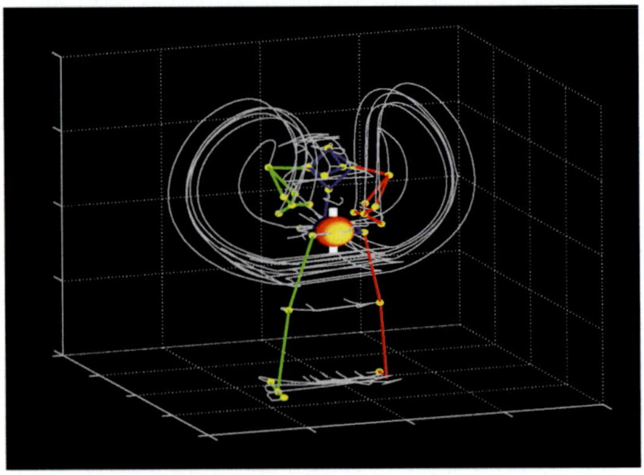

Fig. 12 Snapshot of the stick figure together with marker trajectories after the translations and rotations of the body center have been removed. Such transformed time series are much better suited for an extraction of the relevant features of the other body parts, e.g. arms and legs, than the original data

When the dancer performs the choreography, the movement can be split into a movement of her body center and movements of her body and limbs internal degrees of freedom relative to the center movement. We define the dancer's center as the center of gravity of a triangle given by the markers at her left and right hip, and a marker at her back close to the lowest lumbar vertebra (L5). The location of these points in the 3-dimensional space of the laboratory are given by three vectors we call $\boldsymbol{h}_l$, $\boldsymbol{h}_r$ and $\boldsymbol{l}_5$, respectively, and determine the vector for the body center

$$\boldsymbol{c} = \frac{1}{3}\{\boldsymbol{h}_l + \boldsymbol{h}_r + \boldsymbol{l}_5\} \tag{5}$$

Next, we define a coordinate system for the dancer's body with the x-, y- and z-axis given by

$$\boldsymbol{x} = \frac{(\boldsymbol{h}_r - \boldsymbol{l}_5) + (\boldsymbol{h}_l - \boldsymbol{l}_5)}{|\boldsymbol{h}_r + \boldsymbol{h}_r - 2\boldsymbol{l}_5|} \qquad \boldsymbol{y} = \frac{(\boldsymbol{h}_r - \boldsymbol{l}_5) \times (\boldsymbol{h}_l - \boldsymbol{l}_5)}{|(\boldsymbol{h}_r - \boldsymbol{l}_5) \times (\boldsymbol{h}_l - \boldsymbol{l}_5)|} \qquad \boldsymbol{z} = \boldsymbol{x} \times \boldsymbol{y} \tag{6}$$

After subtracting the translation of the center $\boldsymbol{c}$ from all marker coordinates and applying a rotation that transforms the laboratory coordinate system into the body frame (6), all movements are relative to the body center location and orientation in space. A snapshot of the trajectories for such a coordinated movement is shown in Fig. 12.

The time series after the center translations and rotations have been removed are much better suited for extracting the relevant features of the movement of other body parts. For instance the movement of the torso and the head can now be described in terms of two additional rigid bodies as translations and rotations relative to the center. Here we will restrict ourselves to the dynamics of the arms and legs, which have 14 and 8 markers attached, and therefore their trajectories in 3-dimensional space are given by 42 and 24 time series ($a_n(t)$, $n = 1 \ldots 42$ and $l_m(t)$, $m = 1 \ldots 24$), respectively. To extract the basic movement patterns from these time series we perform a principal component analysis (PCA) separately for the markers on the arms and the

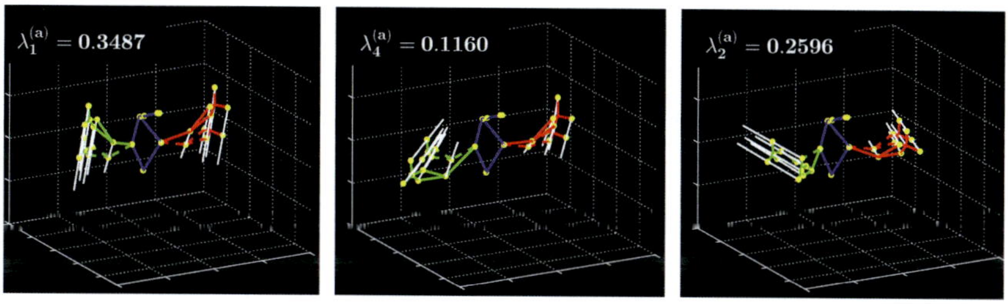

Fig. 13 Three examples from the first six modes for arm movement. The dominating pattern is an in-phase up and down movement (*left*); the corresponding anti-phase pattern is the forth mode (*middle*); the movement corresponding to the second largest eigenvalue is in-phase back and forth (*right*)

markers on the legs. To this end we build the covariance matrix $C^{(a)}$

$$C_{ij}^{(a)} = \frac{1}{T}\int_0^T \{a_i(t) - \bar{a}_i\}\{a_j(t) - \bar{a}_j\}\, dt \tag{7}$$

for the arms, where $\bar{a}_{i,j}$ represents the mean of the corresponding time series. In the same way $C^{(l)}$ for the legs is found; both are symmetric matrices that have real non-negative eigenvalues and orthogonal eigenvectors. The eigenvectors represent basic movement patterns, whereas the corresponding eigenvalues are a measure of how much a given pattern contributes to the variance in the original time series. Specifically, the eigenvectors describe the deviation from a given state, i.e., each triple of components represents the magnitude and direction of the movement for a certain marker. Therefore, each of the eigenvectors represents a movement pattern of all markers for either the arms or the legs. Interestingly, for both the arms and legs, the vectors corresponding to the six largest eigenvalues show coordination patterns that are either in-phase or anti-phase movements. Three examples from the first six modes for the arms and legs are shown in Figs. 13 and 14, respectively. The dominating arm pattern, accounting for almost 35% of the variance is an in-phase up and down movement (Fig. 13 left). The corresponding anti-phase pattern covering about 12% corresponds to the fourth largest eigenvalue (middle). On the right is the second mode, an in-phase back and forth movement covering about 26% of the variance. In total the six dominating eigenvectors cover more than 97% of the arm movements.

For the legs, shown in Fig. 14, the two dominant coordination patterns are an anti-phase (56%) and in-phase movement (18%) in the *yz*-plane (left and middle). The sixth mode is an anti-phase walking pattern in the *x*-direction and the fifth mode (not shown) is the corresponding in-phase movement. The total variance captured by the first six modes for the legs sums up to more than 98%.

It is most intriguing that the complex arm and leg movement by a professional ballet dancer can be captured almost completely by only six basic coordination pattern for both the arms and the legs. Moreover, these patterns fall into groups and can

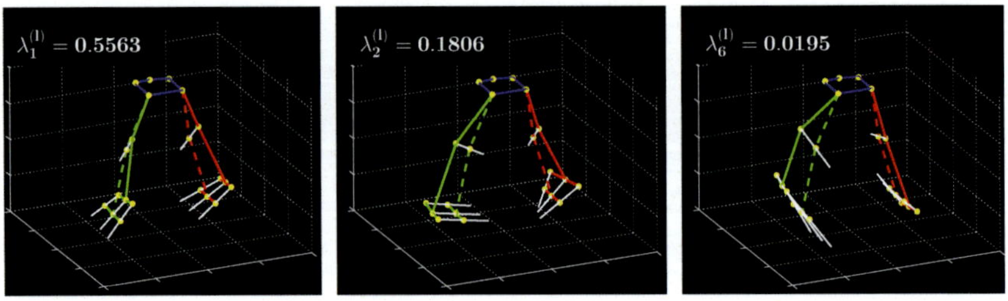

Fig. 14 Three examples from the first six modes for leg movement. Modes one and two (*left* and *middle*) are an anti-phase and an in-phase pattern in the *yz*-plane. The sixth mode (*right*) is an anti-phase walking movement in the *x*-direction; the corresponding in-phase pattern is the fifth mode (not shown)

be classified as anti-phase or in-phase movements along three different directions in 3-dimensional space.

5 Summary and Conclusion

We have tried to point to some of the highlights from more than 30 years of our research in coordination dynamics and its relation to synergetics, where the systems' components form entities and the interaction of such entities, modes or patterns gives rise to self-organization and macroscopic structures. On the macroscopic level, the coordination pattern is not described by the dynamics of the individual finger movements but by their relative phase. The transition in the behavior of this order parameter is also found in the brain signal of the dominating pattern. The way coordination patterns are established (or lost) in social settings leaves footprints in EEG recordings and may allow conclusions regarding the relationship between individuals. The results from the analysis of the ballet dancer in a certain sense take us back to the question we raised at the beginning of Sect. 2, which we can now rephrase as: In what ways do you have to be able to move your limbs in order to perform movements as complex as the choreography of a classical ballet dance? The answer, as we have seen here, is: In-phase and anti-phase in three different directions. Our results attest to a quite remarkable aspect of complex biological systems: They compress their high-dimensionality into lower-dimensional, fundamental patterns that are context-specific. These functional synergies or coordinative structures [21] constitute the building blocks of coordinated behavior in living things. There are good reasons why all healthy humans have these coordination patterns in their repertoire.

Acknowledgements One of us (AF) would like to convey special thanks to the organizers of the meeting and their staff for their great hospitality and for a wonderful time in Schloss Herrenhausen and Hannover. We are grateful to Emmanuelle Tognoli for kindly providing the data analyzed in Sect. 3. Some of the research described herein was supported by NIMH Grant MH080838 and the

FAU Foundation (Eminent Scholar in Science). We thank Makaila Wallace for her performances and Vyacheslav Murzin, Craig Nordham and Lee-Kuen Chua for their help with the collection and preprocessing of the data for the dancing project.

References

1. J.A.S. Kelso, D.A. Engstrøm, *The Complementary Nature* (MIT Press, Cambridge, MA, 2006)
2. H. Haken, J.A.S. Kelso, H. Bunz, A theoretical model of phase transitions in human hand movements. Biol. Cybern. **51**, 347–356 (1985)
3. J.A.S. Kelso, On the oscillatory basis of movement. Bull. Psychon. Soc. **18**, 63 (1981)
4. J.A.S. Kelso, Phase transitions and critical behavior in human bimanual coordination. Am. J. Physiol. **246**, R1000–R1004 (1984)
5. G. Schöner, H. Haken, J.A.S. Kelso, A stochastic theory of phase transitions in human hand movement. Biol. Cybern. **53**, 247–257 (1986)
6. J.A.S. Kelso, G. Schöner, J.P. Scholz, H. Haken, Phase-locked modes, phase transitions and component oscillators in coordinated biological motion. Phys. Scripta **35**, 79–87 (1987)
7. G. Schöner, J.A.S. Kelso, Dynamic pattern generation in behavioral and neural systems. Science **239**, 1513–1520 (1988)
8. B.A. Kay, J.A.S. Kelso, E.L. Saltzman, G. Schöner, Space-time behavior of single and bimanual rhythmical movements: data and limit cycle model. J. Exp. Psychol. Hum. **13**, 178–192 (1987)
9. G. Schöner, W.Y. Jiang, J.A.S. Kelso, A theory of quadrupedal gaits and gait transitions. J. Theor. Biol. **142**, 359–391 (1990)
10. J.J. Jeka, J.A.S. Kelso, Manipulating symmetry in the coordination dynamics of human movement. J. Exp. Psychol. Hum. **21**, 360–374 (1995)
11. J.A.S. Kelso, J.J. Jeka, Symmetry breaking dynamics of human multilimb coordination. J. Exp. Psychol. Hum. **18**, 645–668 (1992)
12. R.C. Schmidt, C. Carello, M.T. Turvey, Phase transitions and critical fluctuations in the visual coordination of rhythmic movements between people. J. Exp. Psychol. Hum. **16**, 227–247 (1990)
13. J.A.S. Kelso, J. DelColle, G. Schöner, Action-perception as a pattern formation process. In: *Attention and Performance XIII*, ed. by M. Jeannerod (Erlbaum, Hillsdale, NJ, 1990) pp. 139–169
14. A. Fuchs, J.A.S. Kelso, H. Haken, Phase transitions in the human brain: spatial mode dynamics. Int. J. Bifurcat Chaos **2**, 917–939 (1992)
15. J.A.S. Kelso, S.L. Bressler, S. Buchanan, G.C. DeGuzman, M. Ding, A. Fuchs, T. Holroyd, Cooperative and critical phenomena in the human brain revealed by multiple SQuIDs. In: *Measuring Chaos in the Human Brain*, ed. by D. Duke, W. Pritchard (World Scientific, Singapore, 1991), pp. 97–112
16. J.A.S. Kelso, S.L. Bressler, S. Buchanan, G.C. DeGunzman, M. Ding, A. Fuchs, T. Holroyd, A phase transition in brain and behavior. Phys. Lett. A **169**, 134–144 (1992)
17. A. Fuchs, J.M. Mayville, D. Cheyne, H. Weinberg, L. Deecke, J.A.S. Kelso, Spatiotemporal analysis of neuromagnetic events underlying the emergence of coordinative instabilities. Neuroimage **12**, 71–84 (2000)
18. J.A.S. Kelso, E. Tognoli, Towards a complementary neuroscience: Metastable coordination dynamics of the brain. In: *Neurodynamics of Cognition and Consciousness*, ed. by R. Kozma, L. Perlovsky (Springer, Heidelberg, 2007), pp. 39–60
19. E. Tognoli, C. Magne, G.C. DeGuzman, B. Tuller, J.A.S. Kelso, Brain rhythms underlying international social coordination. In: *SfN Itinerary Planner* (2007), p. 304.24

20. E. Tognoli, J. Lagarde, G.C. DeGuzman, J.A.S. Kelso, The phi complex as a neuromarker of human social coordination. Proc. Natl. Acad. Sci. USA **104**, 8190–8195 (2007)
21. J.A.S. Kelso, Coordination dynamics. In: *Encyclopedia of Complexity and System Science*, ed. by R.A. Meyers (Springer, Heidelberg, 2009), pp. 1537–1564

The Human Dynamic Clamp: A Probe for Coordination Across Neural, Behavioral, and Social Scales

Guillaume Dumas, Aline Lefebvre, Mengsen Zhang, Emmanuelle Tognoli and J.A. Scott Kelso

Abstract Humans (with their brains, bodies and behaviors) are complex dynamical systems embedded in an environment that includes a multitude of other conspecifics. Moving beyond previous brain- centered views of the human mind requires to develop a parsimonious yet integrative account that relates neural, behavioral, and social scales. Social neuroscience has recently started to acknowledge the importance of relational dynamics when it extended its purview from social stimuli to human-human interactions. Human-machine interactions also constitute promising tools to probe multiple scales in a controlled manner. Inspired by the electrophysiological method of the dynamic clamp, Virtual Partner Interaction (VPI) allows real time interaction between human subjects and their simulations as dynamical system. This provides a new test bed for operationalizing theoretical models in experimental settings. We discuss how VPI can be generalized into a Human Dynamic Clamp (HDC), a paradigm that allows the exploration of the parameter spaces of interactional dynamics in various contexts: from rhythmic and

G. Dumas (✉) · A. Lefebvre
Human Genetics and Cognitive Functions Unit, Institut Pasteur, Paris, France
e-mail: guillaume.dumas@pasteur.fr

G. Dumas · M. Zhang · E. Tognoli · J.A. Scott Kelso
Center for Complex Systems and Brain Sciences, Florida Atlantic University, Boca Raton, FL, USA

A. Lefebvre
Department of Child and Adolescent Psychiatry,
Assistance Publique-Hôpitaux de Paris, Robert Debré Hospital, Paris, France

J.A. Scott Kelso
Intelligent System Research Centre, Ulster University,
Derry Londonderry, Northern Ireland

S.C. Müller et al. (eds.), *Complexity and Synergetics*,
https://doi.org/10.1007/978-3-319-64334-2_24

discrete coordination to adaptive and intentional behaviors, including learning. HDC brings humans and machines together to question our understanding of the natural and our theory behind the artificial.

1 Introduction

Social neuroscience seeks to bridge the gap between the neural, the behavioral and the social. Such an agenda contrasts with cognitive science and the shortcoming of its brain-centered and individualistic approach to the mind. Recently, several approaches have proposed to go beyond a third person representational account of others by investigating social interaction from developmental, dynamical and relational viewpoints. This departure from a strictly reductionist view calls for new manners of empirical investigation of social systems along with a theoretical account of their various scales of organization. With those advances, one aims to integrate complementary aspects of the problem of social coordination into a coherent, comprehensive and parsimonious whole. In this respect, non-linear dynamical systems theory has already proved a good formalism to relate biological, psychological and more recently social levels [1]. This paper discusses a new experimental paradigm grounded in the framework of Coordination Dynamics [2–4]. We describe the development of Virtual Partner Interaction (VPI), a system allowing to couple a human with a theoretical model of movement coordination in real time [5, 6]. We review its generalization into the "Human Dynamic Clamp" (HDC), a new paradigm for Cognitive Science to study the multiple scales of coordination that govern human brain and behavior.

This novel paradigm pursues an already ongoing grip of Cognitive Science toward multiscale coordination [1, 3, 7]. In the exemplary case of hand movements for instance, social interactions span multiple scales in time: from position, phase and frequency of movements to the turn-taking between people (e.g. [8]). Such social interaction also gives rise to neural coordination within and across brains [9–11]. Multiple scales are also present in space, from the processing of information at synaptic levels to the level of large neural assemblies giving rise to different brain rhythms [12]. Moreover, neurophysiology shows how temporal and spatial dimensions are intertwined: neural oscillations at large time-scales (i.e. low frequencies) tend to cover larger scales in space, whereas shorter time-scales (i.e. high frequencies) appear to be more localized [13]. Thus, both brain and behavior are meshed together across multiple scales of time and space.

Since the present scientific approach aims to combine experimental studies with theoretical models, the key challenge is to connect these observations across scales and levels of organization within a coherent theoretical framework [14]. Coordination dynamics aims at such understanding through the synergetic concepts of self-organization [15] and the mathematical tools of dynamical systems theory [3, 16, 17]. It seeks both general principles and functionally-specific mechanisms of coordination [2] and aims at connecting multiple scales by emphasizing reciprocal coupling between levels, upward and downward [1]. In this perspective, coordination between humans represents an operational playground for experimental investigation at the crossroad of the neural, the behavioral and the social.

Recently, hyperscanning techniques have offered access to the simultaneous recording of brain activity from interacting people and thus to the study of brain and behavior coordination at both intra- and inter-individual scales [11, 18–22]. In doing so, this technique has also reintroduced real social interaction into laboratory studies of human behavior, a key feature that was oddly lacking from earlier work within a (cognitively-inspired) social neuroscience, as it resorted to exposing one subject to social "stimuli" rather than examining interactions [23–26]. Further, the use of reciprocal paradigms and a real second-person approach of social cognition do not necessarily require the presence of two or more subjects in the experimental task [26]. Instead, one of the interacting partners can be substituted with a virtual agent whose design sustains bi-directional coupling between real and simulated partners [6, 27, 28].

2 Human-Machine Interaction as a Research Tool

Meanwhile, in other areas of science and engineering, a plethora of studies was concentrating on subjective perception of artificial agents by humans, with the goal of designing realistic avatars for potential applications to, e.g. video games, cinema, or eLearning assistants [29] to name just a few. In this line of research, the exercise was to mimic facets of human behavior rather than to model foundational neurobehavioral mechanisms. Interestingly, participants' beliefs of realism were influenced by emotionally and behaviorally contingent responses made by the artificial agent [30]; see also [31]; this finding hints at the importance of reciprocal coupling with the human.

The development of realistic artificial agents extended the toolset available to social psychological research [32], with more to come as those agents are embedded in virtual realities that are increasingly indistinguishable from "normal" reality. The breakthrough of virtualization has reconciled ecological validity and experimental control, e.g. in the study of visual perception, spatial cognition and social interaction [5, 33].

A first level of social interaction is the mere presence of someone else [34]. Regarding this issue, virtual reality fits particularly well since it creates a sense of presence through mediated environments carrying dynamic animations of virtual characters [35]. Virtual characters are readily perceived as social agents and are thus capable of exerting social influence on humans [36]. Those virtual characters with strong similarity to real human interactions [37] can easily and valuably be combined with neuroimaging recording [32].

Human-machine interaction was also used to investigate motor coordination: for instance a finger tapping study by Repp and Keller [38] used a simple linear phase correction model to drive a virtual agent. It showed that subjects' behavior was systematically modulated by the computational parameters governing that agent. Reframed in a functional neuroimaging study by Fairhurst et al. [39], the same paradigm uncovered some neural basis for motor synchronization and more importantly, for the socio-emotional consequences of different degrees of entrainment success.

In the following, we describe another paradigm, the Human Dynamic Clamp (HDC), that embraces a continuous, multiscale and nonlinear coupling between a human and a machine. By departing from information processing approaches and design-oriented modeling, the HDC offers: (a) a new way to bridge the gap between theory, experiment and models; and (b) an integrative solution to linking neural, behavioral, and social dynamics. HDC puts well established equations of human coordination dynamics into the machine and studies real-time interactions between human and virtual partners. This opens up the possibility to explore and understand a wide variety of interactions [5, 6, 40]. Ultimately, HDC may prove useful to establishing a much friendlier union of man and machine, based on sound interactional design, and perhaps it will even lead to the creation of a different kind of machine altogether.

3 A Principle-Based Virtual Partner

The study of movement coordination is at the core of coordination dynamics and for the last 30-odd years the catchy phrase "let your fingers do the walking" has opened a rich experimental window into human behavior at both intra-individual and inter-individual levels. In a first move, it is important to clarify what we are looking at [1]. What is the behavior? What are the relevant variables and control parameters? These fundamental issues are addressed by uncovering qualitative changes in collective variables from the system called order parameters [15, 41]. Qualitative changes appear in two main flavors within the formalism of dynamical system theory: phase transitions and bifurcations. Although they are both revealed in the phenomenon of transition in collective dynamics, the first is related to the switch between potential modes of behavior simultaneously accessible to the system, and the second concerns global changes of the system's behavioral landscape. The landscape is usually described with a manifold in phase space (the frame of reference representing the relationship between variables associated with each degree of freedom). The challenge then is to uncover the most parsimonious model that can exhibit these qualitative changes, and fit its parameters according to the experimental data (see the discussion of *Phenomenological Synergetics* in [42]. One key issue to keep in mind lies with the biological constraints that make it possible to link a model to actual physiological mechanisms. In this perspective, it is fundamental to recognize that all models are false by definition. However, dynamical system theory offers good candidates for a universal class of models, given the needed parsimony for elegant theory [43, 44].

Born from this aim was our recently developed Human Dynamic Clamp, a paradigm that took inspiration from the electrophysiological dynamic clamp

[45, 46] to allow real-time interaction between a human subject and a computational model. Using empirically-grounded models not only validated reciprocal and fully dynamical design protocols for experimenters to use, but also provided the opportunity to explore parameter ranges and perturbations that were out of reach of traditional experimental designs with live interactions. The symmetry between the human and the machine and the fact that they carry the same laws of coordination dynamics were keys to our approach [6]. The design of the virtual partner (VP) was grounded in the equations of motion for the coordination of the human neurobehavioral system. These laws were obtained from accumulated studies over the last 30-odd years to describe how parts of the human body and brain self-organize, and to address the issue of self-reference, a condition leading to complexity.

The first version of the Human Dynamic Clamp called Virtual Partner Interaction [6] embodied the Haken–Kelso–Bunz (HKB) model [47]. The original form of HKB describes and predicts the coordination dynamics of two rhythmically moving fingers, with its characteristically complex phenomena such as multistability, phase transitions, hysteresis, critical slowing-down and fluctuation enhancement ([42, 17] for reviews). Since then, the model has also been successfully validated experimentally for the coordination between different limbs (e.g. [4]), between people (e.g. [48]) and even between species [49], within unimodal and multimodal contexts [50]. It has been supported by empirical evidence ranging from brain dynamics within [51, 52] and between brain areas [46, 53–55], to coordination with external stimuli [56] and neural counterparts thereof [57, 58]. This universal characteristic supports HKB as an ideal candidate for the Human Dynamic Clamp.

In its original implementation, the VPI system was composed of a goniometer continuously digitizing the finger position of a human participant; a computational circuit simulating the HKB model; and a screen rendering the virtual partner's behavior (see Fig. 1a–b). The computational circuit calculates the position of VP continuously according to the differential equations of HKB (Fig. 1b), and the resulting dynamics is mapped onto a virtual avatar displayed on the screen.

The HKB model at the collective level describes the equation of motion of the relative phase, a variable that distills the coordination of two oscillatory components. In this form, the HKB model reads:

$$\phi = \mathrm{a} \sin \phi + b \ \sin 2\phi, \qquad (1)$$

where ϕ is the relative phase between human and VP's finger position, and a and b are constants (for more details, see [59]).

However, since computers do not have direct access to the relative phase, the internal dynamics of VP is governed by the HKB model at the component level (see Fig. 1b). In this form, two non-linearly coupled non-linear oscillators represent the interaction between the two fingers. The collective form in Eq. (1) can be derived from the equations at the component level (Fig. 1b). At the component level, variables are no longer the relative phase but the individual finger positions (and velocities by derivation). x and y represent VP's and human's finger positions, α, β and γ are constants associated with the intrinsic dynamics of VP, ω is VP's

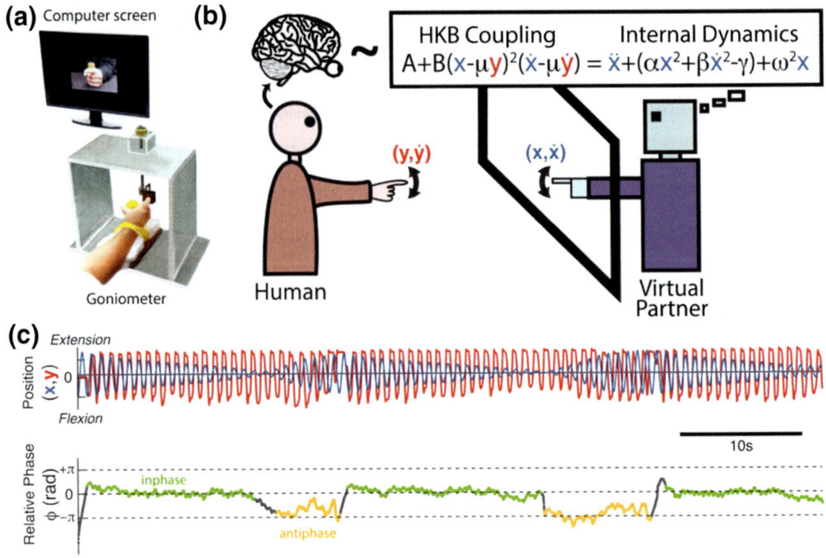

Fig. 1 The VPI system. **a** presents it key components (goniometer to transduce human movement behavior and screen to display Virtual partner's behavior) from the human viewpoint. Task and coupling are outlined in (**b**). Human's behavior is digitized and fed into a computer whose software computes the corresponding position of the VP in real time, following a theoretical model of behavioral coordination—here HKB. The picture of the VP is updated on the screen (**a**) according to the output of the model. Data are stored for further study (**c**) to test hypotheses about the relationship between the agent's intrinsic properties, coupling parameters and emergent collective behavior

pulsation (frequency), A and B are constants associated with the coupling from VP to human, and finally μ is a constant fixed to either +1 or −1, indicating VP's preference for in-phase or anti-phase coordination.

In the original study [6], VP and human behaviors were chosen to be quite simple. Both partners were tasked to coordinate finger movements with one another, the human with the intention of achieving in-phase coordination with the VP (trying to synchronize his/her flexion and extension movements with VP's). On the VP side, the parameter μ was set to −1, inducing a VP preference for anti-phase coordination and thus a goal opposite to human's. Subjects were instructed to maintain a smooth and continuous rhythmic movement with their right index finger (flexion-extension) and to avoid stopping their movement at any time. Visual coupling was experimentally manipulated: from unidirectional in two conditions (VP "perceives" human movement but human does not perceive VP's behavior; or reciprocally), to bi-directional in another (both VP and human have access to each other's finger movement). VPI accommodated the whole set of behavioral coordination modes described by the HKB model. For instance, when VP and human participants did not have the same preferred movement frequency, their relative phase conformed to predictions by the extended version of HKB [56] and exhibited phase wrapping (not shown) or metastability (see Fig. 1c). Pitting machine against human through opposing task demands is a way to enhance the formation of emergent behavior, and also allowed us to examine each partner's individual contribution to the collective behavior. An intriguing outcome of the experiments was that subjects ascribed intentions to the machine, reporting that it was "messing" with them. A later study further suggested that VP elicits emotional experiences in human: subjects' emotional arousal was greatest when VP interactions were (falsely) deemed to be with a human rather than with a machine [31].

In summary, Kelso et al. [6] initial VPI experiment demonstrated the feasibility of the Human Dynamic Clamp in the context of a continuous coordination of rhythmic movements. It uncovered unexpected behaviors, which were theoretically tested afterward. In the following, we show how to explore a new set of behaviors with other theoretical models of human behavior.

4 Expanding the Behavioral Repertoire

Embedding the HKB model in a Virtual Partner demonstrated that the explicit use of non-linear relational dynamics in an experimental paradigm can lead to new observations of emergent phenomena that linear models may miss out on. The Human Dynamic Clamp paradigm is about developing this idea by integrating other principle-based models grounded on canonical behaviors observed in experimental work. More complex behaviors can then be approached through the combination of canonical models in a modular and hierarchical manner [5, 60], see also Fig. 2.

4.1 Discrete Behavior: Phase-Space Sculpture

Although it is undeniable that living organisms rely both on rhythmic and discrete behaviors, the field of motor control has traditionally studied them separately. This led to two different ways of theoretically approaching and modeling them. While rhythmic movements have been extensively studied through the prism of dynamical systems, discrete movements' modeling has focused on equilibrium points or control signals [61]. Unifying rhythmic and discrete movements is often posed to be a key theoretical challenge in behavioral science [62, 63]. However, there is no specific need to invoke two separate mechanisms for discrete and rhythmic behavior [64–66]. For instance, Schöner [17] extended the HKB model to the case of discrete bistable coordination by changing the intrinsic dynamics. Sternad et al. [67] proposed another model for unimanual coordination with two mutually inhibiting subsystems, each of which handled the discrete and the continuous cases respectively.

Along similar lines, Jirsa and Kelso [68] modeled discrete and rhythmic movement based on the phase flow topology of the so-called "Excitator" model (see also [63]). The Excitator defines a universal class of two-dimensional dynamical systems able to exhibit limit cycles for rhythmic movement, and fixed

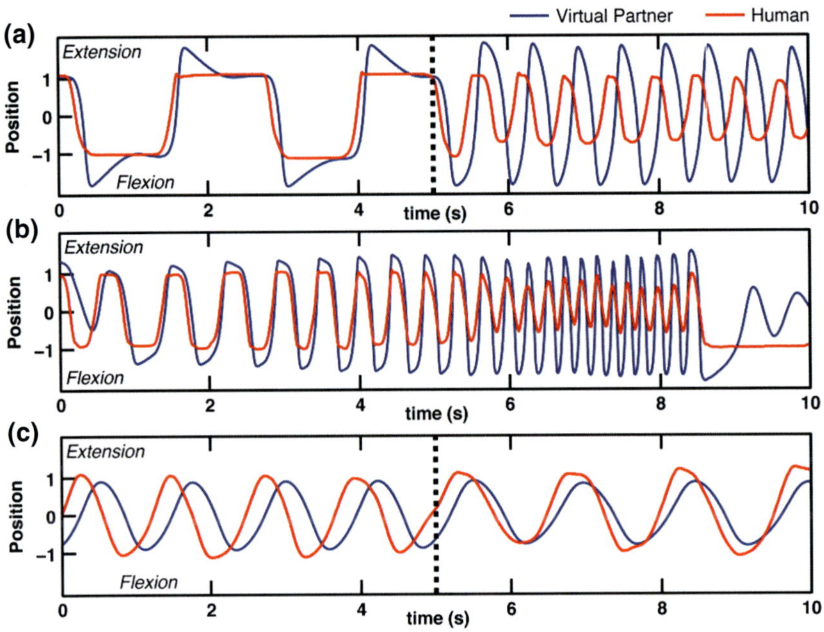

Fig. 2 Examples of interactions between a human participant (*red*) and a VP embedding alternative models of relational dynamics (*blue*). **a** The Excitator model (with parameters $a = 0; b = 0; A = 1; B = -0.2; \tau = 1; \omega = 1$; *dashed line* indicates switch from discrete to rhythmic movement in the human participant); **b** the adaptive Excitator model ($a = 0; b = 0; A = 1; B = -0.2; \tau = 1; \omega = 1; K = 1$); **c** a modified HKB with an intended relative phase of pi/2 ($a = 0.641; b = 0.00709; A = 0.12; B = 0.025; C = 1; \omega = 1$; *dashed line* indicates release of the VPI intentional forcing, i.e. switch to normal HKB model)

point dynamics for discrete movement. This model is based on topological considerations and is a parsimonious way to handle discrete and continuous behaviors simultaneously. Furthermore, in line with the approach of HKB modeling, the Excitator provides predictions regarding false-start phenomena that have been confirmed experimentally [69]. Finally, it is a biologically realistic model since it follows the self-excitable property that the FitzHugh-Nagumo model drew from single neurons [70].

The structure of the model contains three characteristics related to topological constraints: boundedness of the trajectory, existence of a separatrix marking the boundary between two separate regimes in phase space, and existence of a limit cycle for rhythmic movements and of one or two stable fixed point(s) for monostable and bi-stable discrete movements respectively.

The equations read as follows: $$\begin{cases} \dot{x_1} = \omega(x_1 + x_2 - g_1(x_1))\tau \\ \dot{x_2} = -\omega(x_1 - a + g_2(x_1, x_2) - I)/\tau, \end{cases} \quad (2)$$

where x_1 and x_2 are internal variables of the oscillator, ω is the pulsation (frequency) of VP, a the term controlling the position of the separatrix, b the term controlling the angle of the separatrix, I an instantaneous external input, and τ the time constant of the system.

Note that the choice of functions g_1 and g_2 is not fixed but must nevertheless guarantee the boundedness of the system so that the system belongs to the class of self-excitable systems. Here we take

$$g_1(x_1) = \frac{1}{3}x_1^3 \text{ and } g_2(x_1, x_2) = -bx_2 \tag{3}$$

When this is put in unidimensional form, we retain the same coupling terms as HKB model's. The coupling causes either convergence or divergence of the trajectories in phase space depending on initial conditions. Since trajectories are bounded, constraints lead to in-phase or anti-phase modes of coordination (for more details, see [68]).

Implementation of the Excitator dynamics in an HDC is quite straightforward: one only needs to substitute the relevant equations from (2) in the software controlling VP's behavior. Note that these equations introduce a new term of importance: parameter I allows to modify the phase flow according to an external input. An external input can originate from the experimenter or the human partner him/herself. It is a key component for modeling discrete behaviors, which rely on external information and is non-autonomous in a mathematical sense. The introduction of the new variable allows VP and human to coordinate diverse movements that range from simple rhythms to discrete actions. Figure 2a presents an interaction between a human and a VP governed by the Excitator model, and shows a transition from discrete movement (flexions and extensions interrupted by quiescent behavior) to continuous movement.

4.2 *Adaptive Behavior: Parameter Dynamics and Modularity*

The Excitator model shows how a single dynamical system may give rise to different behavioral modes of coordination between human and virtual partner. However, each mode required a different set of parameters. Once those parameters are fixed, the differential equations set the functional structure of the system for a specific behavioral context. But structure, function and dynamics are not separated in nature; everything is constantly evolving on different time scales [7, 3]. In biology, organisms change their own behavior and learn new ones to better face the world, and interact with their peers in a more effective manner. Robert Rosen even associated adaptation as the most characteristic property of living things [71]. The process of adaptation is ubiquitous in so-called complex adaptive systems that may also encompass physical or artificial aspects [72]. In the case of the brain, it is not surprising to observe such ongoing anticipation continuously [73]. Adaptation is especially important in social behavior, for instance mimicry at the morphological level [74] or interactional synchrony during cooperative imitation and skill learning [75].

Coordination may be seen as a subtle blend of reaction and adaptation to the other [76]. Whereas reaction takes place at a given time, adaptation builds up over time. For instance, humans may have a preferred movement frequency but they can adapt to different partners by slowing down or speeding up their movements. In the case of the Human Dynamic Clamp, frequency adjustment is a good candidate to address adaptive behavior in a manner that is fully compatible with the previously described systems, and uses the same formalism. Basically, frequency adaptation requires a new equation in the system of differential equations that manages the rate of change of frequency ω. At a more conceptual level, it fits with the idea that adaptation depends on the system's ongoing intrinsic dynamics. Furthermore, adaptation can enhance the realism of the interaction, by expanding beyond an instantaneous coordination with the position of a finger or the phase of a movement.

Different strategies for modeling frequency adaptation have been proposed. In a pure Artificial Intelligence (AI) tradition, a specific module detects the frequency of the human partner which then controls VP's actual frequency. This shows that it is possible to successfully design an artificial device that is able to do the job. In the Bayesian approach, adaptation is error-based and relies on reinforcement learning [77]. This approach is inspired from real physiological processes. In predictive coding, adaptation of model parameters is associated with Hebbian and synaptic plasticity in the brain [78]. Other bottom-up strategies have been developed in the fields of signal processing [79] and robotics [80]. Here we continue to follow the strategy of Coordination Dynamics and Dynamical System Theory. That approach was shown to better account for frequency adaptation in fireflies [81] and in tempo adaptation to musical rhythms [82]; see also [83]. In contrast with the AI approach, it is worth noting that the equations stay totally continuous and do not relate to an artificial measurement of the human frequency. This illustrates how adaptation relies on parameter dynamics according to the scale of observation [84].

Following Righetti and colleagues [85, 86], we introduce frequency adaptation through the addition of a new dimension—related to ω—in the set of differential equations:

$$\begin{cases} \dot{x} = f_x(x, v, \omega) + KF(t) \\ \dot{v} = f_v(x, v, \omega) \end{cases} \text{and } \dot{\omega} = \pm KF(t)\frac{v}{\sqrt{x^2 + v^2}}. \tag{4}$$

where K is the coupling strength of the adaptation, x and v are variables capable of producing a limit cycle, and $F(t)$ is the coupling part of the system. Figure 2b shows how a VP governed by the extended Excitator equations is able to follow changes in movement frequency. Addition of a third dimension also leads to unstable dynamics, less predictable from the human point of view. This may be associated with the emergence of chaotic regimes that are typical of 3-dimensional nonlinear dynamical systems [87]. Such unpredictability can be associated with a form of intention [88]: a model of intentional behavior could be further designed. That is what we will see in the next section.

4.3 Intentional Behavior: Symmetry Breaking and Forcing

In the case of an adaptive system, we have seen that adding a third dimension renders the dynamics less predictable. The system is nevertheless not random and appears more autonomous while still being governed by deterministic rules. This balance between autonomy and coupling creates successful agency illusion and can trigger an attribution of intention to the human observer [89, 90]. Keeping in mind that the Human Dynamic Clamp aims at operationalizing models for experimental purposes, a teleonomic system is not adequate, because its intention is not directly controllable by the experimenter.

In the initial VPI experiment [6], the control parameter μ (Eq. (1)) modulated intention attribution in some participants. In general, adopting a principle-based modeling requires redefining the boundary conditions of the model. Until now, we were dealing with spontaneous coordination. It has been shown experimentally, however, that intention affects the spontaneous potential landscape by stabilizing and destabilizing specific dynamic patterns [91] including at the brain level [92]. The former empirical findings motivated an extension of the HKB model [17]; see also [5] through the introduction of new term in the relative phase equation:

$$\phi = a \sin\phi + b \sin 2\phi + c \sin\psi - \phi, \tag{5}$$

where ψ is the intended relative phase. By incorporating an intentional forcing term c which stabilizes or destabilizes particular patterns, the model was able to explain experimental observations related to intentional switching between in-phase and anti-phase.

We recently generalized the Schöner and Kelso coupling model [5], so the intended relative phase angle ψ can take on any value between $-\pi$ and $+\pi$:

$$C_{\text{int}} = -C(cos(\psi)(\dot{x}-\dot{y}) + sin(\psi)\omega y). \tag{6}$$

This modification of VP dynamics makes it possible to direct a collective behavior towards any desired pattern of coordination (see Fig. 2c). This offers new experimental perspectives, e.g. to study how new dynamical patterns are learned on top of a subject's spontaneous behavioral repertoire [40].

5 Conclusion

In this chapter, we have seen how a hybrid system called the Human Dynamic Clamp allows for real-time interaction between humans and virtual partners, based on the equations of coordination dynamics built originally from HKB and its extensions. A key aspect is that the human and its virtual partner are reciprocally coupled: the human acquires information about the partner's behavior through perception, and the virtual partner continuously detects the human's behavior through the input of sensors. Our approach is analogous to that of the original dynamic clamp used to study the dynamics of interactions between neurons, but now scaled up to the level of behaving humans. This principle-based approach offers a new paradigm for the study of social interaction. While stable and intermittent coordination behaviors emerged that had previously been observed in ordinary human social interactions, we also discovered novel behaviors or strategies that had never been observed in human social behavior. Those novel behaviors pertained to unexplored regions of the theoretical model and were possible ways of coordination for people to interact with each other. Such emergence of novel behaviors demonstrates the scientific potential of HDC as a human-machine framework. Modifying the dynamics of the virtual partner with the purpose of inducing a desired human behavior, such as learning a new skill or as a tool for therapy and rehabilitation, is one of several applications of VPI.

HDC allows to study social interaction with more experimental control than other recent social neuroscience methods (e.g. hyperscanning); it is also a test bed for theoretical models. HDC moves away from simple protocols in which systems are 'poked' by virtue of 'stimuli' to address more complex, reciprocally connected systems where meaningful interactions occur. Thus, the Human Dynamic Clamp supports the development of a computational social neuroscience where theory, experiment and modeling work hand-in-hand across neural, behavioral and social scales [93].

Acknowledgements The research described herein was supported by grants from the National Institute of Mental Health (MH080838), the Davimos Family Endowment for Excellence in Science and The FAU Foundation (Eminent Scholar in Science).

References

1. J.A.S. Kelso, G. Dumas, E. Tognoli, Outline of a general theory of behavior and brain coordination. Neural Netw. Off. J. Int. Neural Netw. Soc. **37**, 120–131 (2013). https://doi.org/10.1016/j.neunet.2012.09.003
2. J.A.S. Kelso, Coordination dynamics, in *Encyclopedia of Complexity and Systems Science*, ed. by R.M. Meyers (Springer, New York, 2009), pp. 1537–1565
3. J.A.S. Kelso, *Dynamic Patterns: The Self-Organization of Brain and Behavior* (The MIT Press, Cambridge, 1995)

4. J.A.S. Kelso, J.J. Jeka, Symmetry breaking dynamics of human multilimb coordination. J. Exp. Psychol. Hum. Percept. Perform. **18**, 645 (1992)
5. G. Dumas, G.C. de Guzman, E. Tognoli, J.A.S. Kelso, The human dynamic clamp as a paradigm for social interaction. Proc. Natl. Acad. Sci. U.S.A. **111**, E3726–E3734 (2014). https://doi.org/10.1073/pnas.1407486111
6. J.A.S. Kelso, G.C. de Guzman, C. Reveley, E. Tognoli, Virtual partner interaction (VPI): exploring novel behaviors via coordination dynamics. PLoS ONE **4**, e5749 (2009). https://doi.org/10.1371/journal.pone.0005749.t002
7. S.L. Bressler, E. Tognoli, Operational principles of neurocognitive networks. Int. J. Psychophysiol. **60**, 139–148 (2006). https://doi.org/10.1016/j.ijpsycho.2005.12.008
8. O. Oullier, G.C. de Guzman, K.J. Jantzen, J.A.S. Kelso, On context dependence of behavioral variability in inter-personal coordination. Int. J. Comput. Sci. Sport **2**, 126–128 (2003)
9. G. Dumas, J. Nadel, R. Soussignan et al., Inter-brain synchronization during social interaction. PLoS ONE **5**, e12166 (2010). https://doi.org/10.1371/journal.pone.0012166
10. V. Müller, J. Sänger, U. Lindenberger, Intra- and inter-brain synchronization during musical improvisation on the guitar. PLoS ONE **8**, e73852 (2013). https://doi.org/10.1371/journal.pone.0073852.s012
11. E. Tognoli, J. Lagarde, G.C. de Guzman, J.A.S. Kelso, The phi complex as a neuromarker of human social coordination. Proc. Natl. Acad. Sci. U.S.A. **104**, 8190–8195 (2007)
12. G. Buzsáki, A. Draguhn, Neuronal oscillations in cortical networks. Sci. NY **304**, 1926–1929 (2004). https://doi.org/10.1126/science.1099745
13. E. Tognoli, J.A.S. Kelso, The metastable brain. Neuron **81**, 35–48 (2014). https://doi.org/10.1016/j.neuron.2013.12.022
14. H.R. Maturana, F.J. Varela, *The Tree of Knowledge: The Biological Roots of Human Understanding* (New Science Library/Shambhala Publications, 1987)
15. H. Haken, *Advanced Synergetics* (Springer, Berlin, Heidelberg, 1983)
16. A. Fuchs, *Nonlinear Dynamics in Complex Systems* (Springer, Berlin, Heidelberg, 2013)
17. G. Schöner, A dynamic theory of coordination of discrete movement. Biol. Cybern. **63**(4), 257–270 (1990)
18. F. Babiloni, L. Astolfi, Social neuroscience and hyperscanning techniques: past, present and future. Neurosci. Biobehav. Rev. **44**, 1–18 (2012). https://doi.org/10.1016/j.neubiorev.2012.07.006
19. G. Dumas, F. Lachat, J. Martinerie et al., From social behaviour to brain synchronization: review and perspectives in hyperscanning. Irbm **32**, 48–53 (2011)
20. U. Hasson, A.A. Ghazanfar, B. Galantucci et al., Brain-to-brain coupling: a mechanism for creating and sharing a social world. Trends Cogn. Sci. **16**, 114–121 (2012). https://doi.org/10.1016/j.tics.2011.12.007
21. I. Konvalinka, A. Roepstorff, The two-brain approach: how can mutually interacting brains teach us something about social interaction? Front. Hum. Neurosci. **6**, 215 (2012)
22. P.R. Montague, G.S. Berns, J.D. Cohen et al., Hyperscanning: simultaneous fMRI during linked social interactions. NeuroImage **16**, 1159–1164 (2002). https://doi.org/10.1006/nimg.2002.1150
23. G. Dumas, Towards a two-body neuroscience. Commun. Integr. Biol. **4**, 349–352 (2011). https://doi.org/10.4161/cib.4.3.15110
24. R. Hari, L. Henriksson, S. Malinen, L. Parkkonen, Centrality of social interaction in human brain function. Neuron **88**, 181–193 (2015). https://doi.org/10.1016/j.neuron.2015.09.022
25. R. Hari, M.V. Kujala, Brain basis of human social interaction: from concepts to brain imaging. Physiol. Rev. **89**, 453–479 (2009). https://doi.org/10.1152/physrev.00041.2007
26. L. Schilbach, B. Timmermans, V. Reddy et al., Toward a second-person neuroscience. Behav. Brain Sci. **36**, 393–414 (2013). https://doi.org/10.1017/S0140525X12000660
27. J. Mattout. Brain-computer interfaces: a neuroscience paradigm of social interaction? A matter of perspective. Front. Hum. Neurosci. (2012). https://doi.org/10.3389/fnhum.2012.00114

28. U.J. Pfeiffer, B. Timmermans, G. Bente et al., A non-verbal Turing test: differentiating mind from machine in gaze-based social interaction. PLoS ONE **6**, e27591 (2011). https://doi.org/10.1371/journal.pone.0027591
29. J. Scholtz, Theory and evaluation of human robot interactions, in *Proceedings of the 36th Annual Hawaii International Conference on System Sciences*, 2003 (IEEE, 2003), p 10
30. M. Ochs, C. Pelachaud, D. Sadek, An empathic virtual dialog agent to improve human-machine interaction, in *Proceedings of the 7th International Joint Conference on Autonomous Agents and Multiagent Systems* (Estoril, Portugal, 2008)
31. M. Zhang, G. Dumas, J.A.S. Kelso, E. Tognoli, Enhanced emotional responses during social coordination with a virtual partner. Int. J. Psychophysiol. **104**, 33–43 (2016). https://doi.org/10.1016/j.ijpsycho.2016.04.001
32. L. Schilbach, A.M. Wohlschlaeger, N.C. Kraemer et al., Being with virtual others: neural correlates of social interaction. Neuropsychologia **44**, 718–730 (2006). https://doi.org/10.1016/j.neuropsychologia.2005.07.017
33. J.M. Loomis, J.J. Blascovich, A.C. Beall, Immersive virtual environment technology as a basic research tool in psychology. Behav. Res. Methods Instrum. Comput. **31**, 557–564 (1999)
34. C. McCall, J. Blascovich, How, when, and why to use digital experimental virtual environments to study social behavior. Soc. Personal Psychol. Compass **3**, 744–758 (2009)
35. M.V. Sanchez-Vives, M. Slater, From presence to consciousness through virtual reality. Nat. Rev. Neurosci. **6**, 332–339 (2005). https://doi.org/10.1038/nrn1651
36. F. Biocca, C. Harms, J.K. Burgoon, Toward a more robust theory and measure of social presence: review and suggested criteria. Presence **12**, 456–480 (2003)
37. M. Garau, M. Slater, D.-P. Pertaub, S. Razzaque, The responses of people to virtual humans in an immersive virtual environment. Presence Teleoper. Virtual Environ. **14**, 104–116 (2005)
38. B.H. Repp, P.E. Keller, Sensorimotor synchronization with adaptively timed sequences. Hum. Mov. Sci. **27**, 423–456 (2008)
39. M.T. Fairhurst, P. Janata, P.E. Keller, Being and feeling in sync with an adaptive virtual partner: Brain mechanisms underlying dynamic cooperativity. Cereb. Cortex **23**, 2592–2600 (2013). https://doi.org/10.1093/cercor/bhs243
40. V. Kostrubiec, G. Dumas, P.-G. Zanone, J.A.S. Kelso, The virtual teacher (VT) paradigm: Learning new patterns of interpersonal coordination using the human dynamic clamp. PLoS ONE **10**, e0142029 (2015)
41. J.A.S. Kelso, Phase transitions: foundations of behavior, in *Synergetics of Cognition*, ed. by H. Haken, M. Stadler (Springer, Berlin, 1990), pp. 249–268
42. J.A.S. Kelso, G. Schöner, J.P. Scholz, H. Haken, Phase-locked modes, phase transitions and component oscillators in biological motion. Phys. Scr. **35**, 1–9 (1987)
43. M. Golubitsky, I. Stewart, *The Symmetry Perspective: From Equilibrium to Chaos in Phase Space and Physical Space* (Springer Science and Business Media, 2003)
44. K.G. Wilson, Problems in physics with many scales of length. Sci. Am. **241**, 158–179 (1979)
45. A.A. Prinz, L.F. Abbott, E. Marder, The dynamic clamp comes of age. Trends Neurosci. **27**, 218–224 (2004). https://doi.org/10.1016/j.tins.2004.02.004
46. S.H. Strogatz, *Nonlinear Dynamics and Chaos* (Westview Press, Boulder, 2008)
47. H. Haken, J.A.S. Kelso, H. Bunz, A theoretical model of phase transitions in human hand movements. Biol. Cybern. **51**, 347–356 (1985)
48. O. Oullier, G.C. de Guzman, K.J. Jantzen et al., Social coordination dynamics: measuring human bonding. Soc. Neurosci. **3**, 178–192 (2008)
49. J. Lagarde, C. Peham, T. Licka, J.A.S. Kelso, Coordination dynamics of the horse-rider system. J. Mot. Behav. **37**, 418–424 (2005)
50. J. Lagarde, J.A.S. Kelso, Binding of movement, sound and touch: multimodal coordination dynamics. Exp. Brain Res. **173**, 673–688 (2006)
51. J.A.S. Kelso, S.L. Bressler, S. Buchanan et al., A phase transition in human brain and behavior. Phys. Lett. A **169**, 134–144 (1992). https://doi.org/10.1016/0375-9601(92)90583-8

52. J.A.S. Kelso, A. Fuchs, R. Lancaster et al., Dynamic cortical activity in the human brain reveals motor equivalence. Nature **392**, 814–817 (1998)
53. S.L. Bressler, J.A.S. Kelso, Cortical coordination dynamics and cognition. Trends Cogn. Sci. **5**, 26–36 (2001)
54. K.J. Jantzen, F.L. Steinberg, J.A.S. Kelso, Coordination dynamics of large-scale neural circuitry underlying rhythmic sensorimotor behavior. J. Cogn. Neurosci. **21**, 2420–2433 (2009)
55. E. Tognoli, J.A.S. Kelso, Brain coordination dynamics: true and false faces of phase synchrony and metastability. Prog. Neurobiol. **87**, 31–40 (2009). https://doi.org/10.1016/j.pneurobio.2008.09.014
56. J.A.S. Kelso, J. DelColle, G. Schöner, Action perception as a pattern formation process, in *Attention and Performance XIII*, ed. by M. Jeannerod (Erlbaum, Hills-dale, NJ, 1990), pp. 139–169
57. K.J. Jantzen, F.L. Steinberg, J.A.S. Kelso, Brain networks underlying human timing behavior are influenced by prior context. Proc. Natl. Acad. Sci. U.S.A. **101**, 6815–6820 (2004). https://doi.org/10.1073/pnas.0401300101
58. O. Oullier, K.J. Jantzen, F.L. Steinberg, J.A.S. Kelso, Neural substrates of real and imagined sensorimotor coordination. Cereb. Cortex **15**, 975–985 (2004). https://doi.org/10.1093/cercor/bhh198
59. A. Fuchs, J.A.S. Kelso, Movement coordination, in *Encyclopedia of Complexity and Systems Science* (Springer, 2009), pp. 5718–5736
60. R. Huys, D. Perdikis, V.K. Jirsa, Functional architectures and structured flows on manifolds: a dynamical framework for perceptual-motor behavior. Psychol. Rev. **121**, 1–113 (2013). https://doi.org/10.1037/a0037014
61. F. Danion, Do we need internal models for movement control? Nonlinear Dyn. Hum. Behav. **328**, 115–134 (2011). https://doi.org/10.1007/978-3-642-16262-6_5
62. S. Degallier, A. Ijspeert, Modeling discrete and rhythmic movements through motor primitives: a review. Biol. Cybern. **103**, 319–338 (2010)
63. R. Huys, B.E. Studenka, N.L. Rheaume et al., Distinct timing mechanisms produce discrete and continuous movements. PLoS Comput. Biol. **4**, e1000061 (2008). https://doi.org/10.1371/journal.pcbi.1000061.g004
64. R. Huys, V.K. Jirsa, *Nonlinear Dynamics in Human Behavior* (Springer, Heidelberg, 2010)
65. J.A.S. Kelso, K.G. Holt, P.N. Kugler, M.T. Turvey, On the concept of coordinative structures as dissipative structures: II. Empirical lines of convergence. Adv Psychol. **1**, 49–70 (1980)
66. P.N. Kugler, J.A.S. Kelso, M.T. Turvey, On the concept of coordinative structures as dissipative structures: I. Theoretical lines of convergence. Tutor. Mot. Behav. **3**, 47 (1980)
67. D. Sternad, W.J. Dean, S. Schaal, Interaction of rhythmic and discrete pattern generators in single-joint movements. Hum. Mov. Sci. **19**(4), 627–664 (2000)
68. V.K. Jirsa, J.A.S. Kelso, The excitator as a minimal model for the coordination dynamics of discrete and rhythmic movement generation. J. Mot. Behav. **37**, 35–51 (2005)
69. P.W. Fink, J.A.S. Kelso, V.K. Jirsa, Perturbation-induced false starts as a test of the Jirsa-Kelso Excitator model. J. Mot. Behav. **41**, 147–157 (2009)
70. R. FitzHugh, Impulses and physiological states in theoretical models of nerve membrane. Biophys. J. **1**, 445 (1961)
71. R. Rosen, *Life itself: a comprehensive inquiry into the nature, origin, and fabrication of life* (Columbia University Press, New York, 1991)
72. J. Holland, *Adaptation in Natural and Artificial Systems* (MIT Press, Cambridge, MA, 1992)
73. M. Kinsbourne, J.S. Jordan, Embodied anticipation: a neurodevelopmental interpretation. Discourse Process. **46**, 103–126 (2009)
74. T.L. Chartrand, J.A. Bargh, The chameleon effect: the perception–behavior link and social interaction. J. Pers. Soc. Psychol. **76**, 893 (1999)
75. A. Fogel, Two principles of communication: co-regulation and framing. New Perspect. Early Commun. Dev. 9–22 (1993)

76. D.A. Engstrøm, J.A.S. Kelso, T. Holroyd, Reaction-anticipation transitions in human perception-action patterns. Hum. Mov. Sci. **15**, 809–832 (1996)
77. J. Peters, S. Schaal, Policy gradient methods for robotics, in *2006 IEEE/RSJ International Conference on Intelligent Robots and Systems* (IEEE, 2006), pp. 2219–2225
78. Friston KJ (2010) The free-energy principle: a unified brain theory? 11:127–138. https://doi.org/10.1038/nrn2787
79. T. Kohonen, Self-organized formation of topologically correct feature maps. Biol. Cybern. **43**, 59–69 (1982)
80. I. Harvey, E. Di Paolo, R. Wood et al., Evolutionary robotics: a new scientific tool for studying cognition. Artif. Life **11**, 79–98 (2005)
81. B. Ermentrout, Neural networks as spatio-temporal pattern-forming systems. Rep. Prog. Phys. **61**, 353 (1999)
82. E.W. Large, M.R. Jones, The dynamics of attending: how people track time-varying events. Psychol. Rev. **106**, 119 (1999)
83. I. Konvalinka, P. Vuust, A. Roepstorff, A coupled oscillator model of interactive tapping, in *Proceedings of the 7th Triennial Conference of European Society for the Cognitive Sciences of Music* (Jyvaskyla, Finland, 2009)
84. E.L. Saltzman, K.G. Munhall, Skill acquisition and development: The roles of state-, parameter-, and graph-dynamics. J. Mot. Behav. **24**, 49–57 (1992)
85. L. Righetti, J. Buchli, A.J. Ijspeert, Adaptive frequency oscillators and applications. Open Cybern. Syst. J. **3**, 64–69 (2009)
86. L. Righetti, J. Buchli, A.J. Ijspeert, Dynamic Hebbian learning in adaptive frequency oscillators. Phys. Nonlinear Phenom. **216**, 269–281 (2006). https://doi.org/10.1016/j.physd.2006.02.009
87. A.A. Sharp, M.B. O'Neil, L.F. Abbott, E. Marder, The dynamic clamp: artificial conductances in biological neurons. Trends Neurosci. **16**, 389–394 (1993)
88. W.J. Freeman, Consciousness, intentionality and causality. J. Conscious. Stud. **6**, 143–172 (1999)
89. J.-J. Aucouturier, T. Ikegami, The illusion of agency: two engineering approaches to compromise autonomy and reactivity in an artificial system. Adapt. Behav. **17**, 402–420 (2009). https://doi.org/10.1177/1059712309344420
90. X.E. Barandiaran, E. Di Paolo, M. Rohde, Defining agency: Individuality, normativity, asymmetry, and spatio-temporality in action. Adapt. Behav. **17**, 367–386 (2009)
91. J.A.S. Kelso, J.P. Scholz, G. Schöner, Dynamics governs switching among patterns of coordination in biological movement. Phys. Lett. A **134**, 8–12 (1988)
92. C. De Luca, K.J. Jantzen, S. Comani et al., Striatal activity during intentional switching depends on pattern stability. J. Neurosci. **30**, 3167–3174 (2010). https://doi.org/10.1523/JNEUROSCI.2673-09.2010
93. E. Tognoli, G. Dumas, J.A.S. Kelso, A roadmap to computational social neuroscience, in *Advances in Cognitive Neurodynamics (V)* (Springer, Dordrecht, in press)
94. G. Schöner, J.A.S. Kelso, A dynamic pattern theory of behavioral change. Science **135**, 1513–1520 (1988). https://doi.org/10.1126/science.3281253

Design Principle for a Population-Based Model of Epileptic Dynamics

Gerold Baier, Richard Rosch, Peter Neal Taylor and Yujiang Wang

Abstract Epilepsy is defined as the brain's susceptibility to recurrent, hypersynchronous discharges that disrupt normal neuronal function. Over the last decades, progress has been made in using dynamical systems theory and computational analyses to characterise the nature of seizure-like activity. Using simplified models of population dynamics, macroscale features of epileptic seizures can be described as expressions of model interactions. There is a trade-off between complexity of these models and their explanatory power: Models that represent biophysical components of the brain often contain many degrees of freedom and nonlinearities, which can make them challenging to interpret and often means that different model parameterisations can produce similar results. Simple models, on the other hand, do not usually have a direct correlate in brain anatomy or physiology, but rather capture more abstract quantities in the brain. However, the effects of individual parameters are easier to interpret. Here we suggest a design principle to generate the complex rhythmic evolution of tonic-clonic epileptic seizures in a neural population approach. Starting from a simple neuronal oscillator with a single nonlinearity, we show in a step-by-step analysis how complex neuronal dynamics derived from patient observations can be reconstructed.

G. Baier (✉)
Cell and Developmental Biology, University College London, London, UK
e-mail: g.baier@ucl.ac.uk

R. Rosch
Wellcome Trust Centre for Neuroimaging, University College London, London, UK
e-mail: r.rosch@ucl.ac.uk

P.N. Taylor
Institute of Neuroscience, Newcastle University, Newcastle, UK
e-mail: peter.taylor@newcastle.ac.uk

Y. Wang
ICOS, School of Computing Science, Newcastle University, Newcastle, UK
e-mail: yujiang.wang@newcastle.ac.uk

S.C. MÃijller et al. (eds.), *Complexity and Synergetics*,
https://doi.org/10.1007/978-3-319-64334-2_25

1 Introduction

Epilepsy is a primary neurological condition marked by sudden, recurrent episodes of synchronized, pathological neuronal activity causing behavioral disturbances and abnormal movements. It can be conceptualized on several time scales, with rhythms observable in the neuronal states during seizures, irregular recurrence of seizures, diurnal variations in seizure burdens, and changes of seizure patterns during patients' lifetimes [1]. Epileptic seizures are typically characterized by abnormal waveforms of electric potential in extended regions of the brain. Using electroencephalography (EEG), epileptic seizures have been recorded for decades and their observation is an important part of clinical diagnosis and treatment decisions. Yet the underlying pathophysiological mechanisms and their relationship to the variation in observed seizure phenomena are still not well understood. Computational models have been employed to try to describe epileptic rhythms in terms of nonlinear dynamics (see [2, 3] for examples and [4] for a review). These often start from the assumption that epileptic rhythms are the consequence of the biophysical and biochemical interaction of known components, generally firing neurons [5]. However, due to the enormous complexity of the human brain, simulations of large-scale dynamics as an ensemble of individual, coupled neurons is not feasible. Therefore heuristic models at a higher level of abstraction were introduced to capture the essence of the interactions at the level of large neuronal populations [6]. These have the advantage of the phenomenology of EEG recordings, whilst allowing a more comprehensive analysis of their dynamic complexity and can therefore be used to explain a variety of different waveforms associated with different types of epileptic seizure dynamics [7, 8].

Very few if any of the computational model features have been used to undergo serious testing in the clinical setting of observations in humans. This is partly due to the unavailability of patients to scientific experimentation but also due to the nature of the models. None of the variables in the models (nodes of the networks) and none of the interactions between the components (edges of the network) can be studied in isolation and a step-by-step derivation of biophysico-chemical principles is generally not possible. Therefore, alternative approaches have been sought.

An alternative to this approach is to evaluate the dynamic consequences of generic model components (excitation, inhibition, time scales) without direct reference to biophysical circuit elements. In previous work, we have proposed a building block strategy to systematically compose models with desired dynamic behaviors [8]. This has the advantage that through comprehensive exploration of parameter space within a simple framework of interacting components, many different dynamic phenomena can be reproduced. Here we use the approach of minimal interacting components to systematically generate complex rhythms of paroxysmal seizure dynamics. We restrict the study to a single basic type of nonlinearity, the sigmoid transition function in self-regulation and connection between neural populations. As an example we present the stepwise design of a complex rhythm sequence which is representative of the so-called tonic-clonic seizure rhythm. We discuss qualitatively different ways to create the seizure and the potential value of this finding in future studies of epileptiform dynamics in clinical data.

2 Basic 2-Variable Oscillator Model

The model studied in the present contribution uses a neural population approach based on the neural field equations proposed by Amari [9] and was employed previously to describe epileptiform dynamics, e.g. in [10–12]. The set of two ordinary differential equations is based on the canonical feedback loop between the activities of an excitatory and an inhibitory neuronal population and is given as:

$$\begin{aligned} \frac{dEx1}{dt} &= \tau_{ex1}(h_{ex1} - Ex1 + C_1 f[Ex1] - C_3 f[In1]) \\ \frac{dIn1}{dt} &= \tau_{in1}(h_{in1} - In1 + C_2\, Ex1) \end{aligned} \tag{1}$$

where $h_{ex1,in1}$ are input parameters, $\tau_{ex1,in1}$ are time scale parameters, $C_{1,2,3}$ are connectivity parameters and $f[x]$ is the sigmoidal activation function:

$$f[x] = (1/(1 + \varepsilon^{-x}),$$

where $x = Ex1, In1$ and $\varepsilon = 1000$.

These equations implement the set of interactions shown schematically in Fig. 1. The scheme consists of an excitatory population $Ex1$ which can self-excite. The strength of the self-excitation is controlled by parameter C_1. This population also excites the inhibitory population $In1$, control parameter is C_2. In return, the inhibitory population inhibits the activity of the excitatory population, controlled by parameter C_3. The input parameters h_{ex1} and h_{in1} describe constant input from external, unspecified sources. Both populations operate at a time scale adjusted by the two time scale parameters.

Parameter values are selected based on previous studies of the dynamics of epileptiform rhythms. In this approach we adjust parameters to produce certain dynamic

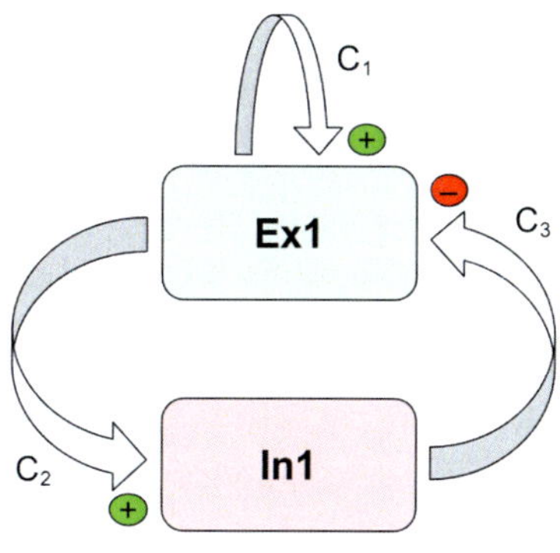

Fig. 1 Scheme of interaction between two neural populations, one excitatory (Ex1) and one inhibitory (In1). Labels to *arrow* denote positive (*green*) or negative (*red*) impact of coupling on a population. The C_i denote the parameter to control coupling strength

Table 1 Parameter values used to produce Fig. 2

Parameter	Interpretation	Fig. 2a	Fig. 2b	Fig. 2c
C_1	$Ex1 \rightarrow Ex1$ connectivity strength	2	2	2
C_2	$Ex1 \rightarrow In1$ connectivity strength	6	6	6
C_3	$In1 \rightarrow Ex1$ connectivity strength	0	2	2
τ_{ex1}	$Ex1$ timescale	2	2	2
τ_{in1}	$In1$ timescale	3	3	3
h_{ex1}	Input $Ex1$	Scanned	Scanned	See caption
h_{in1}	Input $In1$	−1	−1	−1
ε	Sigmoid steepness	1000	1000	1000

features characteristic of a type of epileptic seizure, rather than relating the parameters themselves to match biological realism. The dynamic features we focus on are: (i) an onset of a regular rhythm with characteristic frequency, simple waveform and a temporal evolution where amplitude increases and frequency decreases; (ii) transition of this onset rhythm into a complex waveform consisting of large amplitude spike-wave discharges and further slowing of frequency; and (iii) an abrupt end of the seizure rhythm leading to a non-oscillatory post-seizure state. These dynamic features are typical of tonic-clonic seizures, a common seizure type in a number of different epilepsy syndromes. A clinical recording is shown in Fig. 6c. Investigations of this type of seizure rhythm have indicated that the characteristic frequency slowing and the abrupt ending is consistent with the occurrence of a homoclinic bifurcation in an autonomous oscillator [3]. We therefore adjusted the model parameters of Eq. (1) in the vicinity of a homoclinic bifurcation. The parameter values used to produce the figures are listed in Table 1.

We first look at the dynamics of a single neural population variable in the absence of inter-population coupling. This is done by setting coupling parameter C_2 equal to zero. A single variable can only have steady state solutions, but if we scan h_{ex1} as a control parameter we find that there exists a region of bistability of the activities of both populations. Figure 2a displays a forward and backward scan of parameter h_{ex1} showing the region of bistability or coexistence of two stable solutions in the region between ≈ -1.4 and ≈ -0.5. In that region, the stable, observable state of the system can be either on the upper or on the lower branch, depending on the preparation of the initial conditions. The bistability is created by the only available nonlinearity, the sigmoidal self-excitation.

When inter-population coupling is introduced, $C_2 > 0$, the bistability persists but its nature can change. We chose a setting where there is a bistability of a steady state on the lower branch with an oscillatory solution on the upper branch. The bifurcation diagram in Fig. 2b shows the region of bistability in this case. Technically, the region of bistability is defined by two bifurcations. The upper border of the bistable region is due to a saddle-node bifurcation of the lower branch fixed point. The lower border is due to a homoclinic bifurcation of the limit cycle on the upper branch. Outside of the bistability, there is a region of parameter where the limit cycle exists as the only stable

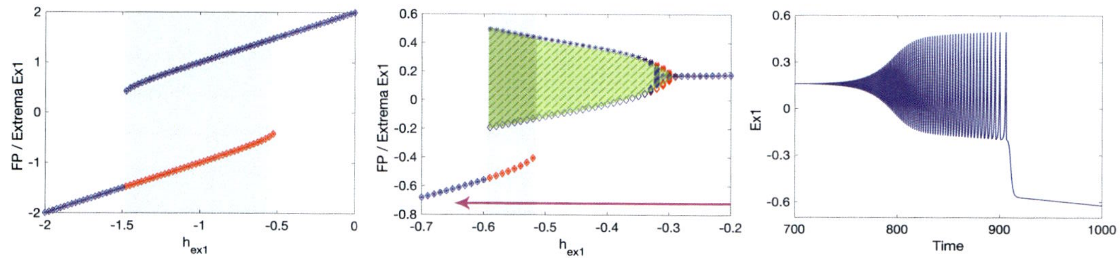

Fig. 2 Bifurcations in the 2V system when scanning parameter h_{ex1}. **a** Bifurcation diagram of Eq. (1) with $C_2 = 0$. The system shows a bistability (*shaded grey*) between an upper and a lower fixed point. **b** Bifurcation diagram of Eq. (1) with $C2 = 2$. The system shows a Hopf bifurcation at about $h_{ex1} = -0.3$. There is additionally a bistability (*shaded grey*) between the oscillations (*shaded green*) and a lower fixed point. **c** Time series of oscillator Eq. (1) with C2 = 2 and continuously changing parameter: $h_{ex1} = -0.15 - 0.0005 * t$. The slow parameter change underlying the time series is additionally indicated in (**b**) as a *magenta arrow*. Parameter values provided in Table 1

solution. It ceases to exist in a supercritical Hopf bifurcation at $h_{ex1} \approx -0.3$ resulting in the stable upper fixed point. Our single-parameter space is therefore structured into four distinct regions.

When interpreting the model output in the context of epilepsy, the oscillatory region is taken to represent the seizure rhythm. The bifurcation diagram then represents specifically what can happen for certain values of constant input to population $Ex1$. To explain the onset and the offset of the seizure rhythm it is commonly assumed that the value of at least one of the model parameters changes continuously on a slow time scale.

To implement this, our oscillator model is prepared in the upper fixed point and parameter h_{ex1} is then slowly decreased and the dynamics recorded (Fig. 2c). The simulation shows a continuous evolution of the dynamics from the upper fixed point via the first (Hopf) bifurcation to fast small amplitude oscillations. It then continues with growing amplitude and decreasing frequency until there is a sudden drop to the lower fixed point where the value of activity slowly decreases further.

If one wanted the model to undergo a second transition into the seizure rhythm one would have to reset parameter h_{ex1} to a value where the activities settle on the upper branch. While this can easily be done in the model equation, there is no known and intuitive, biophysical equivalent of this resetting mechanism in the human brain.

Starting with a simple model containing only two populations, as well as some structured knowledge of neuronal dynamics, we were able to recreate some key features of epileptic rhythms as seen in tonic-clonic seizures. Two nonlinearities (in the self-excitation, and the feedback inhibition, respectively) were used to obtain a region of bistability and oscillations. Introducing slow variation within one parameter, allowed transition between the dynamic states that resembles the natural progression of seizure dynamics as outlined above. In this scheme, repeated changes of the control parameter would result in the recurrent "seizures", a diagnostic criterion for epilepsy.

3 The Oscillator with an Ultraslow Variable

The previous section showed how a two-variable oscillator with a specific bifurcation structure can be turned into a model of recurrent epileptic seizure dynamics by a recurring independent parameter manipulation. This approach was taken in a large number of previous computational studies of seizure rhythms, explicitly or implicitly [13, 14]. Here we present a set of additional possibilities to explain this phenomenon.

The length and appearance of the seizure rhythm in the previous section depends directly on the time scale of the continuous parameter change. Following previous suggestions [3], we now introduce a new variable that helps us dynamically implement these changes. Specifically, the model with a region of bistability such as Eq. (1) can be made to autonomously switch between the two coexisting stable dynamics (fixed point and oscillations) with such an addition.

We expand our oscillator with a variable on a much slower ("ultraslow") time scale to get the following 3 variable model:

$$
\begin{aligned}
\frac{dEx1}{dt} &= \tau_{ex1}(h_{ex1} - Ex1 + C_1 f[Ex1] - C_2 f[In1] - C_{u1} f[Ul)] \\
\frac{dIn1}{dt} &= \tau_{in1}(h_{in1} - In1 + C_3 Ex1) \\
\frac{dUl}{dt} &= \tau_{ul}(h_{ul} - Ul + C_{1u} f[Ex1])
\end{aligned}
\tag{2}
$$

where h_{ul} is the additional input parameter for the ultraslow variable Ul, τ_{ul} the corresponding time scale parameter, and TC and CT are two new connectivity parameters. We label the new variable "ultraslow" to distinguish it from the two "slow" wave populations in the context of spike-wave oscillations below.

The design principle is known from the composition of burst oscillations in the context of neuronal firing [15].The input of one of the activities of a bistable system into the ultraslow variable leads to its continuous modulation. The signs of the coupling terms are such that the excitatory population leads to a decrease of the ultraslow variable while it is on the lower branch. Because of its negative coupling the ultraslow variable leads to an increase of input to the excitatory variable, mimicking an increase in control parameter h_{ex1} as in the previous section. Eventually, the system then reaches the bifurcation point of the lower branch fixed point and the activities of the $Ex1/In1$ subsystem jumps to the upper branch. It starts to oscillate. While on the upper branch, the ultraslow variable starts to continuously increase, thereby diminishing the input to variable $Ex1$, mimicking an increase in control parameter h_{ex1} as in the previous section. Eventually this leads to a sudden jump back to the lower branch where the process starts to repeat itself.

The result is a dynamics that switches spontaneously and periodically between the two branches. In the context of epilepsy the period of time spent on the upper branch of activity of population $Ex1$ would be considered the seizure state. Figure 3 shows one cycle of the resulting periodic switching between the lower steady state and the oscillation on the upper branch of Fig. 2b. The trend of the ultraslow variable while

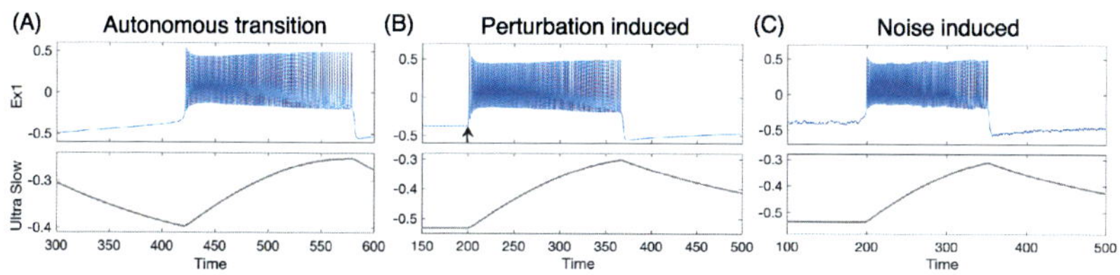

Fig. 3 Transitions to seizures by three different mechanisms. **a** Autonomous, periodic switching to seizure rhythm in Eq. (2). **b** Excitability leading to an induced seizure rhythm in Eq. (2) in response to a pulse stimulation at time 200 (indicated by *black arrow*). **c** Noise-induced transition to seizure rhythm in Eq. (2) in the presence of normally distributed additive noise in the first two variables. Parameter values provided in Table 2

the activity of population $Ex1$ is in the seizure rhythm is similar to the continuously shifting parameter in Fig. 2c, and therefore both the increase in amplitude and the decrease of the frequency are observed during that phase. The evolution of the period between maxima is plotted at the bottom panel. The period first decreases (when settling into the oscillation), then remains nearly constant but increases towards the end. Interestingly, during the final increase there is a an irregularity of the increase of the period. Thus for certain ranges of parameter values a periodic occurrence of the seizure rhythm can be modeled. The period spent in the silent and in the rhythmic state, respectively, depends on the quantitative parameter settings and can be adjusted.

The periodic bursting is limited to a finite range of the chosen control parameter, e.g. h_{ex1}. If the systems is prepared just outside that range the dynamics become excitable. Excitability means that a threshold exists which separates trivial from non-trivial responses to single pulse stimulation. This is similar to excitability in individual neurons. If a neuron is prepared just outside the conditions for periodic (tonic) firing, a single action potential is recorded from a neuron in response to a threshold crossing (supra-threshold) stimulation [15].

Figure 3b shows the response dynamics of the model Eq. (2) following a brief pulse perturbation. The dynamics is prepared in the lower branch fixed point. Following perturbation (arrow) it switches to the upper branch, settles to an oscillation, amplitude increases and frequency decreases until it suddenly drops back to the stable lower fixed point where it remains if no further perturbation is applied. A repetition of seizures is seen when the supra-threshold perturbation is repeated.

In the same setting yet another mechanism is possible. In the absence of parameter variations and specific pulse perturbations it is possible to induce a seizure-like dynamics by adding a noise term to the equations (Fig. 3c). The noise continuously perturbs the system and leads to occasional (unpredictable) crossings of the threshold resulting in a period of seizure-like activity. While the seizure rhythm period depends on the parameter settings of the autonomous model, the frequency of seizures depends on the amplitude of the noise level.

Table 2 Parameter values used to produce Fig. 3

Parameter	Fig. 3a	Fig. 3b	Fig. 3c
C_1	2	2	2
C_2	6	6	6
C_3	2	2	2
τ_{ex1}	1	1	1
τ_{in1}	1.5	1.5	1.5
τ_{ul2}	$\tau_{ex1}/500$	$\tau_{ex1}/500$	$\tau_{ex1}/500$
h_{ex1}	−0.45	−0.49	−0.493
h_{in1}	−1.1	−1.1	−1.1
ε	1000	1000	1000

Note that in Eq. (2) we treat the ultraslow variable formally as an additional neural population although this is not necessary to produce the dynamics described. However, it is unknown what physiological process might underlie this ultraslow process and we therefore emphasize that qualitatively similar results can be generated with other equations as well as long as the time scale and coupling signatures observe the requirements.

4 Two Coupled Oscillator Model

An important feature of tonic-clonic seizures is the transition from tonic to clonic rhythm. Typically, this means that not only does the frequency decrease but in addition a qualitative change of the waveform is observed. The main feature of the clonic part is the "spike-wave" (SW) appearance of the rhythm. The SW oscillation is a typical feature of models with two different time scales and indeed has been modelled using a combination of neural population with different time scale parameters.

To model a system with oscillators at two different time scales we simply duplicate the populations contained within the system in Eq. (1). This leads to the following four-variable system [16]:

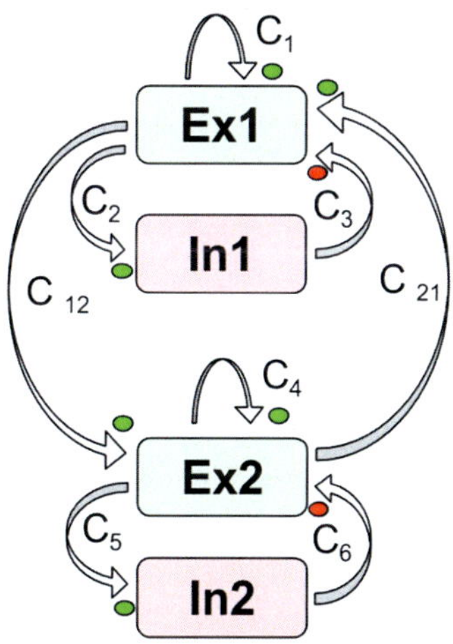

Fig. 4 The two-oscillator scheme

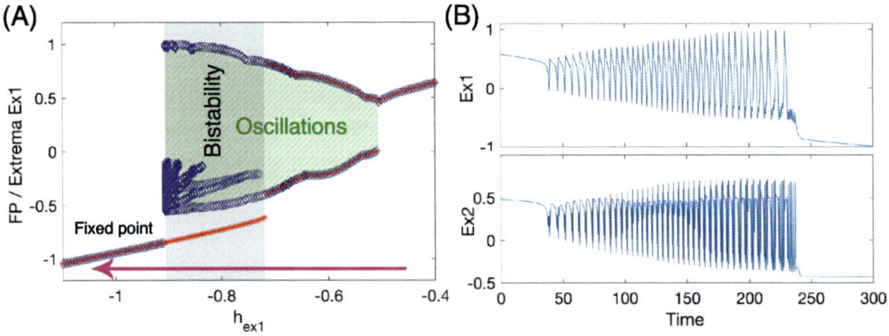

Fig. 5 Bifurcations in the 4V system in the parameter h_{ex1}. **a** Bifurcation diagram of Eq. (3). The system shows a bistability (*shaded grey*) between the oscillations (*shaded green*) and a lower fixed point. Note the between $h_{ex1} = -0.7$ and -0.9 the oscillations are of a (poly-) spike-wave form (hence multiple maxima and minima), see (**b**) for details. **b** Time series of oscillator Eq. (3) and continuously changing parameter: $h_{ex1} = -0.45 - 0.002 * t$. The slow parameter change underlying the time series is additionally indicated in (**a**) as a *magenta arrow*. Parameter values provided in Table 3

$$
\begin{aligned}
\frac{dEx1}{dt} &= \tau_{ex1}(h_{ex1} - Ex1 + C_1 f[Ex1] - C_2 f[In1] + C_{21} f[Ex2]) \\
\frac{dIn1}{dt} &= \tau_{in1}(h_{in1} - In1 + C_3\, Ex1) \\
\frac{dEx2}{dt} &= \tau_{ex2}(h_{ex2} - Ex2 + C_4 f[Ex2] - C_5 f[In2] + C_{12} f[Ex1]) \\
\frac{dIn2}{dt} &= \tau_{in2}(h_{in2} - In2 + C_6\, Ex2)
\end{aligned}
\tag{3}
$$

where $h_{ex1,in1,ex2,in2}$ are input parameters, $\tau_{ex1,in1,ex2,in2}$ are time scale parameters, $C_{1,2,3,4,5,6,12,21}$ are connectivity parameters and f[.] is defined as above.

A scheme of the model of two coupled oscillators is depicted in Fig. 4. The dynamics of this model with its large number of parameters and additional nonlinearities is too complex to be comprehensively evaluated in the present context. We therefore reduce the discussion to a specific setting which relates to the dynamics of tonic-clonic seizures. Specifically, the parameters in the two isolated oscillators are the same with the exception of the time scale parameter. Scanning control parameter h_{ex1} and recording the bifurcation diagram we get the result displayed in Fig. 5a.

We can identify the lower and upper branch of activity in the two oscillator model. Also, as before, the lower branch is a stable steady state whereas the upper branch contains oscillatory and steady state dynamics. In contrast to the diagram in Fig. 2b, the oscillatory dynamics now comprise multiple waveforms and, in particular, complex oscillations. Due to the choice of different time-scale parameters, the oscillations include spike-wave (SW) dynamics where a fast spike or multiple fast spikes are followed by slow waves. In the current simulations the complex oscillations are periodic but for other parameter sets we have also seen quasiperiodic and chaotic mixtures and spikes and waves.

Table 3 Parameter values used to produce Figs. 5 and 6

Parameter	Fig. 5a	Fig. 5b	Fig. 6
$C_1 = C_4$	2	2	2
$C_2 = C_5$	6	6	6
$C_3 = C_6$	2	2	2
$C_{12} = C_{21}$	1.1	1.1	1
C_{u1}	NA	NA	1.5
C_{1u}	NA	NA	1
τ_{ex1}	1	1	1
τ_{in1}	1	1	1.5
$\tau_{ex2} = \tau_{in2}$	5	5	4
τ_z	NA	NA	$\tau_{ex1}/300$
h_{ex1}	Scanned	See caption	–0.6
h_{in1}	−1.1	−1.1	−1.1
h_{ex2}	–0.53	–0.53	–0.53
h_{in2}	−1	−1	–0.5
ε	1000	1000	1000

We can now proceed as above and continuously vary control parameter h_{ex1}. The resulting evolving time course is shown in Fig. 5b. The representative variable $Ex1$ now displays a transition from small fast tonic spiking to slow clonic bursting reminiscent of the tonic-clonic seizure evolution. As before, the onset of oscillations can be attributed to a supercritical Hopf bifurcation, the offset is due to a homoclinic bifurcation, in our case in both oscillators. Due to the coupling of two nonlinear oscillators the waveform is now no longer strictly periodic but shows some variability in the evolution even in the absence of noise.

5 Spontaneous Seizure Model

We extend the model of two interacting oscillators with an ultraslow variable following the design principle in the first results section. That is, we add a ultraslow variable which is fed from one of the oscillators and which feeds back to one of the oscillators. The coupling is again such that spontaneous switching between the states on the lower and upper branch of the bistability region happens.

The model with five variables is:

$$\begin{aligned}
\frac{dEx1}{dt} &= \tau_{ex1}(h_{ex1} - Ex1 + C_1 f[Ex1] - C_3 f[In1] + C_{21} f[Ex2]) - C_{u1} f[Ul] \\
\frac{dIn1}{dt} &= \tau_{in1}(h_{in1} - In1 + C_2\, Ex1) \\
\frac{dEx2}{dt} &= \tau_{ex2}(h_{ex2} - Ex2 + C_4 f[Ex2] - C_6 f[In2] + C_{12} f[Ex1]) \\
\frac{dIn2}{dt} &= \tau_{in2}(h_{in2} - In2 + C_5\, Ex2) \\
\frac{dUl}{dt} &= \tau_{ul}(h_{ul} - Ul + C_{1u} f[Ex1])
\end{aligned} \tag{4}$$

with conventions as above. Note that the ultraslow variable modulates the slow oscillator. The coupling is negative to enable cyclical interaction around the region of bistability, similar to the set-up in the three variable model of spontaneous bursting above.

With the parameters from the four variable model in Fig. 5a it is fairly trivial to adjust the couplings such that spontaneous switching is found. An exemplary time series of the result is displayed in Fig. 6. Parameters can be adjusted to obtain more or less frequent and even rare seizure outbursts. In the case shown the onset of fast tonic activity is spontaneous, i.e. with finite amplitude. This shows that the bistability region encompasses only oscillatory dynamics on the upper branch. If the upper branch included the Hopf bifurcation and upper branch fixed point dynamics, a seizure dynamics with transition from zero amplitude could also be observed. Again, the dynamics of the oscillations in the model involving two coupled oscillators is irregular due to dynamic interaction, even in the absence of noise.

6 Additional Remarks

The presented results show a systematic design principle to synthetically compose (reverse engineer) a complex dynamic process from simple ingredients. The main ingredients are a bistable element, a negative feedback oscillator and an ultraslow variable. The use of a single ultraslow variable requires the presence of a region of bistability to allow for spontaneous switching between qualitatively different states. If a bifurcation diagram includes more than one dynamic state without bistability (for example a saddle-node on invariant cycle bifurcation), it is possible to generate autonomous switching between them if two ultraslow variables are added [15].

We have focussed our model on the presence of a Hopf and a homoclinic bifurcation to generate the seizure rhythm. However, in equations with two variables, four bifurcations are known and in principle each of them could be used to generate complex oscillations either in the presence or in the absence of bistability. Thus an enormous repertoire of waveforms and transitions between them is accessible. This is important when it comes to the interpretation of clinical seizure rhythms. So far, the specific dynamic components of individual seizure types have mostly been

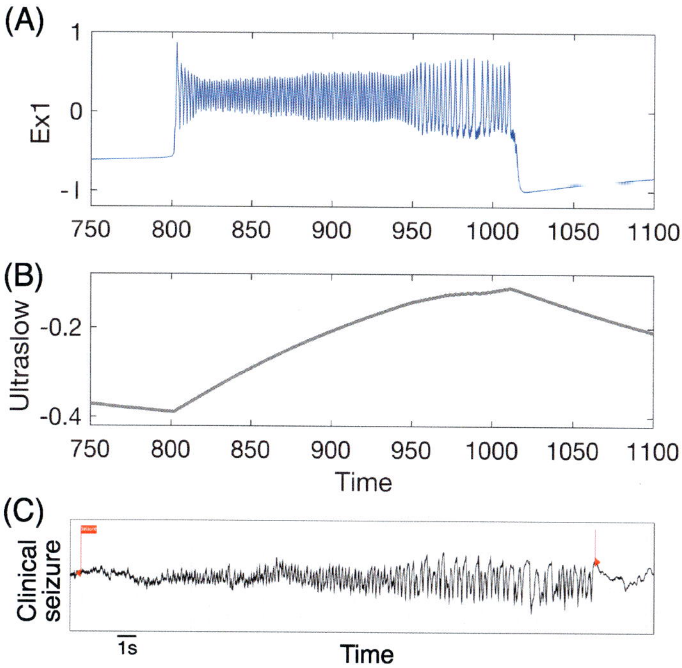

Fig. 6 Time series of spontaneous transitions from tonic to tonic-clonic rhythms. In the full model (Eq. (4)), autonomous transitions to seizures can be observed, where the seizure itself shows a transition from tonic rhythms to tonic-clonic rhythms. Parameters are given in Table 3. *Red* markers in the clinical recording indicate the begin and end of the seizure (data obtained from iEEG.org, Study 031, second seizure, channel RPG35). Parameter values provided in Table 3

ignored. But as shown for example in [3] the careful observation of onset, evolution and offset of the seizure rhythm allows certain conclusions about the dynamics that can be modelled following the suggested design principle.

The model proposed here does not provide insight into the neuroanatomical basis of tonic-clonic seizures in humans. Certain features, such as the spatial spread, the irregular and unpredictable occurrence of seizures, and the co-existence of different oscillatory states (e.g. background activity, sleep, seizure activity), are beyond the scope of this analysis. However, minimal models that generate qualitative features of epileptic seizure rhythms can be used in a number of different ways.

First, it is possible to fit the model dynamics to clinical recordings. Because of the nonlinear mapping from model parameters to neuronal states, this approach may reveal natural categories within dynamically diverse seizure phenotypes, that are not intuitively apparent from visual EEG analysis alone [17, 18]. Second, one can use the model to generate hypotheses about the response to perturbations, e.g. drug administration, electric stimulation, transcranial magnetic stimulation and others. With recent advances in epilepsy surgery and deep brain stimulation, as well as a first in-man trial of chronic implantable recording devices, a better characterised model of the dynamics governing epileptic seizures will be imperative to harness the potential of these novel approaches [19, 20]. Third, while the knowledge of the mammalian neocortex is rather advanced on the neuronal level, large-scale understanding of epileptic seizure rhythms in the human patient is still limited. Epilepsy is still commonly conceptualised on the level of an imbalance between excitation and inhibition, as recently demonstrated in a report on invasive multiscale recordings during presurgical monitoring of patients with focal seizures [21]. This approach is useful, but unlikely to fully explain the qualitatively different dynamic features seen in patients, and their associated neurological comorbidities and prognoses. Similar modelling approaches as used applied have been used as an excellent tool to generate model-based suggestions to differentiate between clinical seizures based on dynamic features, as suggested in the context of focal seizures in [22]. Finally, these building blocks can be coupled laterally to investigate spatio-temporal dynamics of seizures, e.g. seizure spreading in heterogeneous brain networks [23–25] which may have clinical implications on surgery [26, 27] and stimulation [20, 28].

Acknowledgements We thank Otto Rössler, Viktor Jirsa, Ulrich Stephani and Beate Diehl for discussion.

References

1. M.P. Richardson, New observations may inform seizure models: very fast and very slow oscillations. Prog. Biophys. Mol. Biol. **105**, 5–13 (2011)
2. M. Breakspear, J. Roberts, J. Terry, S. Rodrigues, N. Mahant, P. Robinson, A unifying explanation of primary generalized seizures through nonlinear brain modeling and bifurcation analysis. Cereb. Cortex **16**, 1296–1313 (2006)
3. V.K. Jirsa, W.C. Stacey, P.P. Quilichini, A.I. Ivanov, C. Bernard, On the nature of seizure dynamics. Brain **137**, 2210–2230 (2014)
4. F. Wendling, P. Benquet, F. Bartolomei, V. Jirsa, Computational models of epileptiform activity. J. Neurosci. Methods (2015)
5. W. Lytton, Computer modelling of epilepsy. Nat. Rev. Neurosci. **9**, 626–637 (2008)
6. H. Wilson, J. Cowan, Excitatory and inhibitory interactions in localized populations of model neurons. Biophys. J. **12**, 1–24 (1972)
7. F. Lopes Da Silva, W. Blanes, S.N. Kalitzin, J. Parra, P. Suffczynski, D.N. Velis, Epilepsies as dynamical diseases of brain systems: basic models of the transition between normal and epileptic activity. Epilepsia **44**, 72–83 (2003)
8. Y. Wang, M. Goodfellow, P. Taylor, G. Baier, Phase space approach for modeling of epileptic dynamics. Phys. Rev. E **85**(061), 918 (2012)
9. S. Amari, Dynamics of pattern formation in lateral-inhibition type neural fields. Biol. Cybern. **27**, 77–87 (1977)
10. D. Fan, Q. Wang, M. Perc, Disinhibition-induced transitions between absence and tonic-clonic epileptic seizures. Sci. Rep. **5** (2015)
11. P.N. Taylor, G. Baier, A spatially extended model for macroscopic spike-wave discharges. J. Comput. Neurosci. **31**, 679–684 (2011)
12. P.N. Taylor, G. Baier, S.S. Cash, J. Dauwels, J. Slotine, Y. Wang, A model of stimulus induced epileptic spike-wave discharges. IEEE Proc. SSCI **2013**, 53–59 (2013)
13. G.K. Cooray, B. Sengupta, P.K. Douglas, K. Friston, Dynamic causal modelling of electrographic seizure activity using bayesian belief updating. NeuroImage **125**, 1142–1154 (2016)

14. L. Kuhlmann, D.B. Grayden, F. Wendling, S.J. Schiff, Role of multiple-scale modeling of epilepsy in seizure forecasting. J. Clin. Neurophysiol. **32**, 220–226 (2015)
15. E.M. Izhikevich, *Dynamical Systems in Neuroscience: The Geometry of Excitability and Bursting* (The MIT Press, Cambridge, 2007)
16. P.N. Taylor, Y. Wang, M. Goodfellow, J. Dauwels, F. Moeller, U. Stephani, G. Baier, A computational study of stimulus driven epileptic seizure abatement. PLoS ONE **9**(e114), 316 (2014b). https://doi.org/10.1371/journal.pone.0114316
17. A. Holgado-Nevado, F. Marten, M. Richardson, J. Terry, Characterising the dynamics of EEG waveforms as the path through parameter space of a neural mass model: application to epilepsy seizure evolution. NeuroImage **59**, 2374–2392 (2012)
18. F. Wendling, F. Bartolomei, J. Bellanger, P. Chauvel, Epileptic fast activity can be explained by a model of impaired GABAergic dendritic inhibition. Eur. J. Neurosci. **15**, 1499–1508 (2002)
19. M.J. Cook, T.J. O'Brien, S.F. Berkovic, M. Murphy, A. Morokoff, G. Fabinyi, W. D'Souza, R. Yerra, J. Archer, L. Litewka, S. Hosking, P. Lightfoot, V. Ruedebusch, W.D. Sheffield, D. Snyder, K. Leyde, D. David Himes, Prediction of seizure likelihood with a long-term, implanted seizure advisory system in patients with drug-resistant epilepsy: a first-in-man study. Lancet Neurol. **12**, 563–571 (2013)
20. Y. Wang, F. Hutchings, M. Kaiser, Computational modeling of neurostimulation in brain diseases. Prog. Brain Res. **222**, 191–228 (2015)
21. N. Dehghani, A. Peyrache, B. Telenczuk, M. Le Van Quyen, E. Halgren, S.S. Cash, N.G. Hatsopoulos, A. Destexhe, Dynamic balance of excitation and inhibition in human and monkey neocortex. Sci. Rep. **6**, 23,176 EP (2016)
22. Y. Wang, M. Goodfellow, P. Taylor, G. Baier, Dynamic mechanisms of neocortical focal seizure onset. PLoS Comput. Biol. **10**(e1003), 787 (2014)
23. T. Proix, F. Bartolomei, P. Chauvel, C. Bernard, V. Jirsa, Permittivity coupling across brain regions determines seizure recruitment in partial epilepsy. J. Neurosci. **34**, 15,009–15,021 (2014)
24. P.N. Taylor, M. Goodfellow, Y. Wang, G. Baier, Towards a large-scale model of patient-specific epileptic spike-wave discharges. Biol. Cybern. **107**, 83–94 (2013b)
25. P.N. Taylor, M. Kaiser, J. Dauwels, Structural connectivity based whole brain modelling in epilepsy. J. Neurosci. Methods (2014a)
26. F. Hutchings, C. Han, S. Keller, B. Weber, P. Taylor, M. Kaiser, Predicting surgery targets in temporal lobe epilepsy through structural connectome based simulations. PLoS Comput. Biol. **11**(e1004), 642 (2015)
27. L. Martinet, O. Ahmed, K. Lepage, S. Cash, M. Kramer, Slow spatial recruitment of neocortex during secondarily generalized seizures and its relation to surgical outcome. J. Neurosci. **35**, 9477–9490 (2015)
28. P.N. Taylor, J. Thomas, N. Sinha, J. Dauwels, M. Kaiser, T. Thesen, J. Ruths, Optimal control based seizure abatement using patient derived connectivity. Front. Neurosci. **9** (2015)

Discrete Modeling for a Minimal Circuit in the Hippocampus

Anastasia I. Lavrova and Eugene B. Postnikov

Abstract We present a simple discrete model for a minimal circuit in the hippocampal area CA3, which consists of various types of connected cells that differ in morphology and functional properties (pyramidal, basket and O-LM cells). This model allows for reproducing not only basic characteristics of the cells oscillations for all these types (period, amplitude and phase shift) but also to demonstrably explain the key property of switching between different rhythms using only one control parameter. The model results are confirmed via comparison with in vitro experimental results and discussed.

1 Introduction

The hippocampus is a part of the mammalian brain located inside the medial temporal lobe, beneath the cortical surface [1]. It belongs to the limbic system and plays an important role in long-term memory formation and in encoding of new information. The hippocampus consists of several parts (EC, DG, CA1, CA2, CA3). Since different neuronal cell types in each region are organized into networks in the hippocampus, it has been used as a model system for neurophysiological studies.

A.I. Lavrova (✉)
Medical Faculty, Saint-Petersburg State University, Universitetskaya emb. 7/9, Saint-Petersburg, Russia
e-mail: aurebours@googlemail.com

A.I. Lavrova
Saint-Petersburg State Research Institute of Phthisiopulmonology, Lygovsky avenue 2-4, Saint-Petersburg, Russia

E.B. Postnikov
Kursk State University, Radishcheva St., 33, Kursk 305000, Russia
e-mail: postnicov@gmail.com

S.C. Müller et al. (eds.), *Complexity and Synergetics*,
https://doi.org/10.1007/978-3-319-64334-2_26

In particular, it has been shown experimentally that the neuronal network in area CA3 can exhibit two different types of oscillations (so called theta- and gamma oscillations) and dynamically switch between them [2]. Spatial information is coded at theta frequencies (4–12 Hz) by neurons distributed along the longitudinal axis of the hippocampus. Gamma oscillations (30–80 Hz) are generated by the hippocampus and involved in information transmission and storage. It has been proposed that the hippocampus governs memory processes in brain and can switch long-term memory into dynamics.

Thus, the development of mathematical models, which can explain the background of such dynamical regimes is subject of ongoing discussions. Among the first physiologically based work on this topic, the articles [2, 3] should be mentioned. It has been shown that there are at least three types of hippocampal cells, whose dynamics and interconnections determine slow and fast oscillations: two fast oscillating basket cells, two slow oscillating oriens-lacunosum-molecular (O-LM) cells and one two-compartmental pyramidal cell. More recent work [4] has allowed for a simplification of the minimal network by reducing the number of basket cell to one and replacing the cumbersome Hodgkin-Huxley equations by the simplified FitzHugh-Nagumo system.

At the same time, although this simplification allows for reproducing basic dynamical regimes and transitions between them, the strong non-linear character of the coupled non-linear ordinary differential equations used there prevents us from a detailed revealing of the switch mechanism and a choice of a basic parameter (or couple of them). For this reason, discrete modelling is more promising from both point of view, algorithmic computational simplicity and qualitative neurophysiological interpretation.

The main goal of the present work is to reduce the detailed ODE model [2] of basic elements of a hippocampal network to a discrete system. We reproduce period, amplitude and phase shift of the oscillations and analyze the influence of synaptic connections between cells on the mentioned characteristics. In comparison with the data from in vitro experiments [2], we extract a key parameter for the switch between rhythms.

2 Model and Method

To model switching between rhythmic patterns in the hippocampus we have considered a minimal circuit shown in Fig. 1. The biophysical neuronal foundation determines the scheme, in which elements act as follows: (i) O-cells (O-LM-cells) are independent oscillators and activated by an E-cell (pyramidal cell) only at the initial time; (ii) E-cell spiking depends on the applied current and total inhibition by the O- and I-cells (basket cells); (iii) the phase shift between O-cells and their period depends on the total contribution of all synaptic connections between cells; (iv) I-cells are self-sustained oscillators, which inhibit each other.

As a result, the dynamics of all kinds of cells used is simulated by the following set of maps, where each discrete iteration is marked by j.

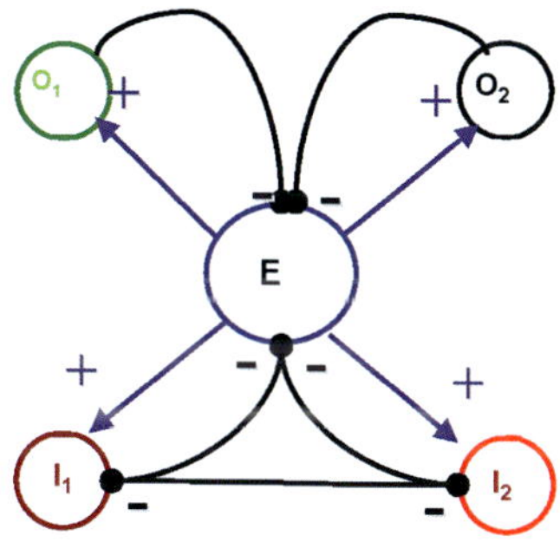

Fig. 1 The network studied represents a system of discrete active elements. The *circles* denote cells with their names inside; *lines* show activator (+) and *inhibitor* (−) directed connections

2.1 Discrete Equation for E-Cell

The spike of an E-cell depends on its potential in the previous discrete step $V_E(j-1)$, on the increment of the potential r, and on the inhibition by the O- and I-cells. Every spike of a basket cell occurs after a spike of a pyramidal cell, which allows for introducing a delay s between I- and E-cell spikes. Concerning O-LM cell spiking, their firing is conditioned by the first spike of an E-cell only. Therefore, the inhibition of an E-cell by the O- and I-cells is the "sum" of the membrane potentials of I-cells with delay s ($V_{I_1}(j-s)$, $V_{I_1}(j-s)$) and the potential of the O-cells in the previous step. For simplicity, we put $s = 1$ in this model. We have to keep in mind that the spike of the second O-cell is shifted relatively to the first by $\Delta\varphi$ and the membrane potential growth of pyramidal cell depends on the phase shift between the O-cells.

To take into account that the influence of O-cells leads to inhibition of E-cells in combination with I-cells, a factor $kV_{I_1}(j-s)/nV_{TE}$ is introduced: it is equal to 1 only if a spike of I-cells takes place and it is equal to 0 otherwise. The parameters k and n are normalization constants and V_{TE} is the maximum potential value for E-cells.

$$V_E(j) = V_E(j-1) + r\Delta\varphi - \frac{kV_{I_1}(j-1)}{nV_{TE}}\left[V_{I_1}(j-1) + V_{I_2}(j-1) - \left(V_{O_1}(j-1) + V_{O_2}(j-1-\Delta\varphi)\right)\right] \quad (1)$$

Figure 2 explains how the model parameters are determined using basic experimental data [2]. Since the principal task of this work is modelling of the period and type of oscillations under various conditions, the scheme realized in Eq. 1 connects the discrete measurable quantities: spike amplitude, phase shift, rate of the membrane potential's growth and period of interspike intervals for E-cells.

Due to the discrete nature of the model, we are not interested in reproducing of the shape of the oscillations. Therefore, the growth of voltage from its minimal value ($\min(V_E)$) to the maximum ($\max(V_E)$) is considered as a linear map $V_E(j) = V_E(j-1) + r_0$. The ratio of two sides (the difference of maximal and minimal voltages, which are standard characteristics of a neuron, and the measured period of oscillations) of the right triangle obtained by this map provides the growth rate r_0. The experiments also show that the period T of E-cell and I-cell spikes and the phase shift between O-cells spikes $\Delta\varphi$ are correlated. Thus, the constant r is finally determined via linear correlation analysis applied to the experimental data.

The third term in Eq. 1 describes the repolarization process due to inhibition from the spiking I- and O-cells. This feedback loop is organized via the coupling described by Eq. (2). The latter states that I-cells fire when the E-cell voltage reaches its maximum, more precisely, during the subsequent iteration. At the same time, since the "time steps" are integer numbers j and the E-cell's intermediate voltage line slope is not, the target gap $\Delta V_E = r$ is slightly extended by the introduction of a small additive phenomenological constant $\varepsilon = 0.1$. It provides more robust calculations.

Finally, the rest constants n and k are chosen in such a way that the voltage intervals of O- and I-cells as well as the repolarization by subtracting them from the E-cell voltage satisfy experimental ranges.

2.2 Discrete Equation for I-Cells

$$\left.\begin{array}{l} V_{I_1}(j) = nV_{TE}, \\ V_{I_2}(j) = nV_{TE} \end{array}\right\} \quad \text{if} \quad V_{TE} < V_E(j-1) < V_{TE} + (1+\varepsilon)r \tag{2}$$

and

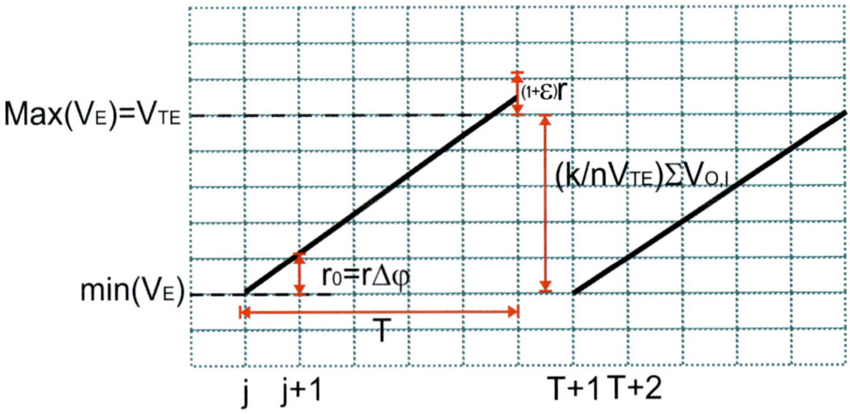

Fig. 2 Diagram demonstrating the interconnection of observable quantities (the minimal and maximal voltages, and the period of oscillations for E-cells) and parameters of the model

$$\left.\begin{array}{l} V_{I_1}(j) = V_{I_1}(j-1) - V_{I_2}(j-1), \\ V_{I_2}(j) = V_{I_2}(j-1) - V_{I_1}(j-1) \end{array}\right\} \quad \text{otherwise.} \tag{3}$$

Equations (2) and (3) describe the excitation of I-cells and the mutual inhibition of two I-cells, respectively; it is expressed as the difference between the potentials at the previous step. The factor $1 + \varepsilon$ multiplied by r is introduced to correct possible small errors emerging from the size of the discrete time steps.

2.3 *Discrete Equation for O-Cells*

$$\left.\begin{array}{ll} V_{O_1}(j) & = nV_{TE}, \\ V_{O_2}(j-T) & = nV_{TE} \end{array}\right\} \quad \text{if} \quad V_E(j) < V_E(j-1)\,\&\,\text{first event} \tag{4}$$

and

$$\left.\begin{array}{l} V_{O_1}(j) = V_{O_1}(j-T), \\ V_{O_2}(j) = V_{O_2}(j-T) \end{array}\right\} \quad \text{otherwise.} \tag{5}$$

Equation (4) describes the excitation of O-cells at the first occurrence of the V_E spike (first event). It is expressed via the conditions $V_E(j) < V_E(j-1)$. The second cell fires with a phase shift $\Delta\varphi$ (the condition $V_{O_1}(j - \Delta\varphi) \neq 0$).

Equation (5) describes a self-sustained oscillator: after the first firing, its current state is simply a repetition of its state T time steps before.

2.4 *Realization of the Simulations*

Equations (1)–(5) are simulated using a simple iteration algorithm in MATLAB. To avoid problems related to the delays in (1) and (5), the first $T + \Delta\varphi + 1$ elements of all arrays are filled with zeros. Thus, the first real simulated output determined by the dynamics of the network corresponds to $j = T + \Delta\varphi + 1$.

The parameters n, V_{TE}, r, T, s were adjusted to the in vitro experimental parameters (amplitude and period of oscillations) of individual cells [2] and kept fixed during all simulations. Thus, the phase shift $\Delta\varphi$ is the only control parameter.

3 Results

The important property of the system (1)–(5) is based on the phenomenological fact [4] of the strict dependence of the E-cell oscillation period on the phase shift between spikes of O-cells [2]. This is the reason for explicitly including this shift $\Delta\varphi$ into the term which provides the linear voltage in Eq. (1). We do not aim speculate regard-

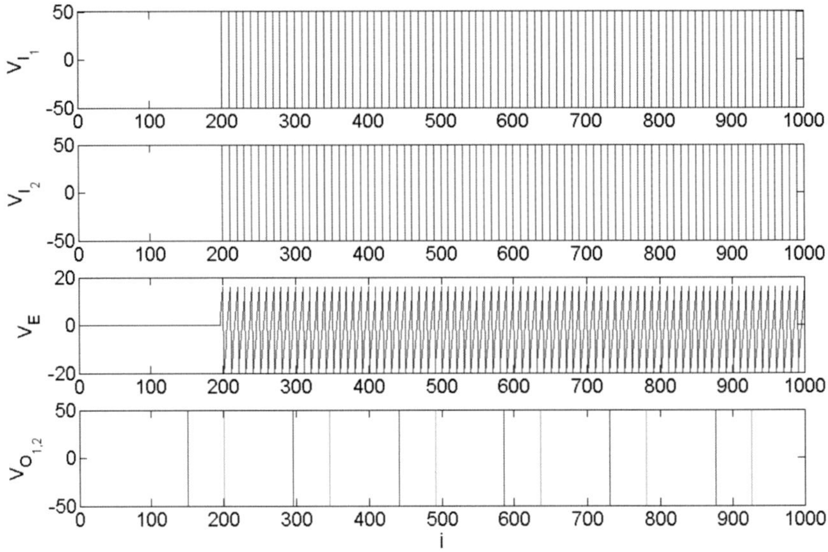

Fig. 3 The regime of gamma oscillations corresponding to the control parameter value $\varepsilon = 0$. Here *light* and *dark* (*green* and *blue* in the color online version) *lines* mark to the spikes of O_1 and O_2 cells, respectively

ing a microscopic mechanism for such a behaviour but to reveal the background for the transitions between theta-, gamma-, and theta-gamma-oscillations by taking this quantity as a key control parameter.

During the simulations, we use a phase shift correction $\Delta\varphi = [\Delta\varphi_0 - \varepsilon n V_{TE}]$, where square brackets denote the integer part (one needs this construction since the phase shift must be equal to an integer number of discrete time steps) and V_{TE} is the maximal voltage available for an E-cell before a discharge. Such a scheme mimics the cell interactions drawn in Fig. 1 and has a certain neurophysiological justification (see the discussion of voltage amplitude and phase properties in linked neurosystems in [5]).

Thus, we show below that various dynamical regimes are regulated by varying the coupling intensity ε. The rest parameters are the same for all simulations: $k = 0.2$, $r_0 = 0.08$, $n = 10$, $\Delta\varphi_0 = 50$, $T = 145$, $V_{TE} = 10$.

Figure 3 demonstrates gamma oscillations typical for the default dynamics of connected cell spikes, i.e. in absence of an additional influence of E-cell voltage amplitude, $\varepsilon = 0$. One can see that the default phase difference between O-cells spikes as well as their intrinsic period is sufficiently larger than the period of E-cell oscillations. As a result, the dynamics of an E-cell is particularly simple: the voltage grows up to the maximum value, discharges and this process repeats periodically. Thus, we see uniform saw-like dynamics for an E-cell and equidistantly distributed bars marking spikes of O- and I-cells.

The case of sufficiently larger values of the phase-amplitude coupling parameter ($\epsilon = 0.4$, see Fig. 4) leads to a decrease of $\Delta\varphi$. This is reflected in the graph where the spikes of the two O-cells are closer to each other than in Fig. 3. Consequently,

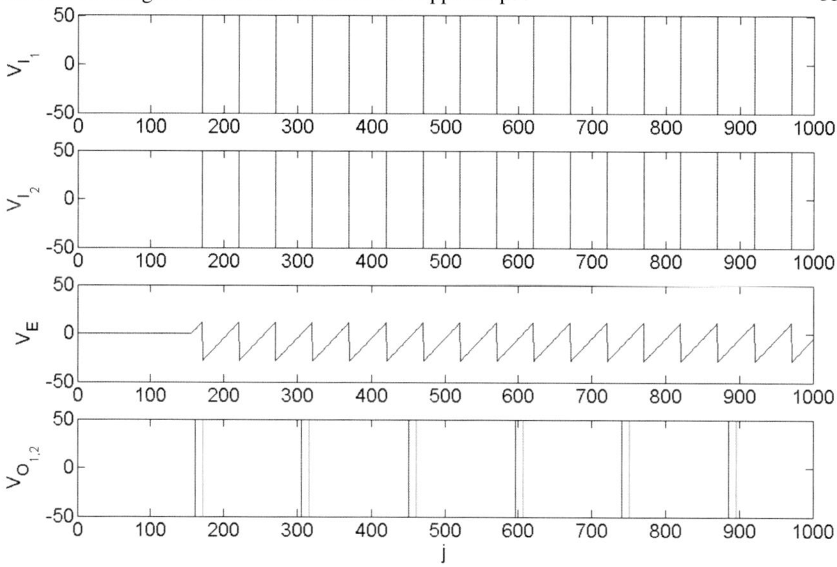

Fig. 4 The regime of theta oscillations corresponding to the control parameter $\varepsilon = 0.4$

it takes more time for E-cell voltage to reach its amplitude value. As a result, theta-oscillations emerge, see Fig. 4. Note also that the periods of I-cells and O-cells are still incompatible. They do not affect each other (see the expression in the brackets in Eq. (1)) and there are no intermittent dynamics in the system.

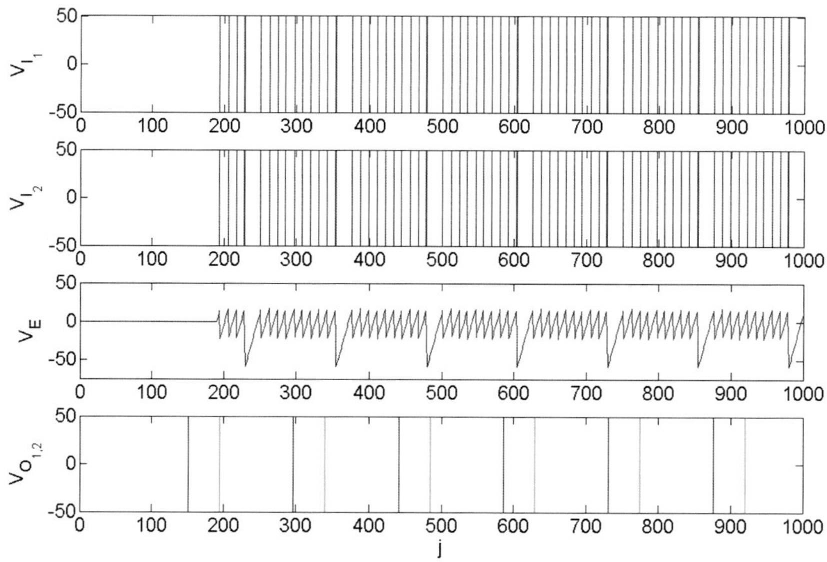

Fig. 5 The regime of theta-gamma oscillations corresponding to the control parameter $\varepsilon = 0.06$

The case of compatible oscillations periods for E-cell and both O-cells results in the modulated theta-gamma regime, see Fig. 5. The value of the coupling parameter, $\varepsilon = 0.06$ is small, thus the basic periodic regime is gamma. Each discharge of an E-cell produces an I-cell spike. Their superposition with the oscillations of O-cells causes a larger deactivation of the E-cell, i.e. the "gaps", the periods of which correspond to theta-oscillations. The fine adjustment of these processes in our model originates from the chosen discrete formulation of Eqs. (2)–(5) leading to a representation of the corresponding spikes as point-wise localized.

4 Discussion and Conclusion

Although the majority of works in computational neuroscience is based on the well-developed theory of non-linear ordinary differential equations varying from detailed Hodgkin-Huxley-like systems to simplified ones (FitzHugh-Nagumo, integrate-and-fire, Morris-Lecar, etc.) [6], the map-based models start to attract the attention of researchers (see the comprehensive review of the recent state-of-art in [7]). However, a sufficient part of the systems reviewed are obtained either via discretization of the standard ODEs or via the introduction of their discrete counterparts (non-linear maps). This technique allows to obtain a variety dynamical regimes provided by the corresponding non-linear maps (regular, chaotic oscillations, bursting, etc.). However, these systems keep a high complexity for analytical study; the absolute majority requires an extensive numerical bifurcation analysis. The consideration of connected lattices and networks of such elements increases complexity and difficulties in devising of systems with prescribed properties and their adjustment to neurophysiological experiments.

On the other hand, the approach proposed in the present work has an advantage in the easier correspondence to measurable properties of neuronal cell systems. In fact, the experimentalists do deal with the timing of firing for different types of neurons that determines an attention to phase resetting approaches in their modelling (see e.g. [8, 9]). At the same time, there is evidence that a pure phase description may not be enough for the correct description of real neurons' dynamics and one needs to take into account its interplay with the voltage amplitudes (see the discussion in the review [5]).

In the present work, we combine both these factors in the most direct way: the basic oscillatory properties of all three types of cells are primarily postulated via linear delayed mapping. Its parameters are determined from experimental data since we do not have the aim to model individual oscillators (it is a known solved problem). Therefore, we can concentrate on the study of the influence of connections on the dynamics of the small network block in its entirety.

We have found that two types of oscillations in the hippocampal area CA3, theta and gamma, as well as the combined theta-gamma regime, are determined by the phase shift between spikes of O-LM cells, which act as independent timers. The inhibitor input of one isolated O-LM cell (the situation with a large phase shift between both) is not strong enough to compensate the self-sustained activation of the pyramidal cell. Thus, the gamma oscillations of the latter are completely determined by its self-activation and the resetting induced by basket cells. If both O-LM cells act simultaneously (small phase shift between them), the voltage growth of the pyramidal cell is stretched and theta oscillations are produced. The most interesting combined theta-gamma regime emerges as a combination of two inhibitor inputs, which coincide in the regime of fractional order phase locking: from fast oscillating basket cells and low oscillating O-LM cells. All these details of phase regimes are confirmed by experimental results (see Fig. 5 in Ref. [2]).

Finally, this simple construction confirms the conclusion [4] about necessity of two O-LM cells in a minimal network mimicking oscillations of constitutive blocks of the CA3 region of the hippocampus. But in contrast to the citied ODE-based model, our discrete system allows further to form larger networks of connected blocks taking in mind the revealed rules for connections determining switching between oscillatory states.

Acknowledgements We thank Prof. I.M. Sokolov (Humboldt-Universität zu Berlin) for the useful comments on this work.

References

1. P. Andersen, *The Hippocampus Book* (Oxford University Press, 2007)
2. T. Gloveli, T. Dugladze, H.G. Rotstein, R.D. Traub, H. Monyer, U. Heinemann, M.A. Whittington, N.J. Kopell, Orthogonal arrangement of rhythm-generating microcircuits in the hippocampus. Proc. Natl. Acad. Sci. USA **102**, 13295 (2005). https://doi.org/10.1073/pnas.0506259102
3. A.B.L. Tort, H.G. Rotstein, T. Dugladze, T. Gloveli, N.J. Kopell, On the formation of gamma-coherent cell assemblies by oriens lacunosum-moleculare interneurons in the hippocampus. Proc. Natl. Acad. Sci. USA **104**, 13490 (2007). https://doi.org/10.1073/pnas.0705708104
4. A.I. Lavrova, M.A. Zaks, L. Schimansky-Geier, Modeling rhythmic patterns in the hippocampus. Phys. Rev. E **85**, 041922 (2012). https://doi.org/10.1103/PhysRevE.85.041922
5. P. Sauseng, W. Klimesch, W.R. Gruber, S. Hanslmayr, R. Freunberger, M. Doppelmayr, Are event-related potential components generated by phase resetting of brain oscillations? A critical discussion. Neuroscience **146**, 1435 (2007). https://doi.org/10.1016/j.neuroscience.2007.03.014
6. G.B. Ermentrout, D.H. Terman, *Mathematical Foundations of Neuroscience* (Springer, Heidelberg, 2010)
7. B. Ibarz, J.M. Casado, M.A.F. Sanjuán, Map-based models in neuronal dynamics. Phys. Rep. **501**, 1 (2011). https://doi.org/10.1016/j.physrep.2010.12.003
8. R.F. Galán, G.B. Ermentrout, N.N. Urban, Efficient estimation of phase-resetting curves in real neurons and its significance for neural-network modeling. Phys. Rev. Lett. **94**, 158101 (2005). https://doi.org/10.1103/PhysRevLett.94.158101
9. S. Achuthan, C.C. Canavier, Phase-resetting curves determine synchronization, phase locking, and clustering in networks of neural oscillators. J. Neurosci. **29**, 5218 (2009). https://doi.org/10.1523/JNEUROSCI.0426-09.2009

Part VII
Additional Topics

I See What You Do Not See

The World Observed with High-Speed Cameras

Kinko Tsuji

Abstract High-speed cameras are very useful and important tools for studying fast phenomena that we cannot follow with our own eyes. Recent developments of photosensitive semiconductors with memory (sensors with on-chip memory) enable us to take images of up to 10 million frames/s without losing spatial resolution. We explain the principles of these sensors (a CCD (charge coupled device) sensor and a CMOS (complementary metal oxide semiconductor) sensor, both with on-chip memory), and their trigger systems for high-speed cameras. As typical applications the following four experiments are shown: 1. material testing: destruction processes of a piece of CFRP (carbon fiber reinforced plastic) by tensile strain observed with two cameras synchronously, 2. cavitation: behavior of a microbubble in various geometries of their environment, 3. shock waves: their propagation observed with a Mach-Zehnder interferometer, and 4. crack formation: the process of glass cracking. Applications to chemical, biological and medical fields are on-going promising projects.

1 Introduction

The human eye can detect changes with a maximum of 10–20 Hz. Therefore, we are not disturbed by fluorescent lamps which work with a frequency of 50/60 Hz. And we can enjoy "continuous" motions in classical movies using a frequency of 16 Hz.

How can we then observe fast phenomena which we cannot directly see with our eyes? Since the middle of the 19th century people have tried to take high speed photographs. The first practical images were taken by Eadweard Muybridge in 1887 [1, 2]. He used 12 cameras and produced a series of images of a galloping horse. At that time, there was an exciting discussion on whether at least one of the four feet touches the ground during galloping. His photographs showed that there is a moment when all four legs are in the air. Moreover, they showed that very complicated coordination of

K. Tsuji (✉)
Shimadzu Europa GmbH, Albert-Hahn-Str. 6-10, 47269 Duisburg, Germany
e-mail: kts@shimadzu.eu

S.C. Müller et al. (eds.), *Complexity and Synergetics*,
https://doi.org/10.1007/978-3-319-64334-2_27

the legs is involved: there is no symmetry between front legs and hind legs, and no synchronized movement either of the two front legs or of the two hind legs.[1]

Various techniques and devices for taking high-speed images have been developed during the following decades: using stroboscopes, rotating prisms, rotating mirrors, streak cameras, high speed mechanical shutters, and others. One excellent example is the observation of cavitation bubbles with a rotating mirror technique by Lauterborn and Bolle [3]. They took a few images of the cavitation bubbles with 300,000 frames/s.

In parallel, the technique of capturing still images was drastically changed from analog (using films with photosensitive substances, e.g., silver halide) to digital (using photosensitive semiconductors). Since then the combination of photosensitive semiconductors and some of the techniques for high-speed imaging (e.g., rotating mirrors) became a fundamental tool for high-speed cameras. Among various semiconductors, CCD (charge coupled device) and CMOS (complementary metal oxide semiconductor) are photosensitive and therefore frequently used.

However, even with this combination, there are difficulties to take a series of images with both high temporal resolution and high spatial resolution (some exceptions: e.g., a huge Brandaris camera, 150 cm (w) $\times$ 150 cm (h) $\times$ 20 cm (d), 140 kg [4]), because the rate determining step of such cameras is the read-out time of data: usually the rate of data transfer from the image sensor to a device for data storage (e.g., a PC). Each camera has a maximum amount of data transferred per second, depending on sensors, electric circuits and storage devices. Therefore, when the number of frames per second (the temporal resolution) is increased, the number of pixels in each frame (the spatial resolution) is reduced. In an extreme case a high-speed image consists of only a few lines, and such images are often not very useful for applications.

In order to avoid sacrificing the spatial resolution at high frame rate (and to remain compact in size), image sensors with a storage function (on-chip memory) have been developed. Different from the conventional image sensors without on-chip storage, however, the total number of frames is limited, because of the available storage space on the chip.

In this presentation we will briefly explain the principles of cameras with on-chip memory (Sect. 2), and show in Sect. 3 what we can see with these cameras: for example, how a material is deformed/damaged, how a cavitation bubble behaves during shrinkage, how a shock wave develops, or how cracking in glass occurs. In the last section (Sect. 4) we will describe some possible applications towards future use in the field of chemical, biological and medical applications. Note that what we describe here is limited to some examples, which are only a small part of works done by many people.

---

[1]Perhaps Gioacchino Rossini knew this acoustically before Muybridge. He composed the William Tell Overture in 1829. The rhythm of galloping horses in this overture reveals how their legs are moving.

2 Principles of High-Speed Cameras with On-Chip Memory

We will now explain two different image sensors (CCD and CMOS) with on-chip memory and their trigger functions in order to understand for which kind of application these cameras are useful (or limited). An appropriate trigger is essential for optimizing the limitation of the number of frames.

2.1 *CCD Image Sensor with On-Chip Memory [5]*

On the CCD chip each photo-sensitive element, P, has storage elements, the number of which corresponds to the number of frames. As shown in Fig. 1a, photons reaching P are converted to electron charges and transferred to the first storage element 1. When photons arrive newly at P during the next time frame, these photons are newly converted to electron charges and are again transferred to the first storage element 1. At the same time the charges which had occupied the first storage element are transferred to the second storage element 2. During the "image capturing" mode these processes are repeated, and charges for each frame are continuously transferred from one element to the next. Charges arriving at the last storage element n are drained (D). When a trigger signal arrives, the charge transfer is stopped: it means that the information of the amount of photons reaching P at every frame is fixed. Then the amount of electron charges in each storage element is read out into an external storage device (e.g., a PC).

2.2 *CMOS Image Sensor with On-Chip Memory [6]*

In this sensor photons are captured by the photo-sensitive element, P, converted to photoelectrons, further to voltage and stored in the storage element. The signal reading-out process from P to on-chip memory takes less than 10 ns. Different from the CCD image sensors no charge transfer from one storage element to the next element occurs. The voltage signal for the first frame is kept in the storage element 1, for the second frame in the storage element 2 and further on (see Fig. 1b). The number of the storage elements for each P corresponds to the number of frames. When the storage element n is occupied, the signal stored in element 1 is overwritten by the next signal (n + 1). During the image capturing mode, the memory is overwritten one by one in each frame. When a trigger signal is sent, the image capture is stopped and the signals in each storage element are read out. A big advantage of the CMOS sensor with on-chip memory compared with the CCD sensor with on-chip memory is less heat dissipation, and subsequently, less energy consumption: on the CMOS sensor only one storage element per P is driven at each frame, while on the CCD sensor all storage elements are driven for transferring electrons from one storage element to the next.

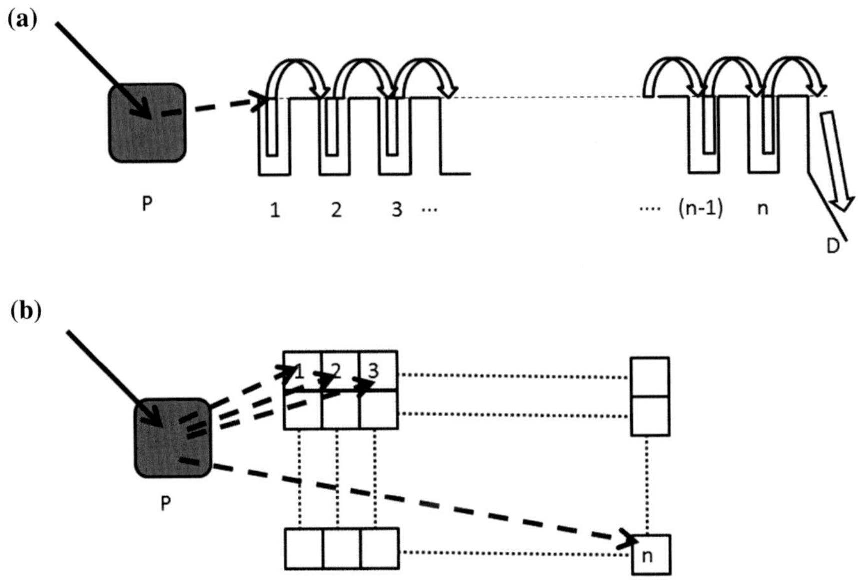

Fig. 1 **a** On-chip memory of a CCD image sensor, **b** on-chip memory of a CMOS image sensor. P, photosensitive element; 1, 2, 3, …, n are memory storage elements; D, drain

2.3 Trigger

For high-speed image capturing, a trigger is important for taking images at the right moment. Especially when the number of frames is limited such as in the case of on-chip memory, an appropriate timing of the trigger is essential. As mentioned above images are captured continuously, till a trigger signal arrives. When this signal arrives, image capturing and overwriting processes are stopped. Afterward the image signals of all storage elements are read out one by one. Usually a TTL (Transistor-Transistor-Logic) signal from electronic, optical, acoustic or other sensors is used as a trigger signal. It is practical to set a variable delay to stop the image capturing process after an input of the TTL signal. It is also possible to set a number of frames (m), during which images are still captured after the trigger signal arrives. For example, if the image capturing is stopped immediately (m = 0), all signals kept in the memory are those for the phenomena happening before the trigger (pre-trigger mode: case (a) of Fig. 2). If the image capturing continues for n frames (m = n) after the trigger, then images of n frames after the trigger are recorded (post-trigger mode: case (b) of Fig. 2). The image capturing can be stopped at any frame within n frames (case (c) of Fig. 2). Thus, the on-chip memory has an additional advantage regarding such flexible trigger modes.

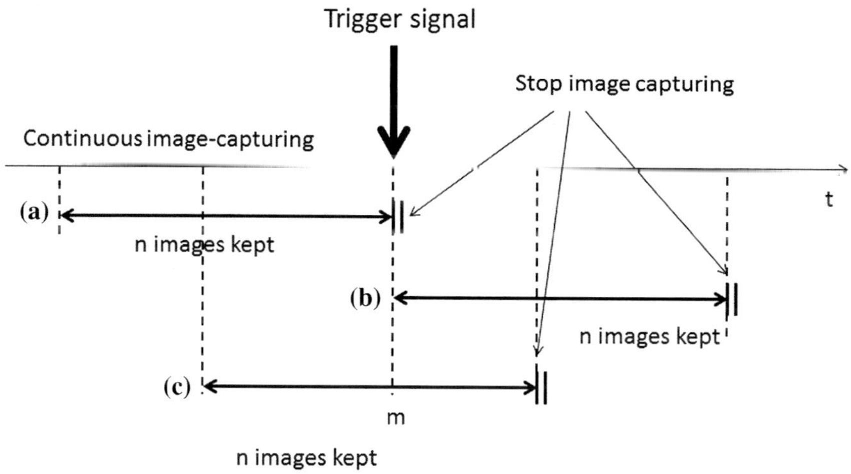

Fig. 2 Flexible trigger modes. **a** Pre-trigger mode, n images before the trigger are kept for read-out, **b** post-trigger mode, n images after the trigger are kept for read-out, **c** intermediate-trigger mode, m images before the trigger and (n-m) images after trigger are kept for read-out

3 Applications

3.1 Why Do We Need a High-Speed Camera?

- Temporal resolution

The human eye can detect changes of 10 Hz. This corresponds to a camera which takes images every 0.1 s. If we like to see the light intensity changes of a fluorescent lamp lighted by a power source with alternating current of 50 Hz, we need a camera which works with a time interval shorter than 0.02 s, as long as the lamp itself does not move. If the change of the light intensity is faster, then we need a correspondingly faster camera: for instance, for observing light intensity changes of 10^6 Hz, we need a camera which records images faster than every 1 μs.

- Spatio-temporal resolution

When an object is moving (relative to the camera), then we have to take a spatial factor into consideration. In order to illustrate such a case, we often show a movie of a fast train passing with 180 km/h a camera fixed on the platform. With a conventional video camera you see a kind of band moving, whereby you maybe detect segments of wagons. Here the question is: With which frequency do you have to take images in order to see faces of passengers in this train?

In our example the train moves in one direction with 50 m/s (=180 km/h). Human eyes with the temporal resolution limit of 10 Hz can observe objects in the train larger than 5 m length along the train. If you have to detect an object of 1 cm length (which is necessary for the recognition of human faces), a frequency 500 times larger (5000 Hz corresponding 0.2 ms) is required.

- Microscopic observation

When a small particle is moving with a velocity of 0.01 m/s, you can observe this movement with your own eyes, as far as the particle is large enough to be seen. However, if the particle is so small that you have to magnify it with a microscope, then the observation of even such a slow movement (0.01 m/s) is not any more so obvious. Under a microscope an apparent velocity is magnified, depending on the magnification factor. If the magnification factor is 100, the apparent velocity of the moving object with 0.01 m/s is 1 m/s. Moreover, the field of view gets small, also depending on the magnification.

3.2 *Material Tests*

One of the most familiar applications of high-speed cameras is material testing. It is essential for new materials to test how stable (or weak) they are under various stresses. A carbon-fiber-reinforced polymer (CFRP), for example, is known to be a strong and light material and, therefore, is used often for airplanes or racing cars. Since CFRP is brittle, fracture occurs instantaneously (for our eyes). A frame rate faster than 100,000 frames/s is required for observing the destruction process. In the experiment shown in Fig. 3, a piece of CFRP (l: 150 mm, w: 36 mm, d: 2.5 mm) with a hole of 6 mm diameter at the center was tested. The CRFP piece was set in the tensile testing machine (AG-X plus, Shimadzu Corp., Japan) and pulled with 100 kN and a speed of 5 mm/min. A front view (Fig. 3a) and a side view (Fig. 3b) are documented synchronously with two cameras (HPV-X2, Shimadzu Corp., Japan) [7]. The direction of the strain is vertical. As shown in Fig. 3a the cleavage

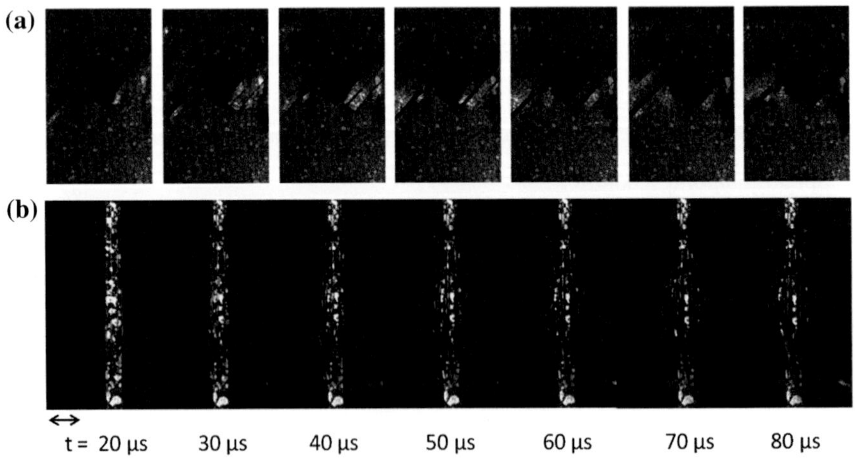

Fig. 3 Fracture observed in the tensile test of a CFRP piece with a hole at the center. Synchronous imaging of **a** front view and **b** side view. Camera speed: 200,000 frames/s, scale bar: 5 mm

starts from the vicinity of the hole along the alignment of the surface layer of fibers (at 45° to the strain direction). The detailed evolution of the cleavage pattern can be seen in the subsequent time series. The side view (Fig. 3b) indicates additionally how the fibers are raveling.

3.3 Cavitation

This is a typical example of high-speed imaging with a microscope. Zwaan et al. studied single cavitation bubbles oscillating in a confined fluid environment [8]. A single bubble is generated in PDMS (polydimethylsiloxine)-based microsystems by using a pulsed Nd-Yag laser (532 nm, 5 μJ, 6 ns pulse duration) in a light-absorbing liquid. The bubble dynamics is recorded with 1 million frames/s (HPV-1, Shimadzu Corp., Japan). The behavior of the flow can be visualized by tracer particles of 2 μm diameter. Figure 4 shows the dynamics of a cavitation bubble in various geometries of its environment. When there is no boundary close to the bubble, the cavitation process is circular symmetric in both expansion and shrinkage (Fig. 4a). If there is a pin in the neighborhood, then the circular symmetry is broken in the vicinity of the pin, and a jet is observed towards the pin (Fig. 4b). When a bubble is created in a triangular or square boundary, the shrinking bubble forms a three-leaf or four-leaf clover as shown in Fig. 4c and d. Jet-like flows emerge from each corner. Another example is the case close to the surface of a single wall (Fig. 4e). A very fast process of expansion (within 4 μs) is followed by a slower collapsing process (~10 μs), where a jet towards the wall is observed. The velocity of the flow here reaches a maximum of 7.4 m/s. It is known that screw propellers of ships are damaged by small bubbles which are formed on the surface during fast rotating motion. The jet towards the surface of propellers could be an explanation of these damages.

3.4 Shock Waves

Processes in transparent media with changing density can be visualized by several methods, e.g., the Schlieren method, the shadow method and interferometry. Figure 5 shows the dynamic behavior of shock waves exiting from an open-end shock tube (inner diameter 20 mm) as a sequence of Mach-Zehnder interferograms, taken with the HPV-1 [9]. The initial shock speed in the tube is 465 m/s. The shock wave quickly attains a hemispherical shape (t = 40 μs) and continues to expand at a velocity of 370 m/s, which is slightly higher than the sound velocity. When the exiting shock diffracts around the corner of the shock tube, it induces a prominent vortex ring. The influence of different exit geometries with various initial shock speeds is under further investigation.

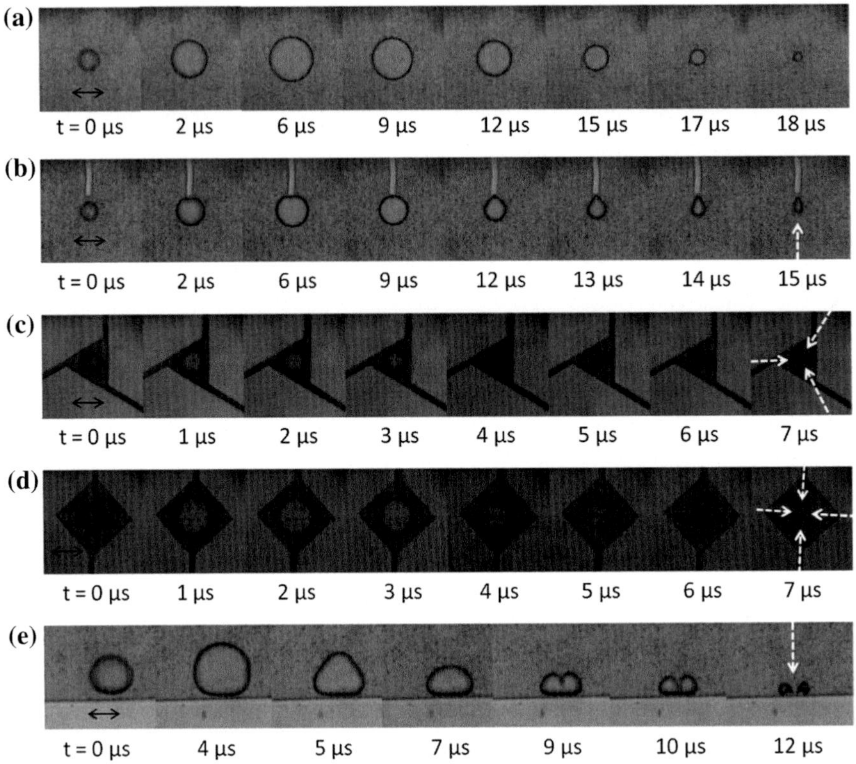

Fig. 4 Microbubble dynamics in various geometries of their environment. A bubble is created **a** in a boundary-free environment, **b** close to a pin, **c** in a triangular wall, **d** in a square wall, **e** close to a wall of a channel. The *dashed arrows* indicate the direction of the jet flows. Camera speed: 1 million frames/s, scale bar: 100 μm

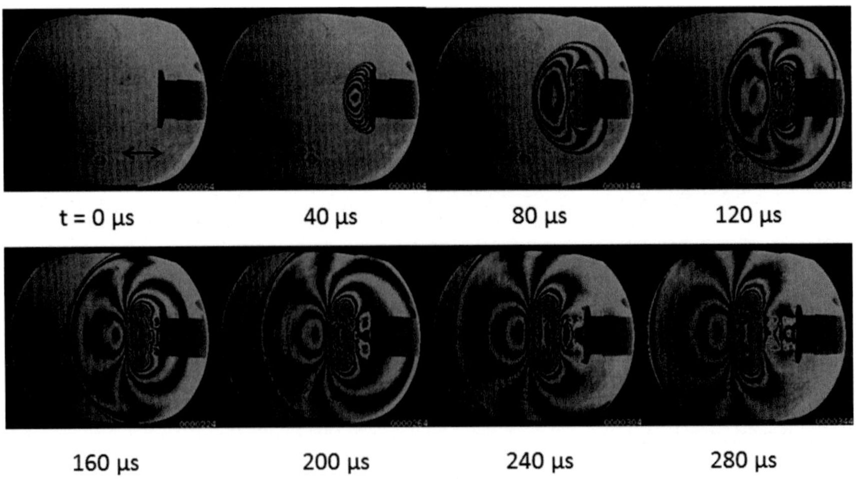

Fig. 5 Shock waves exiting from an open-end shock tube, observed by a Mach-Zehnder interferometer. Camera speed: 250,000 frames/s, scale bar: 20 mm

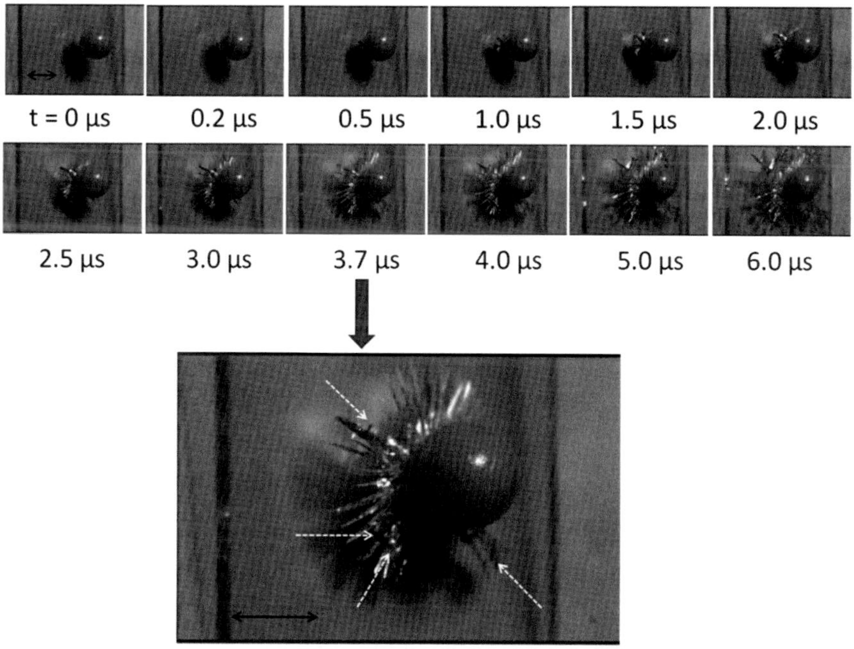

Fig. 6 Observation of glass cracking. *Dashed arrows* point to the start of dendrite formation. Camera speed: 10 million frames/s, scale bar: 5 mm

3.5 Glass Cracking

Crack formation in glass is a very fast process. Figure 6 shows the development of cracks on a slide glass for microscopes, when a polystyrene ball of 6 mm diameter collides with it at a velocity of 50 m/s, taken by 10 million frames/s with a high speed camera (HPV-X, Shimadzu Corp., Japan) [10]. The length of a crack (a selected crack of medium length) and the velocity of its development are plotted against time in Fig. 7a and b, respectively. The graphs were drawn up to 7 μs, because some cracks reach the edge of the slide glass at 7 μs. The average velocity of the crack development is about 1100 m/s. The small minimum of the velocity between 3 and 4 μs could be associated with the formation of dendrites. However, it is necessary to analyze the data statistically for clarifying this point.

4 Towards the Future

Many applications of high-speed cameras, some of which are described in Sect. 3, are in the field of physical changes of non-living objects. Recently, some people challenge to visualize chemical reactions, biological processes and medical

Fig. 7 Length of one of the cracks (**a**) and the velocity of its development (**b**) plotted versus time

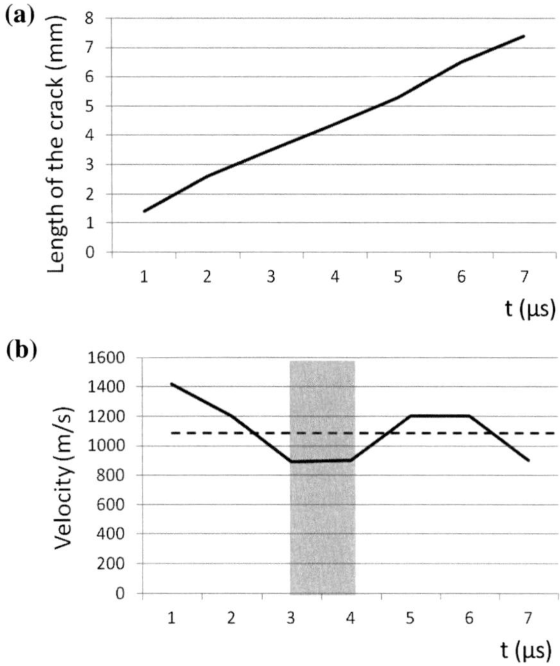

treatments, although this is not so trivial. In order to observe chemical reactions one should choose systems where any kind of optical change is involved. It is further critical how to initiate the reaction in the observation field and how to trigger the camera. Tsuji and Müller investigated an early stage of color change caused by a droplet containing a pH indicator when impinging on alkaline solution [11]. High-speed observations of up to 16,000 frames/s reveal a finger formation at the reaction front, suggesting an instability at the boundary. Le Gac et al. studied how cancer cells are killed with a single cavitation bubble in a microfluidic system [12]. Here one should be careful not to kill the cells by other effects like strong illumination or heat. One example for a medical application is dental root canal cleaning through cavitation and microstreaming [13]. Here the cleaning efficiency of needle irrigation, ultrasonic irrigation and laser-activated irrigation was studied by using artificial glass root canal models. So far it has not yet been achieved to observe such fast phenomena in a living body. It would be interesting to see, for example, how kidney stones are destroyed by ultra sound.

Some of these difficulties could be overcome if techniques of high-speed imaging were developed further, for example by increasing the number of frame sequences, by increasing sensitivity or by triggering with a different concept (e.g., an image trigger—triggered by changes in the image itself).

Finally, we would like to compare the development of high-speed cameras with the development of microscopes: the former is a tool to obtain better temporal resolution, while the latter is a tool to obtain better spatial resolution.

Since traditional optical microscopes can detect only objects larger than the wavelength of visible light, people were not able to see most viruses, the size of which is about 100 nm or less. Therefore, before the electron microscope was invented, there was no way for people to verify the existence of viruses, even though they knew something smaller than bacteria causing diseases had to exist. Only after the invention of electron microscopes, they were able to observe them. For the temporal resolution, the invention of high-speed cameras with a time resolution larger than 1 million frames/s corresponds to the invention of the electron microscope. There would be many fast movements/phenomena/steps which we have not observed yet. It is possible that we will be able to observe something that we never knew exists. We are sure that high-speed cameras contribute to further developments in science and technologies for human beings and the environment.

Acknowledgements We are grateful to Nobuyuki Tokuoka for checking of the technical details on high-speed cameras. We thank Stefan C. Müller and Sho Tsuji for critical reading of the manuscript.

References

1. M. Leslie, The man who stopped time, *Stanford Magazine* (May–June 2001)
2. P. Oridger, *Time Stands Still: Muybridge and the Instantaneous Photography Moment* (Oxford University Press and Stanford University, Stanford, 2003)
3. W. Lauterborn, H. Bolle, Experimental investigations of cavitation-bubble collapse in the neighbourhood of a slid boundary. J. Fluid Mech. **72**, 391–399 (1975)
4. C.T. Chin, C. Lancee, J. Borsboom, F. Mastik, M. Frijlink, N. de Jong, M. Versluis, D. Lohse, Brandaris 128: a digital 25 million frames per second camera with 128 highly sensitive frames. Rev. Sci. Inst. **74**, 5026–5034 (2003)
5. T.G. Etoh, D. Poggeman, G. Kreider, H. Mutoh, A.J.P. Theuwissen, A. Ruckelshausen, Y. Kondo, H. Maruno, K. Takubo, H. Soya, K. Takehara, T. Okinaka, Y. Takano, An image sensor which captures 100 consecutive frames at 1 000 000 frames/s. IEEE Trans. Electron. Dev. **50**, 144–151 (2003)
6. Y. Tochigi, K. Hanzawa, Y. Kato, R. Kuroda, H. Mutoh, R. Hirose, H. Tominaga, K. Takubo, Y. Kondo, S. Sugawa, A global-shutter CMOS image sensor with readout speed of 1–T pixel/s burst and 780-M pixel/s continuous. IEEE J. Solid-State Circ. **48**, 329–338 (2013)
7. Shimadzu Corporation, The fraction observed in the tensile test of open hole CFRP—synchronized imaging with 2 high speed video cameras, *Shimadzu Application News* **V19** (2015)
8. E. Zwaan, S. Le Gac, K. Tsuji, C.-D. Ohl, Controlled cavitation in microfluidic systems. Phys. Rev. Lett. **98**, 254501 (2007)
9. H. Kleine, H. Olivier, K. Tsuji, K. Takehara, K. Etoh, and T.G. Etoh, Time-resolved Mach-Zehnder interferometry, in *Abstracts of 14th International Symposium on Flow Visualization*, EXCO, Daegu, Korea, 20–24 June 2010
10. https://www.shimadzu.eu/applications
11. K. Tsuji, S.C. Müller, Chemical reaction evolving on a droplet. Phys. Chem. Lett. **3**, 977–980 (2012)
12. S. Le Gac, E. Zwaan, A. van den Berg, C.-D. Ohl, Sono-poration of suspension cells with a single cavitation bubble in a microfluidic confinement. Lab Chip **7**, 1666–1672 (2007)
13. B. Verhaagen, *Root Canal Cleaning through Cavitation and Microstreaming*, Dissertation, University of Twente (2012)

Performative Science—Transgressions from Scientific to Artistic Practices and Reverse

Hans H. Diebner

Abstract *Performative Science* is a research practice that is inherently processual and takes up a first person stance on epistemic things. It bears resemblance to phenomenological, that is to say, artistic practices and is most notably characterised by its reflective power. The current era of harsh reification of the meaning of *Being* in the course of an ongoing cybernetisation demands for such a reflection. *Performative Science* is a veritable attempt to turn technology into a hermeneutic practice, and if possible, even as part of an hermeneutics of facticity, that is to say, the inherently temporal understanding of (human) *Being*. A few concrete applications are introduced that partially have self-deconstructive features in order to emphasise the radical openness of *Performative Science*, that is far from being a scientific method. Rather, a method becomes emergent by tuning oneself into the epistemic thing.

1 Enacting the Performative

Performative Science started in 1999 as an embodied epistemology bridging practices from the arts and the sciences in order to come close to an enacting of a performative dimension, i.e., to performativity, within scientific research [1]. Performativity is here understood as opposed to semioticity, i.e., disjunct to any formally signifying concept (the libretto, if you like), thus subject to a first-person-perspective only. Initially, it shared many ideas with other fast-growing art and science move-

[1]"Enframing" refers to Heidegger's notion of "Gestell" with which he characterised the "truth-obstructing" essence of technology.
[2]Heidegger's notion of "Machenschaft" refers to the scientific practice to conceive Being from an inventory, computational or instrumental point of view.

H.H. Diebner (✉)
Faculty of Medicine Carl Gustav Carus, Institute for Medical Informatics and Biometry, Technical University Dresden, Dresden, Germany
e-mail: hans@diebner.de

S.C. Müller et al. (eds.), *Complexity and Synergetics*,
https://doi.org/10.1007/978-3-319-64334-2_28

Fig. 1 Series of film stills showing the interaction of spectators with the reactive installation *Liquid Perceptron* taken during the opera enactment *Einstein on the Beach* [6]

ments as, e.g., the so called artistic research [2, 3]. However, it gradually turned out that the scientific partners within these hybrid practices tend to monopolize the arts, which is in line with the conception of systems sciences (including synergetics) as a unified discipline [4, 5]. Consequentially, *Performative Science* startet to cultivate a critical discourse including a reflection of its own initial constitution.

In his historical and epistemological investigations, Andy Pickering felt compelled to call the constructive approaches of complex systems sciences an "ontological theatre" [7, 8], thus putting them quasi equal to performative practices in the arts. Yet, in such an equalisation the non-propositional aspects of the arts or, using Martin Heidegger's vocabulary, the fundamental ontological basis of the arts, is buried in oblivion [9, 10]. Although Pickering claims that complexity theory escapes from the process of *enframing*,[1] it rather perfects what Heidegger called *machination*.[2] Remarkably, soon after the introduction of cybernetics and the mathematical theory of communication, Heidegger noticed that:

> No prophecy is necessary to recognise that the sciences now establishing themselves will soon be determined and steered by the new fundamental science which is called cybernetics. This science corresponds to the determination of man as an acting social being. For it is the theory of the steering of the possible planning and arrangement of human labour. Cybernetics transforms language into an exchange of news. The arts become regulated-regulating instruments of information [10, p.14, in German].

Within the scope of *Performative Science*, several "ontological stagings" have been designed and presented to the public in enactments (Figs. 1, 2) or in the context of exhibitions (Fig. 3) [1, 11]. No doubt, such stagings caused reflections about scientific results but also meta reflections about art and science including their hybridisations, which all to often mock sciences and at the same time ridicule art [12].

After all, *Performative Science* is understood in a triadic way as (i) a critical philosophy of science [5, 13, 14], (ii) a critical discourse of new media art including the art and science movement [12, 15] and, (iii) an exploration of performative modes of expression of scientific research [1]. From this perspective, the art and science synergy is unlocked neither by a unification of art and science nor by building a bridge, but rather by allowing for leaps from the enframing essence of the sciences down into the abysmal arts and reverse.

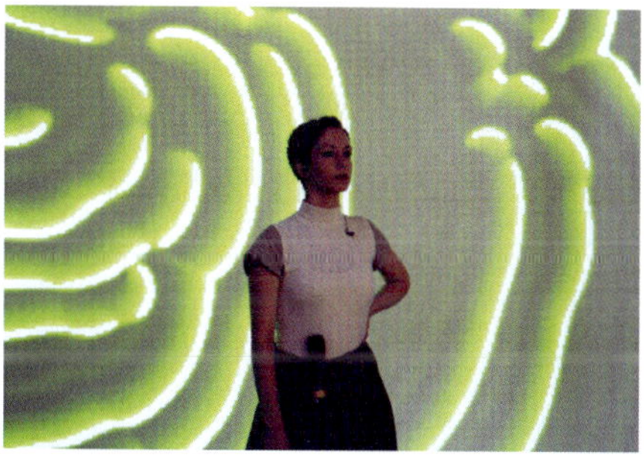

Fig. 2 Opera singer in front of the reactive installation *Liquid Perceptron*. Photo taken during the enactment of the opera *Einstein on the Beach* [6]

Fig. 3 Spectator/user of *Eye Vision Bot* as presented in the ZKM|Media Museum, Karlsruhe [37, 38]

2 Designing Freedom

The love-hate relationship between humanities and sciences has a long history [16]. Recently, the controversy reached an unprecedented peak with an extreme self-referential flavour, that is to say, the discussion is guided more by a vociferous meta dispute of whether or not art is equal to science, rather than by concrete realisations. Prominent forums advocating the art and science community are "Leonardo" [17] and the "Society for Artistic Research [...] that nurtures, connects and disseminates artistic research as specific practices of creating knowledge and insight" [3].

It is beyond the scope of this short essay to touch on all controversial aspects of art and science. Having said this, it is worth mentioning for the needs of the present critique that more and more trained scientists commit themselves to the arts which can be seen from contributions in forums as, e.g., Leonardo [17] and SAR [3] or by visiting relevant exhibitions (see [18] for an example). Surely, it is not the first time that scientists felt that their methodological "corset became to tight". Perhaps Paul Feyerabend's "anything goes" drawn from his "masterpiece" "Wissenschaft als Kunst" is the most prominent historical example. However, it seems that the term "straitjacket" is more appropriate to characterise the current uneasiness.

To give a short analysis of this situation with historical reference, we point to the deeply normative, partially even moralistic nature of early general systems theory:

> Especially the gap between natural and social sciences, or, to use the more expressive German terms, of Natur- und Geisteswissenschaften, is greatly diminished, not in the sense of a reduction of the latter to biological conceptions but in the sense of structural similarities. […] If, therefore, we would have a well-developed science of human society and a corresponding technology, it would be the way out of the chaos and impending destruction of our present world. This seems to be plausible and is, in fact, but a modern version of Plato's precept that only if the rulers are philosophers, humanity will be saved [19, p. 52, 87].

Such scientifically justified moral imperative expressed by one of its founders, Ludwig von Bertalanffy, has been leading motif behind the foundation of general systems theory. Brian Gaines, co-founders of the 1954 established Society for General Systems Research, wrote in 1979 [20] that systems theory

> [...] is also foundational in the extreme and hence philosophical but differs from philosophy in that philosophers are not expected to make things work. [...] Systems theory is a form of philosophical engineering. On these grounds I would certainly claim Plato's Republic as an outstanding example of early work on general systems theory as distinct from philosophy.

Now compare this with the blurb of Pickering's book "The cybernetic brain", where he states that we need an "imaginative model of open-ended experimentation in stark opposition to the modern urge to achieve domination over nature and each other" [21]. All this lead to a temptation that became manifest in the media art community: Art conceived as rescuer. Cybernetics conceived as art [22, 23]. Voilà!

In this context it is worth to recall that both art and cybernetics have been exploited from the end of World War II on by the US government and the CIA to construct and control a "better" society as an answer to the Bolshevist threat (cf. [12] for details). The GDR cybernetician Georg Klaus presented cybernetics as a rigorous mathematical representation of Historical Materialism [24]. Cybernetics has also been used in the Chilenian CyberSyn project to "design freedom" [25, 26]. CyberSyn can be called a megalomaniac surveillance project meant to design an "invisible governing hand" hard to beat for its hubris. The susceptibility for ideologisation is completely comprehensible, for cybernetics can be instantiated in all kind of fields, including social engineering [27]. A moral claim for systems theory/cybernetics in such a presumptuous way entails its *reductio ad absurdum.*

3 Emergency Brake

Some 15 years after CyberSyn, project leader Fernando Flores called for a halt of a pure operational cybernetics that degrades humans to "nodding-through-machines." Together with Terry Winograd he suddenly advocated a hermeneutic perspective and promoted Heidegger's "Being-in-the-World" within systems theory in their book "Understanding Computers and Cognition" [28], as if hermeneutics never were the fundamental characteristic of systems theory [5, 13].

3.1 Pharmakon

These days, a well-intentioned *machine désirante* ("desiring machine", after G. Deleuze), as most anticipating algorithms can be called, functions in an even more obscure and subtle way best described by *φάρμακον* (pharmakon), a notion used in critical theory [29]. A provocative piece by the eminent artist collective Ubermorgen.com gives a pointed explanation. The video installation "Psych|OS" is an entirely confusing recording made by one of the members of the net-actionist group, Hans Bernhard, during his stay in a psychiatric hospital due to a psychosis resulting from troubles with the FBI—such is the narration—after provocative Internet activities. The quarrel with "the net" constitutes a self-therapy for H. Bernhard:

> Hans Bernhard's neuronal networks are connected to the global network, and his mental illness – the bipolar affective disorder that in March 2002 sent him to a mental hospital – is the network's illness. The video called Psych|OS (2005) sums up that experience, in which those two levels – digital and real, bio & tech, nervous system and operative system – merge. This nervous system, infected by the hi-tech, needs a treatment, and the hi-tech society prescribes its remedies, bio-chemical 'agents' which control the internal information flow. [...] The "Psych|OS Generator" (2006) is the literal application of this kind of control: a piece of software that asks the user about the symptoms of her disease and provides her with a remedy, in the form of a 'forged original' medical prescription [30].

I hypothesise that the true cause of such a "systemic disease" lies in a lapse of *Being* (Seinsverfehlung) after Heidegger [9, 10].

Beyond the hyperbolic narrative by Ubermorgen.com, more and more critical appraisals and serious attempts to get rid with an increasing reification are published, in particular with reversions to hermeneutic (as opposed to algorithmic or machine-learning-based) understanding [31–35]. At the same time, scientists acquire a taste for artistic or performative approaches either within their own disciplines or complementing their trained research skills (cf. [36] for an example).

The usage of anticipating machines is of course not immoral per se. However, even in the area of information engineering, the "Midas touch problem" is frequently discussed. This name comes from King Midas in Greek mythology who asked the Gods to turn everything he touches into gold. The Gods' fulfilment of his desire turned his beloved daughter after a warm hug into a golden statue. Eventually he died from starvation after turning all food into gold.

3.2 Eye Vision Bot

To relate this back to our own research, we developed a content-based image retrieval interface called *Eye Vision Bot* (Fig. 3) that uses an eye-tracking device [38]. In the beginning of the search, a collection of about 25 images randomly retrieved from a database or the Internet are shown to the user. With the aid of the eye-tracking device, the attention paid by a user to each image is measured via the gazing times that the images receive. The gazing times are used to anticipate the searched-for category with the help of adaptive algorithms. The collection of images is then progressively replaced. The search progresses by supplying more and more images that come out of the desired category. In a certain sense, the artificial intelligent systems is able to anticipate the user's decisions and takes over the decision making process.

In developing *Eye Vision Bot*, we faced an antinomy within the concept. An efficient adaptation to the most desired category keeps away from any future correction. The converged category can never be disapproved, but only be confirmed through gazing at images out of this desired category. Whether or not this is welcomed by the users depends on their attitudes. However, this is my assessment, they often do not really become aware that the bits are bite-sized tailored, such that they are delivered to the persuasive power due to convenience.

Self-modifying features built into the algorithms are regarded as rescue but are mere window dressing. What nowadays, with the help of complex systems theories, comes under the euphemistic label of evidence-based knowledge management, most likely belongs to the same category of tools that I prefer to call a "prejudice confirming machinery." The deeper cause of the failure of "artificial intelligence" lies in the categorical mistake to define rationality in an operational way (cf. [39]).

4 Liquid Perceptron

Liquid Perceptron simulates a neural network excitable by an external stimulus taken from the live video of spectators moving in front of the video projection (Figs. 1 and 2). The artificial neurons are arranged to a 400×300 array and connected to their neighbours. The neuronal activity is colour coded, where bright areas code for a high activity. In a specific setting, chosen to yield excitability as a result of "playful" interactions, the network reacts such that if no evocative stimulus is present, it reduces its activity to a small background noise, whereas an externally applied stimulus via video signals excites spatially correlated neurons. The activity of the excited neurons spreads over the network due to neuronal coupling and synchronizes to a coherent dynamical pattern that encodes the stimulus in an abstract representation.

4.1 Formal Representation

The (Turing) pattern forming features of *Liquid Perceptron* are well known in complex systems research [40]. The model is used in different physiological contexts as, e.g., calcium oscillations in biological cells [41]. *Liquid Perceptron* obeys the following set of differential equations,

$$\begin{aligned} \dot{x}_{i,j} &= a + x_{i,j}y_{i,j} - \tfrac{c_1 x_{i,j}}{c_2 + x_{i,j}} + D_x \operatorname{Diff}^x_{i,j} + c \cdot \operatorname{Video}_{i,j} \\ \dot{y}_{i,j} &= b - \dot{x}_{i,j}\dot{y}_{i,j} + D_y \operatorname{Diff}^y_{i,j} \end{aligned},$$

with $\operatorname{Video}_{i,j}(t) = \operatorname{pixel}_{i,j}(t) - \operatorname{pixel}_{i,j}(t-1)$ and a coupling constant c in the order of 0.1. For the diffusion terms $\operatorname{Diff}^{\xi}_{i,j}; \xi \in \{x, y\}$, zero-flux boundary conditions are used. The parameters a, b, c_1, c_2, D_x and D_y are constants. A working set of parameter values is given by: $a = 0.056, b = 0.12, c_1 = 0.5, c_2 = 0.1$, $D_x = 0.1$, $D_y = 0.03$. The difference of two subsequent video frames, $\operatorname{Video}_{i,j}(t) = \operatorname{pixel}_{i,j}(t) - \operatorname{pixel}_{i,j}(t - 1)$, is used to excite the neurons. Thus, only motion in the external world leads to a stimulation of the network.

4.2 *Performative Dimension*

The stimulation through the live video leads to the emergence of coherent dynamical spiral patterns. If decoupled from the stimulus, the network relaxes to an equilibrium state of low amplitude random oscillations. Variations of parameter values yield an enormous variability of the qualitative behaviour of the network, the transient time, for example, which is related to "echo effects" of the "virtual reality". Thus, Liquid Perceptron supplies an experiential environment for a hermeneutic understanding of this complex dynamical system.

Although *Liquid Perceptron* is far from capturing most of the characteristics of a real brain, the interaction of the spectators with the model simulation yielded valuable insight in perceptional procecesses. Human brains seem to have a tendency to strive for synchronization. In this regard, two productions of Philip Glass' and Bob Wilson's "Einstein on Beach" directed by Berthold Schneider in 2001 and 2005, respectively, allowed for valuable observations [6]. The spectators move in front of the projection and try to synchronize the motion of patterns, their own movement, and the musical rhythm (Fig. 1). In this sense, the brain is a synergetic system that follows eigen-dynamics according to Hermann Haken's so called "enslaving principle" [42]. However, in my view, human *Beings* (potentially) have the freedom to choose the "order parameters" such that the notion of "enslaving" sounds harsh within this context and certainly serves only as a rough sketch. A performative phenomenological approach can contribute to a deeper understanding, such is the expectation.

The essential point is that the model of the epistemic thing is performatively explored and interpreted from a first-person-perspective. Since the epistemic thing here coincides with a basic experience of *Dasein* (the functioning of the brain), the formal/semiotical articulation of this experience helps to understand it in the sense of a hermeneutics of facticity if fed back to the spectators. Thus, *Liquid Perceptron* also shares characteristics with biofeedback, however, beyond mere conditioning of behaviouristic modi. An *apriori* method how to interact with *Liquid Perceptron* does not exist, rather the methodology itself is emergent. The "field study" during the enactments yielded most valuable performative scientific experiences. Remains to be mentioned that the success of these events is due to a consistent general curating and choreography, including other positions from the arts.

5 At the End: Preliminarity and Openness

Performative Science is a phenomenological practice, designs systemic images [43] and sound [44], and has thus proximity to the arts. Phenomenology, thus *Performative Science*, can remain faithful to itself only under the self-conception of its openness and preliminarity. *Performative Science* is, therefore, determined by its radical self-critical essence that becomes manifest by its readiness to continually scrutinise and revise its own results (e.g., *Eye Vision Bot*). Moreover, *Performative Science* cannot be characterised by the utilization of a specific method. Rather, the method becomes emergent by tuning oneself in to the epistemic thing. It is up to the spectator to relate this back to a scientific context. Having said that, the spectator, seen from the perspective of *Performative Science*, is considered more as co-researcher than as recipient. The largest benefit, for sure, is derived by the *bricoleurs* themselves.

References

1. H.H. Diebner, *Performative Science And Beyond—Involving the Process in Research* (Springer, Wien, 2006)
2. J. Badura, S. Dubach, A. Haarmann, D. Mersch, A. Rey, C. Schenker, G.T. Pérez (eds.), *Künstlerische Forschung-Ein Handbuch* (Diaphanes, Zürich, 2015)
3. SAR. Society for artistic research, (2016), http://www.societyforartisticresearch.org. Accessed 4 April 2016
4. W. Saltzer (ed.), *Zur Einheit der Naturwissenschaften in Geschichte und Gegenwart* (Wissenschaftliche Buchgesellschaft, Darmstadt, 1990)
5. H.H. Diebner, in *Messages and Messengers—Angeletics as an Approach to the Phenomenology of Communication*, ed. By R. Capurro, J. Holgate (Fink Verlag, München, 2011), pp. 229–254
6. B. Schneider, Einstein on the beach. opera by P. Glass and B. Wilson, Parochialkirche Berlin, July 26–Aug 5 (2005), http://www.einsteinjahr.de/page_743.html. Accessed 7 Apr 2016
7. A. Pickering, *The Mangle of Practice—Time, Agency, and Science* (University of Chicago Press, Chicago, 1995)
8. A. Pickering, *Kybernetik und Neue Ontologien* (Merve Verlag, Berlin, 2007)
9. M. Heidegger, *Gesamtausgabe* (Vittorio Klostermann, Frankfurt am Main, 1975)
10. M. Heidegger, *Gesamtausgabe*, vol. 14 *(Zur Sache des Denkens)* (Vittorio Klostermann, Frankfurt am Main, 2007)
11. M. Klein, H.H. Diebner (eds.), *Neue Medien machen Sinn—Kontext INM: Performative Wissenschaft und Virealität* (Institut für Neue Medien, Frankfurt am Main, 2011)
12. H.H. Diebner, in *Gegenwelten. Momente der Wahrnehmung—zwischen Differenz und Reflexion*, ed. by J. Schläder, F. Weber (Henschel Verlag, Leipzig, 2009), pp. 84–121
13. H.H. Diebner, in *The Picture's Image. Wissenschaftliche Visualisierung als Komposit*, ed. by I. Hinterwaldner, M. Buschhaus (Fink-Verlag, München, 2006), pp. 282–299
14. H.H. Diebner, Studia UBB. Philosophia **LVII**, 33–57 (2012)
15. H.H. Diebner, in *Station Rose—20 Digital Years Plus*, ed. by Station Rose (Verlag für Moderne Kunst, Nürnberg, 2010), pp. 15–20

16. C. Snow, *The Two Cultures* (Cambridge University Press, Cambrige, 1998)
17. Leonardo Online, Leonardo—The international society for the arts, sciences and technology (2016), http://leonardo.info/index.html. Accessed 26 Mar 2016
18. Eres Stiftung, Chaos! Komplexität in Kunst und Wissenschaft (2012), http://www.eres-stiftung.de/de/events/chaos_program.html
19. L.V. Bertalanffy (ed.), *General System Theory* (George Braziller, New York, 1969)
20. B. Gaines, General systems: Yearbook of the society for general. Syst. Res. **24**, 1–9 (1979)
21. A. Pickering, *The Cybernetic Brain: Sketches of Another Future* (University of Chicago Press, Chicago, 2010)
22. J. Reichardt (ed.), *Cybernetic Serendipity—The Computer and the Arts* (Studio International, London, 1968)
23. M. Rosen, in *Pask Present—An Exhibition of Art and Design Inspired by the Work of Gordon Pask, Cybernetician and Artist*, ed. by R. Glanville, A. Müller (edition echoraum, Wien, 2008), pp. 131–191
24. G. Klaus, *Kybernetik und Erkenntnistheorie* (VEB Deutscher Verlag der Wissenschaften, Berlin, 1972)
25. S. Beer. Fanfare for effective freedom cybernetic praxis in government. The Third Richard Goodman Memorial Lecture (1973)
26. E. Medina, J. Lat. Amer. Stud. **38**, 571–606 (2006)
27. A. Pickering, in *Interdisciplinarity: Reconfigurations of the natural and social sciences*, ed. by A. Barry, G. Born (Chapman & Hall, London)
28. T. Winograd, F. Flores, *Understanding Computers and Cognition—A News Foundation for Design* (Addison-Wesley, Reading, Mass, 1999)
29. B. Stiegler, *Die Logik der Sorge-Verlust der Aufklärung durch Technik und Medien* (Suhrkamp, Frankfurt am Main, 2008)
30. D. Quaranta. Lilly controls my foriginals (2006), http://www.ubermorgen.com/publications/HMKV_2006/Lilly_controls_my_foriginal_ext.html. Accessed 5 Apr 2016
31. H.H. Diebner, in *Hierarchies of Communication*, ed. by H.H. Diebner, L. Ramsay (Center for Art and Media, Karlsruhe, 2003), pp. 30–57
32. F. Fonseca, J. Martin, JASIST **56**, 46–57 (2003)
33. M. Gessmann, *Was der Mensch wirklich braucht. Warum wir mit Technik nicht mehr zurechtkommen, und wie sich aus unserem Umgang mit Apparaten wieder eine sinnvolle Geschichte ergibt* (Wilhelm Fink Verlag, München, 2010)
34. M. Gessmann, *Die Zukunft der Hermeneutik* (Wilhelm Fink Verlag, München, 2012)
35. S. Gust von Loh, *Evidenzbasiertes Wissensmanagement* (Gabler Verlag, Wiesbaden, 2009)
36. T.J. Senior, M. Meyer-Bohlen, and D. Poggel. Performativity and scientific practice—workshop at HWK Delmenhorst (2012), http://www.h-w-k.de/en/events/conferences-details/perfomativity-and-scientific-practice.html. Accessed 5 April 2016
37. S. Fischer, L. Scherffig, and H.H. Diebner, in *Making Things Public—Atmospheres of Democracy*, ed. by B. Latour, P. Weibel (MIT Press, Cambridge, Mass, 2005), p. 1017
38. L. Scherffig, *It's in your Eyes—Gaze Based Image Retrieval in Context* (Zentrum für Kunst und Medientechnologie, Karlsruhe, 2005)
39. F. Eberhardt, D. Danks, Minds Mach. **21**, 398–410 (2011)
40. M. Golubitsky, D. Luss, S.H. Strogatz (eds.), *Pattern Formation in Continuous and Coupled Systems* (Springer, New York, 1999)
41. G. Baier, S. Sahle, J.P. Chen, A.A. Hoff, J. Chem. Phys. **110**, 3251–3255 (1999)
42. H. Haken, *Synergetics—An Introduction* (Springer, Berlin, 1983)
43. I. Hinterwaldner, *Das Systemische Bild* (Fink Verlag, München, 2010)
44. F. Grond, S. Janssen, S. Schirmer, T. Hermann, in *Browsing RNA Structures by Interactive Sonification*. Proceedings of ISon 2010, Stockholm, 7 April 2010

Dynamics and Synchronisation in Wind Farms

Mehrnaz Anvari, M. Reza Rahimi Tabar, Joachim Peinke and Matthias Wächter

Abstract Renewable energy sources such as wind power and photovoltaics shall account for about 20% of the total energy consumption by 2020 and 60% by 2050 in the European Union. These renewable resources which are spatiotemporally stochastic in nature, can influence the stability of the power grids. Therefore, a detailed modelling and understanding of their statistical behaviour is necessary to achieve an optimal design and operation of future power grids. Here, we present results on the complex statistics of wind power in general and the local phase synchronisation of turbines in a wind farm.

1 Short-Time Stochastic Characteristics of Wind Power

A wind turbine converts wind velocity fluctuations into power fluctuations. Apparently the conversion dynamics take place at short time scale of seconds for wind power [1–3]. The wind speed and power output of the single wind turbine have a similar spectral (and correlation) behaviour at low frequencies $f < 0.1$ Hz, and obey the 5/3 Kolmogorov-law of turbulence i.e. $S(f) \sim f^{-5/3}$, as shown in Fig. 1. Due to the finiteness of the reaction time of wind turbines, the fast wind fluctuations are partly filtered out [2], while slower wind fluctuations are adiabatically converted into power output fluctuations. We should note that for a wind farm consisting of many turbines, because of the geographical averaging, the reduction of the high frequencies by these averaging effect is stronger for the cumulative wind power than for the wind power

M. Anvari (✉) · J. Peinke · M. Wächter
ForWind, University of Oldenburg, Küpkersweg 70,
26129 Oldenburg, Germany
e-mail: mehrnaz.anvari@uni-oldenburg.de

M.R. Rahimi Tabar
Department of Physics, Sharif University of Technology, 11365-9161 Tehran, Iran

S.C. MÃijller et al. (eds.), *Complexity and Synergetics*,
https://doi.org/10.1007/978-3-319-64334-2_29

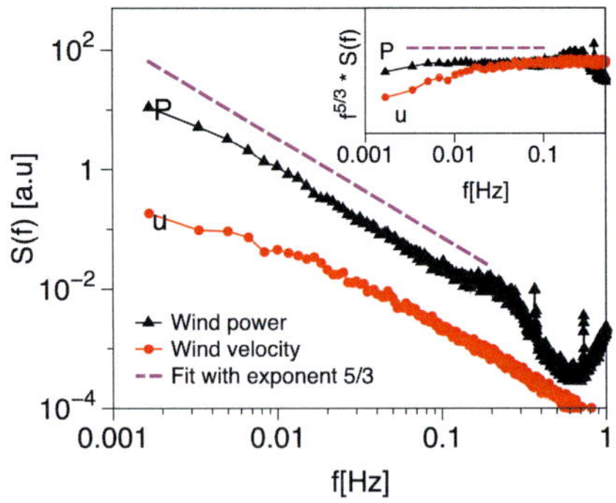

Fig. 1 Power spectra of wind velocity and wind power in log-log scale. The Kolmogorov exponent 5/3 is represented by *dashed lines*. In the *inset*, the log-log plot of the compensated energy spectra $f^{5/3}S(f)$ versus frequency f are plotted. Here, we use the data of Ref. [1]

of a single turbine (see reference [4]). Thus the power output of a wind farm shows weaker high frequency fluctuations than a single turbine.

To understand the observation of the exponent 5/3 in wind power, we point out that wind velocity has the well-known Kolmogorov type of power spectrum, i.e. $S(f) \sim f^{-5/3}$, similar to idealized turbulence. Here, $S_u(f) = |u(f)|^2$ and $u(f)$ is the Fourier transform of the wind velocity time series. Although, the wind power $P(t)$ is proportional to $u(t)^3$ [5], we find that the characteristic power of $-5/3$ does not change significantly for the power spectrum of the corresponding wind power data. This finding was verified by investigating the power spectra of $u(t)^3$, which shows

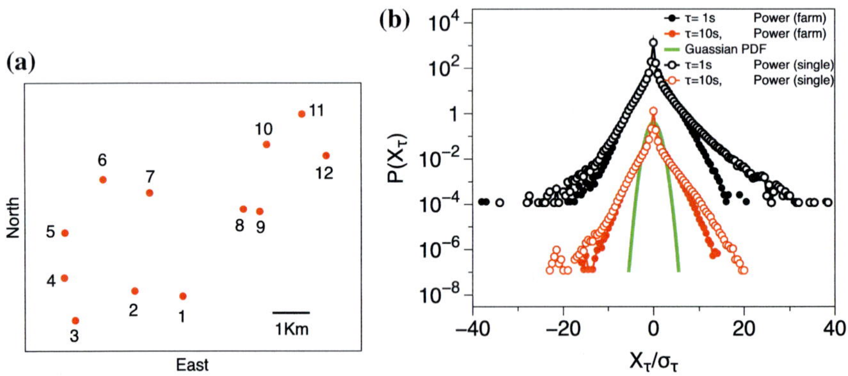

Fig. 2 **a** The spatial position of wind turbines. **b** Probability distribution functions (PDF), $P(X_\tau)$, of wind power increments for time lags $\tau = 1$ and 10 s, in linear-log scale. The PDFs have been vertically shifted for better clarity and a Gaussian PDF with unit variance is plotted for comparison. Clear deviation from a Gaussian PDF is observed for both the single wind turbine and the wind farm; after [1]

the same power law exponent for frequencies $0.001 < f < 0.05$ Hz, of course with different amplitudes, see also [2].

Techniques such as spectral and correlation analysis can uncover only linear structures and correlations of a given time series. To investigate the nature of the wind power fluctuations and their temporal correlations in more details, we consider the statistics of increments [6] for fixed time lags τ, i.e. $X_\tau = X(t+\tau) - X(t)$, where $X(t)$ denotes here the wind power. By considering the complete increment probability distribution functions (PDFs) we are able to capture also higher order two-point correlations.

Here, we present results of wind power data sampled with 1 Hz. These data are from a wind farm in North Germany consisting of 12 turbines. Further details about the data sets are given in reference [1]. Figure 2a shows the spatial arrangement of these wind turbines. The increment PDFs $P(X_\tau)$ of power output for one exemplarily chosen wind turbine and the wind farm for the time lags $\tau = 1$ and 10 s are shown in Fig. 2b. The PDFs are normalised to their standard deviations σ_τ. Note that the standard deviation values of the PDFs for different τ-values are given by the power spectrum. Here we are interested in shapes of PDFs $P(X_\tau)$. As it is clear from Fig. 2b the normalised increment PDFs depart largely from the Gaussian distribution and they possess exponential-like fat tails, which extend to extreme values like $20\sigma_{\tau=1s}$ and more. Such shapes of PDFs are a second remarkable feature of idealised turbulence (cf. [7]), which is called intermittency effect. Most interestingly this effect is present in the power output of a single turbine as well as in that of the whole wind farm.

2 Local Phase Synchronisation in Wind Farms

As Fig. 2b depicts, not only the power output of the single wind turbine deviates largely from the normal distribution but this non-Gaussian behaviour remains also for the aggregated power of the wind farm. Therefore the central limit theorem, predicting a convergence to Gaussianity for independent data sets with defined standard deviation, does not apply. As a consequence, there should exist interdependency between the wind turbines, on what we will focus in the following.

To characterize the interaction between turbines, first we use the cross-correlation method. As an example, Fig. 3 presents the cross-correlation between the power output of turbine 1 and 9 during about 1 h. It is clearly seen in Fig. 3 that the value of the cross-correlation between the turbines changes randomly over time. Taking the average, no significant interdependency between turbines can be detected via the linear cross-correlation method.

A more advanced approach is therefore needed in order to capture the interdependency between turbines. To figure out the time-dependent interaction in the wind farm, here we adopt the phase synchronisation method which is based on the phase locking parameter. The phase locking parameter $|\phi_{n,m}|$ for the power outputs of each pair of wind turbines is calculated as

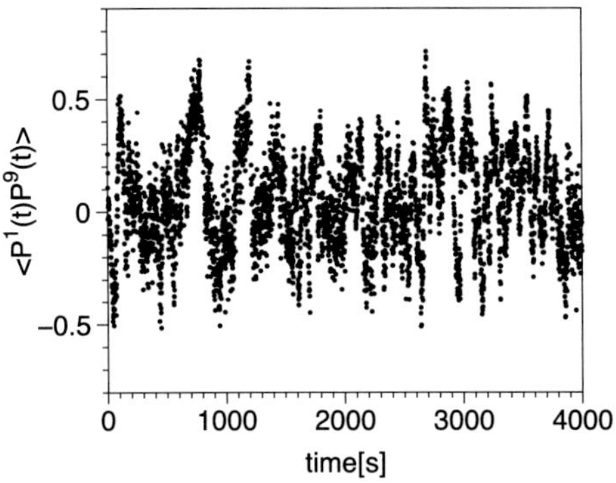

Fig. 3 The cross-correlation, $\langle P^1(t)P^9(t)\rangle$, between the power output of the turbine 1 and 9 in time. The plot shows the value of the cross-correlation during about 1 h

$$|\phi_{n,m}(t)| = |n\phi_i(t) - m\phi_j(t)| < const., \tag{1}$$

where n and m are integer numbers (for simplicity we choose $n = m = 1$), $\phi_{i,j}$ is the phase of the power outputs for turbines i and j. $\phi_{n,m}$ is the respective phase difference. According to the general definition of the phase locking parameter, Eq. (1), $|\phi_{n,m}|$ is bounded. Strictly speaking, the phase difference between two synchronised turbines is not necessarily constant and $|\phi_{n,m}|$ should just remain bounded [8]. Therefore, we put here the same value for the phase locking parameter when it is in a narrow range of the specific phase values.

In order to access synchronisation effects we need to define the phase for our data set properly. Here we use the ideal power curve of each wind turbine, obtained by an averaging procedure after IEC 61400-12-1 [9]. According to Fig. 4, we can divide the power curve to three different parts. The black curve (i.e., the ideal power curve) has two constant value regions, respectively I and III, at the beginning and the end of the diagram. Here, we focus on the middle part of the diagram (i.e. region II) in which the power is proportional to the third power of the wind velocity. In reality, the wind power fluctuates dynamically around the ideal power curve as illustrated by the red curve in Fig. 4. We can now define the consecutive times t_n and t_{n+1} when the power output, or in the other words, the red line crosses the power curve. For details about the dynamics of the power curve, see reference [1], supplemental material. In this way, the power curve can be considered as an attractor of the wind turbine [10]. To finally derive the instantaneous phases of our time series we use the marked events method [8]

$$\phi_i(t) = \pi \frac{t - t_n}{t_{n+1} - t_n} + \pi n, \qquad t_n \le t < t_{n+1}. \tag{2}$$

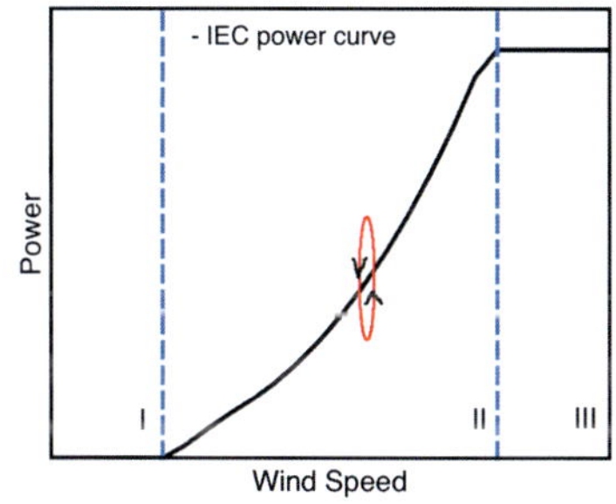

Fig. 4 The *black curve* shows the power curve according to the International Electrotechnical Commission's industry standard IEC 61400-12-1, which is known as the standard power characteristics of a wind turbine. The *red line* illustrates the short-time dynamics of the wind power around the IEC power curve

3 Results and Conclusion

As an example for phase relations between turbines, the phase locking parameter $|\phi_{n,m}|$ of the turbines 1 and 9 is shown in Fig. 5 for about 1 h. As it is visible in the figure, these two turbines are synchronised as the phase difference $|\phi_1(t) - \phi_9(t)|$ has different constant values, each lasting for about 7 min, or is drifting slowly as it can be seen particularly well between 1500 and 3000 s. In contrast, from 3000 to 4000 s the phase difference is highly fluctuating, which means that these two turbines are operating independently.

These results demonstrate that the phase synchronisation approach enables us to quantify the interdependency between the wind turbines. Our analyses show that synchronisation changes in time and location. This means that there are pairs of turbines which are interdependent in specific time intervals.

In this context it remains an interesting question how this synchronisation effects the central limit theorem and how this is related to the intermittent behaviour of the cumulative power output of wind farms. For applications these results may pave the

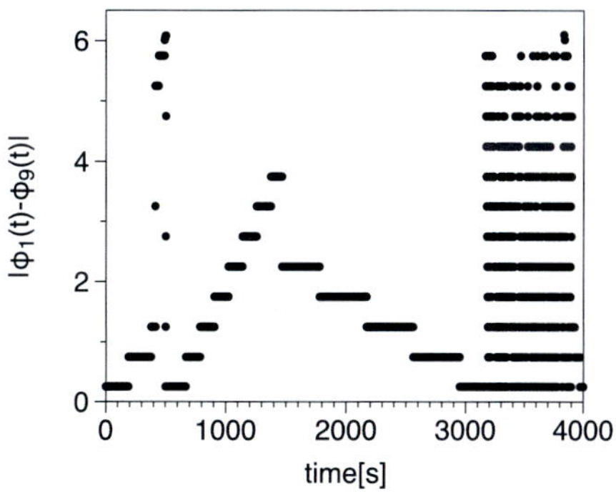

Fig. 5 The phase locking parameter $|\phi_{n,m}|$ of turbines 1 and 9 for about 1 h. As it is visible from the plot, these turbines are synchronized for approximately 30 min where the phase difference does not change in a fast way

way for an optimal design of wind farms in the future which suppresses extreme fluctuations in the aggregated power output. Such fluctuations are known to be a possible cause of instability in power grids with an increased share of wind energy.

References

1. P. Milan, M. Wächter, J. Peinke, Turbulent character of wind energy. Phys. Rev. Lett. **110**, 138701 (2013)
2. M. Reza Rahimi Tabar, M. Anvari, G. Lohmann, D. Heinemann, M. Wächter, P. Milan, E. Lorenz, J. Peinke, Kolmogorov spectrum of renewable wind and solar power fluctuations. Eur. Phys. J. Special Topics **223**, 2637–2644 (2014)
3. G. Lohmann, M. Reza Rahimi Tabar, P. Milan, M. Anvari, M. Wächter, E. Lorenz, D. Heinemann, J. Peinke, Flickering events in wind and solar power, in *EnviroInfo 2014—ICT for Energy Efficiency*, vol. 723, p. 6 (2014)
4. M. Anvari, G. Lohmann, M. Wächter, P. Milan, E. Lorenz, D. Heinemann, J. Peinke, M. Reza Rahimi Tabar, Short term fluctuations of wind and solar power systems. New J. Phys. **18**, 063027 (2016)
5. R. Gasch, J. Twele (eds.), *Wind Power Plants* (Springer, Heidelberg, 2012)
6. A. Morales, M. Wächter, J. Peinke, Characterization of wind turbulence by higher-order statistics. Wind Energy **15**, 391–406 (2012)
7. U. Frisch, *Turbulence: The Legacy of A. N. Kolmogorov* (Cambridge University Press, Cambridge, 1996)
8. A.S. Pikovsky, M.G. Rosenblum, J. Kurth, *Synchronisation* (A universal concept in nonlinear sciences (Cambridge University Press, Cambridge, 2001)
9. International Electrotechnical Commission, Wind Turbine Generator Systems, Part 12: Wind turbine power performance testing, International Standard IEC 61400-12-1 (2005)
10. M. Wächter, H. Heißelmann, M. Hölling, A. Morales, P. Milan, T. Mücke, J. Peinke, N. Reinke, P. Rinn, The turbulent nature of the atmospheric boundary layer and its impact on the wind energy conversion process. J. Turbul. **13**, 1–21 (2012)

Part VIII
Special Lecture

A Primer for Deterministic Thermodynamics and Cryodynamics

Dedicated to the Founder of Synergetics Hermann Haken

Otto E. Rossler, Frank Kuske, Dieter Fröhlich, Hans H. Diebner, Thimo Böhl, Demetris T. Christopoulos and Christophe Letellier

Abstract The basic laws of deterministic many-body systems are summarized in the footsteps of the deterministic approach pioneered by Yakov Sinai. Two fundamental cases, repulsive and attractive, are distinguished. To facilitate comparison, long-range potentials are assumed both in the repulsive case and in the new attractive case. In Part I, thermodynamics—including the thermodynamics of irreversible processes along with chemical and biological evolution—is presented without paying special attention to the ad hoc constraint of long-range repulsion. In Part II, the recently established new fundamental discipline of cryodynamics, based on long-range attraction, is described in a parallel format. In Part III finally, the combination ("dilute hot-plasma dynamics") is described as a composite third sister discipline with its still largely unknown properties. The latter include the prediction of a paradoxical "double-temperature equilibrium" or at least quasi-equilibrium existing, which has a promising technological application in the proposed interactive local control of hot-plasma fusion reactors. The discussion section puts everything into a larger perspective which even touches on cosmology.

O.E. Rossler (✉) · F. Kuske · D. Fröhlich · T. Böhl
Division of Theoretical Chemistry, University of Tübingen,
Auf der Morgenstelle 18, 72076 Tübingen, Germany
e-mail: oeross00@yahoo.com

H.H. Diebner
Department of Medical Information, Technical University Dresden,
Blasewitzerstr. 86, 01307 Dresden, Germany

D.T. Christopoulos
Department of Economics, National and Kapodistrian University of Athens,
Sofokleous 1 str, 10559 Athens, Greece

C. Letellier
Physics Department, University of Rouen CORIA, Avenue de l'Université,
76801 Saint-Étienne du Rouvray, France

S.C. Müller et al. (eds.), *Complexity and Synergetics*,
https://doi.org/10.1007/978-3-319-64334-2_30

1 Introduction

Thermodynamics and its recent new sister discipline, cryodynamics, can be both put on the same deterministic "Newtonian" footing. Hereby the familiar close-range repulsion between the atoms in a gas is formally replaced by a long-range "anti-Newtonian" repulsion so as to thereby facilitate comparison. This "smoothing trick" has surprisingly little impact on the main qualitative behavior [1]. The resulting new mutually anti-symmetric "double reality" is summarized in the following in a tabular format. A third "combined case" is also included with a view to its technological promise. A general, somewhat philosophical discussion is added.

2 Part I. Thermodynamics

FIRST LAW of deterministic statistical thermodynamics

This law is trivial in classical mechanics:

(1) Energy conservation $\left(\frac{dH}{dt} = 0,\ \text{with H the classical Hamiltonian function}\right)$

The Hamiltonian may either be integrable or (in the general case) non-integrable. For the case of a many-particle system at equilibrium, the conservation law of Eq. (1) has an especially intuitive implication:

(1a) Ergodicity

literally "work-path involvement." The term was coined by Boltzmann [2] to describe the fact that the whole phase space is being re-visited arbitrarily closely but not identically everywhere infinitely often across unlimited time (this is one possible interpretation). Ergodicity was demonstrated for deterministic systems for the first time, under the assumption of hard-sphere repulsion, by Yakov Sinai in 1963 [3, 4]. His chaos-theoretic result could later be generalized, both towards soft repulsion [1] and towards far-from-equilibrium conditions [1] (and towards soft attraction, see Part II). In the soft case, so-called Kolmogorov-Arnold-Moser (KAM) tori are automatically included in the phase space [5]. The presence of the latter formally destroys perfect ergodicity but actually has a virtually negligible influence under multi-particle hyperbolic interaction—a fact which does not possess the rank of a theorem, however.

SECOND LAW of deterministic statistical thermodynamics

This law concerns the existence of an attractive (asymptotically stable) equilibrium in the case of gases. This equilibrium was, after its hypothetical generalization towards the whole universe, called "heat death" by Clausius [6]. That dismal fate is being approached in a statistically progressive maximization of phase space volume, cf. [7, 8]. The same tendency in phase space holds true already for very low particle numbers [9, 10]. The second law is sometimes also called the "first arrow" of nature. It implies the fact that no perpetual motion machine of the second kind is possible in nature. It can be described best by focusing on a global observable, the so-called entropy ("inwards-growing"), a term coined by Clausius in 1865 [6] as an onomatopoetic sister term to energy. For the entropy S of a closed system, the following law holds true:

(2) Positive entropy production $\left(\frac{dS}{dt} \geq 0\right)$
Under deterministic conditions, cf. [8, 11], the entropy is equal to the logarithm of phase space volume apart from a scaling factor. A spontaneous increase of disorder in forward time is implicit in Eq. (2). Note that the necessarily positive overall entropy production does not exclude the opposite behavior from occurring in a sub-region of state space or a subsystem, enabling biogenesis.
FIRST SUB-LAW to the second law of deterministic statistical thermodynamics

This sub-law concerns the spontaneous generation of a so-called flow-equilibrium ("Fliessgleichgewicht," von Bertalanffy [12]) which is a point attractor. More generally speaking, "dissipative structures" are being formed [13]. This class includes a progressive chemical evolution [14–16] under realistic natural conditions. In the following, the restrictions that are brought-in by assuming soft (anti-Newtonian) repulsion in place of the usual hard-sphere repulsion between the molecules in a gas and a fluid will be neglected in order to allow for a maximally broad overall picture. The dissipative structures which form in nature are sub-cases to a so-called "**second arrow**" that is implicit in the first arrow of the second law of thermodynamics itself [17]. The second arrow is sometimes called "complexification" following Teilhard [17]. Note that the second arrow forms a sub-arrow of a seemingly inverted, "anti-dissipative" orientation within the first arrow of Clausius. The second arrow implies as its most spectacular special case the **biogenetic arrow**. The latter applies as soon as very many chemical compounds [14] are eligible to being formed (or many nuclear-reactions are likewise enabled on the surface of a neutron star, cf. [18]). Darwin first saw this deductively predictable origin of life as necessarily occurring "in a warm little pond" [19]. Teilhard's parallel law of **complexification** at the same time represents the first complex attractor of history [17]: in the form of his famous

(2a) Point Omega
The deterministic arrow which tends towards a greater and greater and ever more subtle state of order, is as we saw only possible by virtue of the fact that an even larger disorder is simultaneously generated in the "complement" of the dissipative structure in question. It is hereby presupposed that the overall entropy of the whole system can be defined, a question which appears to be new. The entropy no doubt assumes the form of a complicated functional which has never been looked at in mathematical detail. Similarly, the term "complement" appears to be new in the present context. The idea is that the deterministic phase space volume of Eq. (2) is increasing overall, thereby overcompensating the local decrease in subsystems and sub-pockets. "Life feeds on negative entropy" was Erwin Schrödinger's original

phrase [20]. One sees that it all is less trivial than generally assumed but well worth looking at in thermodynamic terms.

SECOND SUB-LAW to the second law of deterministic statistical thermodynamics

This sub-law is not usually considered in a far-from-equilibrium context since it only deals with equilibria. The present sub-law depends on the temperature and the volume as the two decisive parameters, but also on the specific atomic and molecular properties and potentials of the involved particles [21]. It comprises the existence of pressure- and temperature-dependent equilibria which, taken together, can be comprised in a so-called

(2b) Phase diagram or p-V diagram, etc. (cf. also [22] for an example in nuclear chemistry)
The inner energy of the gas is (apart from any remaining potential energies) the kinetic energy of the involved particles including rotational energies.

THIRD LAW of deterministic statistical thermodynamics

This full law, introduced by Walter Nernst [23], concerns the existence of an absolute zero of temperature. It later was complemented by special low-temperature quantum effects which lack a deterministic analog [24–26] and the Bose-Einstein effect [27] and still later, the quantum-Hall effect [28, 29].

(3) Absolute zero, with superfluidity, superconductivity, supercondensation, etc.
which effects have a zero entropy increase each, a property that is essentially retained when they occur at substantially higher-than-zero temperatures. Note that the Cooper pairs in superconductivity have a doubled electric charge and quantum number owing to "grid vibrations" in quantum field theory. But these quantum details exceed the scope of the present survey. The same holds true for the effects of a quantum-based non-deterministic formalism that postulates a negative absolute temperature in excess of any possible positive temperature which nevertheless is then transformable into such [30].

In conclusion of this section, all of the above qualitative facts are well known, even though beautiful open questions remain. Many different ways of looking at the three laws exist. The latter were presented here mostly as a "backdrop" in order to put into a clearer perspective the recently discovered sister discipline to thermodynamics that is to be described next.

3 Part II: Cryodynamics

FIRST LAW of deterministic statistical cryodynamics

This law applies when the potentials which apply between the individual particles are, not short-range repulsive or even long-range repulsive (as suggested above for the sake of comparison), but rather long-range attractive. Such "Newtonian gases" form a new object of discourse even though they could in principle have been considered by Newton himself already. The present law is formally identical to the first law of deterministic statistical thermodynamics (Eq. (1) in Part I above):

(1′) Energy conservation $\left(\frac{\mathrm{dH}}{\mathrm{dt}}=0,\ \text{with H the classical Hamiltonian function}\right)$

just as this held true before in an anti-Newtonian gas. The conserved energy can again assume quite differing forms. Once more, we at equilibrium necessarily have the presence of KAM tori in phase space occupying again only a small ("effectively negligible") subvolume of the phase space of the many-particle system in question. But it actually makes little sense to speak of "ergodicity" applying here, for these formally existing states occur here by definition only in the negative direction of time. Due to their being intrinsically unstable they are predictably unphysical.

SECOND LAW of deterministic statistical cryodynamics

This law concerns the existence-in-principle of an unstable equilibrium (an anti-attractor or synonymously repellor) which can be named "Point Alpha." It is characterized by a progressive (for once not increase but) decrease of overall phase space volume in the forward direction of time. In other words, this is just the opposite of what happens in deterministic thermodynamics as it was considered above. Therefore, we now have for the phase space volume—or rather its logarithm called entropy—the following law:

(2′) Negative entropy production $\left(\frac{dS}{dt}\leq 0\right)$**, with** S **the entropy of a closed system**

This law comes as a real surprise: It describes a spontaneously progressive increase of order in positive time. That is, a positive ectropy production occurs if "ectropy" is defined as minus S. Therefore, the existence of a repellor, an unstable "**Point Alpha**" is formally implicit, named in analogy to Teilhard's first attractor, his **Point Omega**.

FIRST SUB-LAW to second law of deterministic statistical cryodynamics

This law (2′a) is identical to a full-fledged law, the Fourth Law of "negative-time Super-Life" described below as Eq. (4′).

SECOND SUB-LAW to second law of deterministic statistical cryodynamics

This sub-law—much like the analogous second sub-law (2b) of deterministic statistical thermodynamics above—describes equilibrium states that are dependent on the temperature and pressure and density as well as on the specific physical properties of the particles involved. However, these states cannot become manifest empirically by virtue of the repulsive character of the equilibria in question. Nevertheless it makes sense to define here a

(2′b) negative-time phase diagram (being asymptotic in negative time)

This diagram has only a formal character since the equilibria described in it apply in negative time. It has never been investigated and presently it is not clear whether or not it possesses physical significance.

THIRD SUB-LAW to second law of deterministic statistical cryodynamics

This sub-law possesses for once no analog in thermodynamics. It concerns the existence of a family of multi-particle chaotic dynamical regimes which show a certain resemblance to attractors in dissipative systems. For lack of a better term, they can be called pseudo-attractors:

(2′c) Pseudo-attractors of multi-particle systems in astronomy
They apply in celestial mechanics and include:

1. globular clusters of stars (cf. Chandrasekhar [31])
2. spiral galaxy structures
3. elliptical galaxy structures
4. "black" or synonymously "low-surface brightness" galaxies (see [32] for more details about the latter)

The formation of these pseudo-attractors is strongly influenced by the principle of "energetic capitalism" of Sonnleitner [9], see also [33]. Sonnleitner found that under a condition of mutual long-range attraction, a "disproportioning of the kinetic energies" of the involved particles takes place in an inversion of what applies under repulsive Newtonian (as well as the usual close-range repulsive) conditions. An anti-equilibration in which the low-in-kinetic energy particles are "preyed upon" by the high-in-kinetic energy particles was found [9]. After a while, for some reason this anti-entropic feature of systems of mutually attractive particles comes to a standstill and even gives rise to a visible attractor—in the form of galaxies. The latter are a stage in the development of a self-contracting gas cloud. The qualitative dynamics of this very slow process is still ill-understood. There exist billions of specimens before our eyes which all belong into but a few classes. They were, by the way, first named galaxies by a physicist named Immanuel Kant in 1755 (in his "Universal Natural History and Theory of the Heavens," p. 367) who soon after became a famous philosopher. Note that general-relativistic effects play only a minor role in galaxy formation apart from the formation of black holes, including the ultra-heavy ones found located at the galactic centers. Although black-hole theory is still underdeveloped (there appear to exist no finished black holes in finite outer time as a consequence of c-global), the Birkhoff theorem makes sure that something very compact resides inside the invisible attractive region. These heavy residents at the galactic cores by the way are bound to be the products of a very long accretion history.

FOURTH SUB-LAW (non-chaotic) to the second law of deterministic statistical cryodynamics

This sub-law concerns the existence of special virial theorems valid under non-hyperbolic and non-chaotic conditions. It comprises:

(2′d) Kepler ellipses and Mercury-orbit-like Einstein rosettes and long-term stable planetary systems
Note for curiosity's sake that such rosettes occur under many different perturbations of the exactly quadratic Newtonian force law, apart from the familiar general-relativistic one.

THIRD LAW of deterministic statistical cryodynamics

This third law—like its analog in deterministic thermodynamics, Eq. (3) of Part I—comprises low-temperature quantum effects that are valid here under attractive rather than repulsive conditions:

(3′) Absolute zero, with Nernst-type collective quantum phenomena
They predictably show a zero-entropy change again. These predicted equilibrium effects represent an unfamiliar class. A direct physical implementation appears rather unlikely at present.

FOURTH LAW of deterministic statistical cryodynamics

This fourth law possesses no analog in thermodynamics for once, as mentioned. Rather than representing only a second-order phenomenon, as chemical and biological evolution do in thermodynamics (whose dissipative structures do represent only a second-order phenomenon "riding on the back" of the dissipative first-order arrow of Clausius as we saw), the present law represents a **first-order main arrow** —albeit one that is valid in negative time. Hence this arrow represents a full-fledged repellor in positive time. As such, this repellor is **identical** to the *Point Alpha* of the second law of deterministic statistical cryodynamics, Eq. (2′). The implied "**anti-evolution**"—as it can be called—is likely to become a main topic in cryodynamics eventually, but at the present moment in time it is only vaguely perceptible. An interesting open question hereby is: Will a sophisticated subsystem in this class perceive macroscopic time as flowing in the other direction? The present dynamical attractor is only fictitious due to its instability in the forward direction of time. Thus, it is only virtual for us—much like a virtual image. Nevertheless the dynamical routes leading towards it in negative time do represent a very powerful evolution (albeit in negative time). The underlying **anti-attractor** might play a ubiquitous role in the cosmos but its features are totally opaque at present. Nevertheless the new negative-time attractor can already be given a name:

(4′) "Negative-time Super Life"
The proud name is perhaps premature since the properties of this necessarily existing repellor have yet to be unraveled in detail. Owing to the surreptitious numerical instability of Sonnleitner [9] which affects all cryodynamical computer simulations rendering them invalid in detail, it is very hard to say at present how to tackle this negative-time origin problem empirically, that is, numerically. It cannot even be ruled out at present that some cleverly chosen inaccurate numerical simulations will prove useful in coming closer to a solution of this conundrum. The cosmos' own time's arrow thus may or may not be listening to a baton of its own.

FIFTH LAW of deterministic statistical cryodynamics

This fifth law not only is without an analog in thermodynamics but at the same time is maximally counter-intuitive

(5′) Paradoxical Acceleration
suffered by a very tiny particle very rich in kinetic energy, in a cluster of very tiny particles

This law is the natural counterpart to the fact that an energy-poor lighter particle is getting braked by many heavier particles under an attractive interaction (like the photons passing through a gas of galaxies seen by Zwicky in 1929 and independently by Fröhlich in 2003). The present "mirror prediction" is as counterintuitive as the paradoxical braking of Sonnleitner's. It appears to be new.

Ordinarily, the collisions suffered in a gas by a fast heavier particle passing through are non-conducive to acceleration. But if a highly energetic gravitating particle is both markedly heavier and ultra-small, it predictably is bound to undergo an acceleration while traversing a homogeneous gas of much lighter equally small particles. So, for example, in the interior of a planet or a star or even in a maximally dense white-dwarf. This is because the material objects in question are on the finest scale made up from almost point-like particles (electrons and quarks). If for this reason, collisions are sufficiently rare, then the fast heavy mini particle should—under the influence of Newtonian attraction—suffer an acceleration due to the Sonnleitner principle of **energetic capitalism**. This prediction follows in a straightforward manner—just as the previously described braking effect was valid for energy-poor particles. A possible counterargument is that the present effect, while necessarily true at "ordinary" velocities much lower than the speed of light *c*, might cease to apply under relativistic conditions. This objection is probably unfounded though because the interaction between photons and randomly moving galaxies shows that classical cryodynamics remains valid under relativistic velocities.

Why is **paradoxical acceleration**, if it indeed exists, of applied interest? The reason is black-hole theory, of all things. Since micro black holes are necessarily uncharged (as follows from Birkhoff's theorem which is not very well known), they do, if very fast, pass right through massive celestial bodies getting accelerated rather than braked and hence are doing no harm. Hence the onslaught of cosmic ray particles on surface protons of celestial bodies may suffice to produce micro black holes. This is because charged "point particles" like electrons cannot be effectively point-shaped in reality since they would then have to be black holes and hence be uncharged. Therefore, some analog to string theory is bound to be valid empirically. The cosmic-ray borne ultrafast micro black holes (if they exist) will thus leave the earth and other celestial bodies (up to the ultra-dense white dwarf stars) unscathed owing to paradoxical acceleration. The still denser neutron stars are protected by a different mechanism, their superfluidity. Unlike the natural fast micro black holes if they exist, however, any artificial slow micro black holes are *not* innocuous, however.

Now an accident of history wills that ultraslow micro black holes are currently being attempted to be produced on earth. They cannot "Hawking evaporate" as is traditionally believed because a black hole is never finished in finite outer time since nothing including light can ever reach their horizon or come up from it in finite outer time. (Hence their already 19 years old new name "almost black holes.")

The upshot is that the present new phenomenon of paradoxical acceleration which protects celestial bodies from being eaten inside out by micro black holes provided they exist is *inapplicable* to slow micro black holes. This is the situation presently valid on earth. If the experiment is successful in producing at first undetectable uncharged micro black holes on earth, this means that an ultra-slow specimen getting stuck inside earth can gravitationally interact in a non-hyperbolic fashion with a charged micro particle (quark or electron) passed by very close. Such an encounter leads to an "eating event." The on-going interaction (before the final "gulp") then increases the attraction exerted by the momentarily interlocked and hence charged pair on a neighboring charged particle of the opposite polarity by some 30 orders of magnitude. Even a very short "eating time" cannot undo the attendant giant boost in attractive power of the micro black hole. Therefore, an artificial micro black hole inside earth is bound to grow. After an initial phase of linear growth, the growth predictably becomes exponential.

This fact is titillating because the learned official safety report of the mentioned terrestrial experiment, called LSAG [34], goes un-updated for 9 years. The new implication of cryodynamics just shown—that ultra-slow micro black holes are "non-harmless"—will predictably enforce an update of the LSAG soon[1]. This **very indirect side effect of cryodynamics** was perhaps worth mentioning here, too.

4 Some Remarks Leading Over to Part III

Cryodynamics exists as a natural sister discipline to thermodynamics for symmetry reasons, as we have seen. Many of its features are still a *terra incognita*. Does cryodynamics possess an importance that is comparable to that of thermodynamics? In light of the many surprises which the study of thermodynamics has brought over two centuries, an equally rich "bonanza" can be expected to arise in the further development of cryodynamics. As we saw, the defining difference between cryodynamics and thermodynamics lies in the sign change of the deterministic inter-particle potentials in charge.

The fact that in thermodynamics, the inter-particle potentials are actually short-range repulsive rather than long-range repulsive has surprisingly little influence on the broader comparative picture. On going-over from repulsion to attraction, all equilibrium-oriented notions from deterministic thermodynamics lose much of their relevance because the deterministic equilibria are now unstable by definition and hence unobservable. The most directly rewarding sub-task of cryodynamics, perhaps, consists in its explaining galaxy formation as a kind of attractor in forward-time. This task is severely hampered though through the numerical instability governing gravitationally interacting particles, demonstrated by Sonnleitner [9, 33]. Most of the multi-particle numerical galactic simulations performed up until

[1]Equation (5) was triggered by a discussion with Tom Kerwick.

now are automatically put into question as mentioned. Cryodynamics also covers the interactions in a gas that is made up from many galaxies and clusters. And it also governs as mentioned the interaction between the many galaxies on the one hand and the fast-moving cosmic-ray particles and photons on the other. Hereby, relativistic complications automatically arise but they seem not to affect the qualitative features outlined.

In the following, we now turn to a third, once more down-to-earth case: the combined discipline of thermo-cryodynamics as it can be called. Like cryodynamics proper it is a classical subject that previously had gone unnoticed for reasons that are hard to explain in retrospect.

5 Part III: Combined "Thermo-Cryodynamics" or Equivalently "Dilute-Hot-Plasma Particle Dynamics"

FIRST LAW of deterministic statistical Thermo-Cryodynamics

This law concerns the likely non-existence of a globally stable equilibrium in compound systems made up from both mutually repulsive and mutually attractive particles. This situation was historically speaking first looked at by Lev Landau in his proposal to bring into contact two compartments that each contain particles of one of the two types [35]. Next came Donald Lynden-Bell in 1960 [36] (see [37] for details). The more complex "fully mixed" situation is still in a fledgling state today [36, 38]. It nevertheless does allow for some cautious predictions already. The potentially most important one concerns the predictable existence of "partial temperatures" in a hot plasma—not only far from equilibrium where this goes without saying but potentially also at equilibrium proper in case such exists:

(1″) Two-temperature equilibrium (predicted)
Either particle class (electrons; ions) may thus possess a different temperature—at equilibrium. Hereby the electrons should be much "hotter" in general [39]. This prediction comes as a surprise in view of the well-known uniqueness of the thermodynamic equilibrium, the "heat death" of Clausius. While pressure is known to be capable of forming partial pressures at the same temperature and volume in a thermodynamic gas at equilibrium, the prediction of "partial temperatures" existing at equilibrium goes against the grain of the venerable field of high-temperature many-particle physics. And it of course flies in the face of common sense, too. Compare again ref. [38] for some empirical results that may or may not prove compatible with the present conjecture. If the latter is true, the first law of cryo-thermodynamics, Eq. (1″) above, necessarily possesses a corollary:

(1a″) Bistability of temperature (including the formation of hysteresis phenomena)
A consequence of this corollary is a second corollary that is even more foreign: An autonomous "dynamical switching process" of periodic or chaotic type should predictably occur between the two "sub-equilibria" at some parameter values:

(1b″) A non-point attractor at equilibrium (that may even become chaotic)
This hypothetical new attractor applies under "well-stirred" conditions at equilibrium already. Spatial structure formation would then make the situation almost arbitrarily complicated even at equilibrium. A similar complexity is so far only known from the relaxation processes in glass-like solids and spin glasses, cf. [39]. Note moreover that so far, no quantum effects were assumed to be involved. It is important to state once more that the whole First Law of Cryo-Thermodynamics, Eq. (1″), is hypothetical at present. In any new terrain one easily is led astray—as may have occurred here. Fortunately though, experiments in the footsteps of Landau and Lynden-Bell—but of a non-compartmentalized, "mixed" kind—can be conducted in hot-plasma reactors so as to either falsify or confirm Eq. (1″) above.

SECOND LAW of deterministic statistical Thermo-Cryodynamics

The two-temperature equilibriumTwo-temperature equilibrium of the first law of hot-plasma dynamics—Eq. (1″)—is (if it indeed exists) maximally sensitive to an exogenous change of total energy or volume. Hereby, one special case predictably presents a technologically implementable situation:

Experiment: "Concentric multidirectional injection of fast electrons into a hot plasma"

In this way, a momentarily monopolar gas with a higher temperature and density is generated locally. This artificial "spot" can be used in principle to paradoxically cool down the local equilibrium temperature of the ions, if the latter have become too hot locally in a two-temperature hot plasma for it to remain stable and not touch the wall. Because the two temperatures in the plasma possess a fixed ratio at equilibrium (or quasi-equilibrium, respectively) with the electrons being much hotter in accordance with suggestive observations [38], it is predictably possible to exogenously change the electron temperature by active intervention—in order to thereby paradoxically cool-down the local ion temperature in the hot plasma. If so, it is in principle possible also to—by the concentric injection of electrons of higher density with a carefully chosen energy (temperature)—force down the temperature of the "too hot ions" in a developing localized plasma instability that is about to form a tongue that by licking the wall finishes the fusion process. This "non-hydrodynamic" reaction-kinetic effect offers itself for active intervention. This proposal is the prediction of "interventional cooling" to be possible regarding a too hot region in a plasma that is about to get out of control in a continuous Tokamak reactor like the STELLARATOR called the "Wendelstein 7-x" in Greifswald at the Max Planck Institute for Plasma Physics. The latter is the famous continuous-fusion reactor expected to start test operation in the autumn of 2017 [40].

The predicted effect can be achieved interactively by the local generation, through concentric injection from six directions, of a somewhat hotter local partial equilibrium at enlarged density of the lighter particle class (the electrons), as mentioned. This injection method predictably allows one to "interactively cool" the local temperature of the too hot ions, in the case of a local hydrodynamic instability in the plasma going out of control. This new option possesses no analog in the purely PDE-based processes of the so far prevailing plasma reactor models. The new proposal presupposes that the partial pressures and partial temperatures of the two interacting gaseous components are capable of a double-temperature quasi-equilibrium locally. The underlying novel phase diagram has yet to be described. The here proposed new method can be called:

(2″) Adaptively controlled continuous hot-plasma fusion (ACCHPF)
This technological proposal of a "gas-theoretic cooling" represents an approach very different from the traditional approach to an MHD, formulated entirely in terms of local flows and currents in a fluid-dynamical and electromagnetic scenario governed by PDEs. The latter does not lend itself to a point-specific active intervention. The two approaches, the traditional continuous one and the present particles-based localized one, are both legitimate. But whereas the manipulation of local flows and currents is virtually impossible, the dirigible concentric injection method (**dci**) is in principle implementable by using several injection ports for dirigible electron beams [41]. The 6-beam method of Steve Chu, used by him and his coworkers in an entirely different context [42], offers itself for the purpose. The proposed new option is based on the existence of local two-temperature equilibria or quasi-equilibria in hot plasmas. The proposal made can be tested best by "doing in reality" rather than by simulation. This is because in molecular-dynamics simulations, an "inversion of the macroscopic time's arrow" is notorious for intruding surreptitiously as mentioned [9, 33].

Up until now, the high-energy industry did not signal interest in cryodynamics. Trying the new science out empirically is arguably the fastest way to reach the new technological goal of ***interactive*****fusion-reactor-control** (**ifrc**) with its vast economic promise.

THIRD LAW of deterministic statistical Thermo-Cryodynamics

The term "law" is not fully appropriate here because nothing but a technological proposal is at stake. Nonetheless the term "law" was chosen to acknowledge the existence of an alternative technological option to the above-described road towards unlimited "free energy" in the everyday sense of the word. The main technological competitor to the adaptively controlled continuous hot-plasma fusion (acchpf) considered above is the intermittent alternative. The latter consists in the repetitive creation-from-scratch of a hot-fusion plasma ("McGuire reactor" [43]):

(3″) Repetitively installed discontinuous hot-plasma fusion (RIDHPF)
The difference between the second law of thermo-cryodynamics offered above, Eq. (2″), and the present third law of thermo-cryodynamics, Eq. (3″), reminds one of the familiar difference between the continuous *Stirling motor* and the discontinuous-explosion type *Otto motor* in automobile design. In this old technological realm, the intermittent method (*Otto's*) gained the upper hand up until now. At the present moment in time, it again is hard to predict which of the two alternative options for running plasma fusion engines will win out on a long-term

basis: the continuous or the intermittent one (ACCHPF or RIDHPF). However, one thing can already be said for sure: The new fundamental field of cryodynamics is of a high applied interest.

6 Discussion

I

The above tabular presentation of the field of deterministic statistical multi-particle systems founded by Sinai [4] represents a new approach to the qualitative behavior of multi-degree-of-freedom Hamiltonian systems, classical and eventually also quantum. In this regard, the present approach belongs into the footsteps of Paul Dirac who always insisted on "Hamiltonicity" having to be observed also in quantum mechanics (cf. [44]). At the same time, the three-tiered Table offered above takes up the grand tradition of mid-19th-century deterministic billiard ball theory [45] that would later be called "**chaos theory**." This theory possesses an even older connection to the past by reviving van Helmont's first gas theory: the word "gas" stems etymologically from the Greek word "chaos" [46].

Gas theory was for almost two centuries dominated by the assumption that maximally short-range repulsive potentials hold true in between the involved particles. This went so far that the repulsion could often be formally neglected altogether without change of the statistical equations. This statistical approach to the theory of heat remained the royal route for more than a century—right up to the deterministic statistical physics founded by Sinai [3, 4]. The new chaos-theoretic method had to struggle to obtain a niche of its own in the field. The fact that the deterministic approach is strong enough to also cover far-from-equilibrium situations was only discovered in recent years (Fröhlich, Sonnleitner). It then turned out that a "smoothing-out" of the maximally short-range ("hard") repulsive potentials of the billiard particles of a Sinai gas, into "long-range" 1/r repulsive potentials of inverted Newtonian type, constitutes an admissible option [1] (cf. [47] for a first pertinent simulation). The smoothing proves to be admissible without loss of the main qualitative implications of the new deterministic statistical thermodynamics. Gas theory in this way can be re-defined under smooth-repulsive deterministic conditions in the footsteps of the famous KAM theory [5]. (The first author once enjoyed a long encouraging telephone conversation with Jürgen Moser, the "M" of KAM, in mid-1976.) This long-budding development became "fully transparent" only with the "plane-tree alley paradigm," as the generalized (both smoothened and far-from equilibrium) Sinai theory was called [1].

The smoothened repulsive-Newtonian subcase is a direct generalization and outgrowth of Sinai's founding hard-sphere repulsion-based chaos theory. It formed the subject matter of Part I above (with chemical constraints and quantum specifications etc. yet to be added). To this smoothened deterministic case, then surprisingly a twin case arose—valid under mirror-inverted potentials. The latter situation had for intrinsic reasons been impossible to spot in the traditional "hard-core" approach towards thermodynamics: namely the attractive-Newtonian case. Its previous nonexistence is nevertheless surprising in retrospect. It formed the subject matter of Part II. The new physical science termed "cryodynamics" [1] exists in close parallel to the venerable science of thermodynamics (krýos ice is the twin to thermós hot). The new field forms a "second empire" within Sinai's deterministic-chaos based statistical physics.

In this manner, a "parallelization" was achieved between the almost two-centuries-old theory of microscopic Hamiltonian thermodynamics (see [48] for its groping earliest origin) on the one hand, and the brand new theory of Hamiltonian cryodynamics [1], on the other. Finally, Part III dealt in a preliminary fashion with a combination of the two fundamental fields sketched before.

The two parallel anti-Newtonian and Newtonian programs described above could have been initiated more than a century ago—or even three centuries ago by New- ton himself (although the Cartesian billiard theory [48] was only in its earliest infancy at the time). To date, the Newtonian multi-particle theory paradoxically is still in its early infancy. A solid foundation is provided by the theory of deterministic chaos in phase space, originally invented by Henry Poincaré in the late 19th century [49].

II

A first successful numerical simulation of the new surprise tendency towards anti-equilibration valid in the attractive case is due to Klaus Sonnleitner [9] who remained maximally skeptical himself during his long numerical-mathematical voyage. He therefore placed the greatest emphasis on the controls—which fact only made the discovery possible in the end. He achieved the counterintuitive new finding of anti-dissipation while applying his especially transparent and hence accurate fourth-order symplectic simulation method [9] to a formally three-particle but dynamically two-particle system, in a two-dimensional but formally one-dimensional configuration: the T-tube, with a frictionless particle put into either leg [9] (cf. [33]). The immovable third particle (the massless frictionless T-tube itself) was assumed to be held fixed in space by rigid connection to a distant very large mass.

The heavier particle with its mostly larger kinetic energy was placed into the vertical leg of the T-tube, and the light-weight, mostly less energetic, fast particle was placed into the horizontal leg. The Newtonian (or anti-Newtonian, respectively) potential acted across the two-dimensional space in which the T-tube is embedded. All possible mutual relations (as to which particle has more energy initially) were numerically checked with both polarities of the potential (repulsive or attractive, respectively), in two carefully juxtaposed parts (see pp. 99–100 of Sonnleitner's dissertation [9] for a transparent tabular synopsis in the otherwise German text). The not-so-surprising "dissipative" results found in the repulsive case got collected just as painstakingly as the surprisingly emerging "anti-dissipative" results found in the attractive case.

An analytical description of essentially the same 2-D two-particle situation was offered by Ramis Movassagh [50], cf. also [51]. Almost all of the big theorems and sub-theorems which are implicit in the new paradigm of a "Newtonian gas" [1] do wait to be formulated-out in detail on the way towards becoming standard textbook stuff.

The next step planned in the simulation program was to make the heavier vertical particle arbitrarily heavy in a step-wise fashion while leaving its local motion and the effect it locally exerts on the light-weight horizontal particle, unchanged. Eventually, the heavy vertical particle will then act as a mere periodic forcing on the light-weight horizontal particle. If (as can be expected but needs to be verified) the qualitative behavior remains unchanged under this progressive simplification, the resulting final system will be nothing else but a periodically forced Hamiltonian oscillator of the kind first conceptualized by Poincaré, in which the "homoclinic point" and hence chaos was first discovered [49]. The surprise prediction is that essentially the "same" Poincaré map explains, both a temporally directed *dissipative* behavior arising from virgin initial conditions in either direction of time in this Hamiltonian system, and a temporally directed *anti-dissipative* behavior, arising from virgin initial conditions in either direction of time, in this "same" Hamiltonian system following a sign flip in the potential. The very possibility of the latter effect existing had escaped discovery over both the 19th and the 20th century.

If the stepwise simplification works as predicted, a simple "flat" 2-D Poincaré map of two predictably related but at present totally unknown shapes will be found numerically. This 2-D map will then both causally and intuitively, explain either type of qualitative macroscopic behavior in nature—the "Boltzmann" case and the new "anti-Boltzmann" one—in basically the "same" system: namely, in the oldest and simplest chaotic oscillator of history. Boltzmann and Poincaré would then be reconciled at last. The late Sonnnleitner planned to compute this enigmatic "twin" Poincaré cross section in further numerical work. A simplified idealized pair of maps producing the same behaviors can predictably be written down afterwards: to enable the first "real" (that is, qualitative) understanding of the two "miracles" of statistical physics, one called the "time's arrow" and the other awaiting a fitting name ("anti-arrow"?). This simple but not yet existing area-preserving 2-dimensional diffeomorphism with two different mutually nontrivially related but still unknown shapes can—as the oldest playground of chaos theory that it represents—be given a name already as the "double-faced Sonnleitner map." All of this presupposes that the above reasoning did not go astray at some point.

III

So far, the discussion dealt exclusively with the two pure cases—either pure repulsion or pure attraction. The combined theory was touched upon in Part III. This third Part is even farther away from having reached a canonical form. It comes as a surprise that dilute-plasma theory suddenly acquires a fundamental role of its own in science—besides and apart from the traditional formal treatment in terms of idealized smoothened partial differential equations (PDEs) valid approximately under less dilute conditions.

With the here assumed "celestial-mechanics-like" more dilute conditions, the present approach led in Part III to the unexpected (possibly premature) prediction of a two-temperature equilibrium existing in nature. Namely, in hot plasmas when "static interactions" prevail over "Maxwellian interactions." New phenomena that had been overlooked both theoretically and observationally come into focus if no mistake was made. Nature can look amazingly foreign under the influence of the Poincaréan angle of vision later called *chaos theory*. As it is well known, Poincaré also conjured-up *topology*. Whether or not hot-plasma fusion reactors can become a reality much sooner than currently hoped-for, under the new *dilute-plasma paradigm* is an open question of some applied interest.

We have here a down-to-earth application of cryodynamics that offers the prospect of unlimited "free energy" in the everyday sense of the word. This promise renders the new science of cryodynamics an irresistible topic for an energy-thirsty planet. The work-in-progress of Part III needs to be taken with a grain of salt regarding its mathematical rigor. Nevertheless the economic consequences of the new "molecular-dynamical dilute-hot plasma theory" (**mddhp**) have the potential to render the discipline of cryodynamics attractive for applied departments across the world. Never before was an economic bonanza lying closer to one's door steps it appears—provided thermo-cryodynamics keeps its promise. The 6-tiered concentric injection method for electrons was discussed in Part III.

IV

Following this "hardest" (in the sense of technological promise) implication of cryodynamics, we now turn to its "softest": its impact on the only non-experimental since purely observational science—**astronomy**. Celestial mechanics comes into focus again. Sonnleitner's numerical instability implies in the first place that galactic simulations must be taken with a big grain of salt when based on deterministic equations.

The ten billion visible galaxies of but a few morphological types point jointly to the existence of an underlying shape-forming dynamical quasi-attractor (in several versions) in these many-particle Newtonian systems. The empirical spiral shape of galaxies is quite hard to understand apart from the fact that the involved large angular momentum follows in a straightforward fashion from the random asymmetry of the originally present gas cloud (before the latter slowly contracted while also forming local pockets of faster contraction). Binney and Tremaine's learned book [37] gives an excellent survey of many of the current views and open problems. Now Sonnleitner's numerical instability surprisingly stands in the way of a full numerical understanding of galaxy formation.

The bigger structures—clusters and superclusters—are even harder to understand than galaxy formation itself, and so is the fractal structure of the universe at large with its "voids" (first discovered in 1907 by Fournier d'Albe [52] quoted by Mandelbrot[2] [53]). The empirically measured fractal dimensionality of the universe is not much larger than unity [54]. And of course, there is much more mass present everywhere than is optically visible. This is Zwicky's original "dark matter" as he called it in a German-language paper in 1933 in which he proved the existence of the latter with the aid of the virial theorem. This burnt-out, previously non-dark

[2]Fournier is quoted 14 times.

matter is still partly visible on infrared photographs in which ordinary-looking galaxies often appear much larger than their optical counterparts—in conformity with their flat rotation curves first discovered by Rubin [55]. Moreover, there exist seemingly quite small galaxies with "too massive" central structures including a giant black hole—whose outer parts would be worth looking for on deep-infrared photographs. And there even exist fully "low surface brightness galaxies" (formerly called "black galaxies") which look extremely old in the same perspective with even their centers included. The current interpretation, however, is that they must be "very young" [32].

At the farthest distances in the cosmos, we have Riccardo Giacconi's equi-distributed ultra-faint X-ray point sources with but a few photons arriving per day [56]. Their redshift will be very hard to measure in future decades. Thus, one sees that celestial mechanics is still replete with beautiful open empirical, conceptual and numerical problems. These empirical questions are totally neutral regarding a larger overarching picture.

However, in the wake of celestial mechanics, we come via astronomy also in touch with **cosmology**. Cryodynamics possesses direct relevancy for cosmology. This is because a gas of mutually attractive moving particles exists empirically in the sky with the myriad (more than two hundred billion) of visible galaxies. And this "gas" happens to be perfused by a very much larger myriad of ultra-fast tiny particles and photons traversing it.

The Sonnleitner effect of "paradoxical cooling" (jokingly called "energetic capitalism" by him because the rich-in-kinetic-energy are preying on the poor-in-kinetic-energy) allows for a specific prediction to be made here: The energy-poor fast gravitating particles—photons and cosmic rays—traversing the cosmos are bound to lose energy to the energy-rich slow gravitating particles—the galaxies. The same possibility was first raised on intuitive grounds by Fritz Zwicky in 1929 [57] as a competing alternative to Lemaître's expansion postulate for a postulated primordial "cosmic atom." The latter "Big Bang" hypothesis as it came to be called later is a world-famous postulate compatible with (but in no way enforced by) general relativity. This commonly accepted Friedmann-Lemaître postulate now suddenly gives way to a postulate-free first-principles alternative: the comprehensive theory of Newtonian gases called cryodynamics, described in Part II. When applied to the cosmos, cryodynamics directly entails the cosmological redshift phenomenon as an unavoidable qualitative implication.

Cryodynamics thus re-opens the redshift problem in cosmology because it directly implies the existence of a distance-dependent redshift as a first-order effect. So much for a qualitative argument. Quantitatively speaking, however, the exact "percentage question" remains totally open in cryodynamics: How much redshift is Newton based and how much redshift is expansion-based? is a question that remains totally open. This limitation follows from the mentioned impossibility to perform even qualitatively correct numerical multi-particle Newtonian simulations. As is well known historically speaking, a formal error made by Zwicky in his mentioned founding paper of 1929 [57], pointed out to him by Eddington in a private letter immediately made public by Zwicky in his next publication the same year, was destined to make his qualitative insight a laughingstock amongst his colleagues—under the catchy name "tired light theory." A similar fate could now await the claim that the Hubble law along with its famous wiggle at the end follows from cryodynamics in a 100% fashion. Actually, nothing but an unknown percentage value follows up until now from cryodynamics. The modern Big-Bang Cosmology thus only needs to be replaced by a slower-growing, larger and older cosmos of otherwise the same qualitative structure as currently assumed.

This is because as we saw, a quantitative calculation based on cryodynamics that would reproduce the Hubble line along with the famous "wiggle" at the end of its best-studied first portion, is out of the question due to the new instability discovered in the field of numerical mathematics. Cryodynamics therefore only causes an unknown quantitative modification of the by now for 89 years accepted Friedmann-Lemaître cosmology. In this way, everything has only become "more messy" so to speak—or so it appears.

Strange enough, however, there exists an entirely independent proof to the effect that the expansionist model is *fully* ruled out. This unlikely historical coincidence is perhaps worth mentioning here as an insert to "round off" the connection between cryodynamics and cosmology.

Insert: A re-evaluation of the famous "equivalence principle" of Einstein of 1907 reveals a new "fourth" implication. While the final conclusion reached by Einstein remains perfectly true (that light progressing horizontally downstairs appears *slowed* when watched from upstairs), the reason for this fact lies in a previously unaddressed fact: The observable slow-down of a light ray hugging the horizontal floor of Einstein's constantly accelerating rocketship when watched from above reflects, not a proportional relative *reduction in c* as assumed up until now but rather a proportional *increase in size* valid downstairs because the slowed speed of light visible from above is due to a local slant of the horizontally progressing light ray, valid relative to above at every point. If this is true, the speed of light *c* in the vacuum remains a global constant of nature, just as the "younger Einstein" up to the age of 32 had assumed [58] (so it is not just an everywhere locally valid constant, as is commonly assumed). When this retrieved global constancy of *c* is added to cosmology as a new piece of information, only a "zero" cosmic expansion remains possible because any residual velocity of expansion would imply a superluminal mutual recession speed to exist between two sufficiently distant points. This means that, if "*c*-global" is accepted as a "sister result" to cryodynamics, the latter science has become as important for our understanding of the world as thermodynamics has been for the better part of two centuries already.

Zwicky's ridiculed conjecture, proven true by Sonnleitner and Movassagh, revives the insight that the empirical Hubble line is caused by the gravitational interaction between light and matter—that is, by cryodynamics. A *derived empirical prediction* then states that in the direction of very large cosmic voids, especially with several voids aligned in a row, the slope of the Hubble line must be markedly reduced. This prediction is specific enough to warrant an **empirical** investigation. The eventual direction-specific "Hubble surface" (rather than mere Hubble line) will predictably become famous.

If Zwicky will thus be rehabilitated not only qualitatively but also quantitatively, his name belongs into the great historical tradition of the lonely maverick scientist who singlehandedly takes on the whole scientific establishment and wins out. Edwin Hubble, the discoverer of the Hubble line and before it of the existence of galaxies including the Milky Way galaxy (conjectured by the young Immanuel Kant as mentioned) did forego a doubly deserved Nobel Prize for the sympathy he showed towards Zwicky's ridiculous claim. Three centuries before, cosmologist Giordano Bruno still would get burnt at the stakes for picturing an infinite cosmos in the footsteps of Saint Augustine and Heraclitus. Next in line came the first modern chemist, Antoine Lavoisier, who got executed for reasons that may have had to do with his discovery of oxygen which dethroned the ruling heat substance (phlogiston) theory (historical quote: "La république n'a pas besoin de savants"—The republic can do without scientists! [59, 60]). Next in line came Ignaz Semmelweis, the European discoverer and first practitioner of asepsis in the first half of the 19th century, who got driven into suicide for his not being allowed to save the lives of thousands of mothers in childbed [60]. A few decades later, Georg Cantor's transfinite mathematics would owe its existence to the fact that routine drugging had not yet entered psychiatric wards so he could write his best texts there. The 20th-century's Fritz Zwicky was, although widely ridiculed, permitted to continue working at CalTech to make further major discoveries like baryonic dark matter and neutron stars. In the wake of cryodynamics and *c*-global, Zwicky's *dark matter* would remain the only kind of dark matter existing in a non-expanding cosmos. If so, the famous "cold dark matter" introduced in an ad hoc fashion in the late 1990's to explain the wiggle in the Hubble line, becomes the modern "phlogiston." But so of course only unless the new *cc*-based (*c*ryodynamics, *c*-global) cosmology proves to be a mirage. A concerted attempt at falsifying *c*-global (since cryodynamics appears invincible already) is important to confirm the new synthesis.

But cosmology has a second major leg in support of the Big-Bang model: the famous "cosmic microwave background radiation" **CMB**. In the absence of expansion, this leg now instead reflects the "mean cosmic black-body temperature" valid in our galactic neighborhood. The latter radiation was anticipated by subsequent Nobel Prize winner Edouard Guillaume in 1896 already, see [61]. Only its giant intensity and smoothness, found in the late 20th century, remained unpredictable up to the *cc*-based cosmology just discussed. Yet however titillating this may appear after the 90-years reign of the expansionist cosmology, a big drop of water remains present in the wine: Unless cryodynamics becomes the huge commercial success expectable on our energy-thirsty planet, the retrieved Zwicky-type cosmology has no chance to win recognition for another 9 decades. This is because a global consensus built-up over the better part of a century represents much too tightly interwoven a fabric in the carpet of science to be given up without a spectacular—in the present case financial—bonus waiting.

V

Following the most popular pastime on the planet—sky-watching –, the field of **numerics** forms the second big sector of science on which Sonnleitner's numerical result wreaks havoc. Virtually all multi-particle "galactic" and "cosmological" numerical multi-particle simulations are rendered invalid by the previously unrecognized tendency of such numerical simulations to surreptitiously switch-over to the other macroscopic direction of time for numerical reasons. Increasing the accuracy-per-step at the expense of simulation time is to no avail: A transition from a maximally rare trajectory towards a neighboring frequent-in-type trajectory occurs in the manner first conceived by Boltzmann [2], but with the add-on that the new trajectory belongs even to the other direction of macroscopic time. Deterministic galactic-dynamics simulations therefore need to be taken with a large grain of salt.

With the advent of cryodynamics, the chaos community with its mostly young enthusiasts scattered all over the globe is offered a new identity of its own. The chaos-and-fractals movement of the third part of the 20th century with its revived deterministic classical thinking can be expected to get a big boost in the footsteps of Newton, Waterston, Boltzmann, Zwicky and Sinai

VI

The third major application of cryodynamics—for once not in the sky but down on earth—is non-polarizing for once because **plasma theory** is a venerable field. The above-included preliminary Part III offers the unheard-of prediction of two-temperature equilibria with a fixed temperature ratio existing in a hot plasma. This predicted fundamental phenomenon is even more counterintuitive than a recycling eternal cosmos is (*metabállon anapaúetai*—metabolizing it rests—was Heraclitus' phrase by the way). Part III's promise of unlimited "free energy" in the everyday sense of the word endows cryodynamics with a special appeal on an energy-thirsty planet. The work-in-progress of Part III involves several merely conjectural results so far. The economic bonanza promised by deterministic dilute hot-plasma theory (**DDHPT**) makes the discipline of cryodynamics attractive to applied departments across the world. The economic bonanza in case thermo-cryodynamics keeps its promise calls for a critical evaluation by the scientific community soon. An attempt at a direct technological implementation may be the fastest method of evaluation.

VII

Finally, a big disclaimer: The above-presented three-tiered Table based on deterministic chaos theory still contains its giant gaps and no doubt also errors. But it at the same time marks the beginning of a more **unified physics**. The latter will bring a better understanding of macroscopic time whose intrinsic arrow now no longer stands alone. Even a better understanding of the place of human beings in the cosmos, as envisioned by Scheler [62], may arise in the once more infinitely extended picture of our external reality [63–64].

To conclude, the rare case of a new science eager to find its place in physics was sketched in a preliminary, tabular form. We hope that the endeavor was not too premature. Critical help from the part of the chaos community is solicited.

Acknowledgements We thank Yakov Sinai for his long-standing encouragement, and Ali Sanayei, Stefan C. Müller, Valerie Messager, Ralph Abraham, John Kozak, Daniel Stein, Jim Yorke, M.A. Aziz-Alaoui, Cyrille Bertelle, Werner Ebeling, Peter Plath, Boris Schapiro, Ramis Movassagh, Kenzei Hiwaki, George Lasker, Greg Andonian, Peter Weibel, Eric Klien, Andre

Assis, Saurya Das, Günter Häfelinger, Alfred Rieckers, Wolfgang Müller-Schauenburg, Henry Gebhardt, Tobias Winkler, Niels Birbaumer, Walter Ratjen, Günter Radons, Luc Pastur, Dogwon Kim, Jürgen Parisi, Bill Seaman, Joachim Peinke, Rudolf Huebener, Stephen Wolfram, Leon Chua, Niels Schopohl and Matthias Bartelmann for discussions. For J.O.R.

References

1. O.E. Rossler, The new science of cryodynamics and its connection to cosmology. Complex Systems **20**, 105–111 (2011)
2. L. Boltzmann, *Lectures on Gas Theory*, Translated by S.G. Brush (University of California Press, Berkeley, 1964)
3. Y. Sinai, Some remarks on the spectral properties of ergodic dynamical systems. Russ. Math. Surv. **5**, 37–50 (1963)
4. Y. Sinai, Dynamical systems with elastic reflections. Russ. Math. Surv. **25**, 137–189 (1970)
5. A. Kolmogorov, On conservation of conditionally periodic motions for a small change in Hamilton's function (O sohranenii uslovnoperiodicheskhi dvizhenij pri malom izmenenii funkcii Gamil'tona), Dokl. Akad. Nauk. SSSR **98**, 527–530 (1954); V.I. Arnold, Proof of a theorem of A.N. Kolmogorov on the preservation of conditionally periodic motions under a small perturbation of the Hamiltonian. Russian Mathematical Surveys **18**:5, 9–36 (1963); J. Moser, On invariant curves of area-preserving mappings of an annulus. Nachr. Akad. Wiss. Gött. Math. Phys. Kl, 1–20 (1962)
6. R. Clausius, On several convenient forms of the fundamental equations of the mechanical theory of heat (Über verschiedene für die Anwendung bequeme Formen der Hauptgleichungen der mechanischen Wärmetheorie). Ann. Phys. Chem. **125**, 352–400 (1865)
7. J. Gibbs, *Elementary Principles in Statistical Mechanics, Developed with Especial Reference to the Rational Foundation of Thermodynamics* (Yale University Press, New Haven, 1902), republished by Dover in 1960
8. H. Diebner, O.E. Rossler, A deterministic entropy based on the instantaneous phase space volume. Z. Naturforsch. **53a**, 51–60 (1998)
9. K. Sonnleitner, StV4: A symplectic time-reversible Störmer-Verlet algorithm of fourth order for Hamiltonian several-particle systems including two applied examples—gas and T-tube arrangement (StV4: Ein symplektisches zeitreversibles Störmer-Verlet-Verfahren vierter Ordnung für Hamiltonsche Mehrteilchensysteme mit zwei Anwendungsbeispielen (Gas, T-Rohr-Anordnung), (Ph.D. thesis, University of Tübingen 2010)
10. F. Brando, M. Horodecki, N. Ng, J. Oppenheim, S. Wehner, The second laws of quantum thermodynamics. Proc. Natl. Acad. Sci. USA **112**, 3275–3279 (2015). https://doi.org/10.1073/pnas.1411728112
11. O.E. Rossler, An estimate of Planck's constant, in *Dynamic Phenomena in Neurochemistry and Neurophysics: Theoretical Aspects* (Publications of the Hungarian Academy of Sciences, Budapest, 1985), pp. 16–18
12. L. Bertalanffy, *Theoretical Biology (Theoretische Biologie I)* (Gebrüder Bornträger, Berlin, 1932)
13. I. Prigogine, Dissipative structures in chemical systems, in *Fast Reactions and Primary Processes in Chemical Kinetics*, ed. by S. Claesson (Interscience, New York, 1967); I. Prigogine, W. Kestemont and M. Marechal, Velocity correlation and irreversibility: a molecular-dynamics approach, in *From Chemical to Biological Organization*, ed. by M. Markus, S.C. Müller, G. Nicolis (Springer, New York, 1988), pp. 22–26
14. O.E. Rossler, A system-theoretic model of biogenesis. Z. Naturforsch. **26b**, 741–746 (1971)
15. O.E. Rossler, Deductive prebiology, in *Molecular Evolution and Prebiology*, ed. by K. Matsuno (Plenum Press, New York 1984), pp. 375–385. https://doi.org/10.1007/978-1-4684-4640-1 27

16. O.E. Rossler, Is benevolence compatible with intelligence—on the theory of the humane feeling (in German), in *The Theme Park of the Expo 2000, vol. 1: Planet of Visions, Knowledge, Information, Communication* (Springer, Vienna, 2000), pp. 157–163
17. P. Teilhard de Chardin, *The Future of Man* (Harper and Row, New York, 1964)
18. R. Forward, *Dragon's Egg* (Del Rey, New York, 1980)
19. H. Follmann, C. Brownson, Darwin's warm little pond revisited: from molecules to the origin of life. Naturwissenschaften **96**, 1265–1292 (2009). https://doi.org/10.1007/s00114-009-0602-1
20. E. Schrödinger, *What Is Life? The Physical Aspect of the Living Cell* (Cambridge University Press, 1944)
21. J. van der Waals, *On the Continuity of the Gaseous and Liquid States* (in Dutch), Ph.D, thesis, Leiden University, 1873
22. A. Aranda, J.A. Lopez, C.O. Dorso, V. Furci, Mapping the phase diagram of nuclear matter. Bull. Can.-Am.-Mex. Phys. Soc. (1987)
23. W. Nernst, *Über die Berechnung chemischer Gleichgewichte aus thermischen Messungen* (On Calculating Chemical Equilibria from Thermal Measurements), (Nachr. Kgl. Ges. Wiss. Göttingen, 1906), pp. 1–40
24. H. Kammerlingh-Onnes, Further experiments with liquid Helium, D. On the change of electric resistance of pure metals at very low temperatures, etc., V. The disappearance of the resistance of Mercury, Comm. Phys. Lab. Univ. Leiden, **122b** (1911)
25. P. Kapitza, Viscosity of liquid Helium below the Lambda Point. Nature **141**, 74 (1938)
26. J. Bardeen, Theory of non-Ohmic conduction from charge-density waves in NbSe3. Phys. Rev. Lett. **42**, 1498–1500 (1979). https://doi.org/10.1103/PhysRevLett.42.1498
27. S. Bose, Planck's law and light-quantum hypothesis (in German). Zeitschrift für Physik **26**, 178–181 (1924). https://doi.org/10.1007/BF01327326
28. K.V. Klitzing, G. Dorda, M. Pepper, New method for high-accuracy determination of the fine-structure constant based on quantized Hall resistance, Phys. Rev. Lett. **45**, 494–497 (1980). https://doi.org/10.1103/PhysRevLett.45.494
29. R.B. Laughlin, Anomalous Quantum Hall effect: An incompressible quantum fluid with fractionally charged excitations. Phys. Rev. Lett. **50**, 1395–1398 (1983). https://doi.org/10.1103/PhysRevLett.50.1395
30. S. Braun, J.P. Ronzheimer, M. Schreiber, S.S. Hodgman, T. Rom, I. Bloch, U. Schneider, Negative absolute temperature for motional degrees of freedom. Science **339**, 52–55 (2013). https://doi.org/10.1126/science.1227831
31. S. Chandrasekhar, Dynamical friction, I. General considerations: the coefficient of dynamical friction, Astrophys. J. **97**, 255–262 (1943). https://doi.org/10.1086/144517
32. P. Schneider, *Extragalactic Astronomy and Cosmology* (Springer, Heidelberg, 2015), p. 119
33. O.E. Rossler, F. Kuske and A. Sanayei, Deterministic antidissipation. in *Bottom-up Self-Organization in Supramolecular Soft Matter*, ed. by S.C. Müller, J. Parisi (Springer, Berlin, 2015), pp. 271–280
34. CERN 2008, *LSAG* report, review of the safety of LHC collisions, http://lsag.web.cern.ch/lsag/LSAG-Report.pdf
35. L.D. Landau, On the theory of the dispersion of magnetic permeability in ferromagnetic bodies. Physikalische Zeitschrift Sowjetunion **1**, 285–288 (1932)
36. D. Lynden-Bell, Stellar dynamics: Exact solution of the self-gravitation equation. Mon. Not. R. Astron. Soc. **123**, 447–458 (1961). https://doi.org/10.1093/mnras/123.5.447
37. J. Binney and S. Tremaine, *Galactic Dynamics* (Princeton University Press, Princeton, 2008), p. 547
38. M. Hotinceanu, Z. Borsos, O. Dinu, Aspects of thermodynamic equilibrium in plasma. Pet. Gas Univ. Ploiesti Bull. **62**, 97–102 (2010)
39. O.E. Rossler, F. Kuske, A. Sanayei, Cryodynamics can solve the energy problem by stabilizing ITER: a prediction, in *Numerical Analysis and Applied Mathematics ICNAAM 2012: International Conference of Numerical Analysis and Applied Mathematics* (AIP Publishing, Melville, 2012), pp. 642–645. https://doi.org/10.1063/1.4756216; O.E. Rossler,

Hamiltonian chaos: two-temperature equilibrium in hot plasmas predicted, in *Fourth International Conference on Complex Systems and Applications ICCSA 2014*, pp. 24–27
40. P. Seeliger, Perspective contributions of atomic and nuclear processes to a carbon-free energy economy of the future (in German). *Sitzungsberichte der Leibniz-Sozietät der Wissenschaften zu Berlin* **130**, 165–187 (2017), cf. p. 171
41. O.E. Rossler, A. Sanayei, Is hot fusion made feasible by the discovery of cryodynamics? in *Nostradamus. Modern Methods of Prediction, Modeling and Analysis of Nonlinear Systems* (Springer, Berlin, 2013), pp. 1–4
42. C.S. Adams, H.J. Lee, N. Davidson, M. Kasevich, S. Chu, Evaporative cooling in a crossed dipole trap. Phys. Rev. Lett. **74**, 3577–3580 (1995). https://doi.org/10.1103/PhysRevLett.74.3577
43. T. McGuire, The lockheed Martin compact fusion reactor. Thursday Colloquium, Princeton University, August 6, 2015
44. G. Farmelo, *The Strangest Man—The Hidden Life of Paul Dirac, Quantum Genius* (Faber and Faber, London, 2009)
45. J.J. Waterston, On the physics of media that are composed of free and perfectly elastic molecules in a state of motion (1845). Philos. T. R. Soc. Lond. A **183**, 1–79 (1892), (edited after 47 years by Lord Rayleigh). https://doi.org/10.1098/rsta.1892.0001
46. J. van Helmont, *Oriatrike or Physick Refined* (Lodowick Loyd, London, 1662)
47. H. Diebner, *Time-dependent deterministic entropies and dissipative structures in exactly reversible Newtonian molecular-dynamics universes* (in German) (Grauer-Verlag, Stuttgert 1999). Ph.D. Thesis, University of Tübingen
48. R. Descartes, *Principles of Philosophy (Latin original 1644)* (Kluwer, Dordrecht, 1991)
49. H. Poincare, On the three-body problem and the equations of dynamics (in French). Acta Math. **13**, 1–270 (1890), cf. J. Barrow-Green, *Poincaré and the Three-Body Problem*, Vol. 2 (American Mathematical Society. Providence, R.I., 1997)
50. R. Movassagh, A time-asymmetric process in central force scatterings (2013), arXiv:1008.0875[physics.class-ph]
51. O.E. Rossler, R. Movassagh, Bitemporal Sinai divergence: an energetic analog to Boltzmann's entropy? Int. J. Nonlin. Sci. Num. **6**, 349–350 (2005)
52. E. Fournier d'Albe, *Two new Worlds (I) The Infra-World, (II) The Supra World* (World and Longman Green, Longmans, Green and Co., London 1907)
53. B. Mandelbrot, *The Fractal Geometry of Nature* (Freeman, New York, 1977)
54. P.H. Coleman, L. Pietronero, The fractal nature of the universe. Phys. A **185**, 45–55 (1992). https://doi.org/10.1016/0378-4371(92)90436-T
55. V.C. Rubin, N. Thonnard, W.K. Ford, Jr., Extended rotation curves of high-luminosity spiral galaxies, IV: Systematic dynamical properties SA through SC. Astrophys. J. **225**, L 101–111 (1978). https://doi.org/10.1086/182804
56. R. Giacconi, et al., The Chandra Deep field south one million seconds catalog. Astrophys. J. Suppl. **139**, 369–410 (2002)
57. F. Zwicky, On the red shift of spectral lines through interstellar space. Proc. Natl. Acad. Sci. USA **15**, 773–779 (1929)
58. O.E. Rossler, The *c*-global revival in physics. Prog. Phys. **11**, 340–343 (2015)
59. S. Mason, *A History of the Sciences* (Collier-MacMillan, New York, 1968)
60. S. Sambursky, *Physical World of Late Antiquity* (Basic Books, New York, 1962)
61. A. Assis, M. Newes, History of the 2.7 K temperature prior to Penzias and Wilson. Apeiron **2**, 79–84 (1995)
62. M. Scheler, *Man's Place in Nature* (Noonday, New York, 1961)
63. A. Sanayei, O. Rossler, *Chaotic Harmony—A Dialog about Physics, Complexity and Life* (Springer, Heidelberg, 2014)
64. O.E. Rossler, Rolling ball in breathing plane-tree alley paradigm. Eur. Sci. J. **9**, 1–7 (2013)

Index

S.C. Müller et al. (eds.), *Complexity and Synergetics*,
https://doi.org/10.1007/978-3-319-64334-2

Printed in the United States
By Bookmasters